C Programming for Engineering and Computer Science

H. H. Tan

Morrison Knudsen Corporation

T. B. D'Orazio

San Francisco State University

**WCB
McGraw-Hill**

New York   St. Louis   San Francisco   Auckland   Bogotá   Caracas   Lisbon
London   Madrid   Mexico City   Milan   Montreal   New Delhi
San Juan   Singapore   Sydney   Tokyo   Toronto

Vice president and editorial director: Kevin T. Kane
Publisher: Tom Casson
Executive editor: Eric M. Munson
Developmental editor: Holly Stark
Marketing manager: John T. Wannemacher
Project manager: Alisa Watson
Production associate: Debra R. Benson
Senior designer: Cripsin Prebys
Supplement coordinator: Carol Loreth
Compositor: Lachina Publishing Services
Typeface: 10.5/12 Times Roman
Printer: Quebecor Printing Book Group/Fairfield

WCB/McGraw-Hill

A Division of The **McGraw·Hill** Companies

C PROGRAMMING FOR ENGINEERING AND COMPUTER SCIENCE

This book is printed on acid-free paper.

8 9 10   QPF/QPF   0 5 4

ISBN 0-07-016911-X
ISBN 0-07-848164-3 (disk)
ISBN 0-07-913678-8 (set)

Library of Congress Cataloging-in-Publication Data
Tan, H. H.
    C programming for engineering and computer science / H. H. Tan,
T. B. D'Orazio.
        p.   cm. -- (McGraw-Hill's BEST--basic engineering series and
tools)
    Includes index.
    ISBN 0-07-016911-X
    1. C (Computer program language)   I. D'Orazio, T. B.   II. Title.
III.   Series
QA76.73.C15T35   1999
005.13'3--dc21   98-20109

http://www.mhhe.com

McGraw-Hill's BEST—Basic Engineering Series and Tools

To Elizabeth & Buttercup, and to Wei Huang

Engineering educators have had long-standing debates over the content of introductory freshman engineering courses. Some schools emphasize computer-based instruction, some focus on engineering analysis, some concentrate on graphics and visualization, while others emphasize hands-on design. Two things, however, appear certain: no two schools do exactly the same thing, and at most schools, the introductory engineering courses frequently change from one year to the next. In fact, the introductory engineering courses at many schools have become a smorgasbord of different topics, some classical and others closely tied to computer software applications. Given this diversity in content and purpose, the task of providing appropriate text material becomes problematic, since every instructor requires something different.

McGraw-Hill has responded to this challenge by creating a series of modularized textbooks for the topics covered in most first-year introductory engineering courses. Written by authors who are acknowledged authorities in their respective fields, the individual modules vary in length, in accordance with the time typically devoted to each subject. For example, modules on programming languages are written as introductory-level textbooks, providing material for an entire semester of study, whereas modules that cover shorter topics such as ethics and technical writing provide less material, as appropriate for a few weeks of instruction. Individual instructors can easily combine these modules to conform to their particular courses. Most modules include numerous problems and/or projects, and are suitable for use within an active-learning environment.

The goal of this series is to provide the educational community with text material that is timely, affordable, of high quality, and flexible in how it is used. We ask that you assist us in fulfilling this goal by letting us know how well we are serving your needs. We are particularly interested in knowing what, in your opinion, we have done well and where we can make improvements or offer new modules.

Byron S. Gottfried
Consulting Editor

We developed this book to address the difficulty beginning students often find reading computer language texts. We felt that if we could get students involved in a text, hold their interest, and get them thinking about the meaning and uses of C code, we could make the process of learning a first language easier and fun. To accomplish this, we use a question and answer style. The reader's thought processes are stimulated by the same questions about code that students themselves often ask. By answering them directly and clearly, we focus the reader's attention on the important issues of C programming.

We also observed that most computer language texts have very few figures. Because visual images are so useful in teaching we have made a concerted effort to create figures that are accurate yet easy to follow. These figures provide students reinforcement and clarification of concepts illustrated by the actions of the programs. In particular, we have three-dimensional sketches of the actions of loops and if control structures. Students can look at these figures and quickly grasp the flow of these structures. We believe that these figures are an improvement over the standard flow charts. We recognize that one of the most difficult topics for beginning C programmers is pointers. The foundation for our pointer figures is established early in the text with a table of the names, types, addresses, and values of variables. Throughout the text, we use this table with arrows indicating how information at one location in memory relates to that in another location.

Many texts have numerous pages of code that are not adequately explained, and most beginning students are not capable or willing to independently decipher even relatively simple programs without guidance. In this book, we guide students through the code so that they understand both the operations and the thought processes that created the code. The goal is to make students realize what aspects of programming require extra thought and care and the importance of getting the details correct.

The unique style has been enthusiastically accepted among the students who have used early drafts of the book. Faculty unknown and unrelated to the authors have presented this text and asked the students' opinions. When compared with other texts, students have overwhelmingly preferred ours. Given this acceptance, we believe that you will find this book to be a valuable teaching and learning tool.

TEXT ORGANIZATION

The first chapter is an introduction to computing that assumes students have no basic knowledge other than using a computer for simple word processing. It describes hardware, the way information is stored in memory, computer languages, compilers, and software engineering. The purpose of this chapter is to introduce students to the way that computers work and the concepts behind software design.

Chapters 2 through 4 cover the fundamental aspects of a procedural programming language, basic syntax, and control structures. C library functions are described and their use illustrated in these chapters. In Chapter 5, user-defined functions are covered, emphasizing the concepts of modularity and reusable code. Pointers are gently introduced, being integrated with the development of functions that use addresses as arguments in function calls. After this chapter, the effects of using C features with user-defined functions are described. Chapter 6 focuses on numerical arrays.

Chapter 7 describes both strings and pointers. Because strings are most commonly manipulated using addresses, this chapter is well suited to describing working with pointers to modify memory. Chapter 8 covers structures in C and their uses in creating linked lists, stacks, queues, and binary trees. Also, large program design is covered in this chapter. This is included because engineering programs can quickly become very large. The importance of using C features to handle large programs is considered fundamental to preparing students for employment with firms that develop commercial software products in engineering. We have called Chapter 9 (available on McGraw-Hill's web site at http://www.mhhe.com/engcs/general/best/tandorazio. mhtml) an introduction to C++, but it is actually much more. Because of the thorough coverage we have given C we are able to describe many of the core issues of object oriented programming with C++. Classes, encapsulation, and polymorphism are described in simple terms. This chapter is richly illustrated. The simple language and illustrations provide students the background to use many of the fundamental C++ features.

Each chapter is divided into two parts, the Lessons and the Application Programs. A typical chapter begins with eight or nine Lessons and ends with four or five Application Programs. The Lessons teach syntax, form, and basic constructs. The Application Programs illustrate how what is taught in the Lessons can be used to solve real engineering and computer science problems. The Application Programs show the thought processes a developer goes through to create a program. The goal of the Application Programs is to give students the ability to follow a structured methodology in developing their own programs.

FEATURES

The text has the following features:

1. The book uses a simple question and answer approach that students find more friendly and accessible than standard narrative.
2. Each Lesson is focused on a single sample program. Students are guided through the program with descriptions of observations they should make and questions they should attempt to answer. Students gain understanding of C because they are compelled to follow the details of this code. After being guided through the code, it is explained in detail in the Explanation portion of the Lesson.
3. The Application Programs given in the second part of each chapter illustrate the usefulness of the C language for solving engineering and computer science problems. They are comprehensively explained. The examples focus on program design, software engineering, modularity, and creating reusable code.
4. There are more than 200 figures illustrating programming concepts. Many of the figures are unique and give students an ability to grasp concepts quickly and easily. Original pointer illustrations assist students in visualizing pointer fundamentals.
5. A structured four-step method (becoming five steps after introducing strings and more complex data structures) of program development is illustrated in describing the Application Programs. The method includes creating structure charts and data flow diagrams.
6. Numerical method examples, which can be used in courses that combine programming and numerical methods, are included in the Application Programs.
7. The Lessons have annotated code that assists students in understanding program details and flow. The annotations help students focus on the code and highlight the important points that the code is demonstrating.
8. We realize that students typically will not follow multiple pages of code on their own. Therefore, the Application Programs, each of which can cover two or three pages, are explained completely. There are no multiple pages of unexplained code.
9. We know that students struggle with the concept of pointers. We also have found that for students to understand pointers, the figures representing them should be rooted to something that the students can visualize. It is not enough to have a box with an arrow pointing to another box. Using tables and grid-like sketches of memory, we have taken much of the mystery out of pointers. We have found that after reading our text, students "get" the concept of pointers.
10. Modification exercises after the Application Programs can be used for courses that have a laboratory. Instructors can tell the students in advance of the laboratory session to read a particular Application Program. During the lab, students can be guided through some of the changes that need to be made. Further changes can be assigned as home exercises.
11. Beginning students struggle with debugging because the process is new and foreign to them. Students often become frustrated because they must debug their very first program. Recognizing this, we have included a detailed example of

debugging very early in the text (Chapter 2). Beginners also find debugging loops difficult. In this text, we focus on loops and illustrate how values change as loops are executed. Students learn to trace loops and find errors. In addition, common beginners' errors are noted at appropriate locations throughout the text.

12. The True-False questions at the end of each Lesson (with solutions) allow students to quickly assess their progress in grasping the basics.

13. The Application Exercises at the end of each chapter can be used by an instructor for home assignments.

14. A diskette with all of the programs in the book is included with the text. Students can modify and execute these programs to gain insight into how they operate.

15. We have included what is called an introduction to C++ that is available for free on McGraw-Hill's web site (http://www.mhhe.com/engcs/general/best/tandorazio.mhtml). It actually covers more than just the basics. After reading this chapter, students will be able to use many of the fundamental aspects of object oriented programming.

16. Many of the Application Programs give students an introduction to numerical methods. A total of 10 different numerical methods are illustrated.

HOW TO USE THIS BOOK

Students

In Chapter 1, you should focus on understanding what can be stored in memory, how a compiler works, and the steps in software engineering. The rest of the chapters cover programming in C. Each lesson in these chapters begins with a short introduction to the lesson's source code. Use the introduction to guide you through the important points of the code. Then read the code and the annotations in the boxes. After doing this, make sure that you begin to understand the major topics being covered in the lesson. Then read the Explanation and do the True-False and short answer exercises. If you do not do well on the exercises, reread the lesson until you feel comfortable with it, and then proceed to the next lesson.

After reading all the lessons in a chapter, begin the Application Programs. The purpose of these is to illustrate the thought processes that you would typically go through when you write your programs and to show practical uses of C. You will find as you write your own programs that you will be addressing many of the same issues raised in the Application Programs. In these, focus on learning the methodology and understanding the logic of each program. Remember, the details are very important in programming. Every statement of your programs must be correct. By fully understanding each Application Program, you will know how to design and create an accurate and efficient program. Use this knowledge in writing the programs assigned by your instructor.

Instructors

For a very rigorous 3- or 4-unit course in C programming with an introduction to C++, we recommend that you cover the entire text in the order presented. However, it is possible to cover the material in an order different from that given. For instance, Lesson 3.9 (Single Character Data) can be covered not as the last lesson of Chapter 3 but immediately before the first lesson of Chapter 7. Also, Lessons 8.11 (Creating Header Files), 8.12 (Use of Multiple Source Code Files and Storage Classes), 8.19 (Function-like Macros—on McGraw-Hill's web site), and 8.20 (Conditional Inclusion—on McGraw-Hill's web site) can be covered with the material in Chapter 5 (Functions) if so desired.

It is also possible to postpone some of the last lessons in Chapters 7, 8, and 9 and cover them as time permits. For instance, Lesson 7.9 (Pointer Notation vs. Array Notation) can be postponed until immediately before Lesson 8.15 (Pointers to Functions and Functions Returning Pointers).

For 2-unit courses, we recommend that you cover through Lesson 7.8 and Lessons 7.10, 8.1, 8.2, 8.3, 8.4, and 8.5. For a 1-unit course, students will be able to write valuable and sophisticated C programs if you cover just the first six chapters.

The book is filled with exercises. After each lesson are True-False and short answer exercises. The students should do these on their own. Some simpler programs are also assigned at the end of some of the lessons. These can be assigned as homework. One week is probably a sufficient amount of time for students to complete one of these programs.

The Modification Exercises at the end of the Application Programs can be used for courses that have a laboratory. Students should prepare for the laboratory by studying the pertinent Application Program. During the lab, students can be guided to make the modifications required by the exercises. The diskette at the back of the text has the source code for all the programs. Students simply can edit these programs on diskette. Some of these exercises are straightforward while others are difficult. Ones that are difficult can be assigned as home exercises.

At the end of each chapter are Application Exercises. These tend to be the most challenging exercises in the book, and therefore are most appropriate as home assignments. Two to four weeks is an appropriate amount of time for doing them depending on the difficulty level. McGraw-Hill will furnish a solutions manual to selected exercises.

In addition, the book can be used as a reference for much of ANSI C. Reference tables are distributed throughout the book and some are given in the appendices.

ACKNOWLEDGMENTS

We would like to thank Eric Munson and Holly Stark of McGraw-Hill for their interest, support, encouragement, and insightful comments regarding this project. It has been a real pleasure working with them. We would also like to thank Byron Gottfried, the consulting editor of the BEST series, for both his support and valuable criticism, and Alisa Watson and her production team at McGraw-Hill for putting together what we believe is a very nice-looking text.

We also had a number of very thoughtful reviewers. Betty Barr, University of Houston; Raymond Bell, University of Texas, El Paso; Tat W. Chan, Fayetteville State University; Bart Childs, Texas A&M University; Chris J. Dovolis, University of Minnesota; Janet Hartman, Illinois State University; Elden W. Heiden, New Mexico State University; Elias Houstis, Purdue University; Joseph Konstan, University of Minnesota; Jandelyn Plane, University of Maryland; and Matthew Ward, WPI, all made quite helpful comments and suggestions. Their input is greatly appreciated.

Our early computing careers began at the University of Michigan and U. C. Berkeley. We would like to thank Professors J. M. Duncan, John Lysmer, and Raymond Canale (the instructor in TBD's first freshman year computing class), for inspiring (and pushing) us to tackle creating some very complex programs. The satisfaction in developing these programs motivated us to write this book.

We would also like to thank Suzanne Lacasse and Kaare Hoeg, Director and former Director of the Norwegian Geotechnical Institute, respectively. Their confidence in us and their financial support in writing geotechnical application programs formed the basis for extending our computing backgrounds and established the foundation for further developing our programming skills. We thank Hui Xian Liu, the previous director of the Institute of Mechanical Engineering in Harbin, China, for his unending support, guidance, and encouragement of HHT.

Lastly, we would like to thank our families. TBD's wife, Elizabeth, although very busy herself, found the time to be encouraging and supportive during the entire process. With her, this endeavor has been a joy. HHT's wife, Wei Huang, has been wonderful with her warmth, tolerance, and intelligence. Sijing Tan, HHT's daughter, drew most of the figures in the first four chapters. We very much appreciate her effort and skill in doing these.

CONTENTS

*Located at McGraw-Hill's web site—http://www.mhhe.com/engcs/general/best/tandorazio/mhtml.

*Located at McGraw-Hill's web site—http://www.mhhe.com/engcs/general/best/tandorazio/mhtml.

Computers and Computing Fundamentals

1.1 HISTORY OF ELECTRONIC COMPUTERS

Electronic computers have a relatively short history. The Atanasoff-Berry computer (ABC) was developed in the 1930s at Iowa State University for the sole purpose of solving large numbers of simultaneous equations. The ENIAC (electronic numeric integrator and calculator) was a military computer built shortly thereafter for general computations, but the wiring had to be reconfigured, in a manner similar to an old-time switchboard, each time a new task was to be performed. For both of these computers, once the electronic wiring was set, only the data (that is, numbers to be added or other information with which to work) were input to the computer's memory.

A major advance in computing theory was made by the mathematician John von Neumann, who proposed an alternative to the process of reconfiguring the wiring. He introduced the concept of storing a computer's instructions in its own memory. The instructions, among other things, would dictate the directions and locations to which electronic pulses would flow in much the same way that wires dictated the flow. For computers built on von Neumann's concept, both the data and the computer's instructions were input to the computer's memory.

Figure 1.1a schematically illustrates a computer with direct wiring. Data is stored in cells A–F and is directly connected to the output cells, with A, B, and C going to X and D, E, and F going to Y, giving $X = A + B + C$ and $Y = D + E + F$. If we wanted to create the equations $X = A + C + E$ and $Y = B + D + F$, we would have to rewire the computer.

In contrast, Fig. 1.1b shows input cells for both data and instructions and wiring going to both X and Y from each input cell. One can envision that it would be easy to put other instructions into the instruction cells to change the output. For instance, if the desired results were to be $X = A + C + E$ and $Y = B + D + F$, it would be straightforward to change the input instructions and thereby get the desired result. Thus, the computer illustrated in Fig. 1.1b is superior to the one shown in Fig. 1.1a,

FIG. 1.1

Conceptual drawing of hardwired computer and computer with stored instructions

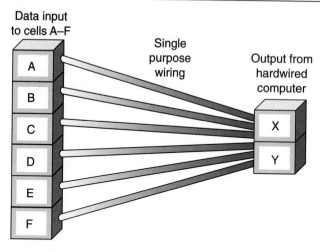

(a) Hardwired schematic with X = A + B + C
Y = D + E + F

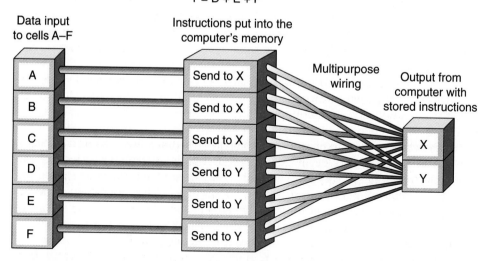

(b) Schematic showing output from computer using stored instructions with X = A + B + C
Y = D + E + F

because no rewiring is required in solving new problems. Only the instructions need to be changed to solve new problems for the computer of Fig. 1.1b. This represents the situation of modern day computers, in that both instructions and data are supplied to the computer. The programmer supplies the instructions to the computer.

The stored program concept led to the development of multipurpose computers because only the instructions needed to be changed to change the type of problem to be solved. It was found that each instruction input, although simple in itself, when combined correctly with other simple instructions could lead to getting the results of quite complex problems. For instance, one set of instructions could be put into the computer's memory to get the computer to perform long division and another set of instructions could be put in to get the computer to perform square root calculations. Then many of these sets of instructions could be combined to solve more difficult problems. As computers have become more complicated, developing computer instructions has become a science. In this book you will learn how to create instructions using the C language.

Electronics has changed significantly since the early days of computing. Vacuum tubes were used in the ABC and ENIAC. In the late 1950s, large transistors were the primary logic units for computers. In the middle 1960s, integrated circuits were developed. Integrated circuits actually contain large numbers of very small transistors, resistors, diodes, and capacitors. Today, integrated circuits on silicon chips form the heart of most computers. Technology improvements in materials and manufacturing processes have made the electronics smaller and faster. The processing power of a room full of 1950s computing equipment today can be put into a microprocessor chip the size of a small coin.

Advances have been made in the way that people communicate with computers as well. Manually moving wires, punching holes in cards, and typing on keyboards were some of the first methods used. Mice, pens, and voice are more recent developments.

Unlike the early days, today many different categories of computers exist. The boundaries between the categories sometimes are not distinct, but the following list describes some of the computers in use today:

- Supercomputers—The supercomputers are the largest and fastest of all computers. They are used primarily for large scientific or military calculations. Like the computers of old, they occupy a room or rooms of space. They are so expensive that universities do not own a supercomputer by themselves but instead share the use of a supercomputer, possibly located in another state or city. In your advanced studies, you may be allowed to use a supercomputer.
- Mainframe computers—These computers are not as fast or as large as supercomputers. Most universities own one or more mainframe computers. These computers usually take up a portion of a room and can be accessed from remote locations on campus.
- Minicomputers—These are less powerful than mainframes and inexpensive enough to be owned by small businesses.
- Workstations—Workstations can be as powerful or nearly as powerful as minicomputers. They are small enough to be able to fit on a desktop or next to a desk.
- Personal computers—These also can fit on a desktop. They are meant to be inexpensive enough for individuals, even students, to own. They have also been called *microcomputers.*

- Laptops—These can fit into a briefcase and are as powerful as personal computers.
- Palmtops—These are small enough to fit into a pocket and are less powerful than personal computers and laptops.

Amazingly enough, today's laptops are more powerful than the very large ENIAC.

In summary, over the years, computers have become smaller, faster, and easier to use. Originally developed for scientific and military applications, computers have evolved to being used in everyday life for many individuals.

1.2 ARCHITECTURE

The architecture of a typical computer, from mainframe to palmtop, usually consists of a central processing unit (CPU), main memory, controllers, and peripheral devices as illustrated schematically in Fig. 1.2. The power of the computer comes primarily from the CPU and the main memory, and in essence these two together are what people are referring to when they speak of "the computer" or the "brain of the computer."

Hardware is a term that refers to electronic and mechanical devices integrated into or connected to a computer. As such, all the devices mentioned in this section are considered to be hardware.

FIG. 1.2
Schematic of the architecture of a typical computer

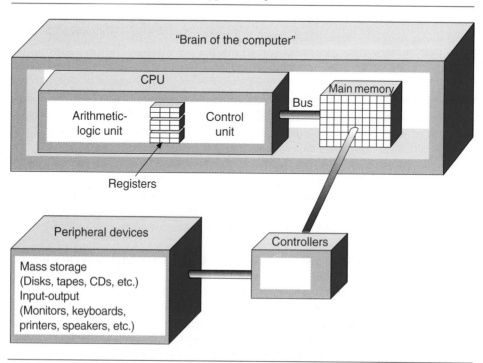

1.2.1 Main Memory

The purpose of the main memory is to store information that is to be processed or store instructions on how to process information. Information is stored in main memory in bits, which is short for "binary digits." Each bit is capable of being in one of two states. These two states are described in numerous ways:

On or off
Charged or discharged
One or zero
True or false
Set or clear
High or low

In this text we will describe the state of a bit as 1 or 0. Why is a two-state bit used in computers? Primarily because it is easy to create one of two states electronically. Figure 1.3 illustrates an electronic circuit storing to and reading from a bit in memory. The position of the switch in the figure determines the value that is to be stored or whether the memory is to be read.

In Fig. 1.3a, on the left, the value 1 is put into the bit location because the switch is set into the position that causes the capacitor to be charged. When we move the switch into the read position (on the right), we know that the capacitor had been previously charged because we can see the bulb light. We interpret the illuminated light as meaning that the bit has the value 1. This contrasts with Fig. 1.3b. On the left, the switch is in the neutral position, not charging the capacitor. After moving the switch to the read position (on the right), the light does not illuminate, meaning this bit has the value 0.

It is typical in memory for bits to be grouped into packets called *cells* or *words*. Very large numbers of cells form the memory. In Fig. 1.4 each cell has 8 bits (8 bits is typically called 1 byte). Each cell has its own address. The address serves a function much the same as your home's address. Your home's address indicates where letters to you are to be sent. A cell's address indicates where in memory a bit pattern is to be sent.

The bit patterns in each of the cells are a type of code that is explained later in this chapter. As a brief example, though, in one type of computer code the capital letter G is represented by the bits 01000111. You can see that cell B5 in Fig. 1.4 contains the code for G. When all of the bits of this cell are read, the letter G is interpreted. We shall see that symbols such as commas and dollar signs and numbers also can be stored in a code represented by bit patterns.

Also capable of being stored in memory are instructions, which, like letters and numbers, are represented by a type of code. For instance, a certain bit pattern stored in cell C3 may be such that it says to take the contents of cells C4 and C5 and add them together. Should a code be needed that is more than 8 bits long, it is possible to use two or more consecutive cells.

The size of memory often is given in terms of megabytes (MB) or gigabytes (GB). As we will see when we look at bit representations, the power of 2 is fundamental.

FIG. 1.3

Conceptual illustration of storing to and reading from memory. The capacitor represents one bit of a cell.

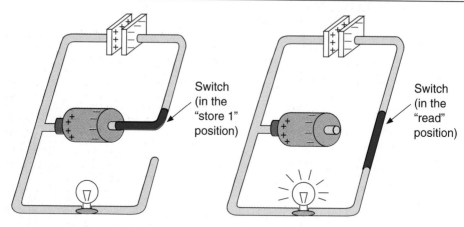

Switch connected to battery, charging capacitor—analogous to storing 1 as a bit in a cell.

Switch connected to bulb with the capacitor having been charged. Bulb lights—analogous to reading the bit value 1.

(a) The bit has a value of 1

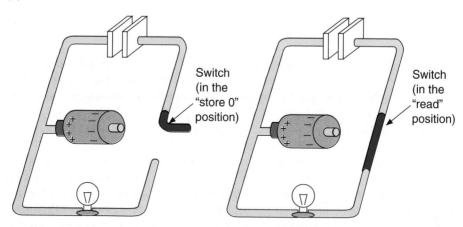

Switch in neutral position, not charging capacitor—analogous to storing 0 as a bit in a cell.

Switch connected to bulb with the capacitor not having been charged. Bulb remains dark—analogous to reading the bit value 0.

(b) The cell has a value of 0

FIG. 1.4

Conceptual image of cells and addresses

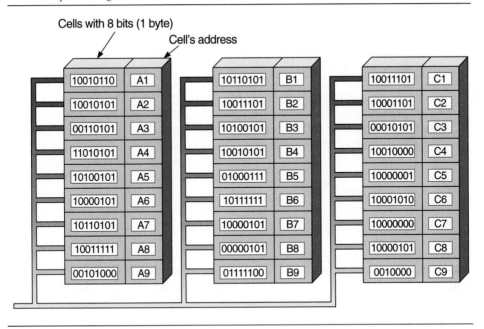

The term *mega,* although it normally means exactly 1 million, when referring to memory it is actually 2^{20}, which is 1,048,576. Therefore, 300MB of memory means $300 \times 1,048,576 = 314,572,800$ bytes of storage. Similarly, giga is not exactly 1 billion but 2^{30}, which is 1,073,741,824.

The main memory is connected to the CPU through a bus. A bus is conceptually a group of wires with one wire connected to each bit of memory. Random access memory (RAM) is a type of memory that allows access to each of the cells in no particular order. It contrasts with serial access, which indicates a sequence of cells that must be followed to extract the desired information. Read-only memory (ROM) represents bits of information that are permanent in memory, meaning the bits cannot be modified by the user because they have been set during the manufacturing process. ROM usually contains instructions and information considered to be fundamental to the computer's performance.

1.2.2 Central Processing Unit

The central processing unit consists of an arithmetic-logic unit (ALU), a control unit (CU), and registers. Each of these is described here.

Registers

Registers are memory cells within the central processing unit. Their purpose is to allow rapid access of information by the control unit and the arithmetic-logic unit.

Control unit

The control unit controls the activities of the CPU. It copies both data and instructions from memory. It decodes the instructions and obeys them. In other words, the control unit is responsible for copying information of interest from the main memory and putting that information into the registers. The unit communicates with the arithmetic-logic unit, informing it which cells contain the information and which cells should contain the results of any manipulations on the information. The control unit, acting on instructions received from memory, directs which circuitry should be utilized within the arithmetic-logic unit for operations to be performed. The control unit under normal conditions reads instructions from memory cells sequentially; however, it can do such things as jump to different cells when it receives instructions to do so.

Arithmetic-logic unit

The arithmetic-logic unit contains the circuitry for performing fundamental data manipulations. In addition to the fundamental arithmetic operations of addition, subtraction, multiplication, and division, it can handle so-called logical operations. Without explaining them in detail, these include the operations of COMPARE, AND, OR, SHIFT, and ROTATE.

1.2.3 Peripheral Devices

You probably are familiar with the peripheral devices, as many of these are the things you see, touch, and hear when you use a computer. Peripheral devices can be divided into two categories:

Mass storage (also called *secondary storage* or *secondary memory*)
Input-output (I/O)

Mass storage devices

The function of mass storage devices is to store information. They include units that operate such things as tapes, hard disks, floppy disks, and CDs. In these devices, the information actually is stored on tape, disk, or CD. The devices that rotate, write to, or read from the disks or tapes are called *drives:* tape drives, hard disk drives, floppy disk drives, or CD-ROM drives.

Information is stored on either disks or tape in a series of "on-off" markings. The "on-off" may be done with magnetic spots, where a spot on a disk (for hard disks or floppy disks) or tape is either magnetized or not magnetized. On CDs, the "on-off" is represented by either a pit or a smooth area. These "on-off" markings are put onto the disks in the same type of code briefly described in the section on main memory. The computer can read and write this code, making these mass storage devices useful for holding information.

Main memory also stores information; however, there are primarily four major differences between mass storage devices and main memory:

1. Accessing information in mass storage is slower than accessing information in main memory. This is because the information from mass storage must be transferred into main memory before it can be processed. Also, mass storage devices typically require mechanical motion. The required mechanical motion slows the access of information considerably.
2. The media of the mass storage devices (disks, tapes, etc.) can be portable and therefore used as a means of conveying information to other computers.
3. Mass storage devices usually have a greater capacity than main memory and serve the function of holding large amounts of information.
4. The information stored on mass storage devices remains even after power is cut off from the computer.

Input-output (I/O) devices

Many different input and output devices can be used by computers. For instance, a small computer in a microwave oven may have a temperature sensor as an input device and its ouput device would be the on-off switch of the oven—causing the oven to turn off when the temperature reaches a certain value. For a standard computer devoted to manipulation of information, the most common input devices are keyboards, scanners, microphones, and pointing devices such as mice. The most common output devices are monitors, printers, and speakers.

1.2.4 Controllers and Communication to Peripheral Devices

The controllers actually are miniature computers within themselves. Their function is to coordinate the actions of the peripheral devices connected to them with the actions of the computer.

One particular design for the connection of computers to peripheral equipment is illustrated schematically in Fig. 1.2. For this design, the central processing unit communicates with peripheral devices through controllers via the main memory. With this design, a controller can act very much like a CPU. For instance, the CPU can write a large amount of information into memory and then instruct that it be sent to a printer. Then the controller can take over and extract the information as needed from the memory and send it to the printer. This process also can work in reverse when, for instance, information is to be extracted from a disk. Based on instructions it has received from the CPU, the controller can take information from the disk and put it into the main memory.

Information constantly flows back and forth between the peripheral equipment and the controller to which it is connected. Why? There are many reasons. One is that the controller needs to know roughly what the peripheral equipment is doing. For example, a typical printer cannot print information as rapidly as it receives it. When the printer falls behind it must signal this to the computer so that the information can be saved and the flow of information to the printer can be slowed if necessary. Because the usefulness of a computer depends on its ability to coordinate peripheral devices, controllers are an important part of the computer system.

1.3   NETWORKS

We include a very short discussion of networks here because you likely will be using a network in doing your classwork. Networks are groups of individual computers and peripheral devices linked together to share information and resources. They can be broken down into two categories: wide area networks (WANs) and local area networks (LANs).

The Internet is an example of a wide area network because it spans such a vast geographical region. It links machines throughout the world. Because of the large number of resources and area encompassed by the Internet, a variety of types of machines exists within it. Part of the challenge of getting the Internet to work efficiently involves getting diverse machines to communicate efficiently with each other. Much of the shared information is sent over telephone lines. When you access a site on the World Wide Web, you may be receiving information from a mass storage device located in another part of the world. Should you so desire, you may have the information transferred to your own mass storage device for use at a later time.

Your university or the company you work for most likely has a local area network. It may be closed, in that it is meant to allow communication primarily among the computers within your organization (although one of the shared devices that may be on the LAN is a modem for communicating with other networks). Often, the type of equipment within a LAN is of a similar type. This simplifies the communication between the machines. Since the communication usually is not done over long telephone lines and the connections are much simpler, the response is much faster than for the Internet. Within a small company, a LAN is very useful for sharing printers, scanners, mass storage, and other devices that are expensive but not used continuously by a single user. Shared machines in a network are meant to appear to be devoted to a single user. A network server is a computer on the network that holds much of the network software and controls shared devices. For instance, network software on the server handles such things as setting the queue (pronounced "cue," it is a sequential list of users) for the printer. It allows each user to use the printer one after another. It also handles electronic mail and sharing mass storage devices.

Networks can be connected in various ways (called *topologies*). A few popular types are shown in Fig. 1.5. The Internet is an example of a network with irregular topology. Local area networks may have other types of topologies.

■ **FIG. 1.5**
Some computer network topologies

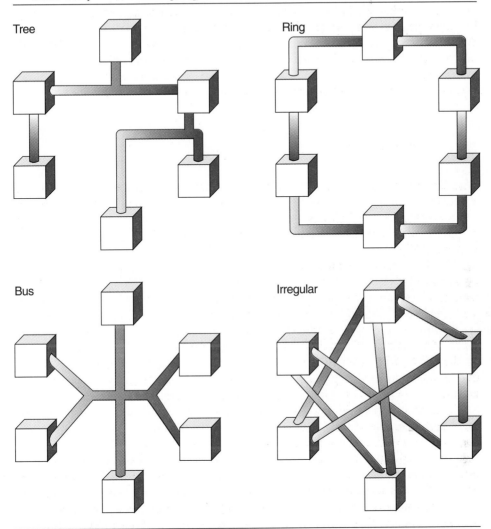

Tree

Ring

Bus

Irregular

■ **1.4   USING BITS TO REPRESENT CHARACTERS AND SYMBOLS, INTEGERS, REAL NUMBERS, ADDRESSES, AND INSTRUCTIONS**

1.4.1 Characters and Symbols

In the early days of data transmission, Morse code was used as a standard for dashes and dots to represent characters and symbols. The advent of computers led to the development of other codes that are more complex but serve essentially the same

■ **TABLE 1.1**
   Code samples

Symbol	ASCII code	EBCDIC code
A	01000001	11000001
B	01000010	11000010
C	01000011	11000011
D	01000100	11000100
E	01000101	11000101
F	01000110	11000110
?	00111111	01101111
+	01001110	01001110
(	01001101	00101000

function. Instead of dashes and dots, the customary notation for computers is 0 and 1. The EBCDIC (pronounced "ebb-sid-dikk" for extended binary-coded decimal interchange code) and ASCII (pronounced "ask-ee" for American standard code for information interchange) are two codes in large use for representing characters and symbols. EBCDIC commonly is used on IBM mainframes and ASCII commonly is used on personal computers.

Table 1.1 lists the codes for the capital letters A–F and a few symbols (a listing of the 128 standard characters for these codes is given in the appendix). Note that each of the characters and symbols is represented by 8 bits, which we will refer to as 1 byte.

1.4.2 Integers

Integers are represented in memory by a type of a binary or base 2 system. Base 2 is used because only 1s and 0s are needed to represent digits, and this fits in well with our bit representation of information. The binary system contrasts with our everyday number system, which is decimal or base 10. In base 10, recall that each placeholder represents a power of 10 with the rightmost placeholder representing 10^0 (which is 1). Each succeeding placeholder to the left represents a successively greater power of 10. For example, the five digit number 78326 is interpreted as being $(7 \times 10^4) + (8 \times 10^3) + (3 \times 10^2) + (2 \times 10^1) + (6 \times 10^0)$.

In binary, placeholders all represent powers of 2. The rightmost placeholder represents 2^0 (which is 1), and each succeeding placeholder to the left represents a successively greater power of 2 as illustrated next:

Base 2 placeholders	2^7	2^6	2^5	2^4	2^3	2^2	2^1	2^0
Decimal value of placeholders	128	64	32	16	8	4	2	1

EXAMPLE. Determine the base 10 representation of the 8 bit base 2 number 10010110.

Solution. Using the preceding table of placeholders, the value can be calculated by

Base 2 placeholders	2^7	2^6	2^5	2^4	2^3	2^2	2^1	2^0
Decimal value of placeholders	128	64	32	16	8	4	2	1
Binary digits	1	0	0	1	0	1	1	0
Binary digit times decimal value	128	0	0	16	0	4	2	0

The base 10 representation then is the sum of all of the numbers in the last row, which is $128 + 16 + 4 + 2 = 150$.

EXAMPLE. Using the previously described representation of binary numbers, determine the largest and smallest values that can be represented by an 8 bit cell.

Solution. The largest number is with all of the digits being 1 and the smallest is with all of the digits being 0. Therefore the largest number is $128 + 64 + 32 + 16 + 8 + 4 + 2 + 1 = 255$ (which is also $2^8 - 1$). The smallest number is 0. Note that negative numbers cannot be represented with this scheme.

Suppose we wanted to represent an equal number of negative and positive integers with 8 bits. How might it be done? One method would be to let one of the bits be a sign bit so that each bit would represent the following:

$$\text{Sign } 2^6 \ 2^5 \ 2^4 \ 2^3 \ 2^2 \ 2^1 \ 2^0$$

We could set the convention to be that 1 in the sign bit would indicate a negative number (0 would be positive). If so, the following range of numbers could be represented:

$$\text{largest } = 0111111 = +(64 + 32 + 16 + 8 + 4 + 2 + 1) = 127$$
$$\text{smallest} = 11111111 = -(64 + 32 + 16 + 8 + 4 + 2 + 1) = -127$$

The effect of using the sign bit is to shift the numbers being represented from all positive to half positive and half negative. This signed scheme is not exactly what is used in most computers to represent integers (the scheme commonly used, called *2's complement,* will not be described in detail in this book); however, it does represent many of the principals of representing integers:

- The number of bits used directly determines the size of the integers that can be stored.
- A sign takes up a bit and therefore a representation has a different range with and without a sign.

We will see that with the C language, we can specify the number of bits we want to use for integers (thereby controlling the size of the integer that we may use) and whether or not we want to use a sign in one of the bits.

1.4.3 Real Numbers

Real numbers are stored in binary format just like integers, characters, and symbols. The binary code for real numbers, though, is different from that used for integers.

The method for converting between binary and decimal for real numbers is very similar to that for integers. The placeholders to the right of the decimal point (called the *radix point* for bases other than base 10) are powers of 2 in the following order:

Base 2 placeholders	2^{-1}	2^{-2}	2^{-3}	2^{-4}	2^{-5}	2^{-6}	2^{-7}	etc.
Decimal value of placeholders	0.5	0.25	0.125	0.0625	0.03125	0.015625	0.0078125	

The next example illustrates the conversions.

EXAMPLE. Determine the base 10 representation of the 8 bit base 2 number 100.10110.

Solution. Using the preceding table of placeholders, the value can be calculated by

Base 2 placeholders	2^2	2^1	2^0	2^{-1}	2^{-2}	2^{-3}	2^{-4}	2^{-5}
Decimal value of placeholders	4	2	1	0.5	0.25	0.125	0.0625	0.03125
Binary digits	1	0	0	1	0	1	1	0
Binary digit times decimal value	4	0	1	0.5	0	0.125	0.0625	0

The answer is the sum of the numbers in the bottom line, which is 4 + 1 + 0.5 + 0.125 + 0.0625 = 5.6875.

To make efficient use of computer memory, though, real numbers are stored in a form of scientific notation. You may recall that in decimal, for example, 15230000 is represented in scientific notation as 1.523×10^7. The 1.523 is called the *mantissa,* 10 is called the *base,* and 7 is called the *exponent.*

We also can use scientific notation in binary. For example,

$$101.01100 \qquad = 1.0101100 \times 2^2$$
$$-0.0001011101 = -1.011101 \times 2^{-4}$$

We will not go into the details of exactly how real numbers typically are stored. However, from these simple examples, we can see that to store real numbers we must

• Store both the mantissa and the exponent
• Have space in memory for a sign for both the mantissa and the exponent

Unlike integers, where the number of bits used to store the integer restricts only the integer size, with real numbers the number of bits restricts both the size and precision (that is, the number of digits after the decimal point) of the numbers that can be stored. This is because, in storing real numbers, the total number of bits allowed must be apportioned between the mantissa and exponent.

1.4.4 Hexadecimal and Octal Notation

At times, humans must write bit patterns using a pencil and piece of paper or some other form. While a computer, in general, has no difficulty in dealing with a large number of 1s and 0s, humans find large numbers of 1s and 0s cumbersome. For instance the 32 bit representation of 635,163,077 is 00100101110110111101000111000101. If you had to deal with these long strings of numbers on a daily basis, you would find your work to be quite tedious and you easily could err. Also, the lack of direct correspondence between decimal number and binary number representations makes decimal numbers difficult to work with when dealing with bits and cells. As a result, hexadecimal (or base 16) and octal (or base 8) notation are used as a human shorthand for representing large strings of bits. Table 1.2 shows decimal, hexadecimal, octal, and bit patterns.

Note that capital letters are used in hexadecimal to represent 10 to 15 (decimal), because it is necessary to have a single symbol as a placeholder. The table can be used to create the octal and hexadecimal representations of long bit strings. The octal digits (0–7) are used to represent 3 bit groups (ignoring the left-most 0 of the bit pattern of Table 1.2) and the hexadecimal digits (0–F) are used to represent 4 bit groups. For instance, the bit pattern 101001100110011000111011 is represented in octal as 51463073, as shown next:

Bit pattern	101	001	100	110	011	000	111	011
Octal digit	5	1	4	6	3	0	7	3

TABLE 1.2
Comparison of notations

Decimal	Hexadecimal	Octal	Bit pattern
0	0	0	0000
1	1	1	0001
2	2	2	0010
3	3	3	0011
4	4	4	0100
5	5	5	0101
6	6	6	0110
7	7	7	0111
8	8		1000
9	9		1001
10	A		1010
11	B		1011
12	C		1100
13	D		1101
14	E		1110
15	F		1111

And 00100101 1101 1011 1101 0001 11000101 is represented in hexadecimal as 25DBD1C5, as shown next:

Bit pattern	0010	0101	1101	1011	1101	0001	1100	0101
Hexadecimal	2	5	D	B	D	1	C	5

You can see that both the octal and hexadecimal representations clearly are more wieldy than the long binary bit patterns.

1.4.5 Addresses

As we have seen to this point, anything that can be coded in binary can be stored in memory. Look at Fig. 1.6. Note that each memory cell has an address associated with it. Although we have not been precise in the way we have indicated the addresses, the concept is that a binary code can be used to represent the address of a cell. For instance, one can assume that the addresses listed in the figure are hexadecimal. One easily could convert these to binary. Then, one memory cell could be used to hold the address of another memory cell. In this figure, the address B5 is contained in cell A7.

FIG. 1.6

Conceptual image of cells and addresses

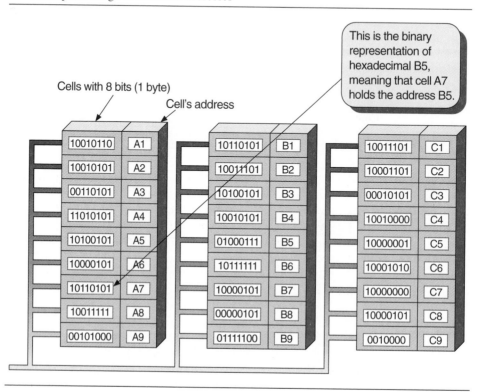

1.4.6 Instructions

Instructions need to be stored in a code, too. In this book we will not go into it in detail. However, you should be aware that the instructions you write in the C language must be converted into a binary type code and stored in memory. This binary code is referred to as *machine language*. Section 1.8 describes how you will convert your C instructions to machine language.

1.4.7 Comment

As described, all these can be stored in memory cells:

> Characters and symbols
> Integers
> Real numbers
> Addresses
> Instructions

All of them are stored in a binary code, but the code used for each one is different. Therefore, for the computer to interpret the information in a cell correctly, it must know what type of information is in the cell. In other words, if a cell contains an integer, the computer must somehow know or be told that an integer is in that cell. Otherwise, it would not know whether to treat the binary code as a real number, an instruction, an address, or whatever. In your C programs, you will indicate to the computer the type of information you will be storing in each cell.

In addition, the computer must know whether or not the information spills over into more than one cell. For instance, in your C programs, you will indicate whether a real number is to be represented by 4 bytes or 8 bytes.

1.5   PROGRAMMING LANGUAGES

Machine language was mentioned briefly in the previous section. It is the only language that a computer can understand. The language consists of instructions in binary code (or hexadecimal, for shorthand) that is specific to the particular processor for which it is to be used. Every step that the computer is to take must be written in these instructions. Because machine language is so cumbersome and tedious to write, most programs are written in other languages and translated into machine language. Many languages that can be translated into machine language have been written. Some of them are described here.

1.5.1 Assembly Language

Assembly language is considered to be one level above machine language. In assembly language, all of the instruction steps needed for the instruction list included for a program written in machine language are necessary. In other words, all instructions for moving information from memory to registers and back again must be included to successfully write a program in assembly language. The main advantage of assembly language over machine language is that the instructions in assembly language are not in a binary code but in English words. The English words are translated into machine language code with a language translating program. One can envision that it would not be difficult to convert English type words directly to corresponding binary code.

1.5.2 High Level Languages

High level languages are meant to further simplify the commands needed to be written by human programmers. For instance, to add together two numbers in machine language, a number of steps is required to transfer information from memory cell to memory cell. A simplified method might be simply to write "a + b"; and after the language has been translated, all of the series of instructions necessary to add two numbers would be written in machine language and stored in memory. Unlike machine language, high level languages allow programmers to write programs with far less concern about the internal design of the machine on which the program is to be used. It is necessary, however, that the translator be compatible with the computer. This makes programs portable from one computer to the next. If you write a program in a high level language on your home computer, you should be able to use that program on your university's mainframe, provided that the university has a translator for the language you are using.

High level languages all have rules that must be followed to get an accurate translation into machine language. Such languages are designed to simplify the writing of programs to solve particular types of problems. For instance, some languages are made to write programs that solve scientific problems while others may be meant to deal with such things as business accounting.

Languages (summarized in Table 1.3) can be broken down into four types:

Procedural (or imperative)
Functional
Declarative
Object oriented

At this point it is not worthwhile describing the differences among the various language types. However, it is worth noting that C is a procedural language. As implied in its name, a procedural language requires the programmer to lay out a procedure for solving a problem. As we will see in the Application Programs part of this text, we focus first on developing a procedure for solving a particular problem and then work on writing a program capable of carrying out the steps of the procedure.

Also note from Table 1.3 that C++ is an object-oriented language. Despite this difference, C is a subset of C++, meaning that everything you learn from this text about C can be applied later if you learn C++.

TABLE 1.3
A summary of some high level languages

Language name	Language type	Year developed
Fortran	Procedural	middle 1950s
Basic	Procedural	middle 1960s
Lisp	Functional	late 1950s
Prolog	Declarative	early 1970s
Ada	Procedural	middle 1970s
Smalltalk	Object oriented	middle 1970s
Pascal	Procedural	early 1970s
C	Procedural	middle 1970s
C++	Object oriented	middle 1980s

1.6   SOFTWARE

Recall that *hardware* is the name for all of the tangible equipment involved in a computer's operations. Software is not tangible. Software is a set of instructions that (after being translated into machine code, if necessary) can be read into a computer's memory and later executed on demand.

Software can be broken down into two categories: system software and application software. The vast majority of software on the market is application software, and this is the type of software that you will learn to write from this text. However, to write application software, you will need to use system software. Therefore, you should be familiar with some of the fundamentals of system software. Software categories are shown in Fig. 1.7. The relationship between users and the different categories of software is illustrated in Fig. 1.8.

1.6.1 System Software

System software includes operating systems, utility programs, and language translators.

Operating systems

The operating system is software (that is, a set of instructions) written into memory upon startup of a computer. The operating system gives instructions to the computer essentially to "watch for" and respond to messages given to it from the keyboard, a mouse, or other input device. An operating system establishes the instructions for utilizing such things as peripherals, memory, and registers, thereby freeing users and many programmers from worrying about certain details in dealing with these devices. The operating system creates the "look and feel" of the computer for the user. Usually the user is dealing with the operating system when he or she begins a session on a computer.

FIG. 1.7
Categories of software

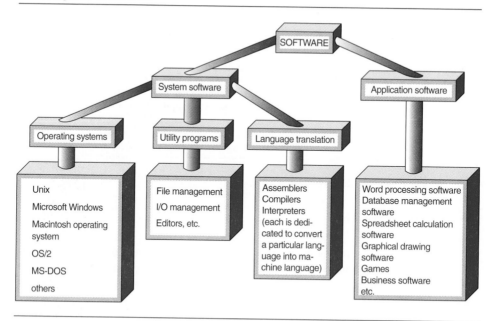

Because the operating system is the fundamental software used on a computer, there needs to be a match between a computer's circuitry and its operating system. In other words, one type of computer may or may not be capable of using a particular operating system.

Operating systems can be broken down into two categories: those for multiple user computers and those for single user computers. The names of operating systems are well-known among users and in the computing industry. Here are a few:

UNIX—multiple user
Microsoft Windows—single user
Macintosh operating system—single user
OS/2—single user
MS-DOS—single user

Multiple user operating systems. Supercomputers, mainframes, minicomputers, and workstations are computers to which large numbers of users can be connected simultaneously, all of the users vying for use of the CPU, printers, and other peripheral devices. One of the operating system's responsibilities is to coordinate all of the computer's activities with all of its users. It needs to do this in such a way as to satisfy the needs of all users—not making them wait too long before their commands are executed, if possible. Typically, an operating system will share the CPU among users by executing at least a portion of each users' instructions before moving to the next user. By doing this, no one user monopolizes the entire computer's resources.

FIG. 1.8
An onion skin type diagram illustrating the relationship between the users, software, and CPU. For instance, a user interacts with the application software, which interacts with the operating system, which interacts with the CPU and memory.

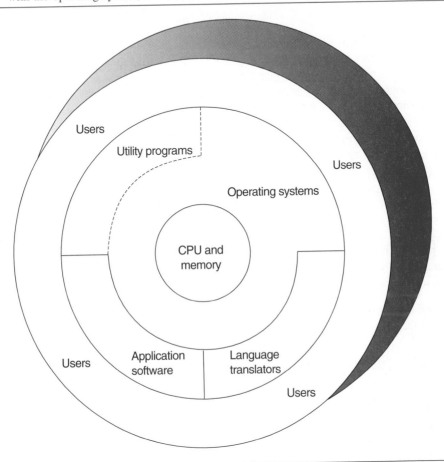

This procedure does not work for sharing printers, for example, because it would make no sense to print a few pages for one user and then a few pages for another, mixing them all up. Therefore, the operating system must share different resources in different ways and do so in an efficient manner.

Single user operating systems. These operating systems are focused on managing just one computer's memory, registers, and peripherals. Although only one user may be on a computer, input to the system can come from disks, the printer, the keyboard, or other input devices. The operating system is responsible for, for instance, allocating space in main memory for files read from disk. It also keeps track of information regarding files in mass storage.

Much has been said about the look and feel of operating systems for personal computers. The trend is to make computers easier to use by making interfaces more user friendly. Popular interfaces in current use are graphical user interfaces such as those used by Microsoft Windows and Macintosh operating systems.

Utility programs

Utility programs are software that perform the basic operations necessary for the fundamental performance of the computer system. Utility programs commonly are included in a package with an operating system. They perform such activities as creating, copying, saving, deleting, merging, and sorting files.

Utility programs encapsulated with operating systems are becoming more sophisticated, so much so that the boundaries between utility software and application software are becoming less distinct. So, you may find that one piece of software that is classified as utility software in one text may be classified as application software in another.

Editors are an example of such software. An editor is a program that enables a user to create and modify the contents of a text file. Early editors were very simple. They allowed a user to put information into a file but had limited capabilities for such things as moving around in the file, copying sections of files, and exchanging data between files. Today, however, such programs can have a variety of features that make them very versatile. Thus, although editors perform fundamental computer tasks, they can still be considered to be application software.

Language translators

Language translators are programs that create machine language instructions (or object code) from instructions written in assembly or a high level language. Your computer programs written in the C language will need to be translated into machine language using a language translator. Three types of language translators exist: assemblers, compilers, and interpreters.

Assemblers convert programs written in assembly language to object code. Since assembly language parallels machine language very closely, assemblers are less complicated in nature than interpreters and compilers.

Compilers work by taking an entire program written in a high level language and converting it to machine instructions. In the conversion process, errors in the high level language code may be detected by the compiler. The compiler looks for violations of established language rules such as improper punctuation or conflicting declarations. However, the compiler is not capable of detecting all possible errors, such as errors in logic. Often, beginning programmers are under the mistaken assumption that a program that compiles correctly is one that gives correct answers. However, a program that compiles with no error messages given by the compiler is much like an English paper written with no grammatical mistakes. It may conform to certain rules, but it does not necessarily make sense.

Compilers also are capable of linking together different translated modules. By linking numerous modules, one large program can be made out of many small ones. Some of the modules may be generated automatically by the compiler, provided statements requesting them are given in the code from a different module. Because modules may be added and because each instruction given in a high level language represents many machine language instructions, the file containing the machine language that is created by the compiler usually is considerably longer than the file containing the high level language code created by the programmer.

You will be using a C compiler to translate instructions you have written in the C language into machine language code. The compiler will search for grammatical errors and inconsistencies in your C code and describe them to you. After you have modified your program sufficiently to satisfy the compiler, you will need to execute and test your program to ensure that it performs the tasks it is intended to perform. If your program does not perform correctly, you will need to modify, add to, or delete some of your C language instructions; compile the program again; and execute it. You will need to repeat this process until you get satisfactory performance from your program.

Numerous C compilers are on the market. Most likely your university will have one that it will allow you to use. Your professor will give you instructions on how to use your particular C compiler. You also can purchase a C compiler from a software outlet. The instructions accompanying the compiler will tell you how to translate your C code into machine code and how to operate other features of the compiler.

Interpreters also are used for high level languages. Interpreters contrast with compilers in that they translate and execute instructions one after another. In other words, an interpreter takes an instruction given in a high level language, converts it to a machine language instruction, and executes it. Then the interpreter moves on to the next instruction and repeats the process until all of the instructions have been executed.

1.6.2 Application Software

Almost innumerable types of application software are in existence. The most commonly used types of application software involve such things as word processing, database management, drawing, graphing, and games. Specialty software for business, science, government, and education is developed every day. An organization that you work for may ask you to write application software for its internal use or to sell on the open market.

In this text, you will learn to write application programs. This book contains examples and illustrates the methodology for writing reliable, understandable, and efficient programs using the C language. In writing large application programs, it is necessary to follow a rigorous software design procedure to assure that the finished product lives up to expectations. The next section outlines various "software engineering" methods.

■ 1.7 SOFTWARE ENGINEERING

Software engineering is a term used to describe the process of software development. It indicates that software is not meant to be created haphazardly, but thoroughly thought out, planned, constructed, and tested. The term reflects the parallel between creating software and creating machines, buildings, and other such things that traditionally are thought of as engineered.

For instance, in the development of a large building, first, the functional requirements of the building must be established. Is the building meant to be used for office space, warehousing material, individual residences, hospital facilities, or maybe a combination of these? Once the purpose or purposes of the building have been established, the shape and layout of the building can be sketched out. At this stage, the expected users, the building's owner, architects, and engineers all can comment on the plans. Modifications can be made to satisfy all of the participants. After the general outline has been agreed to, the individual components of the structure can be addressed. Columns, beams, and walls can be designed in some detail, giving the sizes and exact placement of each individual member. Once as many details as possible have been laid out, construction can begin. However, the design process is not finished. Unexpected changes always need to be made, because as construction proceeds it becomes obvious that certain units do not perform as they are intended. Modifications to the initial design are implemented. Modifications to the design do not stop until a trial period has been completed and the occupants are satisfied. Note that all modifications are thoroughly thought out and detailed so that they fit efficiently and functionally in the building. Engineering drawings are maintained for reference should information about the building be needed to efficiently use the building or make future modifications.

Developing software is a similar process. The function of the software is defined first. An initial sketch of the layout is developed and input from all of the parties (users, owners, programmers, and others) is solicited. Modifications are made, and the design of individual components is addressed. It is recognized in the design that all of the components must fit together properly. Then, like a building, piece by piece, the software is constructed. Modifications are planned out and made as the software is assembled and tested for functionality. After completion, the software is comprehensively tested and modified as necessary. Documentation about the software is maintained carefully so that the software can be used efficiently or modified in the future. Along the way, cost and time estimates for each step are made to keep the project economically feasible and on schedule.

Thus, software design and development is a process that involves many steps and continues throughout the life of the software. In other words, a piece of software does not stop being developed after it is on the market. It continues to be developed and improved until it is found no longer to be useful.

Many of the steps involved in the design and development of software are illustrated in Fig. 1.9. From this figure it is important to note the "loops" shown. The loops are the steps repeated; that is, the testing and modifying steps. These are important because software continually is undergoing testing and modification.

In this text, you will learn how to sketch an initial program layout, write program code in the C language, follow a modular design scheme, test a program, make modifications, and write some of the documentation. Although this does not encompass all of the steps of software development, it is enough to allow you to plan and construct useful C programs.

FIG. 1.9
The steps involved in software development

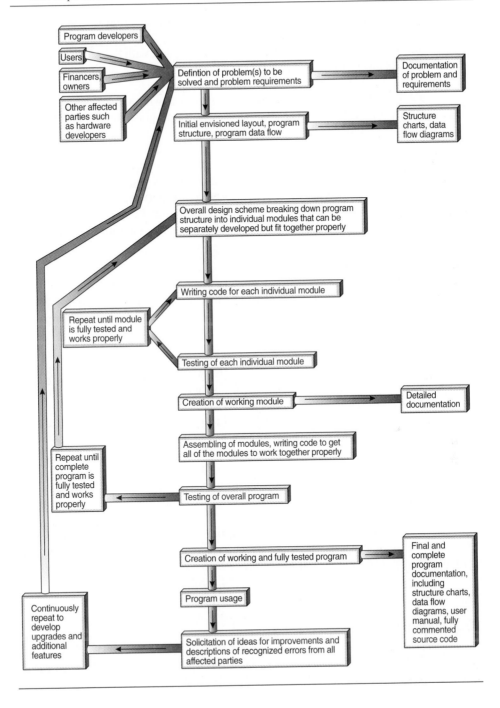

1.7.1 Top-down Modular Design, Structure Charts, and Data Flow Diagrams

A top-down design scheme begins by defining the total function of the software. Then subfunctions are developed, which are parts of the whole. Each part is less complex than the whole and can be designed separately but with the requirement that each of the parts must fit together properly. Each part that can be separately designed is considered a module.

A structure chart for a program illustrates the top-down modular design for that program. At the top of the structure chart is the overall program function. Below that are sequences of modules that perform subtasks.

A data flow diagram for a particular program illustrates the paths that data follow to get into the modules. It reflects the important data each module needs to do its computations, where the data comes from and where it goes.

For example, suppose you are required to write a program that handles the registration of classes at your university. A structure chart for this program might look something like that shown in Fig. 1.10. The modules perform the following functions:

Accept the student's class requests (Module 4)
Collect the student's class history (Module 5)
Accept the department restrictions on class size (Module 3)
Determine whether or not the student has met the prerequisites (Module 2)
Determine the student's class status—being admitted, on waiting list, or rejected (Module 1)
Prepare the official list of students for the class (Main Module)

Structure charts are software development tools. At this point you probably do not have the experience needed to develop efficient structure charts. However, with time, you will develop the skill.

Another type of software development tool is a data flow diagram. The data flow diagram for the registration class list program might look like that shown in Fig. 1.11, which illustrates the data transferred into and out of each module. Clearly, each module performs some manipulation on the data. It shows the external sources (such as the databases) that contribute information to the program. This particular data flow diagram does not rigorously follow the rules of drawing such diagrams; however, it illustrates the concepts.

With a structure chart and a data flow diagram for a large program, the tasks for program development can be given to various parties; for instance, a different person may develop each module. It is critical, however, that on completion, the modules be able to communicate accurately and efficiently with each other.

1.7.2 Functions

In C, the modules are called *functions.* The functions are primarily of two types: library functions and user-defined functions. Library functions are modules that

FIG. 1.10
Structure chart illustrating top-down modular design for a program that prepares the official list of students in a class

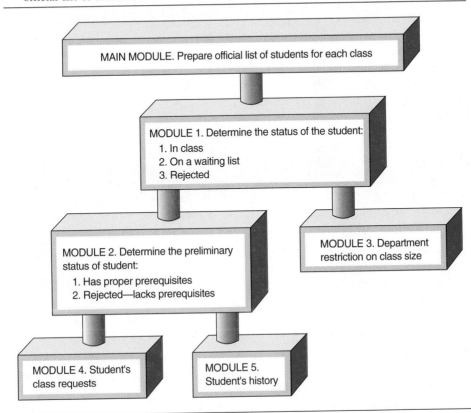

already have been developed and are included with the C compiler. They perform things like standard mathematical operations, such as the sine function, which is capable of finding the sine of an angle. That these functions are available saves the user from developing the specific instructions for performing many standard operations.

User-defined functions can be custom-made by a programmer. They perform tasks that are not available from the function library of the C compiler. In the class registration list program, for instance, a user-defined function would be used to compare a student's list of class requests to the available spaces in each class.

Data flows into and out of both library and user-defined functions. It is up to the programmer to assure that the data flow is correct and efficient. In this book you will learn how to write user-defined functions and how to make use of library functions.

FIG. 1.11
Data flow diagram for the class list program

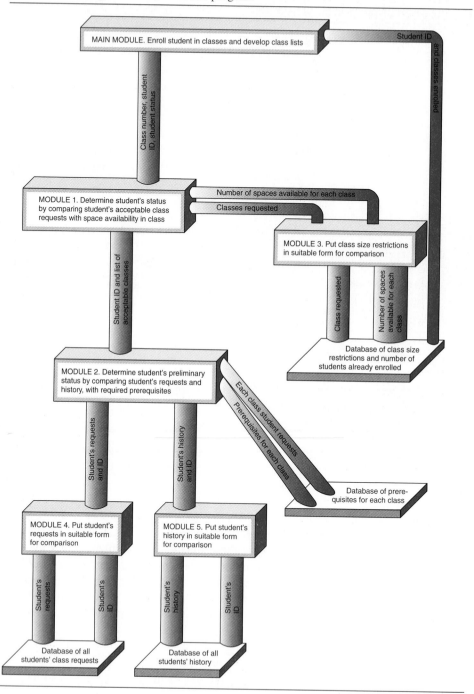

■ 1.8   THE C LANGUAGE, ANSI C, AND C COMPILERS

1.8.1 C and ANSI C

The C language was developed in the early 1970s at Bell Laboratories (now a part of Lucent Technologies, Inc.) by Dennis Ritchie. The language initially was developed to replace assembly language for creating system software. It is a high level language yet capable of controlling low level operations. It is highly portable and machine independent because of its structure that utilizes library functions to a great extent.

To assure that the C language remains standard, in 1989, the American National Standards Institute (Committee X3J11) approved a version of C that is meant to be machine independent (American National Standard X3.159-1989), ANSI C. The intent is that developers of C compilers make their compilers capable of handling correctly all programs written in ANSI C code. Thus, a program you write in ANSI C should be capable of being compiled and executed on any machine with a compiler that is ANSI C compatible. In this text, we follow the ANSI C standard. Wherever we show anything that varies from ANSI C, we note it.

In 1990 (and amended in 1994), an international C standard was adopted (ISO/IEC 9899:1990 with Amendment 1 1994), ISO C. Except for very minor editorial type changes, ISO C and ANSI C are alike. In this book we use the term *ANSI C* quite frequently. However, in whatever statement we use it, the term *ISO C* would be equally valid.

1.8.2 C Compilers

The primary goal in writing the programs we discuss in this text is to create what is called an *executable file*. An executable file is a set of machine language instructions (in binary form) that is ready to be executed by a user. When you buy commercial software, you purchase the executable file (among other things) that you load onto the hard disk of your computer. Once there, you can run it when you need it. You will need to create executable files from the programs that you produce through reading this book. Once you have created these executable files, you can store them on your hard disk and use them when you need them, just like commercial software.

Most of this text is devoted to showing you how you can create your source code. This, generally speaking, is the most difficult part of the process. The process of converting correct source code into an executable file, in general, is much simpler than writing the source code itself, because the compiler does most of the work in converting C statements to machine language. The steps involved in converting a correct source code into an executable file are shown in Fig. 1.12.

In general, modern C compilers perform four different types of operations:

1. Editing text to assist in creating source code
2. Preprocessing source code
3. Compiling source code and code attached by the preprocessor
4. Linking object code generated by compiling with other object code

FIG. 1.12

Steps involved for a programmer to create an executable file. Not all compilers work exactly like this; however, most compilers follow this general form.

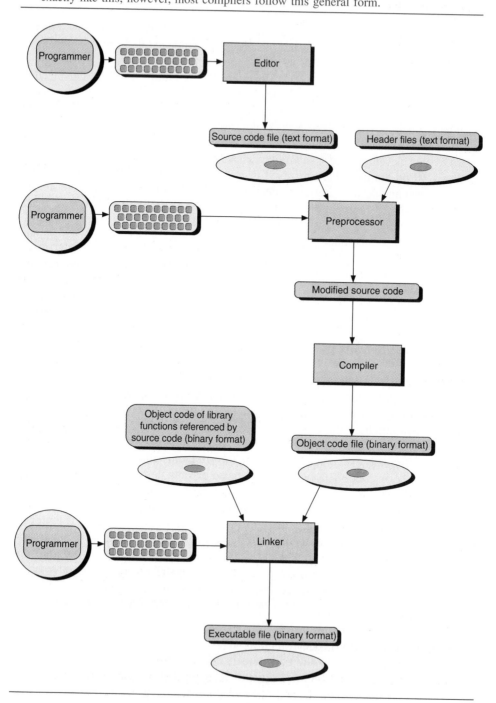

Note that, although it is called a *compiler*, it performs more than compiling (that is, translating the C code into machine language or object code). These four operations are indicated by rectangles in Fig. 1.12.

To generate an executable file,

1. The programmer uses the editor (a particular portion of the C compiler) to create a source code file. The programmer can simply type in the text from the keyboard. This text is written according to the rules described in this book. You will spend many hours learning how to write this C code properly.

2. Once this code is properly written, the programmer can direct the compiler to begin the compilation process. This begins with preprocessing. The preprocessor portion of the C compiler does a number of things. Among them, it attaches source code that has been previously written and built in by the compiler to the source code written by the programmer. It also modifies the programmer's source code following instructions that the programmer has given it.

3. The primary portion of the compiler takes the modified source code passed to it by the preprocessor and translates it into machine language instructions. It creates what is called *object code* and stores this object code in another file.

4. Then the programmer calls on the linker portion of the C compiler (or the compiler does it automatically) to complete the machine language instructions and to make sure that all of the instructions are assembled in the proper manner. In most cases, the source code written by the programmer will call on functions in the C library to perform certain operations. These functions are stored in binary form in the library. The linker determines which functions are called on and attaches them to the object code created by the translation of the source code. Once the linker has finished, compilation is complete and an executable file has been generated. This file is stored on the hard disk.

Compilers on the market have different characteristics. Some have editors built into them, making it very simple for you to create your source code and perform the necessary steps for you to create your executable file. Some of them automatically will perform all of the described compiler operations with just a single keystroke command or click of the mouse. Some compilers require more effort, and there may be times when you may want to manually get them to go through all of the steps. In your class, most likely you will be given access to a C compiler or you will choose to buy one for your personal computer. Because of the number of C compilers available and publishing constraints on this text, space does not permit a description of each compiler. Your course instructor should be able to give you the information needed to utilize the C compiler made available to you.

If you have purchased a compiler, to get started with this book, please refer to the user manual for information on

Naming source code files
Creating source code files (perhaps with a built-in editor)
Preprocessing, compiling, and linking files
Executing programs (also known as *running* programs)
Redirecting input and output (that is, changing the standard input and output
   devices to something different from the keyboard and the screen)

◼ 1.9 ABOUT THIS TEXTBOOK AND HOW TO GET THE MOST OUT OF IT

Each chapter of this book is divided into two parts: the language lessons and the practical applications. The chapters begin with a series of lessons. These lessons are meant to familiarize you with characteristics of the C language. After all of the lessons of the chapter have been completed, the second part of each chapter illustrates practical applications and specific programming techniques that you will find useful in developing your own programs.

1.9.1 The Lessons

To benefit most from the lessons in this book, you should follow this step-by-step procedure:

1. Read the introduction to each lesson and follow the instructions as it guides you to examine the lesson's source code. Look at the source code of the example program and the program's output and make the observations directed in the introduction. Try to answer the questions posed about the source code. This is an important step, and you should try to glean as much information as you can from looking at just the program and its output. In other words, treat this portion of the book as something of a puzzle to be solved. You will learn the material best if you try to figure out on your own what each lesson's program is doing and why it is doing it. We have deliberately designed this portion of the book to guide you and give you enough information to deduce the meaning of the C code by yourself.
2. Read the lesson's explanation portion to gain further insight into programming and to clarify points that may have been confusing in step 1. During this step, you frequently should refer back to the lesson's source code. It is important that you read and interpret as much C code as you can. After finishing this book you should be "fluent" in the C language—unafraid of facing many pages of pure C code given to you by your instructor or employer.
3. Do the exercises at the end of the lesson and satisfy yourself that the concepts have been learned sufficiently well to prepare yourself for the next lesson.

1.9.2 The Application Programs

In reading the Application Examples pay close attention to the methodology used for developing them. In illustrating the development of the application programs, we utilize a multistep procedure. We recommend that you follow the same procedure in developing your own programs.

Beginning in Chapter 3, the procedure is

1. Assemble the relevant equations.
2. Do a hand calculation of an example problem.

3. Write an algorithm (sometimes called *pseudocode*) that uses the equations and follows the pattern of the hand calculation. We recommend that you write an informal algorithm, which is roughly a line-by-line description of what the program does. It should be written in plain English. As you become more advanced, you may be able to make this portion of your preparation look more like the source code that you will write.

4. Use the algorithm to write the actual source code.

As the programs become more complex, we add steps for such things as developing structure charts and data flow diagrams and for planning the data structures to be used in the programs. Although we recommend that you follow our procedure in writing your own programs, we recognize that, as you become more adept at programming, you may be able to skip some of the steps, develop a method that suits your own style better, or use a method preferred by your instructor. While it is not necessarily important whose methodology you follow, it is very important that you follow a formal procedure rather than develop programs haphazardly.

■ EXERCISES

1. Convert the following binary numbers to decimal:

    1001001001,   101010.101010,   1111100000

2. Convert the following hexadecimal integers to decimal integers:

    F8,   3D5,   A5BE

3. Convert the following octal integers to hexadecimal integers:

    12,   345,   7654

4. Convert the following octal integers to bit patterns:

    12,   345,   7654

5. Convert the following hexadecimal integers to octal integers:

    F8,   3D5,   A5BE

6. Convert the following hexadecimal integers to bit patterns:

    F8,   3D5,   A5BE

7. List at least five high level computer languages developed after World War II.

8. List three input devices and three output devices to be used for computer programming.

9. List three types of application software that you can buy on the market.

10. List three popular computer operating systems.

11. List the ASCII codes for the three capital letters—A, B, and C—using binary and decimal integers.

◼ APPLICATION EXERCISES

1. Part of the field of transportation engineering involves developing efficient routing schemes. In this exercise, we describe a computer program that can be used to assist a transportation engineer in developing routes.

Many computer applications involve the use of graphs or meshes. A mesh consists of a set of nodes and a set of connectors that connect them. For example, the freeways that connect San Francisco and New York can be considered a form of a mesh, where different segments of the freeways are the connectors and the cities at the junctions of the connectors are nodes. Assume you are asked to write a program to find a route to drive between Los Angeles and New York.

a. Read all the cities that are in the mesh.

b. Assign a unique number (node) for each city using any sequence that you like. For example, in Fig. App 1.1a, the number 0 is assigned Seattle, 1 to San Francisco, 2 to Los Angeles, 3 to Salt Lake City, and so forth. In total there are eight nodes.

c. Make a table that shows whether a freeway directly connects two cities. The table consists of eight rows and eight columns. Row numbers 0 through 7 represent nodes 0 through 7. Similarly, columns 0 through 7 represent nodes 0 through 7. If a freeway directly connects two cities, we mark 1 in the table; otherwise, we mark 0. For example, there is a freeway between Denver (node 5) and Richmond (node 7). Therefore, we write 1 at the intersection of column 5 and row 7. Similarly, we write 1 at the intersection of column 7 and row 5. Altogether, there are 16 nonzero numbers in the following table, representing eight connectors in the mesh shown in Fig. App 1.1b.

	Row 0	Row 1	Row 2	Row 3	Row 4	Row 5	Row 6	Row 7
Column 0	0	1	0	0	0	0	0	0
Column 1	1	0	1	1	0	0	0	0
Column 2	0	1	0	0	0	0	0	0
Column 3	0	1	0	0	1	1	0	0
Column 4	0	0	0	1	0	0	1	0
Column 5	0	0	0	1	0	0	0	1
Column 6	0	0	0	0	1	0	0	1
Column 7	0	0	0	0	0	1	1	0

d. Read the data in the table.

e. Starting from Seattle, ask the computer to exhaustively explore all the connectors (i.e., possible corridors between two cities), visiting each city (node)

FIG. APP 1.1A
United States interstate highways (partial, not to scale)

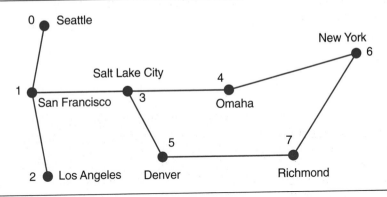

as many times as necessary, until a path between Seattle and New York is found. For each try, we have to make a table to allow the computer to remember cities that have been visited. For example, if the path is from Los Angeles to San Francisco to Seattle, the table should contain nodes 2, 1, and 0. When the computer finds a dead-end route, such as connector 01, which represents the route between San Francisco and Seattle, your program must be smart enough to change the value of connector 01 from 1 to 0 so that, in the next search, connector 01 will not be searched again and again.

f.   Tell the driver when a path is found; for example, path 213576.
g.   Exit the program.

FIG. APP 1.1B
Graph with nodes and edges

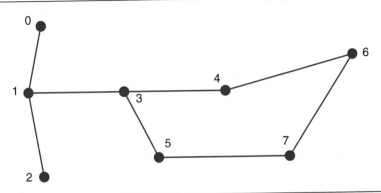

Your assignment is to

h. Create a structure chart and data flow diagram for this program.

i. Add modules to the structure chart and data flow diagram that make the program very user friendly. For example, add a module and ask if the driver wants to find an alternative path. If the answer is yes, then search the next possible path. For example, when path 21346 is found, display it on the screen.

j. Add at least three more modules that will make your program user friendly.

2. We use this exercise because you likely are familiar with computer games. Many computer games use a labyrinth on the screen. The story usually is simple. The hero or heroine enters a demon-haunted labyrinth, captures the demon, and rescues the captive. The labyrinth as shown in Fig. App 1.2 looks different from Fig. App 1.1a. However, the physical form of the labyrinth can be modeled as a mesh as shown in Fig. App 1.1b. The trick is that each junction in the labyrinth is marked with a number that represents a node, the corridor between two junctions therefore becomes a connector. Your assignment is

a. Given the labyrinth in Fig. App 1.2, draw a mesh that models the labyrinth.

b. Given the labyrinth in Fig. App 1.2, make a table that models the labyrinth.

c. Draw a structure chart with at least ten modules to make your game interesting.

FIG. APP 1.2
A labyrinth game

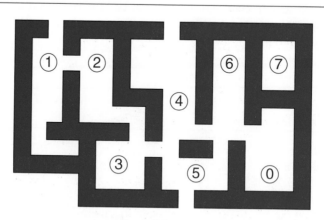

Getting Started—Program Structure, Printing, and Comments

In this chapter you will learn how to write simple C programs that can display words, numbers, and keyboard symbols on the screen. In themselves, the example programs described in this chapter have limited practical use. However, portions of these programs form integral pieces of many valuable C programs.

▨ LESSON 2.1 BASIC STRUCTURE

Topics

- Writing a simple C program
- Using the printf( ) function to display text on the screen
- Structure of a simple C program
- Basic rules for writing a C program

The program that follows illustrates the basic structure of a C program. When you execute the program, the statement

This is C!

appears on the screen and remains there until other tasks eliminate it.

Examine the program and the output carefully before you read the explanation. The program is written in what is called *code*. You may understand why it is called code because it looks difficult to understand. In this book, we want you to try to interpret the code before we tell you what it means. Since at this point you know nothing about C, you probably will find the first few lessons difficult. As you learn more, though, you will find that indeed you can interpret much of the code before we explain it to you.

For this program look most carefully at the line containing the word `printf` and the output from the program. Can you reason out specifically what this line does? If you wanted to display the statement "I am in the process of learning a programming language" on the screen, how would you write the `printf` statement?

Source code

```
#include <stdio.h>
void main(void)
{
    printf("This is C!");
}
```

Output

This is displayed on the screen after the source code has been compiled and the program has been executed:

```
This is C!
```

Explanation

1. What does `#include <stdio.h>` *mean?* As mentioned in Chapter 1, C programs are divided into modules or functions. Some functions are written by programmers such as you. Others are stored in the C library and can be accessed by you. If you want to access library functions, it is necessary to give the C compiler information about the functions to be accessed. At this time, we will not go into the details. However, we can tell you that the file named `stdio.h` (which is included with the C compiler and is called a header file) contains information about the library function, `printf`, used in the program just shown. The directive `#include<stdio.h>` tells the preprocessor portion of the C compiler (meaning the part of the compiler that performs actions prior to the translation of the code into machine language instructions) to attach the `stdio.h` file to the file with the source code shown. This action is illustrated in Fig. 2.1. Because the line of code `#include <stdio.h>` causes the preprocessor to act, it is called a *preprocessor directive.* We use, at most, only a few preprocessor directives for each program that we write. Therefore, writing preprocessor directives will not be a major part of the programming process for us.

After the preprocessor has attached the file `stdio.h`, all of the information necessary to use the function printf is included, and the program can be translated successfully into object code. Before using any library functions in your programs you will need to determine the names of the files that contain the necessary information for each library function you want to use. If you do not include the proper files, the C compiler will send you an error message when you attempt to compile your program, and your program will not compile successfully.

As an example, if you were to use a math library function in your program, the directive would be `#include <math.h>`. A complete list of library functions and

FIG. 2.1

The preprocessor portion of the C compiler attaching the file, stdio.h, to the Lesson 2.1 source code. This is caused by the directive, #include<stdio.h>. After this action, the code can be successfully translated into machine language because it has enough information to properly utilize the function printf, which is used in the source code.

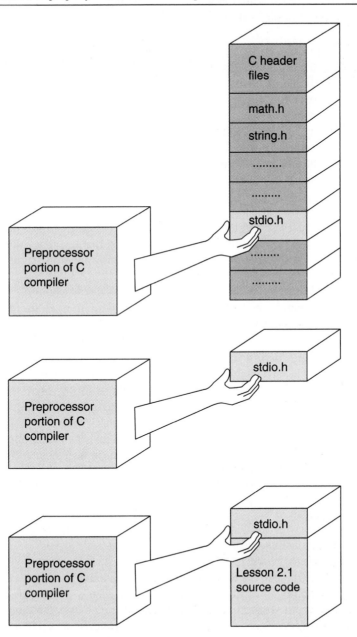

the files with the necessary information to include is given later in this book. At this time you need to know only the form of the directive and that stdio.h is the correct file for the function printf.

You will be able to recognize preprocessor directives in other programs in this book, as they always begin with # and usually are written as the first lines of the programs. The preprocessor also can perform other useful tasks prior to code translation.

2. What does `void main(void)` *mean?* This line gives the name of the function that we have written and information about what goes into and out of the function. In this case, the name of the function is `main`.

C not only allows you to use library functions, it allows you to write your own functions. Your first task in programming is to write the function `main`. Every C program has a *primary function* that must be assigned the name `main`. The name `main` is mandatory and cannot be altered by you, the programmer. In other words, even if your program had the purpose of printing your address to the screen you could not name the primary function `printmyaddress` and have the line be `void printmyaddress(void)`. The function must be called `main`.

We will not describe in detail the meanings of the *voids* in this line. However, we can say that the first `void` in the line indicates that no information is passed from main to the operating system (which is considered to be the unit that calls the function main) and that the second `void` indicates that no information is passed from the operating system to main.

In this book we use `void main(void)` until Chapter 5, where we will cover this topic more thoroughly. For now, you should simply memorize the form of this line as you will use it in all of your programs.

3. Why does this function need to be called main? The C compiler needs to know where execution is to begin. By definition, the primary function, main, is the first function to be executed. For the first several chapters of this book we use only main and library functions. In Chapter 5, we show how to define functions other than main. Other functions can be named almost anything that you want. A typical program has a large number of functions that you have written. The C compiler searches for the function named main and compiles the program in a manner to ensure that main is the first function executed.

4. What is the meaning of the braces? After the line that contains the function name is the function body. The function body has the following features:

- It begins with an *opening brace* {.
- It ends with a *closing brace* }.
- The pair of braces, {}, are used to enclose what is called a *block of code*. We use braces quite frequently to form blocks of code. Sometimes we use blocks within blocks. In this case, the braces enclose the block of code that is the function body.
- The function body consists of C declaration(s) and statement(s). The structure of a typical C main function is as follows:

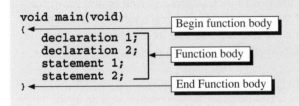

```
void main(void)
{                    ┌─── Begin function body ───┐
    declaration 1;
    declaration 2;   ┌─── Function body ───┐
    statement 1;
    statement 2;
}                    ┌─── End Function body ───┐
```

5. What is the meaning of `printf ("This is C!");`**?** It is a statement that calls the function printf from the C library.

6. What is the form of a call to a library function? A library function is called by giving the function name followed by a left parenthesis and then the information to send to the library function followed by a right parenthesis, such as

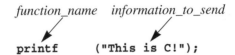

function_name   *information_to_send*

printf   ("This is C!");

The statement `printf("This is C!");` gives the function name, printf, and the information to send "This is C!". The phrase between the double quotes is called a *string literal* or *string constant*. It is a type of parameter that can be transferred to the function printf. The printf function takes the string literal (which it recognizes to be what is called a *format string*) and causes it to be printed on the screen. Figure 2.2 illustrates the sequence of invoking the printf function.

▨ **FIG. 2.2**
Calling the printf function from the C library. For this lesson's program, only the printf function is directly linked to main.

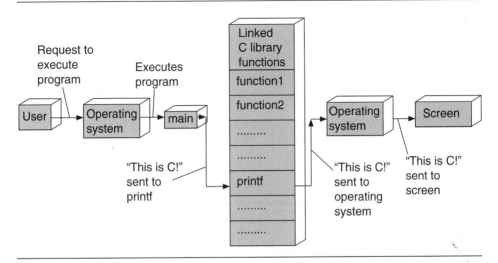

The string literal is sent from the main function and received by the printf function. The printf function is one of many *standard functions* contained in the library of the C compiler. These functions are available to you as a programmer and may be called or invoked by using a statement in your program body.

7. Why does the line `printf("This is C!");` ***end with a semicolon?*** The line `printf("This is C!");` is a *C statement.* C statements appear in the body of functions. C statements must be terminated by a *semicolon,* which is also known as a *statement terminator.* The semicolon at the end of a C statement acts much like a period at the end of a sentence. A typical program has many C statements, each terminated with a semicolon.

8. Will the following C program work correctly?

```
main()
{
        printf("This is C!");
}
```

It may. However, this depends on your compiler. The reason it may work is that

1. The file stdio.h is used so frequently in C that, for some compilers, it is attached automatically to programs despite no specific direction to do this.
2. Even though you have not written "void main(void)"—that is, the voids are missing—C assigns default types for the voids. The word *default* is used commonly in computing. A default value is a value that is used when none is specified. A default type is a type that is used when none is specified. Without going into detail, because C assigns default types for the missing voids, this program may work. However, we do not recommend that you write your programs in this manner. For clarity, we prefer the form shown at the beginning of this lesson. You may see other texts use the preceding form. If you see this, you should simply be aware that defaults are being used in the program.

9. Can we use both uppercase and lowercase letters to write C code? The C language distinguishes between lower- and uppercase letters. Thus, printf is different from PRINTF, Printf, or PrIntF. It is said, therefore, that C is considered to be case sensitive. In this lesson, both `main` and `printf` must be written in lowercase letters. For naming self-developed functions, you may use whatever case you consider appropriate. However, C traditionally is written primarily in lowercase letters. We describe situations when other than lowercase letters commonly are used, as the need arises.

10. Where are blank space(s) permitted in C code? C code consists of a number of *tokens.* A C token is the smallest element that the C compiler does not break down into smaller parts. A token can be a function name, such as main, or a C reserved word, which we discuss later in this chapter. All C words should be written continuously. For example, the expression

```
void ma    in(void)
```

is not legal because no blank characters are allowed between the characters `a` and `i` in the word `main`.

Between tokens, *white-space characters* (such as blank, tab, or the carriage return) can be inserted, but this is optional. For example, the line

```
void main(void)
```

is equivalent to

```
void main (   void      )
```

or

```
void     main     ( void)
```

In general, it is acceptable to add blanks between tokens but not acceptable to add blanks within tokens.

11. Is it necessary to use different lines in writing code? A white-space character, such as that created when you press the Enter key, used between tokens is invisible to the C compiler. Therefore, you have the freedom to write C code at any row or column you like. The C compiler, for example, allows you to rewrite and pack program L2.1.C into one line

```
#include <stdio.h>void main(void){printf("This is C!");}
```

or rewrite it as

```
#include<stdio.h>void
main(   void            ) {
   printf
(   "This is C!"  )    ;        }
```

However, these styles will make your program more difficult to understand and should not be used.

There is no required form for spacing within your program. However, your instructor or employer may want you to adhere to certain standard accepted styles. Our example programs are meant to illustrate acceptable style, however, at times, publishing constraints do not allow us to follow any one accepted style rigorously. Indentation and spacing are considered important for the look of a program, even though they do not affect performance. To make your program readable, do such things as write one statement per line, line up your braces, and add blank lines where there are natural breaks in code instructions.

12. Why is the look of a program important? The look of a program is important because (as was described in Chapter 1) programs continually are undergoing change. A program that is neat and organized is easier to understand and therefore easier to modify. As a result, the likelihood for error is reduced in programs that follow a certain visual and organizational style compared to those that do not.

13. What are the most important things to remember from this lesson? We realize that we give you a large amount of information in this first lesson. Some of it may be confusing because, to prevent digressing too far, we have deliberately not described some of the details. We will cover the details later when you will be able to understand their value. However, from this lesson you should have learned the following:

1. C programs consist of functions and the primary function is main.
2. Information can be transferred from the calling function to the function being called and vice versa. For the program shown, we have not used any information transferred from printf to main, and no information was transferred between main and the operating system. However, a string literal was passed from main to printf.
3. C has a function library that can be accessed by your programs. To use library functions you need to attach a file that has information about the library function you are going to use. The form of the directive attaching a file is

```
#include <filename>
```

4. You can define your own functions. To define a function you need to state the function name and types of parameters passed to and from the function. Until Chapter 5, we will begin the definition of the main function with

```
void main(void)
```

The voids mean that no information is passed directly from main to the operating system or from the operating system to main.
5. A function body is enclosed in braces, { . . . }. The braces form a block. In the block are declarations and statements.
6. You can call a library function from the body of another function. To call a library function, you write the name and the information to be transferred to the function.
7. At this point you do not have to memorize the form or syntax of all the code we have illustrated. However, as you program more, you should commit to memory the syntax of the types of C code that you use frequently.
8. A summary of the components of this lesson's program are shown in Fig. 2.3.

14. What will we cover next? The next lesson will cover how to write comments. After that, in the rest of this chapter, we focus on the use of the printf function. We focus on this function because you will use it repeatedly in your programs. Also, other functions use information similar to that used by the printf function. Therefore, learning printf will give you experience and knowledge in the C language. However, printf is just one of many functions that you will use in your C programs.

EXERCISES
1. True or false:
    a. In general, C statements are case sensitive.
    b. By default, any C statement is location sensitive.
    c. A C statement must be terminated with a period.

■ **FIG. 2.3**
Components of Lesson 2.1

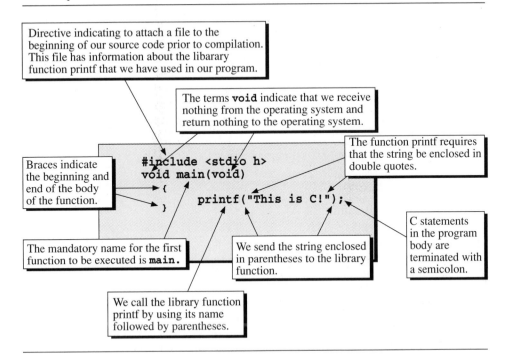

Directive indicating to attach a file to the beginning of our source code prior to compilation. This file has information about the libarary function printf that we have used in our program.

The terms **void** indicate that we receive nothing from the operating system and return nothing to the operating system.

The function printf requires that the string be enclosed in double quotes.

Braces indicate the beginning and end of the body of the function.

```
#include <stdio h>
void main(void)
{

    printf("This is C!");
}
```

C statements in the program body are terminated with a semicolon.

The mandatory name for the first function to be executed is **main.**

We send the string enclosed in parentheses to the library function.

We call the library function printf by using its name followed by parentheses.

---

    d.   The name of the primary function of a C program must be Main. *main*

    e.   main(){} is a complete and correct C program.

    f.   printf and main are C tokens.

2.   Find the error(s), if any, in each of these statements:

    a.   void main(void);

    b.   printf ("Do we need parentheses here?");

    c.   printf("Where do we need blank space? ") ;

    d.   printf("Is any thing wrong here?")

3.   Type, compile, and run this program:

```
main(    void         )
{ printf ("There is no class tomorrow!"  )  ;  }
```

Correct any errors you may find.

4.   Type, compile, and run this program:

```
ma  in() PRINTF *, ('What is wrong?' }
```

Correct any errors you may find.

5. Modify this lesson's program so that it prints your name and address to the screen.

    Solutions

1. a (true); b (false); c (false); d (false); e (true); f (true)

2. a. ```void main(void)```
    b. No error
    c. No error
    d. ```printf("Is any thing wrong here?");```

LESSON 2.2 WRITING COMMENTS

Topics

• Reasons for writing comments
• Structure of comments
• Location of comments
• Continuing comments
• Style of writing comments

If you thumbed through later chapters in this book, you probably noticed that C computer code (that is, the actual C statements) looks somewhat cryptic. Sometimes even the most experienced programmers find it difficult to understand the tasks that a program or a portion of a program are to perform. You very likely will find as you go on that, if you pick up code that you wrote only a few days earlier, you have trouble understanding how it is supposed to work. Also, others may be working with your code and find that what you have written is difficult to understand.

Fortunately, C allows you to write "comments" within the code of your programs. Comments are notes describing what a particular portion of your program does and how it does it (or anything else that you would like to write in the middle of your code). They are a very important addition to programs because they convey information that is difficult to convey through the code itself. When written well, comments reduce the likelihood for error, because a programmer making changes to a program can read the comments to understand how a program operates. Comments serve as an important part of the program documentation and therefore are required by employers and instructors.

If not written correctly, though, the C compiler may think that the words written in the comments are functions, such as printf, or other C expressions or statements. If this occurs, the compiler will indicate that your program has errors when you compile it.

The program that follows is not meant to be an example of good commenting but only an illustration of the mechanics of commenting. It is meant to illustrate a program that performs a task identical to that performed by the program in Lesson 2.1. It prints a line of code to the screen. However, unlike Lesson 2.1's program, it is filled with comments. Comments properly written provide no action instructions to the computer. Compare this code to Lesson 2.1's code. Which lines have been added here? Look at the beginning and end of each new line. Can you tell the way that you signify to the C compiler that a statement is a comment statement? What symbols indicate the beginning and end of a comment? Is it necessary for the beginning and end of a comment to be on the same line? Remember that this source code is not meant to be an example of good commenting style (in fact, it is poor style); however, it does illustrate some characteristics of comments.

Source code

```
/*L2_2.C - In this book, the source for Lesson x_y is Lx_y.c */
 #include <stdio.h>      /* This is an include directive */
void main(void)
{/*The purpose of this program is to print one
   statement to the screen */

    printf("How do we write comments in C?"); /*
   The printf function prints to the screen */

 /*This program performs the same task as
   program 2_1.C */
}/* A closing brace is needed to end the program*/
```

Output

```
How do we write comments in C?
```

Explanation

1. What is the structure of a comment? The syntax of a C comment is

```
/*     Any text, number, or character     */
```

where there should be no blank(s) between the slash and the asterisk. In addition, the /* and */ must form a couple. The /* and */ are called *comment delimiters.*

The /* and */ must form a couple, but they need not be on the same line. Therefore, a comment line may occupy more than one line. A multiline comment starts with /*, followed by multiple lines of text consisting of numbers, characters, or symbols. The multiline comment terminates with */. See the first comment in this lesson's code for an example.

Examples of incorrect comments:

```
/* Wrong comment 1, no end asterisk and slash
/* Wrong comment 2, no end slash *
/ *Wrong comment 3, there is a blank between /and * */
```

2. Can we write comments at the very beginning and end of a program? Yes, a comment line can be written in the very first line of a program. It also can be written on the very last line of a program.

3. Can a comment appear in a C statement? That depends. The C compiler treats comments like a single white-space character. Therefore, comments can appear in a C statement only where a white-space character is allowed. This means that comments may be placed between tokens but not within a token. (Note: a string literal is a token.)

Examples of correct comments:

```
printf /*This comment is legal */("Welcome to C!");
printf ( /*This comment is also legal*/"Welcome to C!");
```

Examples of incorrect comments:

```
pr /* This comment is illegal because it splits a C function
name (i.e., a token) */ intf("Welcome to C!");

printf("This is not a comment and /*will be displayed*/ ");
```

The text contained in the `/*` and `*/` is not a comment because it becomes part of the string literal and will be displayed. For this incorrect comment, the compiler will not indicate an error, it will simply print the phrase:

```
This is not a comment and /*will be displayed*/
```

on the screen. For the other comments, the compiler will indicate an error when you try to compile the program.

4. Can we write a nested comment? No, comment statements cannot be nested (meaning that we cannot write a comment within a comment) in C. For example,

```
/*/* This is an illegal comment because it is */ nested */
```

5. Why would we want to write a nested comment? We never want to write a nested comment, but it is very easy to create one accidentally. A common technique for isolating operations of source code is to "comment them out"; that is, remove some of the operations temporarily by converting them into comments to see the effect on the performance. For instance, suppose we were working with this lesson's program (meaning we were modifying and running it repeatedly to create a new program). We could prevent the printf statement from executing by putting comment delimiters `/*` and `*/` around it to give

```
/* printf("How do we write comments in C?");*/
```

If we compile and run the program with the line of code like this, the compiler will think that it is a comment and not convert it into machine language instructions.

It then is a simple task to later re-create the line of code by removing the comment delimiters. In other words, if you had deleted the line of code to prevent its execution, you would have had to retype it to get it back. Commenting the line out saves you typing and easily gives you exactly what you had before. For this program, such action would not help us very much, but for other programs, you will find this technique quite useful for finding errors in your programs.

However, if we make a mistake in where we put our comment delimiters, we accidentally could enclose a comment with the statements we want to comment out. For instance, if, without thinking carefully, we typed the delimiters on lines above and below the printf statement:

```
/*
    printf("How do we write comments in C?"); /*
    The printf function prints to the screen */
*/
```

we would have created a comment within a comment, which is illegal.

6. How do we write valuable comments? Be wise in your use of comments. Remember that comments are used to enhance the understandability of your programs. Make them pleasing to the eye and clear. With practice, you will develop a style of writing comments that is of benefit to you. In this text, we do not use many comments, primarily because we want you to interpret our code yourself. Your programs should use comments far more frequently than we use them.

We recommend that you avoid writing comments on the same line as other C code unless you can clearly distinguish the comments from the code. It often looks much better if your comments are on lines separate from other statements. There is no standard for writing comments, but we like a style that highlights comments and separates them from other C text. The reason for this is that, if not highlighted, comments tend to blend in with the rest of the code and make following the logic of the code confusing. In other words, do not hide your comments. Make them stand out— they are there to help you and others. We like a style in which each comment line begins with two stars, and the comment block begins and ends with a line of stars. For instance, we strongly recommend that you add a *banner* at the beginning of your program. A banner is a set of comments that describe such things as the name, parameters used, history, author, purpose, and date of the program. A better look for the program for this lesson is with a banner as follows:

```
/********************************************************
** Name: L2_2.C                                        *
** Purpose: Learning how to write comments in C        *
** Date: Written on 11/22/98                            *
** Author: Joe Kelly                                    *
** Reference: None                                      *
********************************************************/

#include <stdio.h>
int main(void)
{
    printf("How do we write comments in C?");
}
```

The disadvantage of making comments stand out is that it takes time to type them in this way. When you are in a hurry, you will tend to skip the comments. Do not let this style of programming continue for very long. Plan ahead to set aside half an hour or so each day to do nothing but write comments. In the long run, the time that you spend writing comments will save you considerable frustration in finding errors in your programs. As you gain experience with programming, you will become aware of how many and what types of comments are useful.

Take writing comments seriously. You will be regarded as a better programmer and your programs will have fewer errors if you write comments properly. Your coworkers and employer will appreciate your good comments.

EXERCISES

1. True or false:
   a. A C comment line may appear on the first line of a program.
   b. A C comment line may not appear on the last line of a program.
   c. We may write a comment at the end of a C statement.
   d. We may write a comment line that contains 120 characters.
   e. At times, C allows us to write a nested comment.

2. Compile and run this program (file LE2_2_2.C) that executes correctly but is not good style:

```
/* Comment before the program */
main /* We learn how to write comments*/ ()
{ printf /*This is not a nested*/ /*comment*/
    ("Let's fly to Paris" );
} /* Comment after the program is OK */
```

3. Correct, rewrite the program neatly, and then compile and run the program file LE2_2_3.C:

```
ma/* This comment is illegal */ in()
{printf("Let's fly to Paris");
/*This /* is a nested*/ comment */ }
/* This comment does not have the closing '*' and '/'
```

4. Modify this lesson's program so that it has a banner reflecting you as the programmer and other pertinent information.

   Solutions
1. a (true); b (false); c (true); d (true); e (false)

▨ LESSON 2.3 FORMATTING OUTPUT

Topics
- Formatting output
- Line feeding

The first example program of this chapter showed how to print a single line to the screen. However, in most cases you will want to print multiple lines to the screen, and you will want to display these lines in such a way that they have proper spacing. Proper spacing can be achieved by what is called *line feeding*.

Look at the first two printf statements in the source code that follows and the output from them. The output looks strange. Even though we used two printf statements, the output appears on just one line and without proper spacing. What does this tell you about how the printf function works?

Within the string literals used in printf statements we can insert symbols that are not printed to the screen but are interpreted by the printf function to be commands to move the cursor (also called the *insertion point*) around the screen. Look at the third printf statement in the code and the output from it. What combination symbol and character in the string literal does not get printed on the screen? What does this symbol-character combination cause the cursor to do?

What do you think the fourth printf statement does? Does the fifth printf statement give the output you expect?

Source code

```
#include <stdio.h>
void main(void)
{
    printf("Welcome to");
    printf("London!");
    printf("\nHow do we\njump\n\ntwo lines?\n");
    printf("\n");
    printf("It will rain\ntomorrow\n");
}
```

Output

```
Welcome toLondon!
How do we
jump

two lines?

It will rain
tomorrow
```

Explanation

1. Suppose we want to display

```
Welcome to
London!
```

in two lines on the screen. Can we use two printf() function calls

```
printf("Welcome to");
printf("London!");
```

to reach our goal? No, this is because the function, printf(), does not automatically advance a line to the next line each time it is called. Therefore, the output from the first printf() function call will be connected to the output of the next printf() function call and that is the reason why "Welcome to" and "London!" are printed on the same line.

2. How do we linefeed a line? The linefeed operation can be done easily with the printf() function using the linefeed symbol \n in the string literal. The symbol \n consists of two characters, \ (backslash, not to be confused with slash, /) and n, with no blank in between. In C, the two character symbol \n is one of many character *escape sequences.* The C compiler considers an escape sequence within a string literal as one character (not two). The importance of this will be seen later. The escape sequence \n causes the cursor to move to the next line and will not be displayed on the screen. Any data behind this symbol is written at the beginning of the next line. You can use \n at any location in the string literal. The \n can be at the beginning, in the middle, or at the end of a string. The number of \n can be more than one. For example, in the statement

```
printf("\nHow do we\njump\n\ntwo lines? \n");
```

the program uses the first \n to move the cursor to a new line, displays "How do we," uses the second \n to jump to a new line, prints "jump," uses the next two \n to jump another two lines, prints "two lines?", and uses the last \n to jump one more line.

3. Can we use the linefeed symbol by itself? Yes, the symbol \n can be used by itself to advance a line as long as it is contained in double quotes. The fourth printf statement in this lesson's code demonstrates this.

EXERCISES

1. True or false:
    a. The statement printf (\n\n\n); will create 3 blank lines. ✗
    b. The statement printf ("\nnn"); will create 3 blank lines. ✗
    c. The statement printf ("\n\n\n"); will create 3 blank lines. ✓
    d. The statement printf ("\n \n \n"); will create 3 blank lines. ✓
    e. The statement printf ("\ n\ n\ n"); will create 3 blank lines. ✗
    f. The escape sequence \n represents two characters. ✗

2. Find the errors, if any, in each of these statements:
    a. printf("I \n Love \n California \n");
    b. printf("I \ n Love \ n California \ n");
    c. printf("I n Love n New York \ n");

3. Compile and run the program file LE2_3_3.C:

```
main()
{printf("I \n Love \n California \n");
 printf("I \ n Love \ n California \ n");
 printf("I n Love n California n");
}
```

How can you correct this program to have the output make sense?

4. Write a program to display a ten-line story on your screen.

5. *Modification exercises.* Modify this lesson's program so that it
    a. Prints all of the text on just two lines.
    b. Prints the following output:

```
Welcome to London! How do we
jump
two lines?
```

```
It will rain    tomorrow
```

    c. Prints the following output using just two printf statements:

```
Welcome to London! How do we jump
```

```
two lines? It will rain
```

```
tomorrow
```

Solutions
1. a (false); b (false); c (true); d (true); e (false); f (false)
2. a. No error
    b. No error but will not linefeed, character n also will be displayed
    c. No error but will not linefeed, character n also will be displayed

▓ LESSON 2.4 MORE ESCAPE SEQUENCES

Topics
- Generating sound
- Concatenating a C string literal

The \n escape sequence is one of many that can be used within a string literal. All of the escape sequences begin with the backslash symbol. Escape sequences shown in the next program are capable of generating a beep and moving the cursor to different locations on a line.

Look at the second printf statement in the source code that follows and the output from it. What does \b do? Look at the third printf statement and its output. What does \r do? What does \t do in the fourth printf statement?

The fifth and sixth printf statements have string literals that carry over to more than one line. Remember, white space is meaningful within a string literal. How have we connected the two lines without printing extra space?

Now that you have seen that double quotes have special meaning in printf statements, you may wonder how we can print double quotes to the screen. Look at the last printf statement and the output from it. Can you see what we need to do to print double quotes? What is within the string literal that has not been printed?

Source code

```
#include <stdio.h>
void main(void)
{
    printf("Listen to the beep now. \a");
    printf("\nWhere is the 't' in cat\b?\n\n");

    printf("I earned $50 \r Where is the money?\n);
    printf("The rabbit jumps \t\t two tabs.\n\n");

    printf("Welcome to\
 New York!\n\n");

    printf("From "        "Russia \
with "        "Love.\n");
    printf("Print 3 double quotes   -\" \" \" \n");
}
```

Output

```
Listen to the beep now.
Where is the 't' in ca?

Where is the money?
The rabbit jumps                    two tabs.

Welcome to New York!

From Russia with Love.
Print 3 double quotes   -"  "  "
```

Explanation

1. How do we generate beeps? The escape sequence \a in the printf() string literal

```
        printf("Listen to the beep now. \a");
```

generates a beep after "Listen to the beep now." has been displayed.

2. How do we backspace? The escape sequence \b in the printf() format string

```
        printf("\nWhere is the 't' in cat\b ?\n");
```

moves the cursor back one space after the *t* in the cat has been displayed, so we do not see the *t*.

3. How do we move the cursor to the beginning of the current line? The escape sequence \r in the printf() format string

```
printf("I earned $50 \r Where is the money?\n");
```

will not display any character before \r. The escape sequence \r represents a carriage return and moves the cursor to the beginning of the current line.

4. How do we concatenate (which means to connect) a C string literal? Here, we illustrate three methods. In method 1, we use a backslash at the end of a line to indicate that a string literal has not finished and continues on the next line. Since the C compiler disregards all blank characters behind a statement, the connection to the next line will start right at the end of the preceding statement. If you want to include blank characters in a statement that occupies two lines, either place them before the backslash in the first line or at the beginning of the second line. For example, the statement

```
printf("Welcome to New       \
York!");
```

is equivalent to

```
printf("Welcome to New       York!");
```

but not

```
printf("Welcome to New York!");
```

In method 2, we enclose each unfinished string literal in double quotes; for example, the statement

```
printf("From " "Russia "
    "with" " love.\n");
```

is equivalent to the statement

```
printf ("From Russia with love\n");
```

Method 3 is a combination of methods 1 and 2. For example, the preceding statement is equivalent to

```
printf("From " "Russia \
with " "love.\n");
```

5. How do we use printf() to display double quotes? Because double quotes are special symbols that could be misinterpreted if used alone within a string literal, we must put a backslash immediately in front of them to display them on the screen. No space is allowed between the backslash and the double quotes following it. Thus, the statement

```
printf("Print 3 double quotes   -\" \" \" \n");
```

produces the following output:

```
Print 3 double quotes   -" " "
```

6. Summary. Character escape sequences consist of a backslash followed by a letter, symbol, or a combination of digits. Each represents a character that has special meaning or specifies an action. The printf() and other output functions may use the following character escape sequences in a format string. (Note: Table 2.1 is a complete list of escape sequences, which you can use as a reference later. At this point you do not need to understand the meanings of all of them.)

Figure 2.4 summarizes the escape sequences used in this lesson's program.

TABLE 2.1
Character escape sequences

Escape sequence	Meaning	Result
\0	Null character	Terminates a character string
\a	Alert/bell	Generates an audible or visible alert
\b	Backspace	Moves the active position (e.g., for the console, this is the current cursor location) back one space on the current line
\f	Form feed	Moves the active position to the initial position at the start of the next logical page (e.g., ejects printer page)
\n	New line	Linefeeds to the initial position of the next line
\r	Carriage return	Moves the active position to the initial position of the current line
\t	Horizontal tab	Moves the active position to the next horizontal tabulation position on the current line
\v	Vertical tab	Moves the active position to the initial position of the next vertical tabulation position
\0ddd	Octal constant	Represents an integer constant using base 8 where ddd represents a sequence of digits 0–7 only
\xddd \Xddd	Hexadecimal constant	Represents an integer constant using base 16, where ddd represents a sequence of the decimal digits, and the letters a–f or A–F represent values of 10 through 15 respectively
\\	Backslash	Displays a backslash
\'	Single quote	Displays a single quote
\"	Double quote	Displays a double quote
\%	Percent	Displays a percent character
\?	Question mark	Prevents the misinterpretation of trigraphlike character sequences; e.g., trigraph sequence ??= will display the character #, but \?\?= will display ??=

FIG. 2.4

Escape sequences of this lesson's program

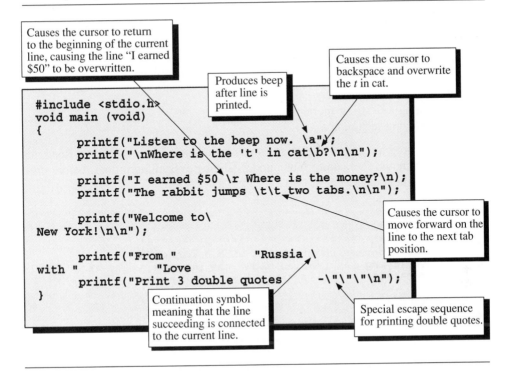

Causes the cursor to return to the beginning of the current line, causing the line "I earned $50" to be overwritten.

Produces beep after line is printed.

Causes the cursor to backspace and overwrite the *t* in cat.

```
#include <stdio.h>
void main (void)
{
        printf("Listen to the beep now. \a");
        printf("\nWhere is the 't' in cat\b?\n\n");

        printf("I earned $50 \r Where is the money?\n);
        printf("The rabbit jumps \t\t two tabs.\n\n");

        printf("Welcome to\
New York!\n\n");

        printf("From "          "Russia \
with "          "Love
        printf("Print 3 double quotes    -\"\"\"\n");
}
```

Causes the cursor to move forward on the line to the next tab position.

Continuation symbol meaning that the line succeeding is connected to the current line.

Special escape sequence for printing double quotes.

EXERCISES

1. True or false:
   a. The statement `printf ("ABC\a\a");` will display ABC and generate two beeps.
   b. The statement `printf ("ABC\b\b");` will display ABC only.
   c. The statement `printf ("ABC\r\r");` will display A only.
   d. The statement `printf ("ABC\t\t");` will display ABC.

2. Compile and run the program file LE2_4_2.C. What do you expect to see on the screen?

   ```
   main()
   {printf("\n1. I Love \a\a\a California \n");
    printf("\n2. I Love \b\b\b Chicago \n");
    printf("\n3. I Love \r\r\r Hawaii \n");}
    printf("\n3. I Love \t\t\t Nevada \n");}
   ```

   Solutions
1. a (true); b (false); c (false); d (true)

▨ LESSON 2.5 REVIEW OF CHAPTER AND COMMENTS ABOUT ERRORS AND DEBUGGING

Topics
- Basic structure of a C program
- Writing comments
- Using character escape sequences
- Displaying special characters
- Concatenating a C string literal
- Basic debugging techniques

The next program uses the techniques you have learned in this chapter. Read the program. You should feel comfortable, understanding everything it does, before proceeding to the next chapter.

Source code

```c
/*********************************************
** Your programs should begin with a banner
** displaying information about the program
*********************************************/

#include <stdio.h>
void main(void)
{
    /***********************************************
    ** The program body of every C main function
    **    must start with an opening brace {
    ** Comments are an important part of C programs.
    ***********************************************/

    printf("Review" " Chapter 1 --- Getting\
started!\n\n");

    /***********************************************
    **  printf() is a standard C function. The
    **  string to be displayed must be enclosed by
    **  two double quotes. The symbol \n means
    **  linefeed and carriage return.
    **
    **
    **  To call a standard C function, we start with
    **  the function name, followed by a pair of
    **  parentheses. The arguments to be
    **  passed are placed inside the parentheses.
    **
    **  Please note the semicolon ';' at the end of
    **  printf() statements. A statement in C must be
    **  terminated with a semicolon.
    ***************************************************/

    printf("We have learned\n\n\
\r 1. How to write comments \n\
```

```
        \r 2. How to linefeed \n\
        \r 3. How to continue a line \n\
        \r 4. How to use tab \\t, beep \\a, and\n\
        \r  carriage return '\\r' \n\
        \r 5. How to print special characters %%, \\,\"\n");
}

    /*********************************************
    ** Braces form and enclose code blocks.
    ** The program body of a C function must end
    ** with a closing brace '}'
    *********************************************/
```

Output

```
Review Chapter 1 --- Getting started!

We have learned
1. How to write comments
2. How to linefeed
3. How to continue a line
4. How to use tab    \t, beep \t, and
   carriage return \r
5. How to print special characters %, \ ,"
```

Explanation

1. What are the important features of a C program?

- The main function name must be `main`.
- The program body must start from {.
- The program body must end with }.
- A C statement must end with ;.
- A C statement is case sensitive.
- A C statement is location insensitive.
- In general, it is acceptable to add blank(s) between tokens in a C statement but not acceptable to add blank(s) within a token.
- C uses character escape sequences that consist of a backslash followed by other character(s) in string literals to represent special characters and actions.
- Make your comments stand out. Do not hide them. Adopt a style that is neat and orderly. Set aside time for writing comments in your programs.

2. What is debugging? In your source code, looking for and correcting errors or mistakes that cause your programs to behave unexpectedly is called *debugging*. In general, there are three types of errors in a C source code: syntax errors, run-time errors, and logic errors.

Syntax errors are mistakes caused by violating the "grammar" rules of C. They easily can be caused by typographical mistakes or a lack of knowledge of the forms of statements required by C. These errors often can be diagnosed by the C compiler as it compiles the program. If your compiler indicates errors when you try to compile a program, it will not translate your code into machine instructions. You must fix the errors before the compiler will translate your code. Therefore, when you have syntax errors you will not generate any output, even if your syntax errors are very minor and located in the very last lines of code.

Run-time errors, also called *semantic errors* or *smart errors,* are caused by violation of the rules during execution of your program. The compiler does not recognize them during compilation. However, the computer displays a message during execution that something has gone wrong and (usually) that execution is terminated. If a run-time error occurs near the end of execution, you may get some of your results. The error message given by the computer may help you locate the source of error in your code.

Logic errors are the most difficult errors to recognize and correct, because the computer does not indicate that there are errors in your program as it does with syntax and run-time errors. It is up to you to identify that there is a problem at all. It is up to you to look at your output and decide that it is incorrect. In other words, your program may have appeared to have executed successfully, perhaps giving very reasonable results. However, the answers may be completely wrong. You must recognize that they are wrong and correct code in the program. (Be careful, though. We will see, as we go further in this book, that the problem may be your input data. Many hours have been spent looking for bugs in programs only to find out that the program is correct, but the input data is incorrect.)

3. How can you reduce the number of bugs that you have in your programs? To reduce the number of bugs in your programs, you need to make sure that you develop good habits and set up an antibug strategy. This includes such things as

Writing your programs neatly
Adding blank lines at natural locations
Lining up your opening and closing braces
Adding comments properly

Following these steps will get you started in avoiding bugs. In essence, you should try to work in an organized and structured manner. Remember, a computer is not forgiving. Any error that you make will not be ignored by the computer. Throughout this text, we will note common errors that are made, to keep you aware of certain issues so that you can focus on these issues and avoid bugs.

4. How do you debug a program? If your program does not run, do not get frustrated. Be confident and calm. Getting frustrated will make you irrational and may cause you to incorrectly change something that is right. Debugging a program is very similar to finding out what is wrong with your car when it will not run. When your car does not start, you usually walk around it, look under the hood for obvious loose wires, check the battery, and other such things. You generally do not begin by disassembling your engine. Unfortunately, many inexperienced programmers do disassemble their programs and randomly change their code when it does not run.

Think globally, look at the whole program, and ask yourself:

Did I type anything wrong? For example, was `printf` typed `print`?

Did I use and follow C punctuation properly? For example, `void main(void)` being typed `void main(void);`.

Are my parentheses and braces in pairs?

In other words look for obvious things first—the common errors. Then, like looking for problems with your car, identify the performance problem and use it as a guide to find the source. For instance, if your car's windshield wipers do not work, you do not look for the source in the rear of the car. Instead, you look at the windshield wiper motor and wiring. With a computer program, if a certain calculation is not performing correctly you look at the portion of code where that calculation is performed and the portions of code that connect to that calculation.

Do not rely on the C compiler to locate the errors. For example, you may just miss the closing `"*/"` in one of your comments at the beginning of your program. If this is the case, the C compiler may think the rest of your program is simply a part of an unfinished comment. This minor error may generate 30 error messages. Do not be alarmed. Remember that the typical C compiler is not particularly sophisticated in identifying syntax errors. You may be able to eliminate a hundred error messages by typing a single character.

Be as independent as you can be, and be selective in your attempts at getting help. In general, try not to rely on others to debug your programs for you. You do not want to interrupt colleagues who are trying to get their own programs to run. Start by trying to work through any problems on your own. After you have made an effort to solve your own problem and still have not solved it, seek help. As painful as it may be, you will learn the most and become a better programmer by solving problems relying just on your books and computer.

EXERCISES

1. True or false:
   a. A C program must begin with a MAIN() statement.
   b. We may use any name as the main program name.
   c. A C statement is location insensitive but case sensitive.
   d. No end mark is required to end a C statement.
   e. We may use double quotes to enclose a string literal.
   f. We use `printf("'/n'");` to linefeed.

2. Use a printf function to display the following message:

   `Can you display "" %% ''\\ // and '' in a line?`

   If you cannot do this, please review the chapter.

Solutions

1. a (false); b (false); c (true); d (false); e (true); f (false)

```
2.  void main(void)
    {printf("Please display \"\" \%\%'' \\\\ // and ''\
     in a line?");
     }
```

APPLICATION PROGRAM 2.1 PRINTING ENGINEERING LOGO TO THE SCREEN

Problem statement

Write a program to print to the screen the logo and name of Sunset Engineering, Inc.:

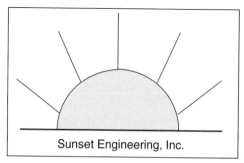

Sunset Engineering, Inc.

Solution

At this point, you lack the skills to duplicate this logo exactly, but you can use what you learned to produce a reasonable facsimile.

Because this particular program is not very complicated, we need not do much planning before we start writing. We will follow the procedure outlined in Chapter 1 for developing more complex programs in Chapter 3. Here, you can immediately start writing on paper or on the computer. In either case, you should begin writing your programs with the fundamental form that you will need for all of your programs until we get to Chapter 5. You can immediately write the following:

```
#include<stdio.h>
void main(void)
{


}
```

Memorize this form and be able to write it within a few seconds of sitting down to write your source code. For this program, the next step is to write the printf statement(s). We show the statement in the source code. Look at it and make sure that you understand exactly what all of the symbols do. We finish the program by writing a short comment at the beginning of the code.

SOURCE CODE

```
/*************************************************
** This program creates a letterhead and
**  border for notes
*************************************************/
#include <stdio.h>
void main(void)
{
printf("\n\n\n\n"
```

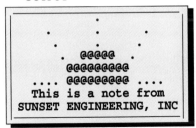

```
\n\n\n\n");
}
```

OUTPUT

COMMENTS

Note that, as with writing all programs, many different ways are acceptable. You could write a very different program from this one to accomplish the identical task of printing the logo. Assuming that your program works and is neat, logical, and well commented, it is an acceptable program.

APPLICATION PROGRAM 2.2 DEBUGGING

This is an important example that you should go through thoroughly. It illustrates a typical session a beginner goes through to get a simple program to run. By understanding this example, you will be more confident and composed as you work to get your own programs to execute correctly.

Problem statement

Create a program that can print the following to the screen:

```
DEBUGGING EXAMPLE:

This is an example of debugging. By following the step-by-step procedure
that we describe in this application example, you will begin to develop the
skills you need for successful and efficient debugging.
```

Envision that you have written the following program as a first attempt to do this (note that we have included line numbers for this program so that we can reference them in the description of the errors):

```
Line number
1      #<include stdio.h>;
2      void main(void);
3      (
4          printf('DEBUGGING EXAMPLE:');
5          printf("This is an example of debugging.\;
6          By following the step-by-step procedure ");
7          printf("that we describe in this application example,
8          you will begin to develop the);
9          printf("skills you need for successful and efficient debugging.");
10     }
```

Use the compiler to help you modify the code. Follow the step-by-step procedure outlined in the following methodology.

Methodology

In this application we illustrate the procedure involved in debugging a program. First, we outline the steps involved in debugging. Then, we use the steps to guide us to debug the preceding program. The procedure we describe is not the only one that can be used. Your instructor may have other suggestions that you will find helpful. However, at this point in your programming career, follow these steps to correct your program errors:

1. Compile your source code.
2. Look at the location in your source code that is indicated by the first error message. At this point, do not try to understand what the error message means. Only the location indicated is important to us right now.
3. Examine five statements at this location—two or three statements above the indicated location, the statement at the location, and two or three statements below the indicated location.
4. Use this text and your reference books to check your syntax (that is, "grammar, notation, punctuation, and form") for all of the lines in the region indicated by step 3.

5. Correct the syntax error(s) you see. At this point, do not attempt to correct your program for every error the C compiler has printed out. Try to fix only the first one.

6. Repeat steps 1–5. Note that each time you compile your program you will likely get a set of error messages that are completely different from the ones the compiler printed previously. This is one reason why we fix only one error at a time. Remember, just one error may cause 100 error messages to be printed. There is no need to read and try to interpret all 100 messages! Fix only the first one and then recompile your program.

   You may need to repeat steps 1–5, even 10 or 15 times. Do not get discouraged, this is not unusual. Eventually, you will fix all of these errors (which are *syntax errors*) indicated by your compiler, and your program will begin executing when you try to run it. However, even after you have fixed all the syntax errors, your program still may have run-time or logic errors.

7. Execute your program and look at the output. Does the output have statements printed that are not like any you have used in your printf statements? You may see such words as *overflow* or *execution terminated.* If you have these, then these are run-time errors. From this message, you may get an idea of the location in your program where the run-time error occurred. Also, your program may have printed some of your printf statements before the run-time error is printed. This means that the run-time error is located after these printf statements in your program. Go back to your source code and look at the statements in the region indicated by the run-time error. Correcting run-time errors is somewhat similar to correcting logic errors. We describe correcting both of these in step 8.

8. You can identify logic errors in your program by noticing that the output is not what you want or expect. In other words, suppose you were expecting the output

<code>This is my output.</code>

but the program printed

<code>This is myoutput.</code>

From this you know which printf statement has an error and you must go back to the code and modify it.

   Sometimes the location and cause of the error is not as obvious as in this example. You often can get an idea of the location of the error, though, by looking at your output. For instance, suppose your program has ten printf statements in it. If your program produces errors after the first five have been printed correctly, then the error is located in your source code after those first five printf statements.

   As we did for syntax errors, look at five or six statements in the region where you feel that the error has occurred. Do not look at just one statement, even if you are sure that execution stopped at that statement. At this point you should ask yourself, "Why is the program not doing what I am trying to tell it to do?" You can begin by making small adjustments to these statements and seeing the effect on the output. You should make adjustments and rerun the program repeatedly. This is how you get experience programming—change something and see its effect. It is a major part of the learning process. Do not get discouraged by the computer telling you that you have made errors. Do not make changes blindly, though. Think about what you are doing and what effect you expect the changes will have. Later in this book, we discuss techniques for helping recognize the source of logic errors and the changes to correct them.

9. After running, making changes, and rerunning your program repeatedly you will have developed a working program. Congratulations! However, before you put this program in your program library, make sure that it is well commented so that you can understand it later or others can easily interpret what you have done. If you have not commented it well, go back at this point and put in comments before you forget what you have done and create whatever other documentation is necessary. Resist the temptation to quit and celebrate. You will thank yourself later for spending a relatively small amount of time at this point to properly document your program. A summary of debugging is presented in Fig. 2.5.

FIG. 2.5
Debugging in a nutshell

- Compile your source code. Look at five or six lines within the region of the location of the first syntax error indicated. Correct the syntax error(s) you see.
- Repeat the first step, correcting only in the region of the first syntax error indicated by the compiler until your program begins to execute.
- Look at your output. If it is not correct, look for logic errors in your program. Make changes in your code and run the program. Repeat making changes and running the program until the results are correct.
- Make sure your program is properly commented and documented. Add comments and create other documentation if necessary.

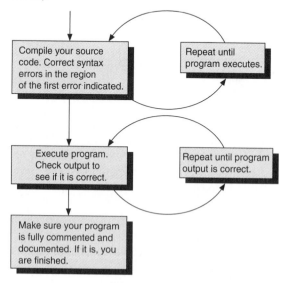

Debugging the given program

1. We begin by compiling the program in the problem statement. We recommend that you now use the diskette that comes with this book and compile this program yourself, if it is possible for you to do so. Each compiler gives slightly different error messages, so your compiler may not give exactly what we show here. When we compile the program, our compiler lists eight messages. The first message points to line 1 of the code (please refer to the source code shown in the problem statement) and says "Unknown preprocessor directive". As stated in the methodology, we will not worry about what this means but simply look at the line indicated and a few lines above and below for an error. Since this is the first line of code we focus on this line. When we do this, we see that we typed:

    ```
    #<include stdio.h>;
    ```

    Comparing this to other examples in this chapter, we see that we should have typed:

    ```
    #include <stdio.h>
    ```

    Note that we had two errors in this statement, the < was in the wrong location and we had a semicolon at the end of the line. At this point we ignore the other seven errors, correct just this one line and recompile the program. Also, refer to Fig. 2.6 for a summary of all of the errors that we describe here.

2. On compiling the next time we get an error indicated at the first printf statement (line 4) with the message, "Character constant too long." Again, we do not try to interpret the meaning of the message. Instead, we look a few lines above our location and see that we typed:

    ```
    void main(void);
    ```

    when we should not have put a semicolon at the end of this line. Thus, we remove the semicolon and recompile the program.

3. We now get one warning and seven errors. Warnings are not fatal errors. In other words, your program can execute correctly and completely with many warnings being given by the compiler. Warnings are exactly those; they are messages from the compiler to the programmer to indicate that something unusual is happening and an error may be caused by this. While you are still trying to locate syntax errors in your programs, we recommend that you ignore warnings and focus on the errors. When your program begins to execute, read and evaluate the warning messages given. For this example, the warning is, "Style of function definition is now obsolete".

    The first error message is ", expected", which is read "Comma expected." The location indicated is again the first printf statement in the program (line 4). Again, we look above this location and see that line 3 has a left parenthesis, (, instead of a left brace,{. So we change the parenthesis to a brace and recompile the program.

FIG. 2.6
Summary of syntax errors in Application Program 2.2

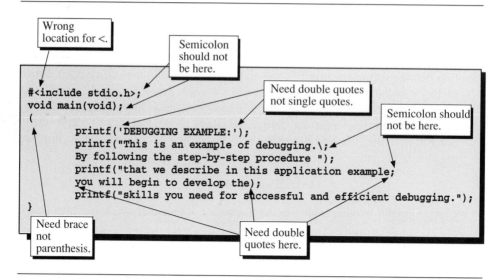

4. On recompiling we find that the number of errors is not reduced, instead it increases to 18 errors! We should not be alarmed by this. Instead, we simply work with the first message that reads, "Character constant too long in function main." The location indicated by this message is once more the first printf statement (line 4). Since we thought we fixed the errors prior to this statement, we look at this statement and realize that we have used single quotes and not double quotes. We, therefore, change the single quotes to double quotes and recompile the program.

5. Our compiler now gives 16 error messages. The first one is indicated by the compiler to be on the second printf statement (line 5) in the program and says, "Unterminated string or character constant in function main." We look at this printf statement and see that we should not have used a semicolon at the end of the line. So we remove it and recompile.

6. We now get seven error messages. The first one points to the third printf statement (line 7) and says, "Unterminated string or character constant in function main." At this point we can describe some of the rationale behind the error messages. We note that the written description of the error for step 5 is the same as that for step 6. Recall that another name for string literal is *string constant*. Our string constants are enclosed in double quotes. The error message says that we have an "unterminated string constant." Remember, white space is significant in string literals. The compiler says that because it is looking for a second pair of double quotes or a continuation character, \, on the same line as our third printf

statement and does not find it, it thinks the string is unterminated. Of course, we meant to terminate it on the next line, but this does not work. Therefore, we add " to the end of the third printf line to terminate the string.

Note that step 5 gave us the same error message. However, in that case our action was to remove a semicolon rather than add double quotes. C interpreted the semicolon to mean the end of the line. Since no matching double quote was at the end of the line, C believed that the string was unterminated. As you get more experience you will be able to make some use of the actual error messages. However, even experienced programmers often find the messages not to be particularly helpful because they do not necessarily point out the immediate problem. Rather than spend time trying to interpret the message, it is frequently more efficient to simply focus on the syntax in the region of the error.

We now recompile the program.

7. We get six errors with the first being, "Undefined symbol 'you' in function main" at line 8. We realize that this line was meant to be part of the printf string literal and we put quotes at the start and end of this line. We recompile the program.

8. We get no errors and therefore attempt to run the program. The program executes completely! However, the output is

```
DEBUGGING EXAMPLE:This is an example of debugging a program.   By following the
step-by-step procedure that we describe in this application example, you will be
gin to develop theskills you need for successful and efficient debugging.
```

Our program executes but is incorrect because the spacing is wrong. Because the compiler sent no error message, we find that we do not have any run-time errors in this program. However, we have logic errors that now must be addressed. At this point, the compiler can no longer aid us in finding or correcting our errors. We must do it completely on our own by looking at the source code and the output and figuring out what is incorrect and changing it. In your own programs, if you are completely stumped at this point, before you ask for help, make small changes in your code and recompile and run the program. By doing this, you will understand the effect changes make, and it may inspire you to find the source of your error.

In this example, we must get the spacing correct. We add \n after EXAM-PLE:, after procedure, and after develop the. We make these changes from the knowledge we developed reading this chapter.

9. After having made the changes indicated in step 8 and rerunning the program we find that the resulting output is

```
DEBUGGING EXAMPLE:
This is an example of debugging.         By following the step-by-step procedure
that we describe in this application example, you will begin to develop the
skills you need for successful and efficient debugging.
```

This is still not quite correct because we want double space after DEBUGGING EXAMPLE: and because we have too much space before the word By. So, we add another \n after DEBUGGING EXAMPLE:. Eliminating the space before By is more difficult. We have extra space because the continuation character, \, on the printf line above By causes the indentation in the source code to be interpreted as extra space. Therefore, we choose to eliminate the continuation character, replace it with double quotes, and put double quotes before By.

10. After having made these changes, we get the correct result when we execute the program. However, we are not finished because we must comment our program. Had our program been longer, we would have written comments earlier. However, for this short program now is a good time to write comments to the program. Our end product is

```
/*****************************************************************
** This is a program that prints a message about debugging
*****************************************************************/
#include <stdio.h>
void main(void)
{
    printf("DEBUGGING EXAMPLE:\n\n");
    printf("This is an example of debugging. "
    "By following the step-by-step procedure\n");
    printf("that we describe in this application example,"
    " you will begin to develop the\n");
    printf("skills you need for successful and efficient debugging.\n");
}
```

11. We run our program after commenting it to make sure that it still works. If this were a larger program, we would need to write supplementary documentation as well. However, for this program we are now finished. Tables 2.2 and 2.3 and Fig. 2.6 summarize the steps and actions taken.

COMMENTS

You should notice the following from Tables 2.2 and 2.3:

1. The line indicated by the compiler may not be the line with the error. You must look above and below that line to find the error.
2. The error messages given by the compiler may or may not be helpful. As you gain experience, you will be able to use the error messages more effectively.
3. When you recompile the program after having corrected an error, the number of error messages indicated by the compiler may actually increase. This does not mean that the program is becoming more incorrect.
4. The compiler does not indicate logic errors. You must observe that logic errors have occurred on your own. Then you must correct them without help from the compiler.

Even though the procedure we have described here is valuable, we cannot guarantee that if you follow it you will definitely eliminate all your errors. For instance, the method does not work if you were to simply forget the terminating / on your comment. Try doing this. You will find that the compiler indicates an error on the very last line of the program when the error is really located on the third line! Therefore,

▨ TABLE 2.2
Syntax error correction summary

Step	Total number of errors indicated by compiler	First line to which the compiler points indicating a syntax error	First error message	First line with a syntax error	Action required to correct the syntax error
1	8	1	Unknown preprocessor directive	1	change `#<include stdio.h>` to `#include <stdio.h>`
2	6	4	Character constant too long	2	change `void main(void);` to `void main(void)`
3	7	4	Comma expected	3	change `(` to `{`
4	18	4	Character constant too long in function main	4	change `'DEBUGGING EXAMPLE:'` to `"DEBUGGING EXAMPLE:"`
5	16	5	Unterminated string or character constant in function main	5	change `debugging.\;` to `debugging.\`
6	7	7	Unterminated string or character constant in function main	7	change `example,` to `example,"`
7	6	8	Undefined symbol 'you' in function main	8	change `you` to `"you`

▨ TABLE 2.3
Logic error correction summary

Step	Indication of error	Action required to correct the logic error
8	Output spacing is incorrect	add `\n` after `EXAMPLE:` after `procedure` after `develop the`
9	Output spacing is incorrect	1. add another `\n` after `EXAMPLE:` 2. replace `\` after `debugging.` with `"` 3. add `"` before `By`

looking only a few lines above and below the indicated error location would not uncover the error. Although the method we presented does not work in this situation, we believe that you will find the method helpful for locating many of your programming errors. As we go further in this book, we describe other techniques for recognizing and correcting errors. The method described here simply is to get you started in the process of debugging.

Last, we urge you to learn the debugging feature of your compiler. Right now you may have difficulty understanding it, but, by the end of Chapter 3, you should have made an effort at learning how to use it. See your compiler's documentation or ask your instructor about how to use it. Learning to use the debugger will save you many hours of programming frustration. We will not discuss it more as each compiler's debugger is different. We simply encourage you to learn this device on your own.

▪ APPLICATION EXERCISES

1. Write a program capable of displaying your name and address and a border on the screen.

2. Design your own logo. Write a program that will display that logo on the screen.

3. Write a program that can print out the first four letters of your first name in this form:

```
NNNN   NN          A        MM        MM    EEEEE
NN NN  NN         A A       M M      M M    E
NN  NN NN        AAAAA      M  M    M  M    EEEEE
NN    NNN      A      A     M   M M   M     E
NN     NN      A      A     M    M    M     EEEEE
```

4. Write a program that can print this shape using % and ". Note that some compilers may allow you to use % without using the escape sequence shown in Table 2.1.

5. Write a program that can display this following shape using \:

```
\\\                                    \\\
\\\\\\\\\                          \\\\\\\\\
\\\\\\\\\\\\\\\\\\\\\\\\\\\\\\\\\\\\\\\\\\\\\\\\\\
\\\\\\\\\                          \\\\\\\\\
\\\                                    \\\
```

6. Debug the following program using the method described in Application Program 2.2. Correct the program until it produces this output shown below:

```
/// This is a program to help you practice "debugging".\\\
///There are a few errors in this program. You should be able
to fix them.\\\
Make your output neat and orderly.
```

```
*******************************************************
**   Comments are valuable additions to programs
*******************************************************
#include<stdioh>
void Main(void)
{printf("/// This is a program to help you practice "debug-
ging"\\\");
printf("///There are a few errors in this program\n
you should be able to fix them.\\\")
print("Make your output neat and orderly")}
```

7. Debug the program until it produces the following output.

```
This program has a large number of errors.
Use the method we
described to fix it.
As you program more and more, you will
get better at debugging.
```

```
/*********************************************
** Put a banner on your programs
*********************************************
#include(stdio.h)
void<main>void
[
printf("This program has a \n large number of errors.\n');
printf("\nUse the method we \
described to fix it.");
pintf(As you program more and more, you will \\
get better at debugging.)
]
```

8. Debug the program shown until it produces the following output.

```
As you program more, you will
realize that you are spending a
considerable amount of time
debugging programs.
You will be able to do it
more quickly as you
get more experience.
```

```
/**************************************
** Put your own comment here
**************************************
#include<stadio.h>
void(main)void
Printf('As you program more, you will');
print{"realize that you are spending a\n
"considerable amount of time
debugging programs.\n");
printf("You will be able to do it"
"more quickly as you"
"get more experience.");
```

9. Debug the program shown until it produces the following output.

```
This program, as it is shown,
executes without syntax errors!
However, it does not produce
the output desired.
```

```
/***********************************
** Another debugging challenge
***********************************
include<sdio.h>
{void main(void)
printf("This program, as it is shown,
executes without syntax errors!");
printf{"However, it does \n not produce \
the output desired.')
}*/
```

10. Debug the program shown until it produces the following output.

```
If you have successfully debugged all
of the programs in this lesson, you
are ready to move on to Chapter 3.
```

```
/*********************************************
** You should feel comfortable with debugging.
*********************************************/
#include stdio.h
{void main(void)
{
printf("If you have successfully debugged all")
printf("of the programs in this lesson, you);
printf('are ready to move on to Chapter 3.);
}
```

The Basics of C—Variables, Arithmetic Operations, Math Functions, and Input/Output

In the second chapter you learned the basic structure of a C program and how to display words and numbers on the screen. In this chapter you will learn how to handle variables and to perform arithmetic calculations.

LESSON 3.1 VARIABLES: NAMING, DECLARING, ASSIGNING, AND PRINTING VALUES

Topics
- Naming variables
- Declaring data types
- Using assignment statements
- Displaying variable values
- Elementary assignment statements

Variables are crucial to virtually all C programs. You have learned about *variables* in algebra, and you will find that, in C, variables are used in much the same manner.

Suppose, for instance, that you want to calculate the area of 10,000 triangles, all of different sizes. And suppose that the given information is

1. The length of each of the three sides
2. The size of each of the three angles

To write an algebraic equation to determine the area, you need to make up your own variable names. You might choose as variable names

1. Lengths: a, b, c
2. Angles: α, β, γ

Or you could name the variables

1. Lengths: l_1, l_2, l_3
2. Angles: $\theta_1, \theta_2, \theta_3$

75

Or you could name the variables something completely different. It is entirely up to you what to name them, and you most likely would choose variable names that for some reason are comfortable to you.

For programming in C, the situation is quite similar. You choose the variable names, and it is best for you to choose names with which you are comfortable. A major difference between typical C programs and typical algebraic expressions is that the variables in most algebraic expressions consist of just one or two characters, maybe with a subscript or superscript. Variables in C programs often consist of entire words rather than single characters. Why? Because, as you will find, programs can get to be quite long, and there simply are not enough single characters to represent all of the necessary variables. Also, you will find that it will be easier for you and others to understand your programs if you have given very descriptive names to each variable.

For instance, for the triangle area program you may use the variable names

1. **Lengths:** `length1, length2, length3`
2. **Angle:** `angle1, angle2, angle3`

Or, if you wanted to be even more descriptive, you could name your variables

1. **Lengths:** `side_length1, side_length2, side_length3`
2. **Angles:** `angle_opposite_side1, angle_opposite_side2, angle_opposite_side3`

These variable names are much less ambiguous than their algebraic counterparts. Unfortunately, expressions using these variable names look much more cumbersome than the ones using simple algebraic notation. However, this is a disadvantage with which we simply must live.

C has rules that you must follow in choosing your variable names. For instance, ANSI C does not allow you to use more than 31 characters for one variable name. This and other rules will be discussed in the lesson. In addition, you must "declare" all your variable names near the beginning of your program, which means essentially to list your variables and indicate of what type they are. For instance, variables can be integers or reals. Also, we can give numerical values to the variables using what are called *assignment statements*. Assignment statements have an = sign with a variable on the left side of it.

The program for this lesson creates three variables: `month`, `expense`, and `income`. It assigns numerical values to these variables and prints them out along with text. To simplify the program description, we list the observations you should make about the program before you read the explanation. You should also attempt to answer the questions given. We will use this approach in a number of lessons throughout this book.

What you should observe about this lesson's program.
1. The first two lines of the body of main use are the keywords `int` and `float`.
2. These two lines declare the variables `month`, `expense`, and `income`.
3. The `=` sign is used in five statements in the program to assign numerical values to the variables.

4. The variables `month` and `expense` are assigned numerical values at the beginning of the program and then assigned different values later in the program.

5. In the printf statements we have used the `%` sign in the string literals a number of times.

6. We have used `%2d` and `%9.2f` among others in the string literals of the printf statements.

7. After the string literal in the printf statements we have listed the names of some variables.

8. The number of variables listed is equal to the number of `%` signs used in the string literals of the printf statements.

9. The output shows a number of spaces between `$` and `111.10`.

10. The output has no spaces between `$` and `82.10`.

Questions you should attempt to answer.

1. What type of variable is month?

2. What types of variables correspond to the `%d` type specifications?

3. What types of variables correspond to the `%f` type specifications?

Source code

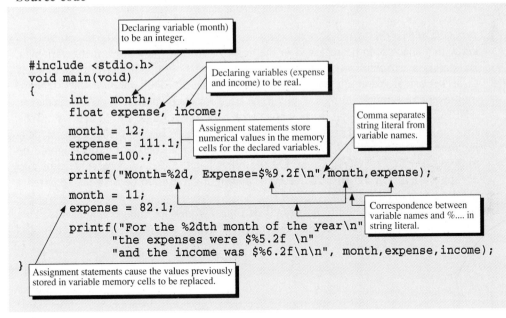

```
#include <stdio.h>
void main(void)
{
      int    month;
      float expense, income;

      month = 12;
      expense = 111.1;
      income=100.;

      printf("Month=%2d, Expense=$%9.2f\n",month,expense);

      month = 11;
      expense = 82.1;
      printf("For the %2dth month of the year\n"
             "the expenses were $%5.2f \n"
             "and the income was $%6.2f\n\n", month,expense,income);
}
```

Declaring variable (month) to be an integer.

Declaring variables (expense and income) to be real.

Assignment statements store numerical values in the memory cells for the declared variables.

Comma separates string literal from variable names.

Correspondence between variable names and %.... in string literal.

Assignment statements cause the values previously stored in variable memory cells to be replaced.

Output

```
Month=12, Expense=$    111.10

For the 11th month of the year
the expenses were $82.10
and the income was $100.00
```

Explanation

1. How do we declare variables? Variable names in C must be declared. The statement

`int month;`

declares the variable month to be of the `int` type (which means integer and must be typed in lowercase letters). An `int` type data contains no decimal point.

2. How do we declare more than one variable? Variables of the same type may be declared in the same statement. However, each must be separated from the others by a comma; for example, the statement

`float expense, income;`

declares the variables `expense` and `income` to be of the `float` (which must be typed in lower case) type (see Fig. 3.1). Float type data contain a decimal point with or without a fraction. For example, 1., 1.0, and 0.6 are float type data. When data without a decimal point are assigned to a float type variable, the C compiler automatically places a decimal point after the last digit.

3. What is the effect of declaring variables? This causes the C compiler to know that space is to be reserved in memory for storing the values of the variables. By stating a variable's type, the C compiler knows how much space in memory is to be set aside. Although not explicitly set by the ANSI C standard, the standard implies the minimum number of bits to be used for each variable type. For instance, ANSI C requires that type int be capable of handling a range of -32767 to 32767 for `int` type variables. This requires 16 bits or 2 bytes of memory. Therefore, declaring the variable month as `int` indicates that 16 bits or 2 bytes of memory should be reserved for this variable's value. On the other hand, a `float` type value typically occupies 4 bytes or 32 bits. Therefore, declaring a variable to be a `float` requires 4 bytes of memory to be reserved.

In addition, C uses different types of binary codes for integers and reals. This means that, for example, the bit pattern for 32 stored as an `int` is completely dif-

FIG. 3.1
Declaring variables

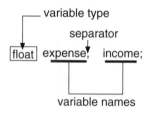

ferent from the bit pattern for storing 32. as a `float`. It is important that we keep this in mind. Forgetting this fact will lead to errors. For instance, if printf attempts to read a memory cell that contains an `int` but it expects a `float` to be in the cell it will misinterpret the cell contents and print out something completely wrong. Since we, the programmers, tell printf what to expect in a memory cell, we must tell it correctly. Later in this lesson we will describe how we indicate to printf what to expect in a memory cell.

4. How do we name variables? Variables in C programs are identified by name. Variable names are classified as *identifiers*. Therefore, the naming convention for variables must obey the rules used by identifiers. For instance, the first character of an identifier cannot be numeric. For the first and other characters the requirements are

Component	Requirement
The first character in identifier	Must be nondigit characters a–z, A–Z, or _
Other characters in identifier	Must be nondigit characters a–z, A–Z, _, or digit characters 0–9

In addition, Table 3.1 lists other constraints on creating valid identifiers.

TABLE 3.1
Some constraints on identifiers

Topic	Comment
The maximum number of characters in an internal identifier (i.e., identifier within a function)	ANSI C allows a maximum number of 31 characters for names of internal identifiers
Use of C reserved words, also called *keywords*, as identifiers	Not allowed; i.e., do not use these words: auto break case char const continue default do double else enum extern float for goto if int long register return short signed sizeof static struct switch typedef union unsigned void volatile while
Use of standard identifiers such as printf	Standard identifiers, such as the function name printf, can be used as variable names. However, their use is not recommended because it leads to confusion.
Use of uppercase or mixed-case characters	Allowed; however, many programmers use lowercase characters for variable names and uppercase for constant names. Differentiate your identifiers by using different characters rather than different cases
Use of blank within an identifier	Not allowed, because an identifier is a token

Examples of illegal variable names:

```
1apple      interest_rate%      float        In come       one.two
```

Examples of legal variable names:

```
apple1      interest_rate       xfloat       Income        one_two
```

5. What is a keyword? A keyword is an identifier type token for which C has a defined purpose. Keywords used in this lesson's program are int, float, and void. Because these have special meanings in C, we cannot use them as variable names. The purposes of the other keywords listed in Table 3.1 are described throughout this book.

6. As a program begins to execute, what conceptually is happening in memory? Conceptually, a table is created internally. This table contains variable names, types, addresses, and values. The names, types, and addresses are first established essentially during compilation, then as execution takes place space is reserved in memory and the variable values are put into the memory cells reserved for the variables. For instance, after the first three assignment statements have been executed in this lesson's program, the table looks like the following (note that the memory cell addresses are written in hexadecimal notation—we use hexadecimal notation for memory cell addresses throughout the remainder of this text):

Variable name	Variable type	Memory cell address	Variable value
month	int	FFF8	12
expense	float	FFF6	111.1
income	float	FFF2	100.

After the fourth and fifth assignment statements have been executed, the table becomes

Variable name	Variable type	Memory cell address	Variable value
month	int	FFF8	11
expense	float	FFF6	111.1
income	float	FFF2	82.1

(Note: You need not be concerned about the memory cell addresses at this point. The addresses are automatically set when you compile and execute your programs. Comparing the two tables, though, we can see that the memory cell addresses do not change, but the variable values do change.) You will find that your programs' purpose is to continually change the values in the variable table.

7. What is an assignment statement? An *assignment statement* assigns a value to a variable, which means an assignment statement causes a value to be stored in the variable's memory cell. For example, the statement

$$month = 12;$$

assigns the integer value 12 to int type variable month. It causes 12, written in 2's complement binary type of notation, to be stored in the month memory cell (Fig. 3.2).

FIG. 3.2
Assignment statement

This assignment statement causes the bit pattern representing 12 (in binary type notation) to be stored in the memory cell reserved for the variable, month.

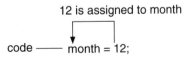

12 is assigned to month

code ——— month = 12;

In C, a simple assignment statement takes the form of

variable_name = value;

where this statement assigns the *value* on the right side of the equal sign to the *variable_name* on the left side of the equal sign. The binary representation of the value is stored in the variable's memory cell after the assignment has taken placed. The *value* can be a *constant,* a variable with a known value, or other, such as a function or an *expression* that returns a value (see the next few lessons for more details). Note that the equal sign in an assignment statement does not really mean equal. As you become exposed to more programming techniques, you will need to interpret an assignment statement by thinking of the table of variable values created and the storing of values in the memory cells rather than the term *equal.*

8. How do we display the value of a variable or constant on the screen? The printf() function can be used to display the value of a variable or constant on the screen. The syntax is

`printf`(*format_string, argument_list*)`;`

where the *format_string* is a string literal that contains three types of elements:

- The first one is referred to by ANSI C as plain *characters.* These are characters that will be displayed directly unchanged on the screen.
- The second one is *conversion specification*(s), which will be used to convert, format, and display *argument*(s) from the *argument_list.*
- The third is escape sequences that the printf function uses to control the cursor or insertion point.

Each argument must have a format specification. Otherwise, the results will be unpredictable. For example, in the statement

`printf("month=%5d \n",month);`

The format string is `"month=%5d \n"`. The plain characters `month=` will be displayed directly with no modification, but the conversion specification `%5d` will be

used to convert, format, and display the argument, month, on the screen. The escape
sequence, \n, moves the insertion point to the next line.

The simplest printf() conversion specifications (also called *format specifications*) for displaying int and float type values have the following forms:

%[*field width*]d   *e.g.,* %5d *for* int
%[*field width*][*.precision*]f   *e.g.,* %9.2f *for* float

where format string components enclosed by [ ] are optional (the characters [and]
are not the part of the format string). The *field width* is an integer representing the
minimum number of character spaces reserved to display the entire argument
(including the decimal point, digits before and after the decimal point, and the sign).
The *precision* is an integer representing the maximum number of digits after the decimal point. For example %5d will reserve five blank spaces for displaying an int type
data; %9.2f will reserve a total of nine blank spaces for a float type data, and two
digits will be displayed after the decimal point. These concepts are shown in Fig.
3.3. If your actual input data contain fewer digits after the decimal point, the *C compiler* will add zero(s) after the decimal point when displaying it. For example, for
the statements

```
expense=111.1;
printf("the expenses were $%9.2f\n",expense);
```

the C compiler adds one 0 to make the precision equal to 2 to give 111.10 for the
output of expense.

FIG. 3.3

Arguments and conversion specifications

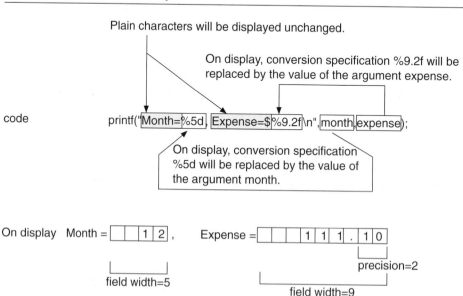

Also, the assignment and printf statements

```
income = 100.;
month = 11;
expense = 82.1;

printf("For the %2dth month of the year\n"
        "the expenses were $%5.2f \n"
         "and the income was $%6.2f\n\n",
        month,expense,income);
```

cause the `int` variable, `month`, to be displayed with the `%2d` conversion specification; the `float` variable `expense` to be displayed with the `%5.2f` conversion specification; and the `float` variable `income` to be displayed with the `%6.2f` conversion specification. Note that both float variables have printed numerals in two places after the decimal. More will be said about conversion specifications in the next lesson.

EXERCISES

1. True or false:
   a. The following `int` type variable names are legal:

      ```
      1cat, 2dogs, 3pears, %area
      ```

   b. The following `float` variable names are legal:

      ```
      cat, dogs2, pears3, cat_number
      ```

   c. The format specifications `5d` or `%8D` are legal for `int` type variables or constants.
   d. The format specifications `6.3f` or `%10.1F` are legal for a `float` type variable.
   e. The two statements that follow are identical:

      ```
      int ABC, DEF;
      int abc, def;
      ```

2. Which of the following are incorrect C variable names and why?

   ```
   enum, ENUM, lotus123, A+B23, A(b)c, AaBbCc, Else, αβχ, pi, π
   ```

3. Which of the following are incorrect C assignment statements and why?

   ```
   year = 1967
   1967 = oldyear;
   day = 24 hours;
   while = 32;
   ```

4. Supposing `year` is an `int` variable and `salary` is a `float` variable, which of the following printf() statements are unacceptable and why?

   ```
   printf ("My salary in 1997 is $2000", salary);
   printf("My salary in 1997 is %d\n",salary);
   printf(In year %d, my salary is %f\n", year, salary);
   printf("My salary in %d year is %f\n, salary,year");
   printf("My salary in %5d year is %10.2f\n\n",year,salary);
   ```

5. The price of an apple is 50 cents, a pear is 35 cents, and a melon is 2 dollars. Write a program to display the prices as follows:

```
*****  ON SALE  *****
Fruit type      Price
Apple           $ 0.50
Pear            $ 0.35
Melon           $ 2.00
```

Solutions

1. a (false), b (true), c (false), d (false), e (false)

▨ LESSON 3.2 CONSTANT MACROS AND MORE ABOUT PRINTING VARIABLE VALUES

Topics

- Using the define directive to define constants
- More about conversion specifications and their components
- Scientific notation
- Flags in conversion specifiers

You will find that sometimes you need to use values in your programs that do not change. For instance, we know that π is approximately 3.14159. For a program that involves areas of circles, it is convenient to simply write the characters PI in the equations rather than the digits 3.14159. This can be done by using what is called a *constant macro*. A constant macro is created by using a preprocessor directive.

Recall that we used the preprocessor directive `#include<stdio.h>` in Chapter 1, and this directive caused the preprocessor to perform actions prior to the translation of the source code into object code. We use the same symbol (#) to indicate a preprocessor directive to create a constant macro.

In the previous lesson we described the use of conversion specifications in the format string of printf statements. In many cases, at the time you are writing the program, you will not know the size of the values you want to display. Therefore, frequently, the conversion specification does not match the value to be displayed exactly. What happens if the conversion specification is too small or too large? In this lesson we answer this question.

When working with very large or very small numbers, scientific notation is convenient. For example, to represent 57,650,000, the scientific notation would be 5.765×10^7, which C would display as 5.765e+007 or 5.765E+007.

By using scientific notation, the printf function decides the value of the exponent, and thus it is possible to display an extremely large number in a small number of spaces. The programmer need decide only on the number of significant digits to display.

This program creates two constant macros, DAYS_IN_YEAR and PI. It prints the numerical values of these constant macros using a number of different conversion specifications. It also prints the value of the `float` type variable, `income`, using a number of different conversion specifications.

What you should observe about this lesson's program.

1. The second and third lines of the program begin with `#define`.
2. White space separates `DAYS_IN_YEAR` and `365`.
3. White space separates `PI` and `3.14159`.
4. Both `DAYS_IN_YEAR` and `PI` are written in all capital letters.
5. A number of different conversion specifications have been used in the printf statements.
6. Each conversion specification is surrounded by double brackets `[[ ]]`.
7. The last conversion specification begins with a negative sign.
8. The output illustrates that different amounts of space are created between the double brackets.

Questions you should attempt to answer.

1. What type of variable is `income`?
2. What types of numbers are printed with the `%d` type specifications?
3. What types of numbers are printed with the `%f` type specifications?
4. What type of conversion specification is used to display scientific notation?

Source code

```
#include <stdio.h>
#define DAYS_IN_YEAR 365
#define PI 3.14159
```

These preprocessor directives cause DAYS_IN_YEAR and PI to be replaced with 365 and 3.14159, respectively, in the source code.

```
void main (void)
{
        float income = 1234567890.12;

        printf ("CONVERSION SPECIFICATIONS FOR INTEGERS \n\n");
        printf ("Days in year = \n"
                "[[%1d]] \t(field width less than actual)\n"
                "[[%9d]] \t(field width greater than actual)\n"
                "[[%d]]  \t(no field width specified)  \n\n\n",
                DAYS_IN_YEAR, DAYS_IN_YEAR, DAYS_IN_YEAR);

        printf ("CONVERSION SPECIFICATIONS FOR REAL NUMBERS\n\n");
        printf ("Cases for precision being specified correctly \n");
        printf ("PI = \n"
                "[[%1.5f]] \t\t(field width less than actual) \n"
                "[[%15.5f]] \t(field width greater than actual)\n"
                "[[%.5f]] \t\t(no field width specified) \n\n",
                PI,PI,PI);
        printf ("Cases for field width being specified correctly \n");
        printf ("PI = \n"
                "[[%7.2f]] \t\t(precision less than actual) \n"
                "[[%7.8f]] \t\t(precision greater than actual)\n"
                "[[%7.f]] \t\t(no precision specified) \n\n",
                PI,PI,PI);
```

Conversion specifications.

```
     printf ("PRINTING SCIENTIFIC NOTATION \n\n");
     printf ("income = \n"
             "[[%18.2e]] \t(field width large and precision small) \n"
             "[[%8.5e]]  \t(field width and precision medium size)\n"
             "[[%4.1e]]  \t\t(field width and precision small) \n"
             "[[%e]]     \t(no specifications)  \n\n",
             income, income, income, income);

     printf ("USING A FLAG IN CONVERSION SPECIFICATIONS \n\n");
     printf ("Days in year= \n"
             "[[%-9d]] \t\t(field width large, flag included)\n",
             DAYS_IN_YEAR);
}
```

Conversion specifications for scientific notation.

Flag.

Output

```
CONVERSION SPECIFICATIONS FOR INTEGERS

Days in year =
[[365]]                    (field width less than actual)
[[     365]]               (field width greater than actual)
[[365]]                    (no field width specified)

CONVERSION SPECIFICATIONS FOR REAL NUMBERS

Cases for precision being specified correctly
PI =
[[3.14159]]                (field width less than actual)
[[       3.14159]]         (field width greater than actual)
[[3.14159]]                (no field width specified)

Cases for field width being specified correctly
PI =
[[  3.14]]                 (precision less than actual)
[[3.14159000]]             (precision greater than actual)
[[3.141590]]               (no precision specified)

PRINTING SCIENTIFIC NOTATION

income =
[[          1.23e+09]]     (field width large, precision small)
[[1.23457e+09]]            (field width and precision medium size)
[[1.2e+09]]                (field width and precision small)
[[1.234568e+09]]           (no specifications)

USING A FLAG IN CONVERSION SPECIFICATIONS

Days in year =
[[365      ]]              (field width large, flag included)
```

Explanation

1. How do we create a constant macro? We use a *preprocessor directive* to create a *constant macro*. As we indicated in Chapter 1, the preprocessor is a system program that is part of the C compiler. It automatically performs various operations prior to the translation of source code into object code. In C, preprocessing directives begin with the symbol # (which must begin the line). A semicolon must not be used at the end of the preprocessing directive. Only the preprocessing directive should be on the line. For example, the line

```
#define DAYS_IN_YEAR 365
```

is a preprocessor directive called a *define directive*.

2. How does the preprocessor work with a define directive? The form of a define directive is

```
#define symbolic_name replacement
```

where *symbolic_name* is the name of the constant macro that we are creating and *replacement* is the value with which we want *symbolic_name* replaced. The word `define` must be completely in lower case. On being instructed by a define directive, the preprocessor replaces any *symbolic_name* (excluding those that appear in comments or in string literals) in the program with the given *replacement*. For example, in this lesson's program, the symbolic name `DAYS_IN_YEAR` in the statement

```
printf("Days in year=%5d\n",DAYS_IN_YEAR);
```

is replaced by `365` before the program is translated into machine language. In other words, the preceding statement will be "rewritten" by the preprocessor to be

```
printf("Days in year=%5d\n",365);
```

prior to the compiler translating the code into machine language. For this example, the constant macro (`DAYS_IN_YEAR`) is replaced with the value `365` throughout the program by the preprocessor before the program is compiled.

Note that only one constant macro can be defined per line. The constant macro cannot be placed on the left side of an assignment statement, meaning that we cannot try to assign a new value to the constant macro at a later point in the program. You can understand why this does not work if you think about the operations involved. For instance, if we wanted to write `DAYS_IN_YEAR = 365.25`, as an assignment statement in our program, the preprocessor would convert this to be `365 = 365.25` before the source code is translated into object code. This statement clearly makes no sense and therefore illustrates why we cannot use a constant macro on the left side of an assignment statement. Therefore, it is said that a constant macro is not an lvalue (pronounced "ell value"), meaning it cannot go on the left side of an assignment statement. A constant macro is considered to be an rvalue (pronounced "are value"), meaning that it can go on the right side of an assignment statement but not the left.

3. What is the convention for naming constant macros? Many C programmers use uppercase characters to name constant macros and lowercase characters to name variables. This book follows this approach in most cases.

4. How does ANSI C convert a float number to scientific notation? It converts a float number to scientific notation using the format

$$[sign]d.ddd \ \ e[sign]ddd$$

where *d* represents a digit, the digits before the decimal point represent the field width, and the digits after the decimal point represent the specified precision. Note that a number in this form is equivalent to

$$[sign] \ d.ddd \times 10^{[sign]ddd}$$

For example, when we use the format `%15.4e`, that is, field width = 15 and precision = 4, to display the number

$$123456789.12$$

in scientific notation, we get

$$bbbb1.2346e+009$$

(where b represents blank), which is equivalent to

$$1.2346 \times 10^9$$

or

$$1234600000.0$$

On display, we lose some accuracy because the specified precision is not high enough. However, this affects the display only. The complete value still is stored in the computer's memory.

5. What is the complete structure of format specifications for int and float type data? The complete structure of format specifications is

$$\%[flag][field \ width][.precision]type$$

where format string components enclosed by [ ] are optional (the characters [and] are not the part of the format string, see Fig. 3.4 for details).

The meanings of these components may vary slightly from compiler to compiler. Therefore, you should check the manual of the compiler you use. A description of some of the flags and types used in ANSI C are given in Table 3.2. You should read the table to get an idea of the features available to you in writing the format specifications for your printf statements.

6. What is displayed if the field width specified for an integer or real is less than or greater than the actual or not specified? As you can observe from the output for this lesson's program, if the field width specified is

FIG. 3.4

Meaning of exponential type conversion specification

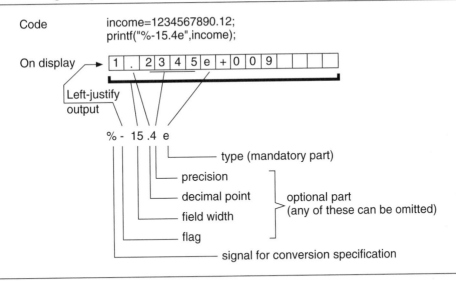

- Less than actual, the printf function displays the actual value with the field width being the same as the actual number of digits
- Greater than actual, the printf function displays the value right justified within the specified field width
- Not specified, the printf function displays the actual value with the field width being the same as the actual number of digits

Figure 3.5 illustrates the effect of the specified width on the display of an integer using a %d format. This figure shows that the number of spaces created is the larger of the number of spaces requested and the number of spaces needed to correctly display the integer. For real numbers (%f or %e format), both the specified field width and the specified precision affect how many spaces will be used in the display. However, for real numbers, the specified field width is somewhat secondary in importance in the display because printf first works with the specified precision.

7. What is displayed if the precision specified for a real is less than or greater than the actual or not specified? From the output you can see that if the precision specified is

- Less than actual, the printf function displays only the number of digits in the specified precision (trailing digits are not lost from memory, they simply are not displayed)
- Greater than actual, the printf function adds trailing zeros to make the displayed precision equal to the precision specified
- Not specified, the printf function makes the precision equal to six (it adds trailing zeros or truncates digits if necessary to get a precision of six; these actions do not change the value stored in memory)

▨   TABLE 3.2
Flags and types in ANSI C

Component	Use
flag = −	This flag causes the output to be left justified within the given field width
flag = +	This flag causes the output to be right justified within the given field width and a plus sign displayed if the result is positive
flag = 0	This flag causes leading zeros to be added to reach the minimum field width; the flag is ignored if the − flag is used simultaneously
field width	This integer represents the minimum number of character spaces reserved to display the entire output (including the decimal point, digits before and after the decimal point, and the sign). If the specified field width is not given or is less than the actual field width, the field width is automatically expanded on print out to accommodate the value being displayed. The field width and precision are used together to determine how many digits before and after a decimal point will be displayed
precision	For floating data types, precision specifies the number of digits after the decimal point to be displayed. The default precision for float type (e, E, or f) data is six. Precision also can be used for integer type data, where it specifies the minimum number of digits to be displayed. If the data to be displayed has fewer digits than the specified precision, the C compiler adds leading zero(s) on the left of the output
type = d	For int type data
type = f	The output is converted to decimal notation in the form of [sign]ddd.dddd, where the number of digits after the decimal point is equal to the specified precision
type = e or E	The output is converted to scientific notation in the form of [sign]d.dddd e[sign]ddd, where the number of digits before the decimal point is one, the number of digits after the decimal point is equal to the specified precision, and the number of exponent digits is at least two. If the value is zero, the exponent is 0.

▨   FIG. 3.5

Action of printf function for printing of integer values (note that the larger of the number of spaces requested and needed is the number created)

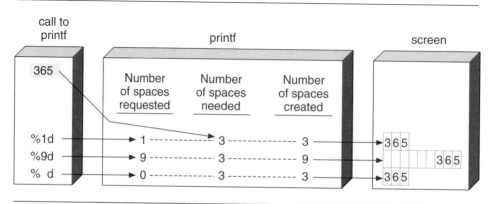

8. What does the flag − do? It left-justifies a value that is put in a field width that is greater than the actual.

9. How would different conversion specifications display 365, 3.1416, and 1234567890.12? Table 3.3, Fig. 3.5, and Fig. 3.6 show the meaning of using different formats for displaying an int constant 365 using type d format, a float type constant 3.1416 using type f format, and a float type constant 1234567890.12 using type e or E format.

Note that if the specified field width is not given or is less than the actual field width, all characters of the value, as long as they are within the limit of precision specification, will be displayed; that is, the field width specification never "truncates" the output value.

FIG. 3.6

Action of printf function for printing real values and scientific notation. Note that printf first works with the "After decimal point number of spaces"—the requested number is created unless it is left blank whereupon a default number, 6 for some compilers, is created. Then the total number of spaces created is the larger of the number requested and the number needed to meet the decimal requirements. For the example 3.14159, the number of spaces "needed to meet decimal requirements" is two more than the "created" after decimal point number of spaces, because "3." occupies two spaces. For the example 1234567890.12, the number of spaces "needed to meet decimal requirements" is six more than the "created" after decimal point number of spaces, because four spaces are used for "e+09" and two spaces are used for "1."

Call to printf		After decimal point number of spaces		Total number of spaces			screen
					(needed to meet decimal requirements)		
		requested	created	requested		created	
3.14159							
%1.5f		5	5	1	7	7	3.14159
%15.5f		5	5	15	7	15	3.14159
%.5f		5	5	0	7	7	3.14159
%7.2f		2	2	7	4	7	3.14
%7.8f		8	8	7	10	10	3.14159000
%7f		0	6	7	8	8	3.141590
1234567890.12							
%18.2e		2	2	18	8	18	1.23e+09
%8.5e		5	5	8	11	11	1.23457e+09
%4.1e		1	1	4	7	7	1.2e+09
%e		0	6	0	12	12	1.234568e+09

TABLE 3.3
Sample displays

Conversion	Flag	Field width	Type	Precision	Display (*b* means blank)	Note
%+5d	+	5	d	none	b+365	Right-justified output, + sign added, total characters displayed is five
%−5d	−	5	d	none	365bb	Flag is −, so output is left justified
%1d	none	1	d	none	365	Specified field width is less than the actual width, all characters in the value are displayed, no truncation occurs
%0.5d	zero	0	d	5	00365	Flag is 0, so output is pre-fixed with zeros, precision is 5, so the number of char-acters to be printed is five
%d	none	none	d	none	365	Field width is undefined, all characters in the value are displayed, no truncation occurs; no blanks are added; value is left justified
%+9.5f	+	9	f	5	b+3.14160	Total digits, including blanks, is nine
%−9.5f	−	9	f	5	3.14160bb	Flag is −, left-adjusted output
%1.3f	none	1	f	3	3.142	Uses precision 3, note the result is 3.142, not 3.141
%f	none	none	f	none	3.141600	Uses the default precision, 6
%+12.4e	+	15	e	4	b1.2346e+009	Flag is +, right-adjusted output, total digits is 12, field width of 15 accommo-dates the e and +, precision is 4
%−12.4e	−	15	e	4	1.2346e+009b	Same as previously, but flag=−, so output is left justified
%5.2e	none	5	e	2	1.23e+009	Precision is 2; field width is too short, so C uses mini-mum field width for output
%E	none	none	E	none	1.234568E+009	Precision is undefined, so C uses default precision of 6; field width is too short, so C uses minimum field width for output

10. Given the same value and using the same format, will programs created using different compilers display exactly the same output? No, because various compilers implement the ANSI C standard differently, so, in general, given the same value and using the same format, the output displayed by programs created with different compilers may be slightly different.

11. What happens if we try to display an int ***with*** %f ***or a*** float ***with*** %d***?*** If you try to do this you will most likely get completely nonsensical values or zeros displayed on the screen. This is a common error that beginners (and even some veterans) make. It is an extremely frustrating error because everything else in your program may be correct but a simple %d instead of %f will make it appear like you have major errors in your program. Then you may spend a lot of time looking at your logic, doing hand calculations and in the end it is a simple conversion specification that has caused the problem. If you get total nonsense or zeros for your output, check your conversion specifications before you investigate other, more difficult to trace, sources of error.

12. For engineering programming, is any danger involved in using a %f ***type display format?*** If your numbers are very small they could be printed as 0. For instance, if you are working with small measurements and you want to print 3.5×10^{-12} meters and you use as a format %f, your output will be 0.000000. Therefore, in working with small numbers, to display meaningful results, you must use an exponential type format such as %e.

13. Why was so much attention given to the printf statement? There are two reasons. One is that you will write printf statements very frequently and other aspects of programming will become easier for you if you feel comfortable writing printf statements. It is a good idea to become proficient at writing them now so that you can easily go on to other programming issues. The other reason is that improperly written printf statements are the source of many errors for beginning programmers. If you understand printf statements, you will substantially reduce your programming errors.

14. Does all this need to be memorized? No. In many cases you simply can use %d, %f, and %e. However, you can use this lesson as a reference when you need to be more exact in displaying values.

EXERCISES

1. True or false:
   a. The statement printf("%-3d",123); displays -123.
   b. The statement printf("%+2d",123); displays $+12$.
   c. The statement printf("%-2f",123); displays 12.0.

  d.   The statement `printf("%+f.3",123);` displays .123.

  e.   The format specification for an `int` type data should not contain a decimal point and precision; for instance, `%8.2d` is illegal.

2.   Find error(s), in these statements:

  a.   `#DEFINE PI  3.1416`

  b.   `#define PI  3.1416;`

  c.   `#define PI=3.14; More_AccuratePI=3.1416;`

  d.   `printf("%f",123.4567);`

  e.   `printf("%d %d %f %f",1,2,3.3,4.4);`

3.   Which of the following are incorrect define directives and why?

```
define speed of light 30000
#define long 12345678901234567890
#define SHORT 0.01;
#DEFINE RADIUS 30
```

4.   Supposing `rate` is a `float` variable and `year` is an `int` variable, correct each of the following statements if you find any errors:

```
printf("The interest rate in %-.d year is %+.F%\n", year,
    rate);
printf("In year %#d5, the interest rate was 8.2f%\n", rate,
    year, rate);
printf("In %+5.8d, the interest rate will be %010.18e%\n",
    year, rate);
```

5.   Write a program to display the following output:

```
12345678901234567890123456789012345
income      expense      Name
+111.1      -999.99      Tom
+222.2      -888.88      Dennis
+333.3      -777.77      Jerry
```

6.   Use four different flags but the same field width and precision, four different field widths but the same flag and precision, and four different precisions but the same flag and field width (i.e., a total of 12 format specifications) to display an int type variable A and a float type variable B, where $A = 12345$ and $B = 9876.54321$.

Solutions

1.   a (false), b (false), c (false), d (false), e (false)

2.   a.   `#define PI  3.1416`

    b.   `#define Pi  3.1416`

    c.   `#define PI  3.14`

         `#define More_AccuratePI  3.1416`

    d.   `no error`

    e.   `no error`

▨ LESSON 3.3 ARITHMETIC OPERATORS AND EXPRESSIONS

Topics

- Operands
- Arithmetic operators and their properties
- Arithmetic expressions

Arithmetic expressions in C look much like algebraic expressions that you write. The first section of the example program for this lesson shows some of the operations that can be performed in C arithmetic expressions. Look at this section of the program and see how addition, subtraction, multiplication, and division are performed. What is the symbol used for multiplication?

Note that in this section of the program are the statements

```
i=i+1
```

and

```
j=j+1
```

Clearly, these two statements would make no sense if you were to use them in a math class. However, in C, not only do these statements (and statements of this type) make sense, they actually are used quite commonly in programs. What do they mean? (Hint: Recall what we said about assignment statements causing the value on the right side of the assignment statement to be put into the memory cell of the variable on the left side of the assignment statement.)

In the second section of this program are expressions with operators whose functions are not quite so obvious. Look at the statements and the corresponding output. The % sign is especially tricky. See if you can figure out what it does. (Hint: It has something to do with division.)

Also the ++ and -- are operators, but there are no equal signs in the statements for these. However, they have an impact on the values of the variables either preceding or succeeding them. Look at the output. What effect do they have on these variables?

Source code

```
#include <stdio.h>
void main(void)
{
        int i,j,k,p,m,n;
        float a,b,c,d,e,f,g,h,x,y;

        i=5;    j=5;
        k=11;   p=3;
        x=3.0;  y=4.0;
        printf("...... Initial values ......\n");
        printf("i=%4d, j=%4d\nk=%4d, p=%4d\nx=%4.2f, y=%4.2f\n\n",
           i,j,k,p,x,y);
```

```
/*-------------- Section 1 --------------------*/
a=x+y;
b=x-y;
c=x*y;
d=x/y;
e=d+3.0;
f=d+3;
i=i+1;
j=j+1;

printf("...... Section 1 output ......\n");
printf("a=%5.2f, b=%5.2f\nc=%5.2f, d=%5.2f\n"
       "e=%5.2f, f=%5.2f\ni=%5d, j=%5d \n\n", a,b, c,d, e,f, i,j);

/*-------------- Section 2 --------------------*/
m=k%p;
n=p%k;
i++;
++j;
e--;
--f;

printf("...... Section 2 output ......\n");
printf("m=%4d, n=%4d\ni=%4d, j=%4d\n"
       "e=%4.2f, f=%4.2f\n",m,n, i,j, e,f);
}
```

Assignment statements. Expressions on right side make use of arithmetic operators.

Special C arithmetic operators.

Output

```
...... Initial values ......
i=    5, j= 5
k=   11, p= 3
x=3.00, y=4.00

...... Section 1 output ......
a= 7.00, b=-1.00
c=12.00, d= 0.75
e= 3.75, f=3.75
i=    6, j=  6

...... Section 2 output ......
m=   2,  n=   3
i=   7,  j=   7
e=2.75, f=2.75
```

Explanation

1. What is an arithmetic expression? An *arithmetic* expression is a formula for computing a value. For example, the expression x + y computes x plus y.

2. What are the components of an arithmetic expression? An arithmetic expression consists of a sequence of *operand*(s) and *operator*(s) that specify the computation of a value. For example, the expression, -x + y, consists of two operands x and y and two operators + and -.

3. What can be an operand? An operand can be a variable, such as x or y, or a constant, such as 3.1416, or anything that represents a value, such as a function (see Chapter 5 for details).

4. What are the meanings of the operators ++, --, *and* %? The operator ++ is an *increment operator,* which can be placed before or after (but not both) a variable. The operator will increase the value of the variable by 1. For example, assuming a variable i is equal to 1, then after the statement

```
i++;
```

or

```
++i;
```

is executed, the value of i is 2 (i++ is not exactly the same as ++i, see Lesson 3.4 for details). Note that the C statement

```
i++;
```

or

```
++i;
```

can be understood as the statement

```
i=i+1;
```

which also causes the value of the variable i to increase by 1. Similarly, the operator -- is a *decrement operator,* which decreases the value of a variable by 1. Also, the statement

```
i--;
```

or

```
--i;
```

can be understood as the statement

```
i=i-1;
```

The operator % is a *remainder operator,* which must be placed between two integer variables or constants. Assuming k and p are two integer variables, the meaning of k%p is the remainder of k divided by p. For example, if k = 11 and p = 3, then k%p is equivalent to 11%3, which is equal to 2 (because 3 goes into 11 three times with a remainder of 2). The operator % is pronounced "mod." So this example would be k mod p. ANSI C states that, if either operand is negative, the sign of the result of the % operation is implementation defined; that is, it is free for the C compiler designer to decide. For example, depending on the compiler you use, the results of -50 % 6 and 50 % (-6) may be 2 or −2.

5. Is an arithmetic expression a complete C statement and how are arithmetic expressions used in assignment statements? An arithmetic expression is not a complete C statement, but only a component of a statement. The value evaluated from the expression may be stored in a variable using an assignment statement. For example, the arithmetic expression x/y is part of a C assignment statement

```
d = x/y;
```

The statement assigns the value obtained from the arithmetic expression on the right to the variable on the left. Thus, the assignment statement

```
i=i+1;
```

although not looking correct algebraically, is a valid C assignment statement. The arithmetic expression i + 1 creates a new value that is 1 greater than the current value of i. The assignment statement then gives i this new value.

Note that we cannot write these two assignment statements as

```
x/y = d;
i + 1 = i;
```

because on the left side of assignment statements we can have only single variables, not expressions. Single variables are allowed to be lvalues (pronounced "ell-values"), meaning they are allowed to be on the left side of assignment statements. Expressions are rvalues (pronounced "are-values") because they are allowed on the right side of assignment statements.

6. Can a single variable also be considered an expression? Yes, for instance, a single variable located alone on the right side of an assignment statement is considered to be an expression. We will see other times when a single variable is considered to be an expression.

7. What happens if we try to divide a number by 0? In general, this causes a run-time error and termination of execution. A common error message displayed by the computer when this occurs uses the word *overflow,* because division by 0 or a number close to 0 produces a very large number. The number being too large to store in the allocated memory causes an interpretation of an overflow problem.

8. What should I do if I get this error in my program? You need to use your debugging skills to trace the source of the problem, and this may not be simple (Table 3.4 summarizes this process). For instance, say, we have in a program:

```
b = c/(x-y);
```

with *x* and *y* being equal values before this statement is executed. This statement will cause a run-time error on execution. However, by simply looking at the code, it is not possible to immediately say that this statement is the source of the problem because nothing is inherently wrong with the statement itself.

Therefore, in your programs, if you get an overflow error message the first thing you need to do is to find the statement in your program at which the division by 0 has taken place.

TABLE 3.4
Finding the source of an overflow error

1. Look at your output. You know the error has been caused by a statement after the printf statements that produced the output.

2. If no printf statements have been executed prior to the overflow error, add printf statements to your program for the sole purpose of helping you find the location of the error.

3. Rerun the program.

4. Use the printf statements executed to locate the error within a few lines.

5. Look at the source code in the region of the lines causing the error.

6. Add more printf statements to print out the values of the variables in these statements. Rerun the program.

7. Use your hand calculator to check the equations used in your program so you can see exactly which statement caused an overflow.

8. At this point, we cannot tell you exactly how to modify your program to correct the error. However, performing these steps should have given you enough insight into your problem to make it easier for you to recognize errors in your equations or other methods for avoiding overflow errors.

9. How do I find this statement? Before the error message saying an overflow has occurred, you may have had some values already printed out. This tells you that the division by 0 has taken place after the printf statements were executed. This starts you in finding the problem statement.

10. What if no printf statements are executed before the error occurs? If no printf statements were executed prior to the error occurring, then one method for finding the statement with the error is simply to put printf statements in the source code at a number of locations for the sole purpose of finding the error causing location. For instance, statements such as

```
printf(" Execution has taken place to statement 5     \n");
```

```
printf(" Execution has taken place to statement 10    \n);
```

can be repeated throughout your program. After writing a large number of these, you can execute your program again (even though you know that it will get an overflow error). On rerunning it, you see which printf statements were executed. If you have spread these throughout your program, then you know that the division by 0 has been caused by a statement shortly after the last printf statement executed.

11. Once I know the approximate location of the error, what should I do? Examine your source code and focus on the statements in the region following the last correct printf statement. Look for divisions that have taken place, paying particular attention to such things as the variables in denominators. Take all of the variables and write a printf statement to have their values printed out. Put that printf statement immediately after the last printf statement executed on the previous run.

Rerun your program again, even though you know that it will still encounter an overflow error. Look at the values of the variables printed out. Use your calculator to perform hand calculations with these variables. Look at your source code and the equations you wrote in the region of the error. By plugging the numbers in your calculator, you should find that one of the denominators in this region calculates to be 0 or nearly 0. It may not work out to be exactly 0 because the computer may be working with numbers slightly different from those that are printed out (one reason is that the computer may be working with 30 digits when you have printed only 5 and use only 5 in your calculator).

By doing this you have now narrowed the problem down to one statement and you understand exactly how the computer is beginning to divide by 0.

12. How do I now correct the problem? This depends on the overall purpose of your computer program, so, we cannot answer the question here. However, in many cases, you will find that the process you have gone through to find the source of the error has made you aware of an error you have made in programming. For instance, maybe you used a subtraction sign where you should have used an addition sign, or you put the wrong variable in the denominator. Recognizing these sorts of things and making the appropriate changes normally will solve your problem. However, you need to address this on a case-by-case basis.

13. Is there an easier way to find the location of the error? Yes, you can learn how to use your compiler's debugger. It will save you from writing all of the printf statements. However, if you do not learn the debugger, you should become proficient enough to be able to write all of the printf statements that you need quickly and easily.

EXERCISES

1. True or false:
   a. The term a+b is a correct arithmetic expression.
   b. The term x=a+b; is a complete C statement.
   c. If a = 5, then a is equal to 6 after a++; is executed, but (with a = 5) it is still equal to 5 after ++a; is executed.
   d. The term 5%3 is equal to 2 and 3%5 is equal to 3.
   e. The operands of the % operator must have integer type.
   f. The meaning of the equal sign, =, in the statement

      ```
      a = x+y;
      ```

      is equal; that is, *a* is equal to *x* + *y*.

2. Supposing a, b, and c are int variables and x, y, and z are float variables, which of the following are incorrect C statements?

   ```
   a+b = c;
   a+x =y;
   c = a%b;
   a/b = x+y;
   ```

```
x = a*3;
z= x+y;
```

3. Write a program to calculate your expected average GPA in the current semester and display your output on the screen.

    Solutions

1. a (true), b (true), c (false), d (true), e (true), f (false)

LESSON 3.4 MIXED TYPE ARITHMETIC, COMPOUND ASSIGNMENT, OPERATOR PRECEDENCE, AND TYPE CASTING

Topics
- Precedence of arithmetic operations
- Initializing variables
- Pitfalls in arithmetic statements
- Mixing integers and reals in arithmetic expressions
- Type casting
- Side effects

Before variables can be used in arithmetic expressions they must first be given numerical values. Giving variables their first numerical values is called *initializing* them. We will learn several different ways to initialize variables. Look at the declarations and the first six assignment statements in the source code for this lesson. What two different ways are shown in this program for initializing variables?

In the program, the arithmetic expressions 6/4 and 6/4.0 are used twice each. The variables on the left side of the assignment statements using these expressions are either float or integer. Look at the output for these variables. Note that only one of the values is 1.5, as one might expect the value to be. Can you guess why the other variable values are not 1.5? Hint: It has to do with the int or float declarations of the variables. What does this tell you about making sure that your declarations are correct and that you understand very clearly how C is performing arithmetic manipulations?

Also included in the program are the compound operators +=, -=, *=, /=, and %=. Look at the code that uses these operators. Compare these statements with the output for k1, k2, k3, k4, and k5. Can you deduce what these operators do? Note: You must also look at the initial values of these variables in order to figure this out.

In this program are assignment statements using the ++ and -- operators. When trying to determine what these statements do, remember that assignment statements take the value of the expression on the right side of the equal sign and give that value to the variable on the left side of the equal sign. Note that, initially, both i and j are equal to 1. Are the values of the expressions i++ and ++j the same? Look at the output for k and h to determine this. What does that tell you about how C computes the values of these types of expressions? Also, note what has happened to the values of i and j after execution of these statements.

You have learned in your math classes that parentheses can be used in arithmetic expressions to control the order in which the operations are performed. Similarly, you can use parentheses in your C code to control the order of performance of operations. Also, C has strict rules about the order of operation of addition, subtraction, multiplication, and division. These rules are established by setting the *precedence* of the operators. Operators of higher precedence are executed first while those of lower precedence are executed later. For two operators of equal precedence, the one that is leftmost in the expression is executed first. Use your calculator to calculate the values of x, y, and z in the program. Compare the values you calculate with the output. Can you determine which operators are of higher precedence—addition, subtraction, multiplication, or division? (Hint: Addition and subtraction have the same precedence and multiplication and division have the same precedence.)

Source code

```
#include <stdio.h>
void main(void)
{
        int    i=1, j=1,
               k1=10, k2=20, k3=30, k4=40, k5=50,
               k, h, m, n;


        float a=7, b=6, c=5, d=4,
              e, p, q, x, y, z;

        printf("Before increment, i=%2d, j=%2d\n",i,j);

        k=i++;
        h=++j;
```

Pre- and postincrement operators on right side of assignment statements. These produce different results.

```
        printf("After increment,  i=%2d, j=%2d\n"
               "                   k=%2d, h=%2d \n\n",i,j,k,h);

        m=6/4;
        p=6/4;
        n=6/4.0;
        q=6/4.0;
```

Right sides of assignment statements have both int and float. Left sides of assignment statements also have both int and float. C has rules for handling these situations.

```
        printf("m=%2d, p=%3.1f\nn=%2d, q=%3.1f\n\n",m, p, n, q);
        printf("Original k1=%2d, k2=%2d, k3=%2d, k4=%2d, k5=%2d\n",
               k1,k2,k3,k4,k5);

        k1 += 2;
        k2 -= i;
        k3 *= (8/4);
        k4 /= 2.0;
        k5 %= 2;
```

Compound assignment operators cause variable on left side to be operated on.

```
        printf("New       k1=%2d, k2=%2d, k3=%2d, k4=%2d, k5=%2d\n\n",
               k1,k2,k3,k4,k5);
```

```
    e= 3;
    x=  a + b -c   /d *e;
    y=  a +(b -c)  /d *e;
    z=((a + b)-c   /d)*e;

    printf("a=%3.0f, b=%3.0f, c=%3.0f\nd=%3.1f, e=%3.1f\n\n",
           a,b,c,d,e);

    printf("x=  a + b -c   /d *e = %10.3f \n"
           "y=  a +(b -c)  /d *e = %10.3f \n"
           "z=((a + b)-c   /d)*e = %10.3f\n", x,y,z);
}
```

> C has precedence rules for operators meaning that some operations in an expression are performed before others.

Output

```
Before increment, i= 1, j= 1
After increment,  i= 2, j= 2,
                  k= 1, h= 2
m= 1, p=1.0
n= 1, q=1.5

Original k1=10, k2=20, k3=30, k4=40, k5=50
New      k1=12, k2=18, k3=60, k4=20, k5= 0

a= 7, b= 6, c= 5
d=4.0, e=3.0

x= a + b -c /d *e  =   9.250
y= a +(b -c) /d *e =   7.750
z=((a + b)-c /d)*e = 35.250
```

Explanation

1. How do we initialize variables? Method 1 uses an assignment statement to *initialize* a variable; for example,

$$e=3;$$

Method 2 initializes a variable in a declaration statement; for example,

$$float \ a=7, \ b=6;$$

2. Assuming that int variables i and j are equal to 1, is the meaning of k = i++; *the same as* h = ++j? No, in the first statement, the value of i is first assigned to the variable k. After the assignment, the variable *i* is incremented by the *postincrement operator* ++ from 1 to 2. Therefore, after executing

$$k=i++;$$

i = 2 and k = 1. However, for h = ++j, the value of j is first incremented by the *preincrement operator* ++ from 1 to 2. After the increment, the new j value, which now is equal to 2, is assigned to the variable h. Therefore, after executing

```
h=++j;
```

$j = 2$ and $h = 2$. In other words, the statement

```
k=i++;
```

is "equivalent to" statements

```
k=i;
i=i+1;
```

However, the statement

```
h=++j;
```

is "equivalent to" statements

```
j=j+1;
h=j;
```

These are the rules for such operators:

1. If the increment or decrement operator precedes the variable, the variable is first incremented or decremented by 1. Then the expression is evaluated using the new value of the variable.
2. If the increment or decrement operator follows the variable, the expression is evaluated first. Then the variable is incremented or decremented.

You must memorize these two rules.

3. How does C interpret and store the number 6 as written in the source code of this lesson's program? Because there is no decimal point, C treats this value as an integer. It therefore stores the number in 2's complement binary form.

4. How does C interpret and store the number 4.0 as written in the source code of this lesson's program? Because a decimal point is written, C treats this as a real type value and stores it in exponential binary form.

5. What is the value of 6/4? Here, we have an integer being divided by an integer. When this occurs, if both operands are negative or positive, the fractional part of the quotient is discarded. Therefore, 6/4 is not equal to 1.5. It is equal to 1. This is illustrated at the center of Fig. 3.7.

6. What is the value of 6.0/4.0? Since both of these values are real, the result is 1.5. This is illustrated at the top of Fig. 3.7.

7. What is the value of 6/4.0? For this operation, one operand is a real type and the other operand is an integer type. When this occurs, C converts the integer to a real type temporarily (meaning that 6 is converted to 6.0) then performs the operation and the result is a real type. Thus, 6/4.0 gives the result 1.5. This procedure is illustrated at the bottom of Fig. 3.7.

FIG. 3.7
Mixed type and same arithmetic operations

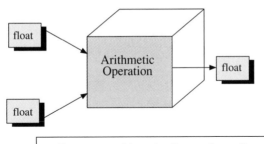

Same type arithmetic: Operands are float.
Result is float.

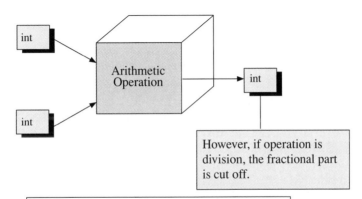

However, if operation is division, the fractional part is cut off.

Same type arithmetic. Operands are int and int.
Result is int but, if operation is division, the
fractional part is cut off.

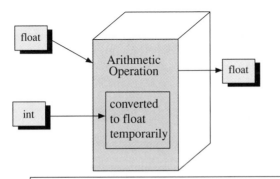

Mixed type arithmetic. Operands are float
and int. Result is float.

8. What happens if we try to assign a real type value to a variable that has been declared to be int? C cuts off the fractional part of the real value, converts the remaining part to int, and stores the result in 2's complement binary form in the memory cell for the variable. For instance, the statement

$$n = 6 / 4.0;$$

takes the real value (6/4.0, which is 1.5), cuts off the 0.5 to give 1.0, converts this to 1 (without a decimal point meaning it is in 2's complement binary form), and stores this in the memory cell for the variable n. The assignment statements for m, p, and q are described in Fig. 3.8.

9. What happens if we try to assign a integer type value to a float type variable? C puts a decimal point at the end and converts the method of storage to exponential binary form and stores the result in the variable's memory cell. Thus,

$$p = 6 / 4;$$

takes the integer value (6/4, which becomes 1 due to the cutting off of the fractional part), converts this to 1.0 (with a decimal point meaning it is in exponential binary form), and stores this in the memory cell for the variable p.

10. Although not used in this lesson's program, what would be the result of the expression, $(-6/5)$? The ANSI C standard says that the result is "implementation defined." This means that ANSI C has no hard and fast rule, therefore, the result depends on the compiler that you use. Check your compiler; it should give either -1 or -2 as a result. Note that the "correct" answer is -1.2. The compiler therefore has a choice of rounding up or down.

11. How does the way that C does arithmetic affect how you need to program? You need to keep in mind a few things while you are writing arithmetic statements in programs:

- When you have a division operation, make sure that it does not involve two integer type variables or constants unless you really want the fractional part to be cut off. Remember, check your variable types before you write a division operation.
- When you are writing your code and a float type variable is on the left side of an assignment statement, to be safe, use decimal points for any constants on the right side of the assignment statement. You may get the correct result without using decimal points, but we recommend that you use decimal points until you feel comfortable with mixed type arithmetic.
- Also, when an int type variable is on the left side of an assignment statement, it is necessary that you make sure that the arithmetic expression on the right side of the assignment statement creates an integer value. If you observe that it creates a real value, you must realize that the fractional part will be lost when the assignment is made.

FIG. 3.8

Same type and mixed type arithmetic operations, assigning real values to int variables and integer values to float variables

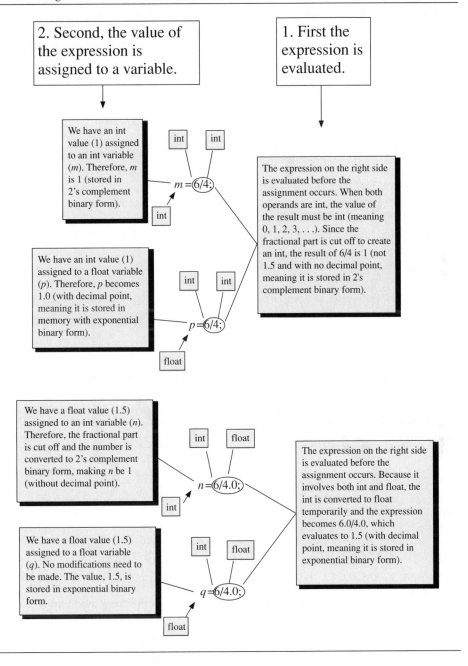

12. Is there a way we can modify the way C uses types in arithmetic operations? Yes, we have not shown it in this lesson's program, but C has *cast* operators. Cast operators can be used to change the type of an expression (recall that a single variable can be regarded as an expression). Thus, we can use cast operators on the right side of an assignment statement to modify the type of an arithmetic expression's result. A cast operator consists of a type name enclosed in parentheses.

For instance, if we have declared the variables

```
int aa=5, bb=2, cc;
float xx, yy=12.3, zz=18.8;
```

then the cast operator can be used in the following way:

```
xx = ((float) aa) / ((float) bb);
cc = (int)yy +(int) zz;
```

To understand how the operations in these statements are to take place, realize that, in performing arithmetic operations, C makes copies of the variable values and works with the copies. After the operation has been completed using the copies, a final result for the expression is obtained. If the expression is on the right side of an assignment statement, the assignment then takes place.

Thus, the cast operators, `(float)` and `(int)`, cause the copy of `aa` to be 5.0, `bb` to be 2.0, `yy` to be 12, and `zz` to be 18. Table 3.5 summarizes these actions. Because the operations are performed with the copies, we can clearly see that the values stored for the variables on the left sides of the assignment statements are

$$xx = 5.0/2.0 = 2.5$$
$$cc = 12 + 18 = 30$$

Had the program statements been written without the cast operators, the results would have been

Expression without cast operators	Stored result
`xx = aa/bb;`	$xx = 5/2 = 2.0$ (because xx is a float while the operands are int)
`cc = yy + zz;`	$cc = 12.3 + 18.8 = 31$ (because cc is an int while the operands are float)

Thus, we can see that the cast operators have changed the results stored for the variables `xx` and `cc`.

▨ TABLE 3.5
    Actions by cast operators

Variable name and type	Initial value	Cast operation	Value of copy used in arithmetic operation
int aa	5	(float) 5	5.0
int bb	2	(float) 2	2.0
float yy	12.3	(int) 12.3	12
float zz	18.8	(int) 18.8	18

13. What is the general form of the cast operator? The general form for use of the cast operator is

(type) expression

where the cast operator must be enclosed in parentheses. The *type* can be any valid C data type. We will learn of more valid C data types later in this text.

14. What are the meanings of operators +=, -=, *=, /=, ***and*** %=***?*** The operators += , -= , *= , /= , and %= are *compound assignment operators*. Each performs an arithmetic operation and an assignment operation. These operators require two operands: The left operand must be a variable; the right one can be a constant, a variable, or an arithmetic expression. In general, the two operands can be of integer or real data type. However, the %= operator requires that its two operands be of integer type.

For instance, the meaning of

```
k1+=2;
```

(not k1 =+ 2;) can be understood to be similar to the statement

```
k1=k1+2;
```

If the original value of k1 is equal to 20, the new value will be 20 + 2 or 22. Similarly, these statements also are valid if we replace the arithmetic operator + with operator - , * , / , or % . For example,

```
k1*=2;
```

is similar to

```
k1=k1*2;
```

15. How do we control precedence in an arithmetic expression? Parentheses can be used to control *precedence*. Arithmetic operators located within the parentheses always have higher precedence than any outside the parentheses. When an arithmetic expression contains more than one pair of parentheses, the operators located in the innermost pair of parentheses have the highest precedence. For example, the + operator in the statement

```
z = ((a+b)-c/d);
```

has higher precedence than the - or / operator and a + b will be evaluated first.

16. What will happen if all operators have the same level of precedence? If all arithmetic operators are of equal precedence in an arithmetic expression, the leftmost operator is executed first.

17. Can we use two consecutive arithmetic operators in an expression? We cannot use two consecutive arithmetic operators in an arithmetic statement unless parentheses are used. For example, x/-y is not permissible but x/(-y) is permissible.

18. What operators can be used in an arithmetic expression? Table 3.6 shows the operators along with their properties that can be used in an arithmetic expression.

19. What are the meanings of the number of operands and their position, associativity, and precedence in Table 3.6? The number of operands is the number of operands required by an operator. A *binary operator,* such as /, requires two operands while a *unary operator,* such as ++, needs only one. Fig. 3.9 shows the concepts of these two operators.

The *position* is the location of an operator with respect to its operands. For a unary operator, its position is prefix if the operator is placed before its operand and postfix if it is placed after its operand; for a binary operator, the position is infix because it is always placed between its two operands. For example, the negation operator in -x is prefix, the postincrement operator in y++ is postfix, and the remainder operator in a%b is infix.

The *associativity* specifies the direction of evaluation of the operators with the same precedence. For example, the operators + and - have the same level of precedence and both associate from left to right, so 1 + 2 - 3 is evaluated in the order of (1 + 2) - 3 rather than 1 + (2 - 3). This concept is shown in Fig. 3.10.

The *precedence* specifies the order of evaluation of operators with their operands. Operators with higher precedence are evaluated first. For example, the operator * has higher precedence than -, so 1 - 2 * 3 is evaluated as 1 - (2 * 3) rather than (1 - 2) * 3. Note that in this example the - indicates subtraction and is a binary operator with precedence 4. The - also can be used as a negative sign,

TABLE 3.6
Arithmetic operators

Operator	Name	Number of operands	Position	Associativity	Precedence
(	parentheses	unary	prefix	L to R	1
)	parentheses	unary	postfix	L to R	1
+	positive sign	unary	prefix	R to L	2
−	negative sign	unary	prefix	R to L	2
++	post-increment	unary	postfix	L to R	2
−−	post-decrement	unary	postfix	L to R	2
++	pre-increment	unary	prefix	R to L	2
−−	pre-decrement	unary	prefix	R to L	2
+=	addition and assignment	binary	infix	R to L	2
−=	subtraction and assignment	binary	infix	R to L	2
*=	multiplication and assignment	binary	infix	R to L	2
/=	division and assignment	binary	infix	R to L	2
%=	remainder and assignment	binary	infix	R to L	2
%	remainder	binary	infix	L to R	3
*	multiplication	binary	infix	L to R	3
/	division	binary	infix	L to R	3
+	addition	binary	infix	L to R	4
−	subtraction	binary	infix	L to R	4
=	assignment	binary	infix	R to L	5

which is a unary operator with precedence 2. For example, `-2 + 3 * 4` is evaluated as `(-2) + (3 * 4)` rather than `-(2 + (3 * 4))`. This concept is shown in Fig. 3.11.

20. How is the cast operator affected by precedence? The cast operator, required to be enclosed in parentheses, has the highest precedence. Thus, the expression

`(float) aa / (float) bb`

is equivalent to

`((float) aa) / ((float)bb)`

because the cast operator has higher precedence than the division operator.

21. How is the use of the cast operator affected by the arithmetic operation rules in C illustrated in Fig. 3.7? The effect of C's arithmetic operation rules is that

FIG. 3.9
Unary and binary operators

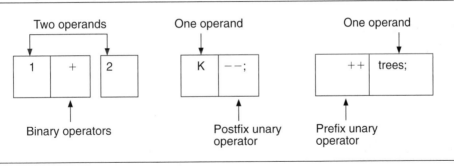

FIG. 3.10
Concept of associativity

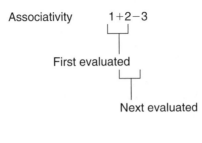

■ **FIG. 3.11**
Concept of precedence

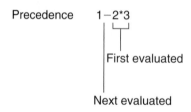

sometimes you may not need to use a large number of cast operators in an expression. For instance, with the declarations

```
int a, b;
float x;
```

the statement

```
x = (float) a + b;
```

gives the same result as

```
x = (float)a + (float)b;
```

because, in the first statement, adding a float and an int is done in C by copying the int in float form and adding the two as floats as illustrated in Fig. 3.7.

22. What is a side effect? The primary effect of evaluating an expression is arriving at a value for that expression. Anything else that occurs during the evaluation of the expression is considered a *side effect*. For instance, the primary effect of the C statement (assuming i originally is 7)

```
j = i++;
```

is that the expression on the right side of the assignment statement is found to have a value of 7. The side effect of the above statement is that the value of i is incremented by 1 (to make i equal to 8). Consider the following C statement:

```
j = (i=4) + (k=3)−(m=2);
```

Its primary effect is to arrive at the value of the expression on the right side of the assignment statement (which is 5 obtained from 4 + 3 − 2). Three side effects occur during the evaluation of the expression:

1. Set i equal to 4
2. Set k equal to 3
3. Set m equal to 2

23. What are the dangers of side effects? At times, side effects can be confusing. For the statement

$$k = (k=4) * (j=3);$$

the result of k will be 12 instead of 4. It is best not to use side effects except in their simplest form, such as

$$i = j++;$$

or

$$i = j = k = 5;$$

Note that, because the associativity of the assignment operator is from right to left, multiple assignment statements such as the preceding one can be written. The order of operation is

1. k = 5
2. j = k
3. i = j

Also, an expression

$$i = j = k = 2 + n + 1;$$

is evaluated in this order:

1. k = 2 + n + 1;
2. j = k;
3. i = j;

because the addition operation has a higher precedence than the assignment operator.

EXERCISES

1. Based on

    ```
    int i=10, j=20, k, m,n;
    float   a,b,c,d,e=12.0;
    ```

    Determine whether each of the following statements is true or false:
    a. The statement i =+ 2; is a valid C statement.
    b. The statement i %= e; is a valid C statement.
    c. The statement i *= (i+j*e/123.45); is a valid C statement.
    d. When k=i/j;   k is equal to 0.5.
    e. When i+=j;   i is equal to 30 and j is equal to 20.
    f. The term k=1/3+1/3+1/3; is equal to 1.
    g. The term d=1/3.+1.0/3+1.0/3.0 is equal to 1.0.
    h. The term a=1/3+1/3+1/3 is equal to 1.0.
    i. The term a=1./3+1/3.+1.0/3.0 is equal to 1.0.
    j. The term i+2/3*3/2 is equal to 11.

k. The term `i+3/2*2/3` is equal to 10.
l. The term `++i=j;` `i` is equal to 21;.
m. The term `++i++;` `i` is equal to 31;.

2. Convert each of the following formulas to a C arithmetic expression:

$$1 + \frac{1}{3} + \frac{1}{5} + \frac{1}{7}$$

$$1 + \frac{1}{2.0} + \frac{1}{3.0} + \frac{1}{4.0}$$

$$\pi R^2$$

$$\frac{(a + b)^2}{(a - b)^3}$$

$$\frac{a + \dfrac{\dfrac{b}{c}}{d + \dfrac{e}{f}}}{gh^2}$$

3. Assume `a`, `b`, and `c` are int variables and have the following values: `a` = 10, `b` = 20, `c` = 30. Find the values of `a`, `b`, and `c` at the end of each list of statements:

a.	b.	c.	d.
`a=c;`	`a=c++;`	`a=++c;`	`a+=c;`
`b=a;`	`b=a++;`	`b=++a;`	`b/=a;`
`c=b;`	`c=b++;`	`c=++b;`	`c%=b;`

4. Suppose that `a`, `b`, and `c` are variables as just defined, write a program to rotate the value of `a`, `b`, and `c` so that `a` has the value of `c`, `b` has the value of `a`, and `c` has the value of `b`.

5. Hand calculate the values of `x`, `y`, and `z` in the following program and run the program to check your results:

```
#include <stdio.h>
void main(void)
{
      float a=2.5,b=2,c=3,d=4,e=5,x,y,z;
      x=  a *  b -  c + d  /e ;
      y=  a * (b -  c)+ d  /e ;
      z=  a * (b - (c + d) /e) ;
      printf("x= %10.3f, y= %10.3f, z=%10.3f",x,y,z);
}
```

6. Calculate the value of each of the following arithmetic expressions:

`13/36,  36/4.5,  3.1*4,  3-2.6,  12%5,  32%7`

Solutions

1. a (false), b (false), c (true), d (false), e (true), f (false), g (false), h (false), i (true), j (false), k (true), l (false), m (false)

LESSON 3.5 READING DATA FROM THE KEYBOARD

Topics

- Using the scanf( ) function
- Inputting data from the keyboard
- The address operator &
- double type data

No programs in any of the previous lessons has had input going into it during execution. These programs had only output, and for them, the output device was the screen (or monitor). Most commonly, your programs will have both input and output. Your program can instruct the computer to retrieve data from various input devices. Input devices include

> the keyboard
> a mouse
> the hard disk drive
> a floppy disk drive

to name a few. The program in this lesson illustrates how input can be retrieved by a C program from the keyboard.

Programs that have input from the keyboard usually create a dialogue between the program and the user during execution of the program. Look at the first printf statement for this program. What is printed to the screen by this statement? Look at the output. Did printf print what you expected? This essentially is the program talking to the user.

Look at the declarations in this program. One of the declarations is of a type that is not int or float. What is the name of the type? We will tell you that this type is for real types of variables but contains more bits (and therefore is capable of representing more digits) than the float type.

What are the names of the variables in this program? Look at the scanf statement following the first printf statement. Is any variable used in this statement? What symbol is used immediately before this variable name in the scanf statement? We also have a string literal in this scanf statement. How can you recognize it? Within this string literal is a conversion specification that we have seen before. What is it? When we used this conversion specification in a printf statement, with what type of variable was it used? What type of variable is in this scanf statement? Do you think there is a correspondence between the conversion specification and the variable type?

This first scanf statement corresponds to the first line of keyboard input. Look at this line of input. The value of which variable has been entered from the keyboard? We can tell you now that scanf essentially has grabbed that value from the keyboard input device and stored it in the memory cells reserved for the variable.

Look at the second scanf statement in the program. What types of variables are these? What are the conversion specifications in the string literal for this statement? Note that one of the variables is of type double. Which conversion specification do you think goes with type double? Look at the second line of keyboard input. How many values have been entered? Does this correspond to the number of variables listed in the scanf statement?

Look at the last scanf statement. What is in the string literal that we have not seen in a scanf string literal to this point? Look at the On-Screen Dialogue. Look at the last line of keyboard input. Has this symbol been written in the input?

Source code

```
#include <stdio.h>
void main(void)
{                              Data type is double
       float income;          (for real numbers).
       double expense;
       int month, hour, minute;    Function scanf reads
                                    values typed on keyboard.
       printf("What month is it?\n");
       scanf("%d", &month);
       printf("You have entered month=%5d\n",month);

       printf("Please enter your income and expenses\n");
       scanf("%f %lf",&income,&expense);
       printf("Entered income=%8.2f, expenses=%8.2lf\n",
          income,expense);

       printf("Please enter the time, e.g.,12:45\n");
       scanf("%d : %d",&hour,&minute);
       printf("Entered Time = %2d:%2d\n",hour,minute);
}
            Function scanf uses the addresses of the declared variables.
```

Output

```
                   What month is it?
Keyboard input     12
                   You have entered month=  12
                   Please enter your income and expenses
Keyboard input     32  43
                   Entered income =  32.00, expenses =  43.00
                   Please enter the time, e.g., 12:45
Keyboard input     12:15
                   Entered Time = 12:15
```

Explanation

1. What is double type data? It is one of C's floating-point (real) data types. The double type usually occupies twice the memory (8 bytes) of the float type (4 bytes). C compilers use the extra memory to increase the range or the number of significant digits as illustrated in Fig. 3.12. Because of the extra precision given by double types, in this text we will use double rather than float to represent real numbers.

2. How do we input data from the keyboard? An easy way to input data from the keyboard is by using the scanf() function. The syntax of the function is

scanf (*format_string, argument_list*) **;**

where the *format_string* converts characters in the input into values of a specific type, the *argument_list* contains the address of the variable(s) into which the input data are stored, and a comma must be used to separate each argument in the list from the others. For example, the statement

scanf("%f%lf",&income,&expense);

will convert the first keyboard input data to a float type value using the %f conversion specification and store the float value in the memory cell reserved for the variable income. Similarly, the second keyboard input will be converted to a double value using the %lf conversion specification and store it in the memory cell reserved for the variable expense. Note that you must precede each variable name with an & when you read a value, because the argument in the scanf() function uses the address of the variable (&income stands for the address of the memory cell for income); similarly, &expense stands for address of the memory cell for expense. This scanf statement is said to pass three arguments to scanf. The first is the string literal that contains the conversion specifications, the second is the address of the variable income, and the third is the address of the variable expense. Fig. 3.13 illustrates the call to the function scanf. If you want to read an int type variable, use %d instead of %f or %lf as the conversion specification. Observe that the "l" in %lf is the lowercase letter "ell."

3. Can you say more about the & symbol? Looking at more C library functions, we will note that some functions need variable values (represented by the names of the variables) passed to them while other functions need the addresses of the variables passed to them. We use the "address of" operator, &, in front of variable names to pass the addresses of variables to functions.

FIG. 3.12
Type double versus type float

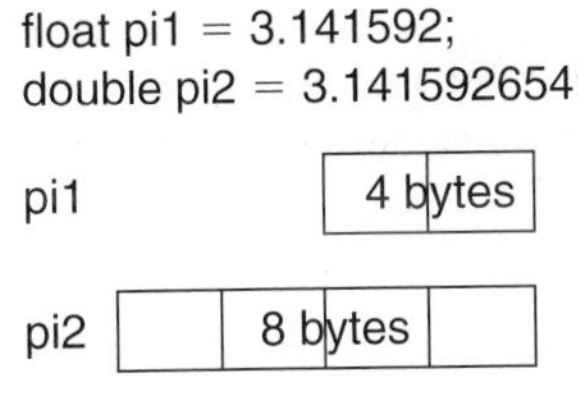

---

float pi1 = 3.141592;
double pi2 = 3.141592654;

pi1 | 4 bytes |

pi2 | | 8 bytes | |

FIG. 3.13
Using scanf( ) for input

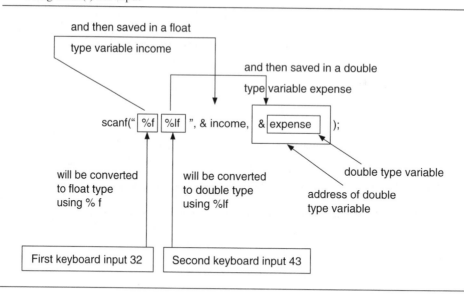

The scanf function needs the address for a memory cell because it needs to know where to put the value typed in at the keyboard. By giving scanf the address, the program knows where in memory to put the value typed. This is illustrated in Fig. 3.14.

4. What are the components of a format string in the scanf() function? The format string may consist of conversion specifications, such as %d or %lf, blanks, and character(s) to be input. If the format string contains character(s), you must match the character(s) when you input information from the keyboard. For example, the statement

```
scanf("%d : %d",&hour,&minute);
```

contains a colon, :, in the format string; if you want to input hour = 12 and minute = 34, the valid input is

```
                           12 : 34
```

If you omit the colon, the data are read incorrectly. In general, the format string in the scanf() function should be kept as simple as possible to reduce errors on input. In most cases in this book we have only conversion specifications in our scanf format strings and avoid using plain characters wherever possible.

4. What errors are common in using scanf? It is common to forget the & in front of the variable name. Remember, with printf do not use &, with scanf use &.

FIG. 3.14
Operation of scanf function

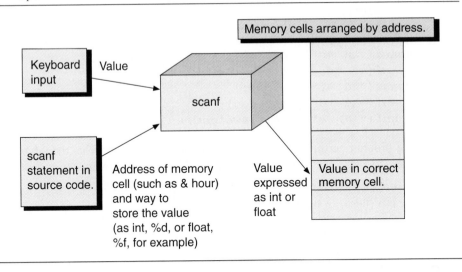

Although not an error, it will cause you problems if you do not have a printf statement to go with each scanf statement you use, because scanf causes program execution to stop. If you do not have a printf statement that has displayed something on the screen telling the user what should be typed in, the computer will just sit there, waiting for keyboard input. Without being prompted by a printf statement, a user will not know when or what to type. So remember, always prompt a user with a printf before using a scanf.

Also important is to pay strict attention to matching your conversion specifications with variable types. In other words, you must use `%lf` with double (`%f` will not work), `%f` with float, and `%d` with int.

EXERCISES

1. True or false:
   a. The statement

      ```
      scanf("%f %lf" &a, &b);
      ```

      will store the first and second keyboard input data to variables `&a` and `&b`.
   b. The statement

      ```
      scanf("%f %lf\n",&a, &b);
      ```

      contains four arguments.
   c. The statement

      ```
      scanf("%d:%f:%lf\n",&a, &b,&c);
      ```

      contains four arguments: `"%d:%f:%lf \n"`, `&a`, `&b`, and `&c`.

2. Based on the statements

```
int cat, dog;
double weight;
```

find error(s) in each of these statements:
a. `scanf("%d %d"),cat,dog;`
b. `scanf(%d %d,cat,dog);`
c. `scanf("%d %f",cat,dog);`
d. `scanf("%d %d",&cat,&dog);`
e. `scanf("%d, %lf",&cat,&weight);`

3. Write a program to input all your grades in the last semester from the keyboard and then display your input and the average GPA on the screen.

Solutions
1. a (false), b (false), c (true)
2. a. `scanf("%d %d",&cat,&dog);`
    b. `scanf("%d %d",&cat,&dog);`
    c. `scanf("%d %d",&cat,&dog);`
    d. No error
    e. No error, but you have to type a comma between the two values.

LESSON 3.6 READING DATA FROM A FILE

Topics
• Opening and closing a file
• Reading data from a file
• Using the fscanf( ) function

If your input data is lengthy and you are planning to execute your program many times, it is not convenient to input your data from the keyboard. This is true especially if you want to make only minor changes to the input data each time you execute the program.

For instance, if your income is the same every month and only your expenses change, it is cumbersome to repeatedly type the same number for each month. It is more convenient to set up a file that has your income and expenses. Your program can read that file during execution instead of receiving the input from the keyboard. If you want to rerun the program with different input data, you can simply edit the input file first and then execute the program.

This lesson's program illustrates how to read data from an input file. In the program, the file name specified is C3_6.IN. You must remember, though, that when you create your input file using your editor that you give that file the same name you have specified in the code for your program. When you execute your program, the computer searches for a file of that name and reads it. If that file does not exist, your program will not execute.

Look at the program. Can you guess the name of the standard function used to read a file? Look closely at the arguments for this function. What is the first argument? How do the arguments differ from the arguments for scanf?

Where else in the program do you see the first fscanf argument? Note that the three locations are associated with fopen, fclose, and FILE. Can you guess what this argument represents?

Source code

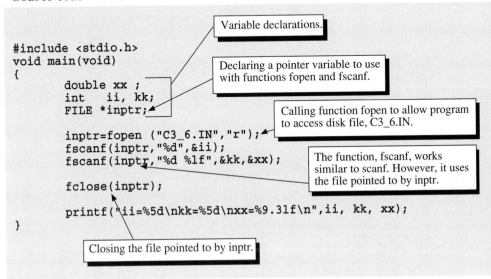

```
#include <stdio.h>
void main(void)
{
        double xx ;
        int    ii, kk;
        FILE *inptr;

        inptr=fopen ("C3_6.IN","r");
        fscanf(inptr,"%d",&ii);
        fscanf(inptr,"%d %lf",&kk,&xx);

        fclose(inptr);

        printf("ii=%5d\nkk=%5d\nxx=%9.3lf\n",ii, kk, xx);

}
```

Variable declarations.

Declaring a pointer variable to use with functions fopen and fscanf.

Calling function fopen to allow program to access disk file, C3_6.IN.

The function, fscanf, works similar to scanf. However, it uses the file pointed to by inptr.

Closing the file pointed to by inptr.

Input file C3_6.IN

```
36
123 456.78
```

Output

```
ii=  36
kk= 123
xx= 456.780
```

Explanation

1. What function is most commonly used to read data from a file? In C, we usually use the fscanf() function to read data from a file (Fig. 3.15). In general, the syntax of the fscanf() function is

$$\text{fscanf}\,(\textit{file\_pointer, format\_string, argument\_list})\,;$$

The fscanf() function reads the contents of the file indicated by *file_pointer* according to the conversion specifications in *format_string*. The contents read are put into the addresses given by the *argument_list*. For example, in the statement

FIG. 3.15

Operations of the fscanf function. This example is for the variable *ii* of this lesson's program.

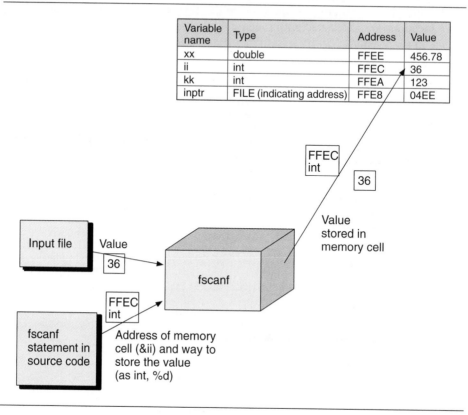

Variable name	Type	Address	Value
xx	double	FFEE	456.78
ii	int	FFEC	36
kk	int	FFEA	123
inptr	FILE (indicating address)	FFE8	04EE

```
fscanf(inptr, "%d %lf",&kk,&xx);
```

the *format_string*

```
" %d %lf"
```

and *file_pointer* inptr indicate that two values are to be read from the input file indicated by inptr. The first is an int (corresponding to %d) and the second is a double (corresponding to %lf). The values read are to be put into the addresses indicated by the addresses in the *argument_list,* &kk and &xx. The functions scanf and fscanf are closely related. What you have learned about the format string and argument list for scanf can be applied to fscanf, and we therefore do not repeat that information.

2. What is a file pointer? *File pointers* will be discussed in Chapter 8. For now, just remember that a *file pointer* is a variable whose memory cell contains an address instead of an int, float, or double value. (Recall that we discussed this in Chapter 1.) This address gives the key to accessing the file stored on disk. To declare the variable that stores the address, the file pointer variable, we must begin the declaration with the word FILE and have an asterisk (*) before the variable name. For example, the statement

FILE *inptr;

declares the variable inptr to be a file pointer, which means that C expects that an address will be stored in its memory cell.

3. What is the convention for naming file pointers? The naming convention for file pointers is the same as the naming convention for other C identifiers. Examples of legal and illegal file pointer names follow:

Legal: **FILE *apple, *IBM93, *HP7475;**
Illegal: **FILE *+apple, *93IBM, *75HP75;**

4. After we have declared the file pointer, how do we make a file available for us to read? We use the C library function fopen. This function gives us the ability to create a link between a disk file and a file pointer. Once this link is created we can work with the file pointer in our program to give us access to the file to which it is linked.

The form for using fopen to create this link is

file_pointer = **fopen** (*file_name, access_mode*);

For example, as illustrated in Fig. 3.16, in the statement

inptr = fopen ("C3_6.IN","r");

the *file_name* as referenced by the operating system is C3_6.IN, the *file_pointer* is named inptr, and the *access_mode* is "r," which means the file is opened for reading in text mode (this contrasts with binary mode, which we will discuss in Chapter 8). You may choose any valid name for *file_pointer.* Note that the *file_name* and the *access_mode* are in string literals. Hence, they must be enclosed by a pair of double quotes.

Observe that this statement is an assignment statement. An assignment statement is used because calling the fopen function creates an expression. The value of the expression is equal to the address through which we can access the file. Thus, the assignment statement assigns this address to inptr, meaning the value of inptr becomes this address. After the assignment has taken place, we can use inptr to access the file, C3_6.IN.

5. Can you say more about FILE? FILE is a type set up for holding information about disk files. At this point, you should realize that a disk *file* is a collection of information in an electronic format. It may contain your personal data, a CIA secret document, or Hollywood's latest video movie. Information in a file is stored in

FIG. 3.16

Operations of the fopen function for this lesson's program. The function, fopen, takes the file name and mode from the source code and creates an address. The assignment statement causes this address to be stored in the memory cell for the variable declared as a FILE pointer.

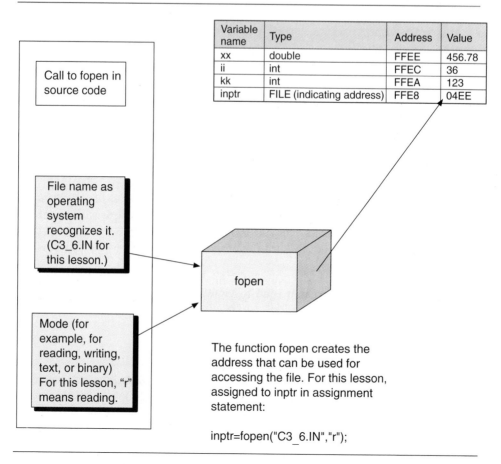

certain section(s) of external device(s), such as disks. Unlike a number, such as +1234, which can be defined by its size and sign, a file is more complicated and contains more features. For example, a file must have a name so that you or a computer can identify it. A file may be opened for reading (that is, getting data from it) or writing (that is, storing data in it). A file can be in text format, such as the source code of this program, or binary format, such as the execution code of this program. In addition, a file needs a temporary storage area to declare its size and other information so that it can be placed correctly by the computer operating system. To keep all of these features in one place, C "invents" a new data type (in reality it is a data structure, this will be explained in more detail in Chapter 8) named FILE that is somewhat similar but slightly more complicated than the other data types you have learned, such as int and float.

FILE is a C *derived data type* defined in the C standard header file stdio.h, which must be included to use FILE in your programs. Without the header file, stdio.h, the C compiler will not understand what FILE stands for and will generate an error message.

When you want to manipulate a disk file, you use the C data type FILE to declare a *file_pointer*, and then use this *file_pointer* to handle your file. This means that there is no direct relation between the C data type FILE and your actual file; that is, you cannot use the following statement

$$\texttt{FILE "L3\_6.IN";}$$

to declare your file. Instead, you must use FILE to declare a *file_pointer*, and then use the *file_pointer* to manipulate your file. The process schematically is

$$\texttt{FILE} \rightarrow file\_pointer \rightarrow actual\_file$$

6. Do we need to close an input file when we are finished using it? It is good practice to close files after they have been used. However, C will automatically close all open files, after execution is completed. To close a file manually, use the fclose() function, whose syntax is

$$\texttt{fclose}(file\_pointer);$$

Note that, the function uses *file_pointer*, not the file name, to close a file. For example, the statement

$$\texttt{fclose(inptr);}$$

uses the file pointer intpr to close the file 3_6.in.

7. How can we specify a file on a particular disk? We can specify, for instance, the C: disk for the file C3_6.IN with the statement:

$$\texttt{inptr = fopen ("C:\\\\C3\_6.IN", "r");}$$

Observe that the two backslashes are required because within a string literal, backslash has a special meaning as described in Chapter 2. Therefore, two backslashes are needed where we would normally use only one backslash.

EXERCISES

1. True or false:
   a. We use the fscanf() function to read input from the keyboard.
   b. We use the fscanf() function to read input from a file.
   c. You must create a link between an external disk file and a file pointer before you can read your input data.
   d. It is good practice to close an input file when you need no further access to the file.

   e. A file pointer is an int data type and can be declared with the other int type variables.

2. Find error(s), if any, in each statement:
    a. `#INCLUDE <stdio.h>`
    b. `file *myfile;`
    c. `*myfile = fopen (C3_6.DAT,r);`
    d. `fscanf("myfile","%4d %5d\n",WEEK,YEAR);`
    e. `close("myfile");`

3. Write a program to read your grades from last semester from an input file named GRADE.REP, which has one line of data consisting of four grades only (no characters); for example,

    `4.0 3.3 2.7 3.7`

    compute your average GPA, and write all the input data and average GPA on the screen.

Solutions
1. a (false), b (true), c (true), d (true), e (false)
2. a. `#include <stdio.h>`
    b. `FILE *myfile;`
    c. `myfile = fopen ("C3_6.DAT","r");`
    d. `fscanf(myfile,"%4d %5d\n",&WEEK,&YEAR);`
    e. `fclose(myfile);`

▨ LESSON 3.7 MATH LIBRARY FUNCTIONS

Topics

- Using standard math header file
- Contrasting the double and float data types
- Other data types

Your calculator makes it very easy for you to perform such operations as sin, log, and square root by having single buttons for them. Similarly, the C compiler makes it easy for you to perform these operations by providing mathematical library functions that you can call from your program. This lesson illustrates the use of some of these library functions. Look at the seven assignment statements in the program. What functions do you recognize? Note that, just like the printf function, these library functions require parentheses for enclosing the argument(s).

Use your calculator to verify the results obtained in the output for this lesson. What are the necessary units for the argument for the sin function? What kind of log does the log function take? Can you guess what the pow function does?

The C compiler has these functions in its library, but you must tell the compiler more information about them. Can you guess which statement in this program tells the C compiler where the extra information is located?

We have not shown it in the source code, but C has numeric data types other than int, float, and double. Again, they either use 2's complement binary or exponential binary form. However, the ones we discuss in this lesson occupy different amounts of memory. We will see that, at times, other numeric types will be useful to us.

In this lesson, we illustrate the difference between double and float. Look at the declarations for c and g and compare their values printed in the output. Observe that they are different. Which is more accurate? Use your calculator to check this. Why do you think one is more accurate than the other?

Source code

```
#include <math.h>  ←─── Including header file for
#include <stdio.h>      math functions.
void main(void)
{
        double x=3.0, y=4.0, a,b,c,d,e,f;
        float g;

        a=sin(x);
        b=exp(x) ;
        c=log(x);                              Standard C
                                               math functions
                                               used in
        d=sqrt(x);                             expressions.
        e=pow(x,y);
        f=sin(y)+exp(y)-log10(y)*sqrt(y)/pow(3.2,4.4);
        g=log(x);

        printf("x=%4.1f     y=%4.1f \n\n\r\
                a=sin(x)      = %11.4f\n\r\
                b=exp(x)      = %11.4f\n\r\
                c=log(x)      = %11.9f\n\n\r\
                d=sqrt(x)     = %11.4f\n\r\
                e=pow(x,y)    = %11.4f\n\r\
                f=sin(y)+exp(y)log10(y)*sqrt(y)/pow(3.2,4.4)= %11.4f\n\n\r\
                g=log(x)      = %11.9f\n",x,y, a,b,c,d,e,f,g);

}
```

Output

```
x= 3.0  y= 4.0

a=sin(x)      =        0.1411
b=exp(x)      =       20.0855
c=log(x)      = 1.098612289

d=sqrt(x)     =        1.7321
e=pow(x,y)    =       81.0000
f=sin(y)+exp(y)-log10(y)*sqrt(y)/pow(3.2,4.4)=    53.8341

g=log(x)      = 1.098612309
```

Explanation

1. How can we see the difference between the double and the float data types?
We already have observed that both of these types cause values to be stored in exponential binary form and that the double data type occupies more memory than the
float data type. When is it a good idea to make sure that we carry a large number of
digits? Carrying a large number of digits may be important when a large number of
calculations are to be done. The drawback in declaring all variables as being of the
double data type is that more memory is required to store double data type variables
than float type variables.

Consider the following example, which illustrates the effect of the number of
digits carried in a calculation. You should try this on your calculator. Suppose you
are multiplying a number by π, 100 times. You essentially will be computing π^{100}.
The influence on the number of significant digits used for π is the following. Using
five significant digits for π gives

$$(3.1416)^{100} = 5.189061599 * 10^{49}$$

while using eight significant digits for π gives

$$(3.1415926)^{100} = 5.1897839464 * 10^{49}$$

Here, it can be seen that the first estimate of π has five significant digits; however,
$(3.1416)^{100}$ is accurate only for the first three digits. This illustrates that accuracy is
reduced after numerous arithmetic operations. Since one computer program easily
can do 1 million operations, one can begin to understand the need for initially carrying many digits.

We can even see a difference in the values calculated for log(3) for this lesson's
program. With double we get the natural log to be

$$c=log(3) = 1.098612289$$

whereas with float we get

$$g=log(3) = 1.098612309$$

If you compare these with your calculator, you will find that the double result
is more accurate. You should be aware that your calculator probably carries 12 or
more digits, whereas float carries only 6 digits. Therefore, you should not use float
if you want to be at least as accurate as your calculator.

2. Are there any other data types used for storing real numbers in C? Yes, in
addition to float and double is long double. The *long double* type occupies even
more memory than double. This means that long double carries more digits and is
capable of storing larger numbers than float and double.

Float, double, and long double data types are compared in Table 3.7 . The ANSI
C standard does not state absolutely how much memory each of the types should
occupy. So, the table lists just what is sometimes used. Check your compiler for the
actual amount of memory it uses for each of these data types. To check your com-

TABLE 3.7
Data types

Item	float	double	long double
Memory used	4 bytes = 32 bits	8 bytes = 64 bits	10 bytes = 80 bits
Range of values	1.1754944E − 38 to 3.4028235E + 38	2.2250738E − 308 to 1.7976935E + 308	Approximately 1.0E − 4931 to 1.0E + 4932
Precision	6	15	19
Simplest format	%f, %e, %E, %g, %G	%lf, %e, %E, %g, %G	%Lf, %Le, %LE, %Lg, %LG

piler, you can look in the file `float.h`. The constant macro, `DBL_MAX_10_EXP`, for instance, may give the number 308 as the maximum exponent for double, as shown in Table 3.7.

Given the low precision of the float data type, we recommend that you use double in your programs. Note that the simplest double format is %lf. This format is used extensively throughout this book. It is also acceptable to use %f with the printf function, however, %f is not acceptable for the scanf function. With scanf you *must* use %lf for double and %Lf for long double.

3. Are there any different data types for integers? Yes, the different integer data types are compared in Table 3.8, again with sizes that are sometimes used.

4. Will we often need to use long int or unsigned long int? Probably not, we usually use integers for counting in programs, not for calculations. Unless you are counting a large number of objects, int probably will be sufficient. However, you should be aware of the limit of roughly 32,000 on int. Later in this book we will give an example of a program that needs an unsigned long int.

TABLE 3.8
Integer data types

Item	int signed int short int signed short int	unsigned int unsigned short int	long int signed long int	unsigned long int
Memory used	2 bytes = 16 bits	2 bytes = 16 bits	4 bytes = 32 bits	4 bytes = 32 bits
Range of values	−232767 to 32767	0 to 65535	−2147483647 to 2147483647	0 to 4294967295
Simplest format	%d, may need to use %hd for short int	%d, may need to use %hd for unsigned short int	%ld	%ld

5. What are the meanings of the functions in this lesson? The meanings of the C mathematical library functions are shown in Table 3.9 (note that the input argument(s) x or y and the return value of each of these functions are of double type). It is very important that you notice that the argument for the sin function (as well as the tan, cos, and other functions that use angles as input) requires that the argument be in radians, not degrees! So, if you want to use the sin of 30 degrees in one of your programs, you must manually write source code that converts degrees to radians by multiplying by $\pi/180$. For instance, this code might be

```
angle = 30.;
x = angle * 3.141592654/180.;
y = sin(x);
```

You should carry a large number of digits for π to maintain accuracy.

Other mathematical C library functions may take different types of data as input and return a different type of data as output. In general, C library functions may vary slightly from compiler to compiler. Check your C compiler manual for details. Table 3.10 lists a few more math library functions:

6. Which header file do we use with C mathematical functions? To use C math functions, you need to add the following statement:

#include <math.h>

or

#include <stdlib.h>   *for* abs *and* fabs

EXERCISES

1. True or false:
    a. The term #include <Math.H> is a correct C preprocessor directive.
    b. A header file must be placed at the first line of a C program.
    c. In C, the value of sin(30) is equal to 0.5.
    d. In C, the value of log(100) is equal to 2.0.

TABLE 3.9
Functions

Function name	Calculating
sin(x)	the sine of x, x is in radians
exp(x)	the natural exponential of x
log(x)	the natural logarithm of x
sqrt(x)	the square root of x
pow(x, y)	x raised to the power of y

TABLE 3.10
More math library functions

Function name	Example	Description
abs(x)	y=abs(x);	Gets the absolute value of an int type argument, x and y are of type int (Note: this function needs #include <stdlib.h> not math.h)
fabs(x)	y=fabs(x);	Gets the absolute value of a double type argument, x and y are of type double (Note: this function needs #include <stdlib.h> not math.h)
sin(x)	y=sin(x);	Calculates the sine of an angle in radians, x and y are of type double
sinh(x)	y=sinh(x);	Calculates the hyperbolic sine of x, x and y are of type double
cos(x)	y=cos(x);	Calculates the cosine of an angle in radians, x and y are of type double
cosh(x)	y=cosh(x);	Calculates the hyperbolic cosine of x, x and y are of type double
tan(x)	y=tan(x);	Calculates the tangent of an angle in radians, x and y are of type double
tanh(x)	y=tanh(x);	Calculates the hyperbolic tangent of x, x and y are of type double
log(x)	y=log(x);	Evaluates the natural logarithm of x, x and y are of type double
log10(x)	y=log10(x);	Evaluates the logarithm to the base 10 of x, x and y are of type double

2. Find `math.h` in your C compiler include directory. Then copy and paste it to the program that follows. Compile, link, and run the program. Do you get the same output as the one from `C3_5.EXE`? Compare the sizes of the source code, object code, and the executable code of the following program and the `C3_5.C` program. Summarize your findings.

```
/* Copy the MATH.H file (without modification)

   below this line */

void main(void)
{
   double x=3.0, y, z;
   y = 4.0;
   z=pow(x,y);
   printf("x=%4.1f, y=%4.1f, pow(x,y)=%7.2f",x,y,z);
}
```

3. The following program can be compiled and linked without error, but you get an error message when you run it. Why?

```
#include <math.h>
main()
{
   float x=-111.11, y=0.5, z;
   z=pow(x,y);
   printf("x=%10.2f, y=%10.2f, z=%10.2f\n",x,y,z);
}
```

4. Write a program to calculate these unknown values:

Alpha(degree)	Alpha(radian)	sin(2*Alpha)
30.0	?	?
45.0	?	?

Solutions

1. a (false), b (false), c (false), d (false)

◾ LESSON 3.8 WRITING OUTPUT TO A FILE

Topics

• Writing data to a file
• Using the fprintf( ) function

Previous programs have displayed all their output on the screen. This may be convenient at times; however, once the screen scrolls or clears, the output is lost.

In most cases, you will want a more permanent record of your output, which can be obtained by writing your output to a file instead of to the screen. Once the output is in a file, you can use a file editor to view it. You can use the editor to print the result on a printer, too.

The program for this lesson illustrates how to print output to a file. Just like an input file, an output file

1. Can have any acceptable file name
2. Must be linked with a file pointer before it is used
3. Must be opened before it is used
4. Should be closed after it is used

As you read the program compare and contrast it to the program in Lesson 3.6, which reads data from a file.

Look at the declarations in the source code. What is the name of the file pointer? Look at the call to the function fopen. In this program, we open the file for writing and not reading. What argument in this function call indicates that we open the file for writing?

The function fprintf is closely related to printf (in a manner similar to fscanf being related to scanf). Look at the calls to fprinf. What is the first argument for each of the calls? Have we closed the file when we finished with it?

Source code

```
#include <stdio.h>
void main(void)
{
        double  income=123.45, expenses=987.65;
        int week=7, year=2006;
        FILE *myfile;

        myfile = fopen("L3_8.OUT","w");
        fprintf(myfile,"Week=%5d\nYear=%5d\n",week,year);
        fprintf(myfile,"Income  =%7.2lf\n Expenses=%8.3lf\n",
                     income,expenses);
        fclose(myfile);
}
```

> The function, fprintf, works similar to printf. Here, it works with the file pointed to by `myfile`.

Output file 3_8.OUT

```
Week=    7
Year= 2006
Income  =123.45
Expenses=987.650
```

Explanation

1. What function do we use to write data to a file? In C, we use the fprintf() function to write data to a file. In general, the syntax of the fprintf() function is

$$\texttt{fprintf}\,(\textit{file\_pointer, format\_string, argument\_list})\,;$$

The fprintf() function writes the values of *argument_list* using the given *format_string* to a file that is linked to the program with the file pointer of *file_pointer* (Fig. 3.17). For example, in the statement

$$\texttt{fprintf(myfile," Week = \%5d\textbackslash n Year = \%5d\textbackslash n",week,year);}$$

the values of *argument_list,* week and year, are written to an external file that has a file pointer named myfile using the format string

$$\texttt{" Week = \%5d\textbackslash n Year = \%5d\textbackslash n"}$$

2. What function do we use to open an output file and to create a link between the file pointer and file? Before data can be written to an external file, the file must be opened using the fopen() function, whose syntax is

$$\textit{file\_pointer} = \texttt{fopen}\,(\textit{file\_name,access\_mode})\,;$$

FIG. 3.17

Operations of fprintf function

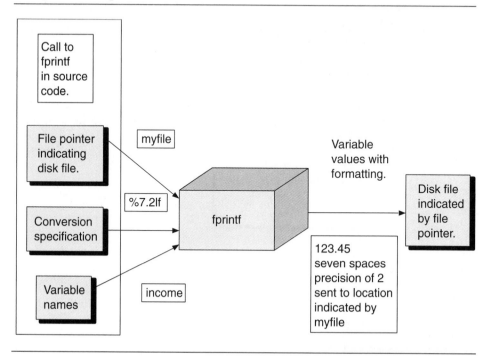

where the definitions of *file_pointer* and *file_name* are the same as those for opening an input file. However, the *access_mode* for writing is "w," which means the file is opened for writing in text mode. For example, the statement

`myfile = fopen ("L3_7.OUT","w");`

opens file `L3_7.OUT` for writing and creates a link between the file pointer, myfile, and the disk file `L3_7.OUT`. If `L3_7.OUT` exists, the contents of the file will be overwritten.

3. Do we need to close the output file? It is good practice to close files after they have been used. However, if no fclose statements are used, C automatically will close all open files after execution is completed. We also use the fclose() function to close an output file.

EXERCISES

1. True or false:
   a. We use the fprintf() function to write output on the screen.
   b. We use the fprintf() function to write output to an external file
   c. You must create a link between an external file and a file pointer before you can write your output in it.
   d. It is good practice to close an output file once you no longer need it.
   e. You must declare a file pointer before you can open an output file.

2. Find error(s), if any, in each statement:
   a. `#include <stdio.h>`
   b. `File myfile;`
   c. `*myfile = fopen (TEST.OUT,w);`
   d. `fprintf(*myfile," Week = %4d\n Year =`
      `%5d\n",&week,&year);`
   e. `fclose("myfile);`

3. Write a program to input all your grades in the last semester from the keyboard, compute your average GPA, and write all the input and average GPA on the screen and in a report file name MYGRADE.REP.

   Solution
1. a (false), b (true), c (true), d (true), e (true)

▓ LESSON 3.9 SINGLE CHARACTER DATA

This lesson is more difficult than the other lessons in this chapter and may be postponed until after Chapter 6 has been completed.

Topics
- The set of characters
- Single character input and output
- Characters treated as integers
- Character input and output functions
- Input buffer
- Flushing the buffer

The single character type, char, can be utilized somewhat similarly to the numeric types (int, float, and double). The term *character,* though, refers to more than just lowercase and uppercase letters. There are graphic characters such as !, #, and ^. Escape sequences (like \n and \r) also are regarded as single characters.

Even the numbers 0–9 can be treated as characters. We will find this helpful when we want to store such things as telephone numbers. Telephone numbers have no arithmetic manipulations on them (it makes no sense to add two of them together). Because of this, a telephone number is more difficult to manipulate if it is stored as an integer. Therefore, we will find it convenient to store numbers that do not require arithmetic operations as characters.

Eleven character variables are used in the program for this lesson. This program simply reads and prints characters.

What you should observe about this lesson's program:

1. The keyword *char* is used to declare 11 single characters.
2. Some compilers contain functions that are not specified by ANSI C. Information about some of these functions is given in the file, conio.h, which is included in this program.

Section 1

1. The first three assignment statements have single quotes surrounding single characters.
2. %c is used as a conversion specification in printf for single characters.
3. The putchar function can be used (as well as printf) to print single characters.
4. The output from the printf statement is exactly the same as the output from the four putchar statements.

Section 2

1. The scanf statement with %c reads single characters typed in at the keyboard.
2. This scanf statement has no spaces between the two %c conversion specifications.
3. The characters xp are typed in and printed out with the scanf and putchar statements.
4. The printf statement uses the %d conversion specification with character variables. The output is numeric.

Section 3

1. The getchar function reads single characters typed in at the keyboard. To call getchar, nothing is enclosed in the parentheses.
2. We can use getchar on the right side of assignment statements.
3. One of the calls to getchar in this section is not on the right side of an assignment statement.
4. The characters qk are typed in and printed out with the getchar and putchar functions.

Section 4

1. The function fflush is used.
2. The argument for fflush is the identifier stdin.
3. The location where stdin is declared is not obvious.
4. The function fflush has been called before and after reading characters typed in at the keyboard using getchar.
5. The characters vs have been typed in and printed out using getchar and printf.

Section 5

1. The getche function, which is not in the ANSI C standard, is called (with no arguments) in the manner we used for getchar.
2. The function fflush has not been called in this section.
3. The characters wq have been typed in and printed out.
4. The return key was not pressed after typing wq.

Questions you should attempt to answer:

1. What does putchar(32) do?
2. What do the numbers mean that are printed out in Section 2 with the printf statement?
3. Why do you think the statement:
    ```
    getchar();
    ```
    was used in Section 3?
4. What do you think fflush does in Section 4?
5. Why is fflush not needed in Section 5?

Source code

```
#include <stdio.h>
#include <conio.h>
```

Header file for function getche. This is not an ANSI C header file. We include this simply to illustrate what functions beyond ANSI C can do. This is one of the few places in this book where we do this.

```
void main (void)
{
    char  c1, c2, c3, c4, c5, c6, c7, c8, c9, c10, c11;
```

Declarations of character variables

```
/***************************************************************
**          SECTION 1
***************************************************************/

    printf("\n***SECTION 1*****************************************\n");

    c1 = 'g';
    c2 = '<';
    c3 = '\n';
    printf ("%c %c\n", c1, c2);
    putchar(c1);
    putchar(32) ;
    putchar(c2) ;
    putchar(c3);
```

Assigning characters to character variables

Printing single characters and more using printf. Note the conversion specifications.

Printing one character at a time using putchar. Note that "space" and "newline" are characters.

```
    printf("\n***SECTION 2 *****************************************\n");

/***************************************************************
**          SECTION 2
***************************************************************/

    printf ("Enter two characters (without spaces), then press return:\n");
    scanf  ("%c%c", &c4, &c5);
    putchar(c4) ;
    putchar(c5);
    printf ("\n%d %d\n", c4, c5);
```

Using scanf to read characters typed in at the keyboard. Note that there are no spaces in the string literal.

```
    printf("\n***SECTION 3 *****************************************\n");
/***************************************************************
**          SECTION 3
***************************************************************/


    printf ("Enter two more characters (without spaces), then press return\n");
    getchar();
    c6 = getchar();
    c7 = getchar();
    putchar(c6);
    putchar(c7);
    putchar('\n') ;
```

Character read from the keyboard but not stored in any variable's memory cell.

Using getchar to read a single character typed in at the keyboard. Note that there are no arguments (that is, the parentheses are empty). Assignment statements cause the characters typed to be stored in memory cells reserved for c6 and c7.

```
    printf("\n***SECTION 4 *****************************************\n");
/***************************************************************
**          SECTION 4
***************************************************************/
```

```
       printf ("Enter two more characters (without spaces) then press return:\n");
       fflush(stdin);
       c8 = getchar();
       c9 = getchar();
       fflush(stdin);
       printf ("%c%c \n", c8, c9);
```

Function fflush clears the input buffer. This allows the next two characters typed to be stored in c8 and c9 without error.

```
       printf("\n***SECTION 5 ***************************************\n");
/********************************************************************
**          SECTION 5
********************************************************************/
```

```
       printf ("Enter two more characters (without spaces) DO NOT press return:\n");
       c10 = getche();
       c11 = getche();
       putchar('\n');
       putchar(c10) ;
       putchar(c11);
```

Function getche operates similar to getchar but it avoids the problem with the input buffer. Because it is not ANSI C, it may not work with your compiler. We show this only to illustrate that you may encounter non-ANSI C functions.

```
       }
```

Output

```
                    ***SECTION 1***************************************
                    g <
                    g <

                    ***SECTION 2 **************************************
                    Enter two characters (without spaces), then press return:
Keyboard input      xpreturn
                    xp
                    120 112

                    ***SECTION 3 **************************************
                    Enter two more characters (without spaces), then press return
Keyboard input      qkreturn
                    qk

                    ***SECTION 4 **************************************
                    Enter two more characters (without spaces), then press return
Keyboard input      vsreturn
                    vs

                    ***SECTION 5 **************************************
                    Enter two more characters (without spaces), DO NOT press return
Keyboard input      wq
                    wq
```

Explanation

1. How do we declare character variables? Character variables are declared with the keyword *char*. The form is `char` *variable1, variable2, variable3, . . .;*

For example, the declaration in this lesson's program:

```
char c1, c2, c3, c4, c5, c6, c7, c8, c9;
```

declares c1, c2, c3, c4, c5, c6, c7, c8, and c9 to be of character type.

2. How do we write an assignment statement with these character variables? To assign a character constant to a character variable, it is necessary to enclose the constant in single quotes, ' '. For instance, the assignment statement

```
c1 = 'g';
```

assigns the character g to variable c1. A common error is to use double quotes " " instead of single quotes. Do not make this error. To correctly deal with single characters and character functions, you must use single quotes.

3. What is a complete list of ANSI C character constants and how does C handle them? A complete list of ANSI C character constants and their ASCII codes/values in decimal is shown in Table 3.11. Note that this list does not include the complete list of ASCII characters, because ANSI C does not support all of the ASCII character set. Therefore, you will notice that the given list does not include such characters as '$' and 'α'. The result of this is that the following statement

```
printf("The first Greek character is %c\n", 'α');
```

may or may not compile with your compiler. Compilers that go beyond the ANSI C standard may be able to compile and execute this statement properly. However, we recommend that you avoid using these characters in your source code. Not using these characters makes your program more portable.

C actually uses the integer value of the characters in its character functions and operations. As we will see, this allows us to use the integer value of characters instead of the actual characters in certain operations.

4. Why are the escape sequences (such as \n, \r*) included in the character set?* C regards the escape sequences as a single character. When you use them, make sure that you do not put a space after the backslash. Because they are regarded as a single character, the assignment statement

```
c3 = '\n';
```

is legal.

5. How do we use the printf function to print characters? We use the %c conversion specification and treat a character variable in the same manner as we have treated integer and double variables. For example, to print the character variables c1 and c2, the statement:

```
printf ("%c %c \n", c1, c2);
```

▨　**TABLE 3.11**
ANSI C characters and their ASCII codes/values (in decimal)

Character	ASCII value	Character	ASCII value	Character	ASCII value
\a	7	<	60	_	95
\b	8	=	61	`	96
\t	9	>	62	a	97
\n	10	?	63	b	98
\v	11	A	65	c	99
\f	12	B	66	d	100
\r	13	C	67	e	101
space	32	D	68	f	102
!	33	E	69	g	103
"	34	F	70	h	104
#	35	G	71	i	105
%	37	H	72	j	106
&	38	I	73	k	107
,	39	J	74	l	108
(	40	K	75	m	109
)	41	L	76	n	110
*	42	M	77	o	111
+	43	N	78	p	112
,	44	O	79	q	113
−	45	P	80	r	114
.	46	Q	81	s	115
/	47	R	82	t	116
0	48	S	83	u	117
1	49	T	84	v	118
2	50	U	85	w	119
3	51	V	86	x	120
4	52	W	87	y	121
5	53	X	88	z	122
6	54	Y	89	{	123
7	55	Z	90			124
8	56	[	91	}	125
9	57	\	92	~	126
:	58	]	93		
;	59	^	94		

accomplishes the task. However, since C treats characters with their integer values, if we were to use a %d conversion specification, we would have received an output showing the integer value for the characters. For instance, the printf statement

```
printf ("\n%d %d\n", c4, c5);
```

causes the integer values of c4 and c5 to be printed. Given that c4 and c5 were input as x and p, the ASCII values are 120 and 112, which are the values printed out.

6. How does the putchar function work? The putchar function prints the character that is its argument to the standard output device (the screen). The form is

```
putchar (character);
```

For example, the statement:

```
putchar (c2);
```

causes the value of the character variable c2 to be printed to the screen. An example of a character constant being printed is

```
putchar ('y');
```

This causes the character *y* to be printed to the screen.

7. What does putchar(32) do? Since C uses the integer value of the character in its functions, putchar(32) causes the character represented by the integer value 32 to be printed to the screen. As can be seen from Table 3.11, the value 32 is the character "space," which means a white-space character is printed to the screen.

So, the printf statement

```
printf ("%c %c\n", c1, c2);
```

(which has a space between the two %c conversion specifications and \n) and the putchar statements

```
putchar(c1);
putchar(32);
putchar(c2);
putchar(c3);
```

perform the same tasks.

8. Will all computers cause putchar(32) **to print a space?** No, only computers that use the ASCII encoding scheme (which is most personal computers) will interpret 32 as meaning space. Another popular encoding scheme, EBCDIC (not in our table) interprets 32 as being a nonprintable control character.

Thus, different computers give different results. To make your programs portable from computer to computer, we recommend you replace putchar(32) with putchar(' ').

9. Which should we use, printf or putchar, to print single characters? Either is acceptable, but many programmers use putchar because it is a little more efficient for printing characters, being designed especially for that purpose.

10. How do we read characters from the keyboard using the scanf function? We use the %c conversion specification. Thus,

```
scanf ("%c%c", &c4, &c5);
```

causes two characters to be read from the keyboard and stored in the memory cells reserved for c4 and c5. Note that there are no spaces in the string literal. We later look at the scanf function more closely and see how spaces, had they been inserted in this string literal, may have caused some difficulties in reading character input.

10. What does the function getchar do? It returns a character that has been input from the standard input device (the keyboard) to the program. The form for it is

`getchar();`

where nothing should be put in the parentheses. Frequently, this function call will be put on the right side of an assignment statement, such as

`c6 = getchar();`

This causes the character sent from the keyboard to be assigned to the variable $c6$, much in the same way that a statement of $y = \log(x)$ assigns the variable y to be the logarithm of x. However, there is no argument in the getchar function call because no information need be sent to the getchar function.

11. Does the getchar function get the characters directly from the keyboard? Not exactly; the getchar function works with the input buffer to get the information typed at the keyboard.

12. What is the input buffer and how does it work with getchar? A buffer is a portion of memory reserved for temporarily holding information that is being transferred. This memory is accessed sequentially, meaning that one memory cell after another (in sequence) is read. A position indicator keeps track of the point at which no further information has been read. On reading a cell the position indicator advances one cell so that the next cell can be read.

The getchar function works with the buffer position indicator to retrieve the next character in the buffer and advance the position indicator. When the getchar function is called, it either reads the item in the next cell or (if there is nothing in the next cell) it stops execution and waits. At this point a character or characters can be input. The getchar function is reactivated when the Return or Enter key is pressed.

In other words, when a key is pressed, the buffer receives and stores the value of the character struck. This continues for as many characters as are struck until the Enter or Return key is pressed. Then the getchar function retrieves the next character (and the next and next if there are multiple calls to the getchar function).

Some difficulty is involved in using the getchar function because getchar is activated by pressing Return or Enter. For example, for this lesson's program, the first keys pressed were

xp~return~

which means the buffer would have (Note: \n is one character since it has its own ASCII code).

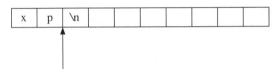

Position indicator

The scanf function has read both x and p so the position indicator is located as shown at the end of Section 2 in this lesson's program.

Then, in Section 3, the next statements are encountered:

```
getchar();
c6 = getchar();
c7 = getchar();
```

The following sequence takes place:

1. The statement getchar(); is executed. The program reads \n from the buffer. However, this character is not stored in a variable location, since this statement is not an assignment statement. The purpose of this statement simply is to advance the position indicator, which occurs after reading a cell in the buffer. The position indicator is advanced to the next position:

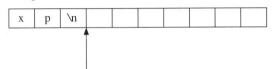

Position indicator

2. Since nothing is in the buffer after the position indicator, the call to getchar in the statement c6=getchar(); causes execution to stop and wait for input from the user. The user types vs to make the buffer

Position indicator

Then the user presses Enter, which has two effects, \n is put into the buffer and execution is reinitiated.

3. Then the statements c6=getchar(); and c7=getchar(); are executed (both can be executed since something is in each of the two cells). The buffer and position indicator become

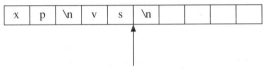

Position indicator

***13. In summary, then, what are the difficulties in dealing with the getchar
function?*** Since the getchar function needs Enter to work, an extra \n always will
be in the buffer and must be dealt with. In addition, the buffer may contain even
more than just an extra \n. For instance, if when the user was prompted to type two
characters in this lesson's program he or she typed three characters by accident and
then pressed Enter, the third character would not be discarded, it would be in the
buffer. As further getchar calls are executed, the extra character would cause execution to be different from what it should be.

14. How else can we deal with this problem? We can flush or empty the buffer
after obtaining the character(s) of interest. This can be done with the function fflush.
To flush the input buffer, the form is

```
fflush(stdin);
```

In this lesson's program, the statements

```
fflush(stdin);
c8 = getchar();
c9 = getchar();
fflush(stdin);
```

have flushed the input buffer before and after the two calls to getchar. Had we not
flushed the buffer before the c8 assignment statement, c8 would have been assigned
\n because (as seen by the location of the position indicator) the next character was
\n. Flushing the buffer after the two calls allows the next calls to getchar to avoid
reading extraneous characters.

15. What is stdin? Recall from the previous two lessons the declarations

```
FILE *inptr;
FILE *myfile;
```

We used inptr as a file pointer to an input file that we created and myfile as a file
pointer to an output file we created. It is not visible in our program, but stdin also
is a file pointer. It is declared in stdio.h, which we have included with the preprocessor directive #include <stdio.h>. In stdio.h, stdin is defined to point
to the standard input stream. Therefore,

```
fflush(stdin);
```

flushes what is pointed to by stdin, which is the standard input stream—the input
buffer from the keyboard. It is possible to use fflush with other file pointer names
as arguments; however, we will not go into the effect of doing so at this point.

Note that, because stdin is defined in stdio.h, you cannot choose to use
stdin as an identifier for a variable in your programs that use stdio.h. If you try
to do it, your compiler will indicate an error.

We will see that it is rare for us to use such an identifier as stdin; that is, an
identifier in lowercase letters that has no obvious source. Therefore, in this textbook,
you can find a declaration for each identifier used in the source codes.

16. Is there another way to avoid the problems of getchar? It is not ANSI C compatible, but the function getche is found on many compilers, especially those made for PCs. This function does not work with the input buffer and, therefore, produces unbuffered input. It deals directly with the operating system and does not require the Enter key to be pressed for the character code to be transferred. As a result, a single keystroke is picked up immediately and the input buffer is not dealt with. Thus, there are no extraneous characters to worry about. Flushing the buffer is not necessary. The form is

```
getche();
```

where nothing should be put in the parentheses.

Using the getche function is a more convenient way to create interactive character input with a user. However, if you use getche, you will lose some portability of your programs, since it is not ANSI C compatible. Be well aware of this loss of portability if you use the getche function! We have included it here only to make you aware that non-ANSI standard functions are available on compilers, and you may work for an employer that wants you to utilize some of them.

17. Why not forget getchar and just use scanf? The function scanf can be used but it has other problems in interpreting character data that we did not have to worry about when we used scanf to read numeric data. At this point, it is worth discussing a little more about how scanf works.

When the scanf function is activated, execution of the program stops. The program user then can type in information, which is transferred to the input buffer. Execution begins again when the user presses the Enter key, causing scanf to read the information in the input buffer.

As we have seen, the scanf function interprets the information in the input buffer based on the string literal used in the function call. The scanf function classifies what is in the string literal (or format string) into three categories:

1. Conversion or format specifiers (begun with a %)
2. White-space characters (which for scanf is not just a space, scanf treats Tab, and Enter [newline \n] also as white space)
3. Non-white-space characters

For example, in this lesson's program, the scanf function call is

```
scanf ("%c%c", &c4, &c5);
```

The string literal here has two conversion specifiers with no white-space and no non-white-space characters. The string literal indicates that the two buffer cells to be read will be interpreted to have characters stored in both of them.

However, it becomes difficult to deal with scanf for characters because of the way that the scanf function deals with white space in both the string literal and the input buffer. This is an issue because, as we have seen, a space can be interpreted as a character (this was not an issue before because a space could not be interpreted as a numeric value).

When scanf has a space in its string literal, that space causes scanf to skip over one or more white-space characters in the input buffer to get to the next non-white-space character. Therefore, when reading characters, you must make sure that you have no white space in your string literal for scanf. Otherwise, you will not read a white-space character that actually is meant to be a white-space character input.

For example, the scanf statement

```
scanf ("%c[]%c", &c1, &c2);
```

will not accept

$$[][][]$$

as input (where [ ] means a single space created by pressing the Spacebar on the keyboard). Try it. Scanf simply will not move forward, even after you press Enter many times, because Enter is white space. Until a non-white-space character is pressed, scanf will not move forward. But,

```
scanf ("%c%c%c", &c1, &c2, &c3);
```

will cause $c1=[\ ]$, $c2=[\ ]$, $c3=[\ ]$ with the same input. Also,

```
scanf ("%c[]%c[]", &c1, &c2);
```

causes a problem because of the trailing space in the string literal. Again, in this case, scanf needs another non-white-space character (after two have been entered) to continue executing. If you enter a third non-white-space character, it would be put into the input buffer but not read by this scanf statement. However, the succeeding scanf statement may unintentionally read this character.

Since newline($\backslash$n) is regarded as white space, you might be tempted to put a space at the beginning of your string literal to get scanf to skip over an extraneous $\backslash$n in the input buffer. We do not recommend this as it may cause you other problems.

If a non-white-space character is used in the scanf format string, scanf expects a matching character in the given location. If scanf gets it, the character is discarded. If it does not find a matching character, scanf is terminated. For example,

```
scanf ("%c-%c", &c1, &c2);
```

expects input of the sort

$$B-2$$

to make the assignment $c1=B$ and $c2=2$. If B2 is input instead, scanf terminates. This feature can be deceptive when you look at it being used in code. For instance, in

```
scanf ("%cd%c", &c1, &c2);
```

the %cd is not a new conversion specifier. It simply means that a d is expected between the characters input. So input would be of the form

$$Bd2$$

to continue executing.

Another feature of scanf allows an unknown item to be read and discarded. If a
* is placed between % and the format code, an item will be read and discarded. For
example,

```
scanf ("%c%*c%c", &c1, &c2);
```

will read

B#2

and assign c1=B and c2=2. The # could be any character, and it would be discarded.

18. What about scanf and numeric input data? White space in the input stream
is treated differently with numeric input. For numeric input data, one or more spaces,
tabs, or newlines is required to separate items. For example,

```
scanf ("%d%d%f", &a1, &a2, &a3);
```

would accept the input

1 2

3.14159

as a1=1, a2=2, a3=3.14159. The input

1 2   3.14159

would produce the same assignments.

19. What is the relationship between scanf and fscanf? The fscanf function
works exactly like the scanf function except the input stream is different (being from
the specified file rather than stdin). Everything we have said about scanf pertains
to fscanf. Therefore, you should have a greater understanding of how your input files
are interpreted by fscanf with both numeric and character input. For instance, you
can see that it is not necessary to put all of your numeric input on one line. The
fscanf function will go to the next line to get more non-white-space characters.

20. Why have we gone into so much detail here about scanf and fscanf? We
have gone into this much detail because it is important that you know exactly what
information your program is getting. A mismatch between data entered and the input
(scanf) statement is a very common source of errors, especially for beginning programmers. In many cases, the rest of the computer program may be completely correct and errors are made only in the input. So, when debugging your programs, do
not forget to check both your input data and the scanf or fscanf statement.

21. Should we learn anything more about character data? Yes, right here we
have just touched the surface of handling single characters. Chapter 7 goes into other
aspects of character handling and manipulation.

EXERCISES

1. True or false, assuming

   `char c1, c2, c3, c4;`

   a. The statement `c1=g`; assigns g to character c1. ✗
   b. The statement `putchar(2)`; causes the number 2 to be printed to the screen. ✗
   c. The statement `getchar(c4)`; gives c4 the value of the next character typed.
   d. The statement `c2=getchar()`; gives c2 the value of the next character typed.
   e. The statement `scanf("%c%c",c3,c4)`; gives c3 and c4 the values of the next two characters typed.

2. Find error(s), in these statements:
   a. `fflush(input);`
   b. `putchar(\t);`
   c. `c4='$';`
   d. `putchar(47);`
   e. `getchar(c4);`
   f. `scanf(" %c %c ", &c1,&c3);`
   g. `printf("%c %c ",c3,  c4);`

3. Write a program to read the following sequence of characters:

   ```
   &gt 891><
   -rew {[]}
   ```

   Print them out as

   ```
   {98we[-gt] -&r1<>}
   ```

4. Write a program to display the following output:

   ```
   ABCDE
    BCDE
     CDE
      DE
       E
   ```

5. Write a program to ask a user to enter five integers between 0 and 255 from the keyboard and convert them to characters. Display both the number and the characters on the screen.

   Solutions

1. True or false:
   a. False, should be `c1='g';`
   b. False, should be `putchar('2');`
   c. False, should be `c4=getchar();`
   d. True
   e. False, should be `scanf("%c%c", &c3, &c4);`

2. Errors

    a. `fflush(stdin);`

    b. `putchar('\t');`

    c. No error, but not ANSI C standard, may not work on all systems.

    d. No error, will display /.

    e. `c4=getchar();`

    f. No error, but scanf will wait for a non-white-space character after c_1 and c_3 are input.

    g. No error.

APPLICATION PROGRAM 3.1 AREA CALCULATION— COMPOUND OPERATORS AND PROGRAM DEVELOPMENT (1)

Problem statement

This application is not meant to be practical but simply an illustration of how to recognize patterns and develop a program based on those patterns. In addition, this program illustrates the program development methodology described in the previous section without introducing new theory that can distract from the focus on patterns and methodology. This example is deliberately simple in concept. Its function is to just introduce you to the logic of writing programs that perform arithmetic calculations.

Write a program that computes the areas of four right triangles. Three of the triangles are shown in Fig. 3.18. You should deduce the dimensions of the fourth triangle from the pattern exhibited by the first three. Use the pattern in writing your program.

Solution

Here, we use the procedure listed at the end of Chapter 1 in developing the application programs. We repeat that, in developing your own programs, it is important

FIG. 3.18
Triangles for the problem

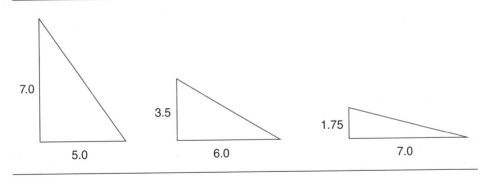

that you follow a formal procedure rather than develop your programs haphazardly. If you choose not to follow the methodology we describe, please follow one recommended by your instructor.

RELEVANT EQUATIONS

Note that there is a pattern to the length of the legs. The lengths of the horizontal legs are 5, 5 + 1 = 6, 6 + 1 = 7, and the vertical legs are 7, 7/2 = 3.5, 3.5/2 = 1.75. Thus, we can see that the fourth triangle has a horizontal leg length of 7 + 1 = 8 and a vertical leg length of 1.75/2 = 0.875.

The horizontal leg length can be computed from the following equations:

$$L_{h2} = L_{h1} + 1$$
$$L_{h3} = L_{h2} + 1$$
$$L_{h4} = L_{h3} + 1$$

where
L_{h1} = horizontal leg length of the first horizontal leg = 5.0
L_{h2} = horizontal leg length of the second horizontal leg
L_{h3} = horizontal leg length of the third horizontal leg
L_{h4} = horizontal leg length of the fourth horizontal leg

Also the vertical leg length is

$$L_{v2} = L_{v1}/2$$
$$L_{v3} = L_{v2}/2$$
$$L_{v4} = L_{v3}/2$$

where:
L_{v1} = vertical leg length of the first vertical leg
L_{v2} = vertical leg length of the second vertical leg
L_{v3} = vertical leg length of the third vertical leg
L_{v4} = vertical leg length of the fourth vertical leg

Note that the area of a right triangle is

$$A = 0.5L_1L_2$$

SPECIFIC EXAMPLE

For this particular program, the results easily can be found using a hand calculator. For most real programs, this is not possible because of the very large number of calculations that are performed by most real programs. The calculations that follow show the lengths and the areas.

Triangle 1

$$L_{h1} = 5$$
$$L_{v1} = 7$$
$$A_1 = (0.5)\,(5)\,(7) = 17.50$$

Triangle 2

$$L_{h2} = 5 + 1 = 6$$
$$L_{v2} = 7/2 = 3.5$$
$$A_2 = (0.5)\,(6)\,(3.5) = 10.50$$

Triangle 3

$$L_{h3} = 6 + 1 = 7$$
$$L_{v3} = 3.5/2 = 1.75$$
$$A_3 = (0.5)\,(7)\,(1.75) = 6.125$$

Triangle 4

$$L_{h1} = 7 + 1 = 8$$
$$L_{v1} = 1.75/2 = 0.875$$
$$A_1 = (0.5)\,(8)\,(0.875) = 3.50$$

ALGORITHM. One purpose of performing a sample calculation is to clearly outline all the steps that are needed to arrive at a correct and complete result. The sample calculation has been used as a guide to writing the algorithm that follows:

```
Begin
Declare variables

Initialize horizontal leg length of first triangle
Initialize vertical leg length of first triangle
Calculate area of first triangle

Calculate horizontal leg length of second triangle
Calculate vertical leg length of second triangle
Calculate area of second triangle

Calculate horizontal leg length of third triangle
Calculate vertical leg length of third triangle
Calculate area of third triangle

Calculate horizontal leg length of fourth triangle
Calculate vertical leg length of fourth triangle
Calculate area of fourth triangle

Print results onto the screen

End
```

SOURCE CODE

The source code has been written directly from the algorithm. Look at each line and make sure that you understand exactly what each is doing. You can compare the code with the algorithm line by line to guide you.

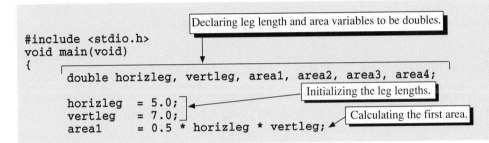

```
#include <stdio.h>
void main(void)
{
        double horizleg, vertleg, area1, area2, area3, area4;

        horizleg  = 5.0;
        vertleg   = 7.0;
        area1     = 0.5 * horizleg * vertleg;
```

Declaring leg length and area variables to be doubles.

Initializing the leg lengths.

Calculating the first area.

```
horizleg    += 1.0;
vertleg     /= 2.0;
area2       = 0.5 * horizleg * vertleg;
```

> Using compound operators and the recognized patterns to calculate new lengths.

> Calculating a new area from the new lengths.

```
horizleg    += 1.0;
vertleg     /= 2.0;
area3       = 0.5 * horizleg * vertleg;
```

> Repeating the form of the preceding statments.

```
horizleg    += 1.0;
vertleg     /= 2.0;
area4       = 0.5 * horizleg * vertleg;
```

```
printf (" \n\
First  triangle area = %6.2f   \n\
Second triangle area = %6.2f   \n\
Third  triangle area = %6.2f   \n\
Fourth triangle area = %6.2f   \n",
areal, area2, area3, area4);
}
```

> Printing the results.

OUTPUT

```
First  triangle area    = 17.50
Second triangle area    = 10.50
Third  triangle area    =  6.13
Fourth triangle area    =  3.50
```

COMMENTS

This program illustrates how patterns are used in programming. One can imagine that it would be very simple to write a program similar to this one that computes the areas of 50 triangles that follow the same pattern. As we illustrate more programming techniques, you will see that you can write such a program with very few statements.

This particular example is somewhat contrived, in that it is deliberately set up to have a pattern to it. You will find, though, that real problems also have patterns and that part of the skill in writing more advanced programs is in recognizing patterns and writing efficient code to take advantage of the patterns.

Modification exercises

Modify the preceding program to

1. Calculate the areas of ten triangles following the same pattern
2. Print the results to an output file

3. Use only three variables (horizleg, vertleg, and area) yet produce the same output
4. Make each vertical leg length double instead of half each time through the sets of equations

■ APPLICATION PROGRAM 3.2 TEMPERATURE UNITS CONVERSION—COMPOUND OPERATORS AND PROGRAM DEVELOPMENT (2)

Problem statement

Write a program that creates a table of degrees Celsius with the corresponding degrees Fahrenheit. Begin at 0°C and proceed to 100°C in 20°C increments. Use no more than two variables in your program.

Solution

First, assemble the relevant equations. The equation converting degrees Celsius to degrees Fahrenheit is

$$F = \frac{9}{5}C + 32$$

where
$$C = \text{degrees Centigrade}$$
$$F = \text{degrees Fahrenheit}$$

SPECIFIC EXAMPLE

Once again, for this simple program, all the following calculations can be done by hand.

$$C = 0$$
$$F = C\left(\frac{9}{5}\right) + 32 = 32$$

$$C = 20$$
$$F = C\left(\frac{9}{5}\right) + 32 = 68$$

$$C = 40$$
$$F = C\left(\frac{9}{5}\right) + 32 = 104$$

$$C = 60$$
$$F = C\left(\frac{9}{5}\right) + 32 = 140$$

$$C = 80$$

$$F = C\left(\frac{9}{5}\right) + 32 = 176$$

$$C = 100$$

$$F = C\left(\frac{9}{5}\right) + 32 = 212$$

ALGORITHM

We use the sample calculations to guide us in writing the algorithm. We added to it the printing of the headings and the results.

```
Begin
Declare variables
Print headings of table
 Set C = 0
 Calculate F
 Print C and F

 Set C = 20
 Calculate F
 Print C and F

 Set C = 40
 Calculate F
 Print C and F

 Set C = 60
 Calculate F
 Print C and F

 Set C = 80
 Calculate F
 Print C and F

 Set C = 100
 Calculate F
 Print C and F

End
```

SOURCE CODE

This source code has been written from the algorithm. Note that this code has used the fact that the values of degrees Centigrade are in increments of 20. Again, read this code line by line and make sure that you understand exactly how the program operates.

```
#include <stdio.h>
void main(void)
{
        double degC, degF;    Declaring just two variables.
```

```
printf ("Table of Celsius and Fahrenheit degrees\n\n"
        "          Degrees              Degrees \n"
        "          Celsius              Fahrenheit \n");

degC    = 0.;
degF    = degC * 9./5. +32.;
printf ("%16.2f %20.2f\n", degC, degF);

degC    += 20.;
degF    = degC * 9./5. +32.;
printf ("%16.2f %20.2f\n", degC, degF);

degC    += 20.;
degF    = degC * 9./5. +32.;
printf ("%16.2f %20.2f\n", degC, degF);

degC    += 20.;
degF    = degC * 9./5. +32.;
printf ("%16.2f %20.2f\n", degC, degF);

degC    += 20.;
degF    = degC * 9./5. +32.;
printf ("%16.2f %20.2f\n", degC, degF);

degC    += 20.;
degF    = degC * 9./5. +32.;
printf ("%16.2f %20.2f\n", degC, degF);
}
```

Printing the table heading. You will find that when you print your results in the form of a table, you must print the table headings before anything else is printed.

Following the algorithm line by line.

OUTPUT

```
Table of Celsius and Fahrenheit degrees

Degrees              Degrees
Celsius              Fahrenheit
   0.00                 32.00
  20.00                 68.00
  40.00                104.00
  60.00                140.00
  80.00                176.00
 100.00                212.00
```

COMMENTS

First, we can see immediately that this program has the same three statements written repeatedly. Had we wanted to display the results for every single degree between 0 and 100 instead of every 20th degree, the program would have been extremely long but with the same three statements written over and over again. Chapter 4 has more advanced programming techniques to allow us to write a program that can accomplish the same task but with many fewer statements.

Second, we could have used the programming technique illustrated in the previous application program, which had a single printf statement at the end of the program instead of one immediately after each calculation of degF. However, this would have necessitated the use of more variables.

For instance, the program could have been

```
#include <stdio.h>                Declaring many Celsius and Fahrenheit variables.
void main(void)
  {
      double  degC1, degC2, degC3, degC4, degC5, degC6,
              degF1, degF2, degF3, degF4, degF5, degF6;

      printf ("Table of Celsius and Fahrenheit degrees\n\n"
              "          Degrees              Degrees \n"
              "          Celsius              Fahrenheit \n");

      degC1    = 0.;
      degF1    = degC1 * 9./5. +32.;

      degC2    = 20.;
      degF2    = degC2 * 9./5. +32.;

      degC3    = 40.;
      degF3    = degC3 * 9./5. +32.;

      degC4    = 60.;
      degF4    = degC4 * 9./5. +32.;

      degC5    = 80.;
      degF5    = degC5 * 9./5. +32.;

      degC6    = 100.;
      degF6    = degC6 * 9./5. +32.;

      printf ("\n"
          "%20.2f %20.2f\n%20.2f %20.2f\n%20.2f %20.2f\n"
          "%20.2f %20.2f\n%20.2f %20.2f\n%20.2f %20.2f\n",
          degC1, degF1, degC2, degF2, degC3, degF3,
          degC4, degF4, degC5, degF5, degC6, degF6);
  }
```

> Because there are many variables, no printf statement is needed after each calculation. Also, the same statements are not repeated as they were in the previous source code.

> A single printf statement can now be used to print all of the variable values.

With this program 12 variables have been used instead of just 2. Variables take up space in the memory of the computer, so the program with 12 variables would occupy more memory than the program with just 2 variables. Efficient programming, in part, means to write a program that takes as little memory as possible. For this very small program, either programming technique could be used on today's computers. However, for very large programs, the memory needed by the program may be very important. So, it is good to develop efficient programming habits while you are learning programming. Reducing memory size is only a part of developing efficient programs. Comments on other ways to make your program efficient will be made throughout this book.

Also note that it is necessary to make your program understandable to someone other than you. The reason for this is that it is common for programs to be developed by teams of people and then undergo several versions. This means that someone who has never seen a particular program may be responsible for modifying it. Your program is more valuable if it is easily understood.

Sometimes you will find a conflict between understandability and efficiency. In other words, efficient programs may not be understandable and understandable programs may not be efficient. You should consult your employer or your course instructor for guidance in determining the more important characteristic for your program.

You can begin to see that there are many ways to write even the simplest of programs. One can argue that there is no right or wrong way, provided the program gives the correct result. However, it is best to write code that is efficient and understandable.

Modification exercises

Modify the program to

1. Calculate the degree conversions every 5 degrees rather than every 20 degrees
2. Print the result to a file called `DEGR.OUT`
3. Do the conversions between -100 and 0 degrees in 20 degree increments

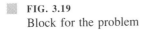

APPLICATION PROGRAM 3.3 FRICTIONAL RESISTANCE—ARITHMETIC OPERATORS, FSCANF, AND CONSTANT MACROS

Problem statement

Write a program that computes the necessary force to move a block across a plane. Friction resists the movement of the block. Consider three different blocks, as shown in Fig. 3.19.

FIG. 3.19
Block for the problem

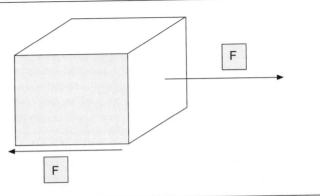

The three different blocks have the following dimensions:

Block	Height (m)	Length (m)	Width (m)
1	0.5	2.0	1.5
2	1.2	0.75	0.2
3	0.8	2.2	1.3

The density of the material of which the blocks are made is 5.7 kN/m^3. The coefficient of friction of each block on the plane is 0.35.

The data file containing the dimensions of each block is FRICTION.DAT. The data file has three lines: each with a block number, height, length, and width as listed. Print to the screen the force required to move the block.

Solution

RELEVANT EQUATIONS

Because friction is the only force resisting movement of the block, the force required to move the block is equal to the frictional resistance. You learned in your physics class that frictional resistance is a function of the coefficient of friction and the normal force on the plane of contact:

$$F = \mu N \tag{3.1}$$

where F = frictional resistance
μ = coefficient of friction
N = normal force on the plane of contact

For this particular case, the normal force is equal to the weight of the block; that is,

$$N = W \tag{3.2}$$

where W = block weight.

The block weight can be computed from the volume and the density, and the volume can be computed from the dimensions as follows:

$$W = Vd \tag{3.3}$$
$$V = hlw \tag{3.4}$$

where V = block volume
d = material density
h = block height
l = block length
w = block width

SPECIFIC EXAMPLE

In many cases you will have to make up the example problem yourself, which means that you will need to create the information that goes into the input file. In this case, however, the input data is given. The following calculation sequence uses the information about the first block:

$$d = 5.7 \text{ kN/m}^3$$
$$\mu = 0.35$$
$$h = 0.5 \text{ m}$$
$$l = 2.0 \text{ m}$$
$$w = 1.5 \text{ m}$$

$V = (0.5)(2.0)(1.5) = 1.5 \text{ m}^3$			from eq. (3.4)
$W = (1.5)(5.7) = 8.55 \text{ kN}$			from eq. (3.3)
$N = 8.55 \text{ kN}$			from eq. (3.2)
$F = (0.35)(8.55) = 2.9925 \text{ kN}$			from eq. (3.1)

ALGORITHM

```
Define constants
Declare variables
Open the input data file

Read first block information
Compute V
Compute W
Compute N
Compute F
Print   F

Read second block information
Compute V
Compute W
Compute N
Compute F
Print   F

Read third block information
Compute V
Compute W
Compute N
Compute F
Print   F
```

SOURCE CODE

The program shown satisfies the requirements of the problem statement. The algorithm has been used to write the source code. Again, go line by line through this program and make sure you understand each statement.

> Because the density and friction coefficient are constant throughout the program, constant macros are used to define them.

```
#define DENSITY          5.7
#define FRICTION_COEFF   0.35
#include <stdio.h>
void main(void)
{
    int     block_number;
    double  weight, height, length, width, volume,
            normal_force, frictional_force,movement_force;
    FILE    *inptr;
```

```
inptr = fopen ("FRICTION.DAT","r");
printf ("\n"
"Block number            Force required for movement(KN)\n\n");

fscanf (inptr, "%d %lf %lf\
%lf",&block_number,&height,&length,&width);
volume = height * length * width;
weight = volume * DENSITY;
normal_force = weight;
frictional_force = FRICTION_COEFF * normal_force;
movement_force = frictional_force;
printf ("\n%10d %30.7f \n",block_number,movement_force);

fscanf (inptr, "%d %lf %lf\
%lf",&block_number,&height,&length,&width);
volume = height * length * width;
weight = volume * DENSITY;
normal_force = weight;
frictional_force = FRICTION_COEFF * normal_force;
movement_force = frictional_force;
printf ("\n%10d %30.7f \n",block_number,movement_force);

fscanf (inptr, "%d %lf %lf\
%lf",&block_number,&height,&length,&width);
volume = height * length * width;
weight = volume * DENSITY;
normal_force = weight;
frictional_force = FRICTION_COEFF * normal_force;
movement_force = frictional_force;
printf ("\n%10d %30.7f \n",block_number,movement_force);
}
```

Opening the input file.

Printing the table headings.

Reading the first line of the data file.

Calculating the movement force.

Reading the second line of the data file.

Calculating the movement force.

Reading the third line of the data file.

Calculating the movement force.

INPUT FILE FRICTION.DAT

```
1 0.5 2.0 1.5
2 1.2 0.75 0.2
3 0.8 2.2 1.3
```

OUTPUT

```
Block number      Force required for movement (kN)
     1                      2.9925000
     2                      0.3591000
     3                      4.5645600
```

COMMENTS

Use your calculator to check these results. Note that once again several statements have been repeated. Also note that every program can be written many ways.

Do you have any ideas on how to change this program to fit your own personal style of programming?

Modification exercises

Modify the program to

1. Calculate the frictional resistance for hollow blocks (read the thickness of the block from input data file; you must develop your own equations for the weight of the block)
2. Read all of the input data from just one fscanf statement, which means you will have to create different variables
3. Print the result to a file called FRICT.OUT

APPLICATION EXERCISES

Use the four-step procedure outlined in this chapter to write the following programs:

3.1. Write a program that creates a table of Olympic competition running distances in meters, kilometers, yards, and miles. The following distances should be used:

100 m
200 m
400 m
800 m

Use the pattern exhibited in these distances to write your program. (Note: 1 m = 0.001 km = 1.094 yd = 0.0006215 mi.)

Input specifications. No external input (meaning no data input from the keyboard or file). All distances are real numbers.

Output specifications. Print the results to the screen in the following manner:

```
Table of Olympic running distances
```

Meters	Kilometers	Yards	Miles
100	--	--	--
200	--	--	--
400	--	--	--
800	--	--	--

Right-justify the numbers in the table.

3.2. Write the program described in Application Exercise 3.1 with the output going to a file called OLYM.OUT. Left-justify the numbers in the table.

3.3. Write a program that computes the length of the hypotenuse of five right triangles based on the lengths of the two legs.

Input specifications. Read the input data from the keyboard by prompting the user in the following way:

Screen output	```Input the values of the leg lengths for five right triangles```
Keyboard input	`leg1  leg2`
Keyboard input	`leg1  leg2`
Keyboard input	`leg1  leg2`
Keyboard input	`leg1  leg2`

All input values are real numbers.

Output specifications. Print the result to the file `HYPLENG.OUT` with the following format:

```
       Hypotenuse lengths of five triangles

Triangle    leg 1    leg 2    hypotenuse
number      length   length   length

   1          --       --        --
   2          --       --        --
   3          --       --        --
   4          --       --        --
```

Right-justify all the numbers in the table.

3.4. Write a program that computes the values of the two acute angles of a right triangle given the lengths of the two legs. Create the input data file `ANGLE.DAT` before executing your program.

Input specifications. Input should come from data file `ANGLE.DAT` with the following form:

line 1	`leg1`	`leg2`
line 2	`leg1`	`leg2`
line 3	`leg1`	`leg2`
line 4	`leg1`	`leg2`
line 5	`leg1`	`leg2`

All the values are real numbers.

Output specifications. The output results should be in degrees, not radians. Make sure that, in your program, you convert from radians to degrees. The output should go to file `ANGLE.OUT` and have the following format:

```
         Acute angles of five triangles

     Triangle      acute         acute
     number       angle 1       angle 2

        1           --            --
        2           --            --
        3           --            --
        4           --            --
        5           --            --
```

3.5. Write a program that is capable of displaying the distances from the sun to the four planets closest to the sun in centimeters and inches given the kilometer distances as follows:

```
Planet      Distance from the sun
            (million km)

Mercury         58
Venus          108.2
Earth          149.5
Mars           227.8
```

Input specifications. No external input is needed. These distances can be initialized in the source code.

Output specifications. Print the results to the screen in the form of the following table:

```
Planet     Distance from the sun
           (million km)       (cm)      (inches)


Mercury        58              --          --
Venus         108.2            --          --
Earth         149.5            --          --
Mars          227.8            --          --
```

Note: To fit the numbers properly in the table, you must use scientific notation.

3.6. The distance that a car (undergoing constant acceleration) will travel is given by the expression

$$S = V_o t + \frac{1}{2} at^2$$

where

$\quad s \;=\;$ distance traveled
$\quad V_o \;=\;$ initial velocity
$\quad t \;=\;$ time of travel
$\quad a \;=\;$ acceleration

Write a program that computes this distance given V_o, t, and a.

Input specifications. The input should come from the file DISTANCE.DAT with the following format:

line 1 v_o t
line 2 a_1
line 3 a_2
line 4 a_3
line 5 a_4
line 6 a_5
line 7 v_o t
line 8 a_1
line 9 a_2
line 10 a_3
line 11 a_4
line 12 a_5

All the above numbers are real numbers. An example data file is

```
10   5
 3
 4
 5
 6
 7
10  10
 3
 4
 5
 6
 7
```

Output specifications. Print the results to the file DISTANCE.OUT in the following form:

```
Car under constant acceleration
```

Initial velocity	time	acceleration	distance
10	5	3	--
		4	--
		5	--
		6	--
		7	--
10	10	3	--
		4	--
		5	--
		6	--
		7	--

3.7. The general gas law for an ideal gas is given by

$$\frac{PV}{T} = \text{constant}$$

where
$$P = \text{pressure}$$
$$V = \text{volume}$$
$$T = \text{temperature (Rankine or Kelvin)}$$

which leads to the equation

$$\frac{P_1 V_1}{T_1} = \frac{P_2 V_2}{T_2}$$

for a given mass of gas.

   Write a computer program that computes the temperature of a gas that is originally at

$$P_1 = 5 \text{ atmospheres}$$
$$V_1 = 30 \text{ liters}$$
$$T_1 = 273 \text{ Kelvins}$$

Input specifications. The input data should come from the file TEMPER.DAT and consist of five lines:

line 1	P_2	V_2
line 2	P_3	V_3
line 3	P_4	V_4
line 4	P_5	V_5
line 5	P_6	V_6

A sample data file is

```
2   40
3   80
6   50
1   15
2   70
```

All of these values are real.

Output specifications. Your output should be to the screen and consist of the following table.

```
The below listed pressure, volume, and temperature conditions
can occur for a given mass of an ideal gas which is originally
at P = 5 atm, V = 30 l, and T = 273 K

   Case    P(atm)  V(l)    T(K)

     1        2      40     --
     2        3      80     --
     3        6      50     --
     4        1      15     --
     5        2      70     --
```

3.8. Ohm's law for a steady electrical current can be written as

$$V = IR$$

where
V = potential difference across a conductor
I = current in the conductor
R = resistance of the conductor

Write a program capable of filling in the blanks in the following table:

Case	V (Volts)	I (Amps)	R (Ohms)
1	10	2	—
2	—	5	7
3	3	—	4

Input specifications. The input data should come from the keyboard and be treated as real numbers. You should prompt the user in the following manner:

```
"For case 1, enter the voltage and current."
"For case 2, enter the current and resistance."
"For case 3, enter the voltage and resistance."
```

Output specifications. Print the completed table to the screen.

3.9. The pressure at depth in water is given by

$$P = h\gamma_w$$

where
p = pressure
h = depth
γ_w = weight density of water

Write a program that determines the pressure at five different depths. Use metric units ($\gamma_w = 9.8$ kN/m²).

Input specifications. Create a data file called PRESS.DAT with your editor. In the data file, list the five depths on one line:

```
depth1  depth2  depth3  depth4  depth5
```

An example data file is

```
10.    15.    828.    1547.    431.2
```

All the data are real.

Output specifications. Print the results to file PRESS.OUT in the following form:

Depth (m)	Pressure (kPa)
--	--
--	--
--	--
--	--
--	--

3.10. The period of one swing of a simple pendulum is given by

$$T = 2\pi\sqrt{\frac{l}{g}}$$

where (in metric units) T = period (sec)
l = length of pendulum (m)
g = gravitational acceleration = 9.81 m/sec^2

Write a program capable of completing the following table:

Length (m)	Period (sec)
0.5	—
1.0	—
—	10.
—	20.
0.32	—

Input specifications. Prompt the user to input the data from the screen in a manner similar to that described in Application Exercise 3.8.

Output specifications. Print the completed table to the screen.

3.11. The kinetic energy of an object in motion is expressed as

$$K = \frac{1}{2}mv^2$$

where K = kinetic energy of object
m = mass of object
v = velocity of object

The work done by a force pushing on an object in the direction of the object's motion is

$$W = Fs$$

where W = work done by the force
F = force on object
s = distance traveled by the object during the time the object is pushed

For an object pushed horizontally from rest, $K = W$, so that

$$Fs = \frac{1}{2}mv^2$$

Assume that one person can push with the force of 0.8 kN and that we have a car of $m = 1000$ kg. Write a program that can complete the following table:

Distance pushed (m)	Final velocity (m/sec)	Number of people required to push
5	10	—
—	10	15
20	—	8

Input specifications. Prompt the user to enter the data from the keyboard.
Output specifications. Print the completed table to the screen.

Beginning Decision Making
and Looping

You will find as you continue learning to program that you will want your programs to make decisions regarding which calculations to perform. For instance, suppose you want to write a program that computes your income tax. Suppose that the percentage of tax is based on your income in the following way:

Income	Percent tax
0–$50,000	20%
$50,000–100,000	30%
>$100,000	40%

For your program to correctly compute your tax, it must be able to decide which percent tax applies to your income. In this chapter you will learn how to get your programs to make decisions of this sort.

The application programs in the previous chapter illustrated many cases where it was necessary to write the same few statements numerous times to perform a calculation repeatedly. In this chapter we describe methods by which it is possible to repeat calculations without writing the statements over and over. This is called *looping*.

LESSON 4.1 IF CONTROL STRUCTURE AND RELATIONAL EXPRESSIONS

Topics

- Simple if statements
- Block if statements
- Controlling program flow
- Relational operators
- Relational expressions

As described in the introduction to this chapter, sometimes you will want your program to make decisions. The if statement often is used for this purpose. The form of the if statement is fairly simple. A relational expression (meaning an expression that compares the values of two variables, for instance) is contained in the if statement; if the relational expression is true, then the statements within the "true" group are executed. If the relational expression is false, then the statements are not executed.

The program for this lesson has four if statements. This program is a simple game. The user is to try to guess the jackpot number. The assumption in this program is that the user does not know the value of the variable, `jackpot`.

Look at the first if statement. The relational expression compares which two variables? For an input guess equal to 3, is this relational expression true or false? Is the subsequent printf statement executed (see the output)? Look at the second if statement. Is the relational expression true or false? Is the subsequent printf statement executed?

Suppose the new guess being read in (by the second scanf statement) is 7. Look at the fourth if statement. Is the relational expression true or false? Is the statement block executed? Look at the relational expression for the last if statement and the program output. From these two, can you guess what the relational expression for this if statement means?

Source code

```
#include <stidio.h>
void main(void)
{
    int i,guess,jackpot=8;

    printf("Try to guess the jackpot number \nbetween 1 and 10!\n");
    printf("Please type a number.\n");
    scanf("%d",&guess);

    if  (guess<jackpot)
        printf("Try a bigger number\n");

    if (guess>jackpot)
        printf("Try a smaller number\n");

    if (guess==jackpot)
            printf("Verify your guess by typing it one more time\n");

    scanf ("%d", &guess);

    if (guess==jackpot)
        {
            printf("You hit the \n");
            printf("JACKPOT!\n");
        }
}
```

true
false

true
false

true
false

true

false

Relational expressions. The result of a relational expression can be regarded as only true or false.

The equality relational operator has *two* equal signs.

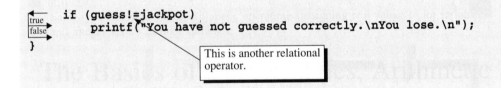

```
if (guess!=jackpot)
    printf("You have not guessed correctly.\nYou lose.\n");
```

This is another relational operator.

Output

	Try to guess the jackpot number between 1 and 10! Please type a number.
Keyboard input	3 Try a bigger number
Keyboard input	7 You have not guessed correctly. You lose.

Explanation

1. What is the meaning of guess<jackpot*?* The expression

guess<jackpot

is a *relational expression* that compares the values of two arithmetic expressions. A relational expression is a type of logical expression and produces a result of either true or false. Here, it checks whether the value of the variable guess is less than the value of the variable jackpot. Its general syntax is

left_operand   relational_operator   right_operand

where the *left operand* and *right operand* can be variables, such as guess in this lesson's program or any arithmetic expression. The *relational operator* is used to compare the values of two operands. C contains six relational operators:

Relational operator	Meaning
<	less than
<=	less than or equal to
==	equal to
>	greater than
>=	greater than or equal to
!=	not equal to

Observe in the table the meaning of !=. It is probably a combination of symbols to which you have not been previously exposed. No ≠ symbol is found on the keyboard, so C uses !=.

2. What does `if (guess<jackpot) printf("Try a bigger number\n");`
mean? It is a *simple if statement* that can be generalized as

if (*expression*) *statement*;

where the *expression,* for instance, `guess<jackpot`, represents a logical expression
and the *statement,* for instance, printf(), is an executable statement. Note that any
executable statement can be used as the *statement,* including if and other control
statements. A logical expression produces a result of either true or false. If the logical expression is true, the *statement* behind the if *expression* is executed. If the logical expression is false, the *statement* behind the if *expression* is not executed. The
expression within an if statement is called a *condition.* (Note: The parentheses are
shown only for clarity. They are not necessary.)

3. What is a block if statement? The statement

if (guess==jackpot)   { *statements...*   }

is called a *block if statement.* If the logical expression is true, the statements in the
"true" block (between two braces) are executed. Otherwise, the entire block of statements is ignored. The general form of a block if statement is

```
      if (expression)
true       {
           executable statement 1;
           executable statement 2;
           ...
false      }
```

Indentation is valuable for readability of your programs containing if statements. The
code block should be indented one tab (at least three spaces) in from the keyword,
`if`. We will see as we go further in this book that indentation is an important way
of making your program understandable to others. Although there are no absolute
rules on indentation (it is not required by ANSI C), it has become common practice
in programming. If you have many block if statements in your program, you can
mark them by hand with arrows as we have done to see if they constitute a valid set
of if statements. You must make sure that no arrows cross.

A conceptual illustration of the fourth if statement for this lesson's program is
given in Fig. 4.1. The figure shows that this control structure causes a block of code
to be either executed or bypassed.

4. What is the difference between = ***and*** ==? The operator = is an assignment
operator and must not be confused with ==, which is a relational operator. A common programming error for beginners is to use = in the relational portion of a relational expression. If you do this you will cause serious errors in your programs.
Remember, in your relational expressions, use ==.

FIG. 4.1
Simple block if statement used in this lesson's code

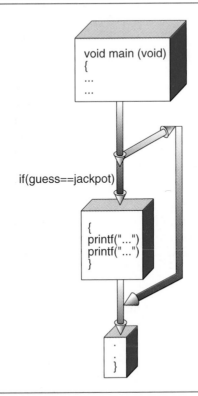

```
void main (void)
{
...
...
```

if(guess==jackpot)

```
{
printf("...")
printf("...")
}
```

```
.
.
}
```

5. Is it recommended to use == to compare the values of double or float type variables? No, in most cases for engineering programming, it is not recommended to compare any of the real type variables using ==, although the C compiler will not indicate an error if it is done. The reason why it is not recommended is that a typical C compiler carries a large number of significant digits for the real variable types (the number depends on the number of bytes used to store the type). If two doubles are compared with == and they differ in only the last significant digit, then the result of a comparison with == is false. For instance, if a = 12.34567891234556789 and b = 12.3456789123456788, then the result of $a == b$ is false. Often in engineering, we perform calculations using slightly approximate values and compare them with other approximate values. In most cases, we are interested only if the numbers are nearly or very nearly equal. Therefore, we want a comparison to evaluate to true when the values are very nearly equal. Because == does not do this, we do not use it.

Also, although we lacked the space to describe it in detail in Chapter 1, the binary representation of decimal numbers often requires an approximation. For instance, we need many binary bits to represent the simple decimal number 5.3 exactly. Because of this characteristic, numbers that we calculate to be exact with decimal arithmetic may not be exact using binary arithmetic with a limited number of bits. A calculation that you believe is exact using a hand calculation in decimal may not be exact using binary with a limited number of bits. We avoid the comparison problems caused by this effect by not using == for real type variables.

6. If we do not use == *with double, long double, or float, how do we compare two real values?* A good way to make the comparison is to use the fabs function and <. For instance, to compare the double values of *a* and *b* as listed previously, the following statement would evaluate to true:

```
if ( fabs (a-b) < 1.0e-10)
```

Here, we have somewhat arbitrarily selected the constant 1.0e−10 as a very small number. In writing your own programs, you will need to decide how small that number should be based on what you require for your problem. In general, though, you will need to use the preceding form to compare real numbers. Remember, to use fabs, you must include stdlib.h.

EXERCISES

1. Find the error(s), if any, in each of these statements:
   a. `if (today = 7) printf("Go to the park");`
   b. `if (today = 7);`
   c. `if (today == 7);`
   d. `if (today ==7) if (Money>100) printf("Dine out!");`
   e. `if (today == 7) j=i;`
   f. `if (Today == 7)`
       ```
       {
          j=i+1; k=100/j;
       }
       ```

2. Write the following text in C:
   a. if $b^2 - 4ac < 0$
   b. if *n* is equal to 0, assign 100 to *x*
   c. if *n* is not equal to 0, calculate 1.0/*n*

3. Write a program to input your grades from the keyboard and average your GPA. If your GPA is less than 2.0, output a 20 line warning message on the screen. If it is higher than 3.9, then produce a beep 10 times to celebrate.

Solutions

1. a. `if (today == 7) printf("Go to the park");`
   b. `if (today == 7);`
   c. No error, but statement does not do anything.
   d. No error.
   e. No error.
   f. No error.

LESSON 4.2 SIMPLE IF-ELSE CONTROL STRUCTURES

Topics
- Simple if-else control structures
- The conditional ? : operator

Another form of the if statement is the if-else form. It is used when a group of statements is to be executed when the logical expression is false.

The program in this lesson computes whether or not you are saving money, based on your income and expenses. Look at the first if line of text. If the relational expression in this line is false, which block of statements is executed (see the output)?

In this lesson we introduce the ? : operator. It is the only operator in C that is a ternary operator, meaning that three operands are needed for it to be used properly. Look at the right side of the assignment statement with the ? : operator. All three operands are on this side of the assignment statement. What are they? The colon separates two of them. Look at the output for the variable interest. The value of one of the operands has been printed out. Can you guess why this operand was printed?

Source code

```
#include<math.h>
 void main(void)
 {
     double  income, expenses, savings, deficit, interest;

     printf ("Enter your income and expenses:\n");
     scanf  ("%lf %lf", &income, &expenses);         If-else control structure.
     printf ("\n\n");

     if (income >expenses)
true        {
            savings = income - expenses ;
            printf ("\nYou are saving money.             \n"
                    "Your savings for this month are: $%8.2f",savings);
            }

        else
            {
false       deficit = expenses - income ;
            printf ("\nYou are running a deficit.        \n"
                    "Your deficit for this month is : $%8.2f",deficit);
            }
                                    Relational expression.

     interest = (deficit > 0.0)  ?  (0.05*deficit) : (0.0) ;
     printf ("\nThe interest you owe on your debt is $%.2f \n",
                interest);
 }
                                 The ? : operator needs 3 operands.
```

Output

Keyboard input	`Enter your income and expenses`
	`3500 4500`
	
	`You are running a deficit.`
	`Your deficit for this month is: $ 1000.00`
	`The interest you owe on your debt is $50.00`

Explanation

1. What is the syntax of a simple C if-else statement? The syntax of a simple C if-else statement is

```
if (expression)
   {
   executable statement 1a;
   executable statement 1b;
...
   }
else
   {
   executable statement 2a;
   executable statement 2b;
...
   }
```

Executable statements 1a, 1b, . . . are part of the "true" block, whereas *executable statements 2a, 2b, . . .* are part of the "false" block. If the expression is true, statements in the true block are executed. If the expression is false, control is transferred to the false block. If the statement block (either true or false) contains more than one statement, the block must be bounded by a pair of braces; otherwise, braces are optional. For example,

```
if (test>=0)
   {
   true block statements...
   }
else
   a single statement
```

where the true block contains more than one statement and therefore must be bounded by a pair of braces. However, the false block contains only one statement, so the braces are optional.

If no statements are to be executed in the false block, then the false block may be omitted. The syntax without the false block is as given in Lesson 4.1.

```
if (expression)
   {
   executable statement 1a;
   executable statement 1b;
   ...
   }
```

FIG. 4.2
Simple if-else control structure in this lesson's program

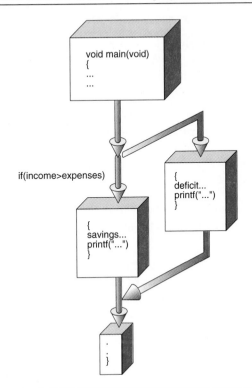

A conceptual illustration of the if-else control structure is given in Fig. 4.2. Note that the impact of using such a control structure is to cause the execution of one block while bypassing another.

2. How does the ? : *conditional operator work?* The ? : operator requires three operands. Its form is the following:

expression1   ?   *expression2*   :   *expression3*

If *expression1* is true, *expression2* is evaluated. If *expression1* is false, *expression3* is evaluated. The value of the entire ? : expression becomes equal to the value of the expression evaluated.

For this lesson's program, *expression1,* deficit>0.0 is true. Therefore, *expression2,* 0.05*deficit, is evaluated and the value of the right side of the assignment becomes equal to the value of *expression2*. Thus, interest becomes equal to 0.05*deficit.

As another example, the ? : operator can be used to find the smaller of two numbers. The statement

$$x = (y<z) \ ? \ y \ : \ z;$$

assigns x the value of the smaller of y and z.

Statements using the ? : operator are good shorthand for longer if-else type control structures. Note that, for the unevaluated expression, no side effects occur.

EXERCISES

1. Find the error(s), if any, in each of these statements:

    a. ```
        if (today = 7) printf("Go to the park");
        else   printf("Go to work");
        ```

    b.  `if (today == 7) ; else  printf("Go to work");`

    c.  `y = ? z>x :a w;`

2.  Write a program to input a number *x* from the keyboard. If the number is larger than 0, find its square root. Otherwise, calculate *x*x*.

*Solutions*

1.  a.  ```
        if (today == 7) printf("Go to the park");
        else            printf("Go to work");
        ```

    b. No error. When today!=7, it displays "Go to work".
        When today==7, nothing will be executed.

    c. `y = z>x ? a : w;`

▓ LESSON 4.3 NESTED IF-ELSE CONTROL STRUCTURES

Topics

• Nested if-else control structures

If-else control structures can be nested, meaning an if-else control structure can be contained within another if-else control structure.

Suppose you are interested in writing a program that tells you what you should be doing on a particular day during the week at a particular time. Your program logic would be very similar to your own thought process if you were to determine what you should be doing on your own.

Say that you have the following schedule that you want to program:

Weekdays
(Days 1–5)
    0:00–9:00   Work
    9:00–24:00   Relax
Weekends
(Days 6–7)
    0:00–8:00   Sleep
    8:00–24:00   Have fun

This schedule can be used to produce the following algorithm, which illustrates the logic that you would use to respond to a question about what you would be doing on a particular day at a particular time:

If day = weekday (1-5), then

    if time = 0:00-9:00   Work

    if time = 9:00-24:00   Relax

If day = weekend (6-7), then

    if time = 0:00-8:00   Sleep

    if time = 8:00-24:00   Have fun

A computer program written in C can duplicate the logic of this algorithm. Examine the program for this lesson to see how if-else statements can be used to mimic this algorithm. Note that these if-else statements are said to be *nested*.

Source code

```
#include <stdio.h>
void main(void)
   {
        int     day;
        float   time;

        printf (" Type the day and time of interest\n\n");
        scanf  (" %d %f ", &day, &time);

        if (day<= 5)
            {
            if ( time<= 9.00 )
                printf (" Work \n\n");
            else
                printf (" Relax \n\n");
            }
        else
            {
            if ( time<=8.00 )
                printf (" Sleep \n\n");
            else
                printf (" Have fun \n\n");
            }
   }
```

true
true
false
false
true
false

If-else control structures contained within an if-else control structure.

Output

	Type the day and time of interest
Keyboard input	3 10.00 Relax

Explanation

1. Can if-else control statements be nested? Yes, in C, different levels of if-else control statements can be nested. However, as if-else control statements become nested, programs become more difficult to understand. As described earlier, under-standability is an important characteristic for programs. Making the program more readable makes it more understandable. The traditional way to make a program read-able is to use indentation. We recommend that you indent each pair of if-else state-ments so that the inner *else* is paired with the inner *if* and the outer *else* is paired with the outer *if*. For example:

```
if (outer)
  {...                        /*if outer is true, execute this block*/
    if (inner_1)
       {... }                 /*if inner_1 is true, execute this block*/
    else
       {... }                 /*if inner_1 is false, execute this block*/

    if (inner_2)
       {... }                 /*if inner_2 is true, execute this block*/
    else
       {... }                 /*if inner_2 is false, execute this block*/
  }
  else
     {... }                   /*if outer is false, execute this block*/
```

A conceptual illustration of the nested if-else control structure used in this lesson is given in Fig. 4.3. Note the branching produced by the nesting.

2. Must the number of "ifs" and the number of "elses" be equal? No, in nested if-else statements, the total number of "ifs" can be greater than or equal to, but not less than, the total number of "elses." By default, an else clause is associated with the closest previous if statement that has no other else statement. Hand-drawn arrowhead braces can be used to clarify the pairing of "if" and "else" clauses.

EXERCISES

1. Find the error(s), if any, in each of these statements:

   a. ```
      if (i>100) printf("Hot\n");
      else   printf("Warm\n");
      else   printf("Cool\n");
      ```

   b. ```
      if (i>100) printf("Hot\n");
      if (i==100) printf("Warm\n");
      else           printf("Cool\n");
      ```

2. Write a program to input these ten incomes from a file:

   ```
   1   187
   2   2768
   3   1974
   4   373
   5   66733
   6   437892
   ```

FIG. 4.3

Nested if-else control structure for this lesson's program. Compare to switch and if-else-if control structure illustration in later lesssons.

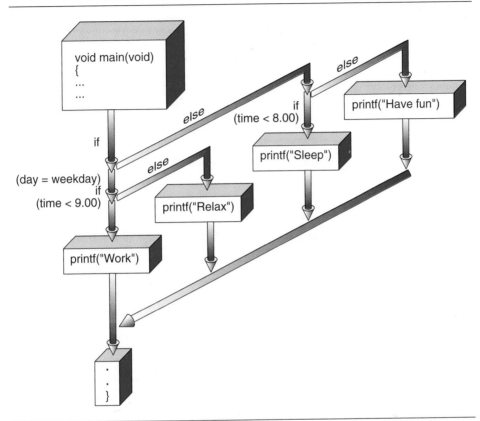

```
7   593
8   8091
9   48903
10  1839
```

and then calculate the tax on each income based on the following assumptions:

a. If income < 1000, no tax.

b. If 1000 < = income < 2000, tax rate = 25 percent.

c. If income > = 2000, tax rate = 500 + 30 percent above 2000.

Solutions

1. a. Too many elses.

b. No error. But when $i > 100$, the program will display both Hot and Cool.

▓ LESSON 4.4 LOGICAL EXPRESSIONS

Topics
- Using logical operators
- Using logical expressions

In the previous lessons you learned that relational expressions can form an integral part of if statements. You saw that whether or not certain statements within an if statement group are executed depends on whether the relational expression is true or false.

Logical operators can be used to connect two relational expressions. For example, the following

$$\text{if } (x == 0 \text{ \&\& } y == 0)$$

is read, "If x is equal to 0 and y is equal to 0."

Here, the logical operator is &&. This operator is read "and." It connects the two relational expressions, x==0 and y==0. The other two logical operators are || and !.

Look at the source code and the output that follow. See if you can determine the meaning of the two other logical operators.

Source code

```
#include <stdio.h>      Logical operators are
void main(void)         commonly used with
{                       relational expressions.
    int x=5,y=0;

    printf("x=%2d, y=%2d\n\n",x,y);

    if (x>0 && y>=0)
        printf("x is greater than 0 and"
               "y is greater than or equal to 0\n\n");

    if (x==0 || y==0)
        printf("x equals 0 or y equals 0\n\n");

    if (! (x==y))
        printf("x is not equal to y\n");
}
```

Output

```
x= 5, y= 0,

x is greater than 0 and y is greater than or equal to 0

x equals 0 or y equals 0

x is not equal to y
```

Explanation

1. How many logical operators are there in C? C has three *logical operators*, &&, ||, and !. The && and || operators are called *binary operators,* because they appear between two relational expression operands. The ! operator is unary, because it precedes a single relational operand. The meanings of these operators follow:

Operator	Name	Operation	Operator type
!	Logical NOT	Negation	Unary
&&	Logical AND	Conjunction	Binary
||	Logical OR	Inclusive disjunction	Binary

The *logical NOT* operator reverses the result of a relational expression. For instance, for $x = 5$ and $y = 0$, x is not equal to y, and the expression x==y is false (which we will see is indicated by the number 0). The logical NOT operator can be used to reverse the false value. Therefore,

$$!(x==y)$$

is true.

The *logical AND* and *logical OR* operators perform much the same way the words "and" and "or" were used in your very early mathematics classes. The C statement

$$if\ (x>0\ \&\&\ y>=0)$$

is read, "If *x* is greater than 0 and *y* is greater than or equal to 0," meaning that the logical expression enclosed in parentheses is true when both $x > 0$ and $y > = 0$ are true. On the other hand,

$$if\ (x==0\ ||\ y==0)$$

is read, "If *x* equals 0 or *y* equals 0." For the logical expression between the parentheses to be true in this case, only one of

$$x==0$$
$$y==0$$

need be true.

Figure 4.4 illustrates the if control structures used in this lesson's program. This figure shows how the printf statements in this program may or may not be printed.

2. What is the result of a logical expression? In the table that follows, we have indicated relational expressions with the symbols *A* and *B*.

A	B	A&&B	A||B	!A	!B
True	True	True	True	False	False
True	False	False	True	False	True
False	True	False	True	True	False
False	False	False	False	True	True

FIG. 4.4
Flow of this lesson's program

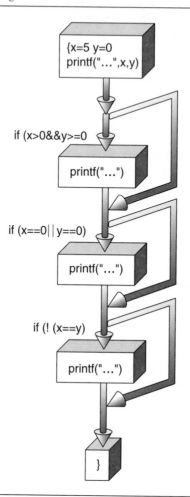

The table is read in the following way: For example, suppose we have the values x = 5 and y = 0. For a relational expression A being x > 0 and a relational expression B being y > = 0, the relational expression A is True, and the relational expression B is True. This means that we use line number 1 in the table. Therefore, A && B is equivalent to True && True and the logical result is True. Looking again at line 1, we can determine the logical result of A||B, !A, and !B.

Similarly, for the relational expressions A being x == 0 and B being y == 0, we have A = False and B = True. This leads us to line 3 of the table. Reading across this line, we get A || B being False. We also can look at it as False || True, which results in True.

Finally, consider relational expression A being x == y (which is False) and relational expression B being !(x == y) (which is equivalent to !(False), which results in True). This also leads us to line 3 of the table. This line indicates $A\&\&B$ is False. Otherwise, looking at this, $A\&\&B$ is True&&False, which is False.

The results of other combinations of relational expressions within logical expressions can be discerned from the table.

One way to remember the table is to realize that, for two expressions being operated on,

1. For &&, if one of the relational expressions is False then the result is False.
2. For ||, if one of the relational expressions is True then the result is True.

Logical expressions commonly are used in if statements. If the logical expression is True, then a set of C statements is executed. If the logical expression is False, then another set of C statements is executed.

EXERCISES

1. Given x = 200 and y = −400, determine whether each of these logical expressions is True or False. Note that, for a and b, it is necessary only to evaluate the first of the expressions to determine whether the result is true or false. Why?
   a. **(x<y && x!=y)**
   b. **(x>y || x==y)**
   c. **!(x>y)**

   (handwritten) a/b >= 100 && a < b

   (handwritten) (a+b) != 200 && b >= 300

2. Write the following statements in C:
   a. If $(a/b) > 100$ and $a < b$
   b. If $(a + b)$ is not equal to 200 and $b >= 300$
   c. If $(a + b) <= 2200$ or $(a − b) \times 4$ is equal to 500

3. Suppose the yearly demand, D, and supply, S, for C programmers in a given area are defined as follows:

   $S=1000+50*(Y-1990)$ $(1990<=Y<=2010)$
   $D=1200$ $(1990<=Y<=1995)$
   $D=1200+60*(Y-1995)$ $(1995<=Y<=2010)$

   Write a program to
   a. Print the yearly demand and supply of C programmers from 1990 to 2010.
   b. Find the years in which there are not enough C programmers for that area.
   c. Find the total number of unemployed C programmers between 1990 and 2010.

   Solutions
1. a. False (because the first expression is false and the operator is &&, the result is false no matter what the second expression is).
   b. True (because the first expression is true and the operator is ||, the result is true no matter what the second expression is).
   c. False

2. a. `if ( (a/b)>100 && a<b)`

b. `if ( (a+b)! = 200 && b>=300)`

c. `if ( (a+b)<=2200 || (a-b)*4==500 )`

▒ LESSON 4.5 PRECEDENCE OF LOGICAL OPERATORS

Topics

• Precedence of logical operators
• Finding the result of a logical expression

To this point, we have not mentioned that C gives relational expressions numerical values just like it gives arithmetic expressions numerical values. If a relational expression is false, C gives it a value of 0. If it is true, C gives it a value of 1 (which you should recognize as being nonzero). The C compiler also operates in the reverse manner. If the value of a relational expression is 0, then it knows the result is false; and if the value of a relational expression is not 0, then the result is true.

Something similar can be done with variables. If the value of a variable is 0, then it can be treated as being false; and if the value of a variable is not 0, then it can be treated as being true. This works this way because (you will recall) that a single variable can be considered to be an expression.

Look at the first three if statements in the source code. Can you rationalize the corresponding output from these if statements by considering the values of the variables? Also, note in the program that the logical NOT (!) can be used on a variable. Can you guess the value of !a without looking at the program output?

We saw earlier that C has an established order of precedence for arithmetic operators. Similarly, C has an established order of precedence for relational and logical operators. From the two compound logical expressions in the source code and the output, can you determine the precedence of the logical operators?

Source code

Single variables can be regarded as true or false depending on their values.

Just like arithmetic operators, relational operators have precedence rules that affect the order of operation.

```c
#include <stdio.h>
 void main(void)
 {
    int   a=4,b=-2, c=0 ,x;

    if(a) printf("a=%2d, !a=%2d\n",a,!a);
    if(b) printf("b=%2d, !b=%2d\n",b,!b) ;

    if(c) printf("Never gets printed\n");
    else  printf("c=%2d, !c=%2d\n\n",c,!c);

    if ( a>b  &&  b>c  ||  a==b )  printf("Answer is TRUE\n");
    else                          printf("Answer is FALSE\n");
```

```
    x=  a>b   ||   b>c   &&   a==b;
    printf("x=%2d,  !x=%2d\n",x,!x);
}
```

> The result of a logical expression can be assigned to an integer variable.

Output

```
a= 4,  !a= 0
b=-2,  !b= 0
c= 0,  !c= 1

Answer is FALSE
x= 1,  !x= 0
```

Explanation

1. What are the precedence and associativity of logical, relational, and arithmetic operators? The precedence and associativity of these operators follow:

Operator	Name	Associativity	Precedence
()	Parentheses	L to R	1 (highest)
++, −−	Postincrement	L to R	2
++, −−	Preincrement	R to L	2
!	Logical NOT	L to R	3
+, −	Positive, negative sign	L to R	3
+=, −=, *=, /=, %=	Compound assignment	R to L	3
*,/	Multiplication, division	L to R	4
+, −	Addition, subtraction	L to R	5
==, >=, <=, >, <, !=	Relational operator	L to R	6
&&	Logical AND	L to R	7
‖	Logical OR	L to R	8
=	Assignment	R to L	9 (lowest)

This table shows that parentheses have the highest order of precedence, followed by unary increment or decrement and logical NOT operators. In general, arithmetic operators, including multiplication, division, addition, and subtraction, have a higher order of precedence than any relational operator. Then come the logical operators, in which the logical AND has higher precedence than the logical OR. The assignment operator has the lowest precedence. The operators are operated (associativity) left to right except the preincrement or predecrement, compound assignment, and assignment operators.

For example, assuming a = 4, b = −2, and c = 0, the expression

$$x=\ (\ a>b\ \ ||\ \ b>c\ \ \&\&\ \ a==b\ )$$

is equivalent to the following expressions (note that the sequence of evaluation is based on the precedence level of each operator; we add parentheses at appropriate locations so that the expression can be grouped and evaluated):

```
x = (   a>b    ||    b>c    &&    a==b   )
x = (  (a>b)   ||   (b>c)   &&   (a==b)  )
x = (  (4>-2)  ||   (-2>0)  &&   4==-2)  )
x = (  TRUE    ||   FALSE   &&   FALSE   )
x = (  TRUE    ||          FALSE         )
x = (              TRUE                  )
```

which results in True. In C, False is defined as 0, and True is defined as nonzero (any positive or negative integer). When we assign the foregoing expression to an integer x as we did in this lesson's program, the value of x is 1, since the result of the expression is True. If x is True, !x is False, and !x is printed as 0.

The effect of the different precedences of the relational and logical operators on the relational expressions in this lesson's program is illustrated in Fig. 4.5.

2. What is the logical value of a single variable? The logical value of a single variable is False if the variable has a value of 0 and True if the value is nonzero. This is illustrated in Fig. 4.6. For example, in this lesson, the logical value of c is False, since c is equal to 0, but the logical values of a and b are True, since a (= 4) and b (= −2) are nonzero. Also, the values of !a and !b are False and therefore printed out as 0, whereas, the value of !c is printed out as 1.

3. Can we use a relational expression of the sort a>b==c? Yes. This expression is evaluated from left to right since all the operators have equal precedence. For

FIG. 4.5

Operation of compound logical expression *a > b* ‖ *b > c* && *a == b,* from this lesson's program.

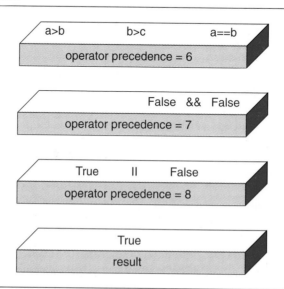

FIG. 4.6
(a) True or false result of integer values. (b) Integer values for !True and !False.

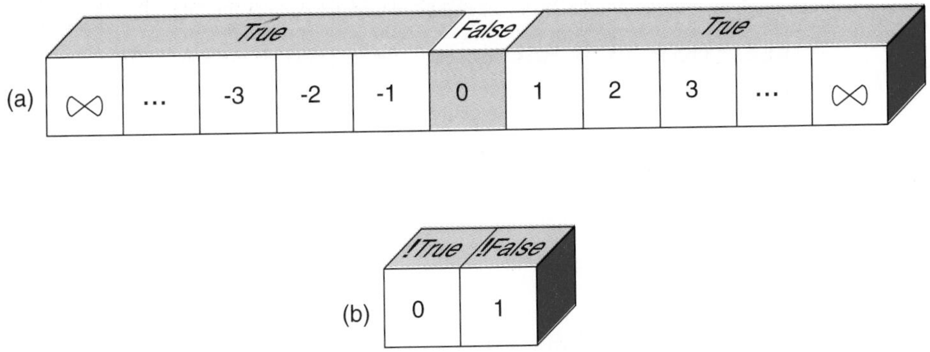

instance, if a = 4, b = −2, and c = 5, this expression evaluates to False. The steps in evaluation are

> a>b is True, giving this expression a value of 1
> 1==c is False.

EXERCISES

1. Assuming a = 100, determine whether each logical expression that follows is True or False.

   a. `a==100 && a>100 && !a`
   b. `a==100 || a>100 && !a`
   c. `a==100 && a>100 || !a`
   d. `a==100 || a>100 || !a`

2. Find the error(s), if any, in these statements, assuming a = 1, b = 2, c = 3, d = 4:

   a. ```
      if (a>b)
        printf("This is an arithmetic if_statement\n")
      ```
   b. ```
      if (a>b)
        {
      printf("This is a block if_statement\n");
      other_statements...
        }
      ```
   c. ```
      if (a>b == c)
        printf("This is a block if_else statement\n");
        other_statements...
      }
      else
      {
      ```
      ```

```
        printf("This is a block if_else statement\n");
        other_statements...
    }
d.  if (a>b) if (c>d)
    printf("This is a nested if_statement\n");
e.  if (a);
        {
          printf("if and else are matched");
          other_statements...
        }
      else
          if(b>a)
            {
              printf("if and else are matched");
              other_statements...
            }
          else
            {
              printf("if and else are matched");
              other_statements...
            }
f.  if (a>b)
    if (c<d)
    {
     printf("More if than else");
     other_statements...
    }
    else
    {
     printf("This else is associated with the last if");
     other_statements...
    }
g.  if (a>b)
    {
     if (c<d)
       {
         printf("More if than else");
         other_statements...
       };
    };
    else
     {
       printf("This else is associated with the 1st if");
       printf("since we use braces to block the last if");
     };
```

3. Create a figure similar to Fig. 4.5 to illustrate why, from this lesson's program, the expression (a>b && b>c || a==b) evaluates to be False.

Solutions

1. a. True && False && False = False
    b. True || False && False = True
    c. True && False || False = False
    d. True || False || False = True

2. a. Needs semicolon at end of printf statement.
    b. No error.
    c. Needs opening brace, {, before first printf statement. Expression is evaluated from left to right.
    d. No error. This is a nested if statement.
    e. Should not have semicolon after if (a).
    f. No error.
    g. Should not have semicolons after the three closing braces, }.

LESSON 4.6 SWITCH AND IF-ELSE-IF CONTROL STRUCTURES

Topics
- Using switch statements
- Using if-else-if control structures
- Comparing switch and if-else-if control structures

Two source codes are given for this lesson. They perform the same tasks but in different ways. The first source code uses an if-else-if control structure, and the second source code uses a switch control structure.

With your knowledge of if control structures you should be able to look at Source Code 1 and the output and trace the program flow. Please do so. Compare this code to Source Code 2.

For Source Code 2, look at the line of code with the keyword switch in it. What type of token follows switch? The keyword case is used three times. Each time it is followed by a different integer constant. What is the relationship between the constants after case and the token after switch? What symbol is used after the constant?

The keyword break is used three times. By following the program flow, can you envision to where break directs program flow? The keyword default also is used. What do you think its purpose is?

Source code 1

```c
#include <stdio.h>
 void main(void)
{
   int option;
```

```
     printf("Please type 1, 2, or 3\n"
            "1. Breakfast\n"
            "2. Lunch\n"
            "3. Dinner\n");
     scanf("%d",&option);

        if(option==1)
             {
             printf("Good morning\n");
             printf("Order breakfast\n");
             }
        else if (option==2)
             {
             printf("Order lunch\n");
             }
        else if (option==3)
             {
             printf("Order dinner\n");
             }
        else
             {
             printf("Order nothing\n");
             }
}
```

if-else-if control structure. Only one of the statement blocks (enclosed in braces) is executed.

Source code 2

```
#include <stdio.h>
void main(void)
{
 int option;

 printf("Please type 1, 2, or 3\n"
        "1. Breakfast\n"
        "2. Lunch\n"
        "3. Dinner\n");
 scanf("%d",&option);

 switch(option)
 {
 case 1: printf("Good morning\n");
         printf("Order breakfast\n");
         break;

 case 2: printf("Order lunch\n");
          break ;

 case 3: printf("Order dinner\n");
          break;

 default:
         printf("Order nothing\n");
 }

}
```

Labels are followed by colon.

Switch control structure.

Break statements cause exiting of switch control structure.

Output from both source codes

```
Please type 1, 2, or 3
1. Breakfast
2. Lunch
3. Dinner
2
Order lunch
```

Explanation

1. How does an if-else-if control structure work? An if-else-if control structure shifts program control, step by step, through a series of statement blocks. Control stops at the relational expression that is True and executes the corresponding statement block. After execution of the statement block, control shifts to the end of the control structure. If none of the relational expressions is true, the final statement block is executed. In this lesson's program the value of option was read in to be 2; so the first statement block was not executed. Because the relational expression option == 2 was True, the second statement block was executed. The third and fourth statement blocks were bypassed and control transferred to the end of the control structure.

The form of the if-else-if control structure is

if (*relational_expression_1*)
       {
       statement_block_1
       }
else if (*relational_expression_2*)
       {
       statement_block_2
       }

  •
  •
  •
  •

else if (*relational_expression_n*)
       {
       statement_block_n
       }
else
       {
       statement_block
       }

Figure 4.7 illustrates the if-else-if control structure for this lesson's program. Note the branching that occurs due to the different values of option.

2. What does the switch statement do? A *switch statement* or switch control structure commonly is constructed similarly to the if-else-if control structure. It also is used to transfer control. Its syntax is

FIG. 4.7

If-else-if control structure for this lesson's program. Compare to switch and nested if-else control structure illustrations.

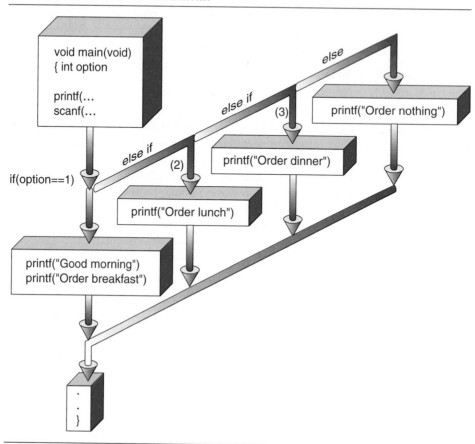

```
switch(expression)
{
case  constant1:
      statement1a
      statement1b
      . . .
case  constant2:
      statement2a
      statement2b
      . . .
      . . .
default:
      statements
}
```

where the expression must be enclosed in a pair of parentheses and must result in an integer type value when the program flow enters the switch block. A switch block must be bounded by a pair of braces. The terms *constant1, constant2,* and so on are integer type constant expressions. Note that all constant expressions are followed by colons. All the constant expressions must be unique, meaning that none can have the same value as another constant expression. Although not required, it is common that the last case type line is the keyword `default`. If no constant matches the value of the expression, the statements in the `default` case are executed. The `default` case is optional. If no default case is given and no constant matches the number, the entire switch block is ignored.

Figure 4.8 shows the switch control structure for this lesson's program. Compare this figure to Figs. 4.7 and 4.3. Note the similarities between the if-else-if, nested if-else, and switch control structures. In all the cases illustrated, the control structure has chosen a single block of code to execute and bypassed the others.

FIG. 4.8
Switch control structure for this lesson's program. Note the importance of the break statement in controlling program flow for the switch control structure. Compare to if-else-if and nested if-else control structure illustrations.

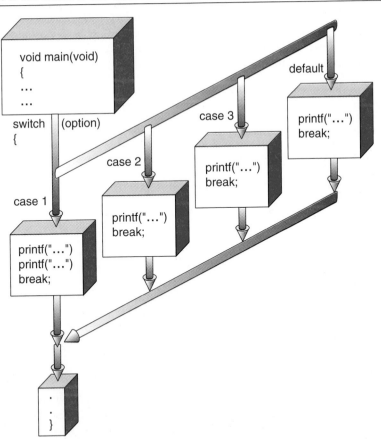

3. What is case*?* It is a keyword that can be used only in a switch control structure. It is used to form a label called a *case label.* A case label is a constant followed by a colon. The label does not affect the execution of the statement that follows it. In switch control structures, C looks for a match between the switch expression and the expression in a case label. C then executes the statement sequence following the matching case label. For instance, for the form shown previously, if the value of the switch *expression* matches *constant1,* then the program flow is transferred to case *constant1* and *statement1a, statement1b,* and so forth are executed. Because the switch control structure can search only for equality, it differs from the if-else-if control structure, which can use other relational operators.

4. What is a break *statement?* A break statement in a switch control structure terminates execution of the smallest enclosing switch statement. The word *terminates* means to send control to the point of the closing brace for the switch structure. We will see that break statements have other uses, which operate similarly in that they cause control to pass to a closing brace.

5. Does a break statement need to be the last statement of a statement sequence following a case label? No, however, often the last statement for each case is the break statement because it terminates the process and exits switch. If no break statement is used, then the statements in the next case are executed. For example, in the following code,

```
switch (option)
{
  case (1): printf("Entering case 1\n");
            break;
  case (2): printf("Entering case 2\n");

  case (3): printf("Entering case 3\n");
            break;
}
```

if option = 1, then "Entering case 1" will be displayed on the screen. If option = 3, "Entering case 3" will be displayed. However, if option = 2, both "Entering case 2" and "Entering case 3" will be displayed because C first finds a match between the switch expression and a case label. Execution then continues, line by line, until a break or the end of the block (indicated by a closing brace) is encountered. This is because a statement label has no effect on the statement that follows it. A statement label serves only as a marker to which control can be sent. The program flow when a break statement is missing is shown in Fig. 4.9.

6. What is default:*?* It is a special label used only for switch control structures. In the event that none of the case label constants agrees with the switch expression, control passes to the default labeled statement sequence. The label default is a keyword and not considered to be a user-defined label.

FIG. 4.9

Flow of program control for the switch control structure of this lesson's program if no break statement were given for the case 2 block. Observe that the break statement causes control to exit the switch structure. With no break statement, control passes directly to the next case.

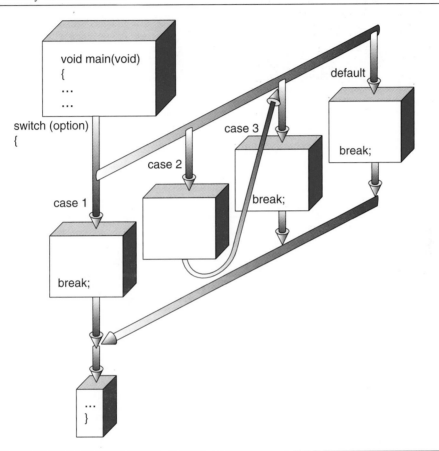

7. Can we nest switch control structures? Yes, a nested switch control structure could take the following form:

```
switch (outer_expression)
        {
        case constant_outer1:
                switch (inner_expression)
                        {
                        case constant_inner1:
                                statement inner_1a
                                statement inner_1b
                                ..
                                ..
```

```
            case constant_inner2:
                    statement inner_2a
                        ..
                        ..

                }
case constant_outer2:
        statement outer_2a
        statement outer_2b
            ..
            ..
case constant_outer3:
        statement outer_3a
        statement outer_3b
            ..
            ..
}
```

An illustration of a nested switch structure is given in Fig. 4.10.

FIG. 4.10
Nested switch control structure with break statements

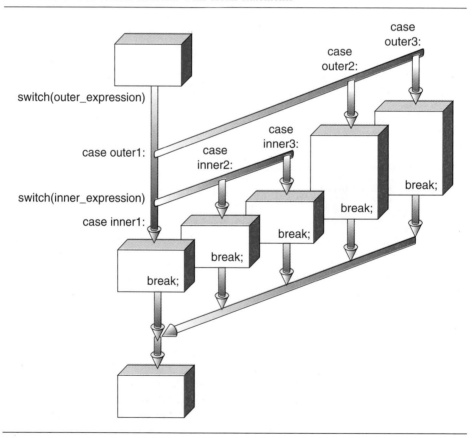

EXERCISES

1. True or false:
   a. The case constants within a switch statement must be arranged in sequence, such as 101, 102, 103, and so forth.
   b. A switch statement can be replaced by an if-else-if control structure.
   c. A switch statement must contain a default case section.

2. Find the error(s), if any, in the following statements (assume a and b are int and a = 1 and b = 2):
   a. **default:**
   b. **switch (a);**
   c. **case 123;**
   d. **switch {a+b}**
   e. **switch (a): {case 1: b=a+2; break;}**

3. Use a switch statement to write a program to calculate the tax based on the following tax code:

income < 1000,	tax = income × 20 percent
1000 <= income < 2000,	tax = income × 30 percent
income >= 2000,	tax = income × 40 percent

4. Use switch statements to program problem 3. (Hint: Introduce an int variable A = income/1000 and use it as a switch variable.)

   Solutions
1. a (false), b (true), c (false)
2. a. No error
   b. **switch (a)**
   c. **case 123:**
   d. **switch (a+b)**
   e. **switch (a) {case 1: b=a+2; break;}**

LESSON 4.7 WHILE LOOP (1)

Topic

• Using while loops

C provides for a number of iterative control structures, known as *looping*. Looping involves the repeated execution of one or more statements. The programmer controls how many times the statements are to be executed and the values that will change with each repetition or iteration. A generic illustration of the effect of looping is shown in Fig. 4.11.

FIG. 4.11
Loops, repeated execution of a block of statements

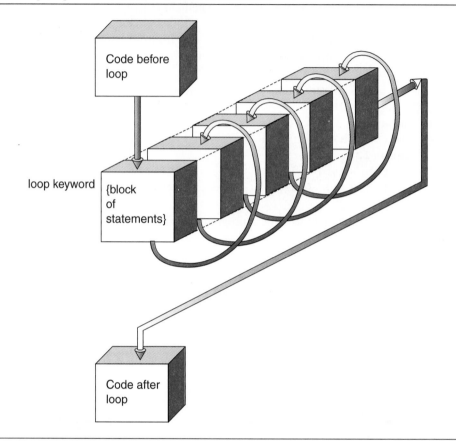

The C language provides several methods for looping. The simplest one is the while loop. A while loop contains just two parts, a test condition part and an execution part. When a program reaches a while statement, the test condition will be checked. If the condition is True, the execution part will be executed and continued to be executed until the test condition becomes False. When the test condition becomes false, the execution part is bypassed and program control is transferred to a point after the end of the while loop.

Look for the line of text in the source code with the keyword *while*. From what you know about statement blocks and relational expressions, can you determine which expression represents the test condition? Which are the statements in the execution part? Look at the output. How many times has the loop been executed? Why did it execute this many times?

Source code

```
#include <stdio.h>
  void main(void)
  {
      int  i;

      i = 1;

      while (i<= 5 )
         {
         printf (" Loop number %d in the while_loop\n",i);
         i++;
         }
  }
```

Test expression.

Statement block is repeatedly executed until test expression becomes False.

Incrementing counter variable.

Output

```
Loop number 1 in the while_loop
Loop number 2 in the while_loop
Loop number 3 in the while_loop
Loop number 4 in the while_loop
Loop number 5 in the while_loop
```

Explanation

1. What is the meaning of while(i<=5) *{statements}?* It means that, while the variable i is less than or equal to 5, the statements between the braces are executed repeatedly. If the variable i becomes greater than 5, the statements between the braces are not executed. In general, the structure of a C *while loop* is

> **while** (*expression*)
> **{**
> *statement1*
> *statement2*
> **...**
> **}**

where the *expression* is a relational expression (the *expression* may be a variable, a constant, or an arithmetic expression) that results in either True or False. If the result is True, the *statements* between the braces are executed. Braces are not required if the loop body consists of only one statement.

The loop body in this lesson's program is executed when i = 1, 2, 3, 4, and 5 and bypassed when i = 6. The i value is increased by 1 for each loop cycle by the statement

i++;

A conceptual illustration of the while loop for this lesson's program is shown in Fig.
4.12. Trace the path indicated by the arrows and observe the changes in the value of
i on each pass through the loop body. Based on the value of i, the test expression
either sends control through the loop body or causes control to pass around the body.
Note that when $i = 6$, the loop body is bypassed.

FIG. 4.12

While loop from this lesson's program. Note that the value i is incremented each time
through the statement block and that the statement block is not executed when $i = 6$.

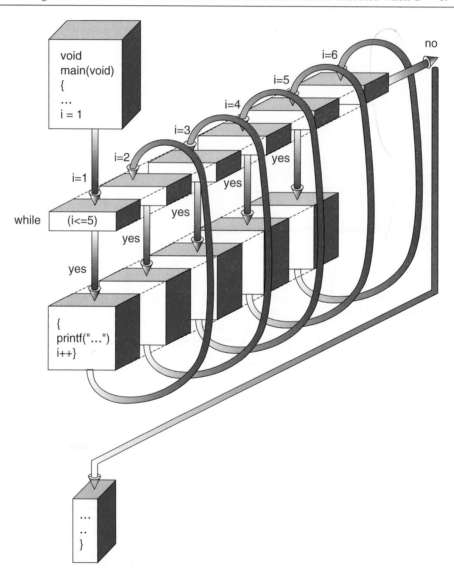

EXERCISES

1. Find error(s), if any, in the following statements (assume a is int and a = 1):
   a. `while (a<5):  {printf("A5%d\n",a);  a11;`
   b. `while (a<5)    {printf("A5%d\n",a);  a22;}`

2. Write a program that uses a while loop to create the following table (X is double type):

   ```
   X          X*X      X+X
   1.0       1.00     2.00
   1.5       2.25     3.00
   2.0       4.00     4.00
   ...
   10.0    100.00    20.00
   ```

3. Use a while loop to show that the following series converges to 2:

$$1 - \frac{1}{3} + \frac{1}{5} - \frac{1}{7} + \ \dots$$

4. Use if and other C statements to determine if a given integer, N, is a prime number or not. (Hint: Use an if statement to test if N can be divided by 2 or any odd integer $<= N/2$.)

Solutions

1. a. `while (a<5) {printf("a=%d\n",a);  a++;}`
   b. No error, but the loop will be executed forever because s never becomes greater than or equal to 5.

LESSON 4.8 WHILE LOOP (2)

Topics

- One variable test expressions
- Compound assignment with increment/decrement
- Avoiding creating an infinite loop
- Early termination of a while loop

The test expression for a while loop does not necessarily have to be a relational expression. Recall that a relational expression that is false is given a value of 0. The C compiler also works in reverse. That is, a variable that has a value of 0 can be treated like a relational expression that is False. Also, a variable that has a value that is nonzero can be treated like a relational expression that is True.

Examine the first while loop in the program. Note that the test expression is not a relational expression but simply the value of i. This loop will stop when i reaches what value?

Look also in this first while loop to see if you can determine how the value of i is changing within the loop. The statement within which the value of i is changing is a compound assignment statement with a decrement. Look at the output and see what happens to the value of i each time the statement is executed. Can you reason out what happens to the value of sum1 in this statement? It is a little tricky; however, you should remember the rules for evaluating expressions that contain increment or decrement operators.

Look at the second while loop and the output associated with it. Recalling what you learned about the pre- and postincrement operators, can you rationalize the values that are printed out for the values of j and sum2?

Look at the third while loop. Determine the values that you should get for k and sum3. Do you see a problem with this while loop? If you see no problem, try running this program on your computer and see what happens. How can you prevent this problem from occurring?

Source code

```c
#include <stdio.h>       Single variable acts as test expression.
void main(void)
{
    int i=4, j=1, k=1, sum1=0, sum2=0, sum3=0;

    while (i)
        {
        printf("old_i=%2d, ",i);
        sum1+=i--;
        printf("new_i=%2d, sum1=%2d\n",i, sum1);
        }
    printf("\n");          Postdecrement operator causes compound
                           assignment to occur first, then decrement.
    while (j<=3)
        {
        printf("old_j=%2d, ",j);
        sum2 += ++j;
        printf("new_j=%2d, sum2=%2d\n",j,sum2);
        }
    printf("\n");          Preincrement operator causes increment to
                           occur before compound assignment.

    while(k)                           Postincrement operator
        {                              causes compound assignment
        printf("old_k=%2d, ",k);       to occur first, then increment.
        sum3 -= k++;
        printf("new_k=%2d, sum3=%2d\n",k,sum3);
        }
}                          We have created an infinite loop because the
                           value of k never becomes False.
```

Output

```
old_i= 4,  new_i= 3,  sum1= 4
old_i= 3,  new_i= 2,  sum1= 7
old_i= 2,  new_i= 1,  sum1= 9
old_i= 1,  new_i= 0,  sum1=10

old_j= 1,  new_j= 2,  sum2= 2
old_j= 2,  new_j= 3,  sum2= 5
old_j= 3,  new_j= 4,  sum2= 9

old_k= 1,  new_k= 2,  sum3= -1
old_k= 2,  new_k= 3,  sum3= -3
old_k= 3,  new_k= 4,  sum3= -6
old_k= 4,  new_k= 5,  sum3= -10
old_k= 5,  new_k= 6,  sum3= -15
. . . .
```

Explanation

1. What is the meaning of `while(i)` *{statements}?* It means that while the variable `i` is not equal to 0 (0 is False and nonzero is True), then the statements between the braces are executed. The variable `i` is called a *flag,* because it is a variable that controls execution of the statements in the while loop body.

In this example, the first while loop begins

<code>while (i)</code>

The loop body is executed when `i` = 4, 3, 2, and 1 (meaning the flag has a value of True) and terminated when `i` = 0 (meaning the flag is False). The values of `i` are summed using the statement

<code>sum1 += i;</code>

The value of `i` is decreased by one for each loop cycle by the expression

<code>i--;</code>

A shorthand for these two statements is

<code>sum1 += i--;</code>

(Please review Lesson 3.4 if you do not understand the shorthand.) This shorthand is a form commonly used and you may see it if you review programs written by others. Note that, in this shorthand, the `--` occurs after the variable on the right-hand side. This is because the value of `i` is to be decremented after it is to be summed.

The second while loop begins

<code>while (j<=3)</code>

The loop body is executed when $j = 1$, 2, and 3 and terminated when $j = 4$. The j value is increased by 1 for each loop cycle by the statement

```
++j;
```

The values of j are summed using the statement

```
sum2 += j;
```

A commonly used shorthand for these two statements is

```
sum2 += ++j;
```

Note that, in this shorthand, the ++ occurs before the variable on the right-hand side. This is because the value of j is to be incremented before it is to be summed.

2. How can we avoid accidentally creating an infinite loop? It often is beneficial to create a counter variable that can be used to prevent the accidental creation of an infinite loop. The counter variable keeps track of the number of times that the loop has been executed; for example,

```
i=1;

while (number)
{
  printf("Please type a number\n");
  i++;
  scanf("%d", &number);
  if (i>50) break;
}
```

The variable i is the counter. In this case, the loop automatically stops after 50 times. If this counter were not used and number never became 0, then the loop would go on indefinitely. If you executed this lesson's program, then you learned that the third while loop is an infinite loop because the value of k never became 0. One way to remedy this is to create a new variable to use as a counter. The variable icount is the counter variable in the loop that follows. It causes the loop to terminate after it has been executed 30 times.

```
icount=0;
while(k)
    {
    printf("old_k=%2d",k);
    sum3 -= k++;
    printf("new_k=%2d, sum3=%2d\n",k,sum3);
    icount++;
    if (icount>30) break;
    }
```

3. What is the function of the break statement? A break statement can occur only in a switch or loop body. The ANSI C standard says that it "terminates execution of the smallest enclosing switch or iteration statement." The line

```
if (icount>30) break;
```

causes control to transfer out of the loop when the relational expression is true. If the counter variable `icount` is larger than 30, the loop will be stopped by the break statement. A C break statement alone is a complete C statement, but usually it is used as part of a control statement to break a process.

4. How can we write the previous while loop using no break statement? We can use a compound relational expression as the test expression. The following loop avoids a break statement.

```
icount=0;
while (k && icount <=31)
    {
    printf("old_k=%2d", k);
    sum -= k++;
    printf ("new_k=%2d, sum3=%2d\n", k, sum3);
    icount++;
    }
```

Note that the bodies of both loops are executed with `icount` = 31.

5. Is it better to write the loop with or without the break statement? In general, it is better to avoid using break statements in loops wherever possible. Break statements make programs less structured because they make program control more difficult to follow.

6. If a while loop's test expression is false initially, will the while loop be executed? No, the while loop will never be executed. For example, the printf() statement in this while loop

```
while (100 <50) printf ("This will never be displayed\n");
```

will never be executed because $100 < 50$ is False.

EXERCISES

1. Find error(s), if any, in the following statements (assume *a* is int and $a = 1$):
    a. `while (5) printf("Good morning\n");`
    b. `while (5<a): {printf("A=%d\n",a); a++;}`
    c. `while (5+1==7) {printf("A=%d\n",a); break;}`

2. Use a while loop to calculate 8 factorial.

Solutions

1. a. No error, but it is an infinite loop. It will print `Good  morning` an infinite number of times because the constant 5 represents True.
    b. `while (5<a) {printf("a=%d\n",a); a++;}`
    c. No error, but the loop will not be executed.

LESSON 4.9 DO-WHILE LOOPS

Topics

- Using do-while loops
- Differences between do-while loops and while loops

To this point we have covered while loops. A second type of loop exists in C, do-while loops. Do-while loops are a slight variation of while loops.

In the program for this lesson, two different forms of do-while loops are used. Examine the first do-while loop and the output. How many times is this loop executed?

The second do-while loop operates similar to the first, even though the form is different. How many times is it executed? Is the test expression for this do-while loop ever True? If this do-while loop were written as a while loop, how many times would it be executed?

Source code

```
#include <stdio.h>
void main(void)
{
    int i=4, j=1;          Test expression.

    do
        {
        printf("old_i=%2d, ",i);
        i--;
        printf("new_i=%2d\n",i);
        } while(i);         Statement block is executed repeatedly until
                            test expression becomes False.

    do ++j
        while(j>999);
                            Test expression is
    printf("j=%2d\n",j);    always False.
}
```

Output

```
old_i= 4, new_i= 3
old_i= 3, new_i= 2
old_i= 2, new_i= 1
old_i= 1, new_i= 0
j= 2
```

Explanation

1. What is the structure of a do-while loop? Where there is only one statement, the *do-while loop* can be written

        do *statement*   **while** *(expression)*;

In general, the structure of a C do-while loop is

```
do
  {
  statement1;
  statement2;
. . .
  }
  while (expression);
```

where the *statements* between the braces are executed at least once regardless of whether the *expression* is True or False. After that, the *expression,* which is a relational expression (including a variable, a constant, and an arithmetic expression) that results in either True or False, is evaluated. If the result is True, the *statements* between the braces again are executed. Braces are not required if the do-while loop body consists of only one statement.

In the first do-while loop of this lesson, the loop body statements are executed once when i = 3, 2, 1, and 0. When i = 0, the while expression becomes False so the loop is terminated. Figure 4.13 illustrates this loop. You should compare this figure with Fig. 4.12. Clearly, the test expression is executed after the loop body. Trace the flow of Fig. 4.13, paying attention to the change in i with each pass through the loop body. In this figure, you can see how the test expression causes the loop to be exited.

In the second do-while loop, the loop body statements are executed once. After that, the while expression j > 999 becomes False because the j value is 2 when the expression is tested, so the loop is terminated.

Remember that, within the statement portion of the do-while loop, it is necessary to increment or modify the variable(s) involved in the expression to assure that an infinite loop is not produced.

2. What are the differences between while loops and do-while loops? While loops and do-while loops are similar. You can use a while loop to replace a do-while loop and vice versa; however, it may take some adjustment in the statements of the loop body to do so. The difference between the two loops is that, in the while loop, the test expression is tested first, and, if the test result is false, the loop body is not executed. However, in the do-while loop, the loop body always is executed once. After that, the test expression is tested; if the test result is False, the loop body is not executed again.

EXERCISES

1. Find error(s), if any, in the following statements (assume a is int and a = 1):

   a. `do printf("Good morning\n"); while(5);`

   b. `do (printf("a=%d\n",a); a++;) while (a<5):`

   c. `Do {printf("a=%d\n",a); break;} while (a>5)`

FIG. 4.13

Do-while loop from this lesson's program. Note that the value i is decremented each time through the statement block and that a portion of the statement block is executed when i = 0 because the decrement occurs in the middle of the block.

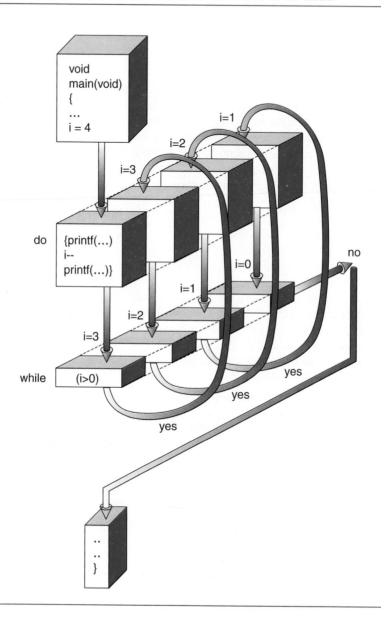

2. Use a do-while loop to make the following table (X is double type):

```
  X       X*X      X+X
 1.0     1.00     2.00
 1.5     2.25     3.00
 2.0     4.00     4.00
 ...
10.0   100.00    20.00
```

3. Use a do-while loop to calculate 8 factorial.

4. Use a do-while loop and the following formula to calculate the value of π. Discuss the accuracy and the rate of convergence:

$$1 - \frac{1}{3} + \frac{1}{5} - \frac{1}{7} + \ldots = \frac{\pi}{4}$$

Solutions

1. a. No error, but it is an infinite loop (meaning that Good morning will be printed an infinite number of times on execution of the program).

   b. `do {printf("a=%d\n",a); a++;} while (a<5);`

   c. `do {printf("a=%d\n",a); break;} while (a>5);`

LESSON 4.10 SIMPLE FOR LOOP

Topics

- The for loop control structures
- Structure of a simple for loop
- Difference between for loops and while loops

This lesson illustrates another looping structure, the for loop control structure. A for loop is very appropriate when you know how many times the operation needs to be repeated. For instance, if an operation is to be performed once for each day of the week then you know that the operation is to be performed seven times. Or if your program involves some sort of countdown from 10 to 0, then you know that the operation is to be performed 11 times. For loops are ideally suited to performing the operations under these circumstances.

In the program, two different for loop control structures are used. Look at the output and see how the printf statements are executed repeatedly. Look also at the three statements enclosed in the parentheses immediately after the word for in the program. These statements control the execution of the loop. Can you see the relationship between these statements and the way the loops are executed? Hint: The first statement indicates how the loop begins, the second statement indicates how the loop ends, and the third statement indicates how the loop goes from the beginning to the end. In the first loop, note where semicolons are located. In the second loop, note the use of braces. Can you speculate on why braces are used in one loop and not the other?

Source code

```
#include <stdio.h>
void main(void)         Initialization.    Test expression.
{
    int day, hour, minutes;
                                            "Increment" expression.
    for(day=1; day<=3; day++)
        printf("Day=%2d\n", day);

    for (hour=5; hour>2; hour--)
        {
        minutes = 60 * hour;                                    Body
        printf("Hour = %2d, Minutes=%3d\n",hour, minutes);      of for
        }                                                       loop.
}
```

Output

```
Day= 1
Day= 2
Day= 3
Hour = 5, Minutes=300
Hour = 4, Minutes=240
Hour = 3, Minutes=180
```

Explanation

1. What is a for loop? A *for loop* is another iterative control structure. For example, the statements

```
for (day=1; day<=3; day++)
        printf("Day=%2d\n", day);
```

cause the printf() function to display the value of day three times; that is, from day equals 1 to day equals 3. The simplest for loop takes the following form:

for (*loop_expressions*)
    single statement for_loop body;

where the *loop expressions,* such as day=1; day<=3; day++, are separated from each other by semicolons (not commas, which is a common error), are enclosed by a pair of parentheses with no semicolon at the end, and must consist of three parts that typically can be described as

1. An initialization expression that initializes the for loop control variable (or counter) and tells the program where to start the loop. For example, day=1; initializes control variable day to 1 and starts the loop at day = 1.
2. A loop repetition condition. This serves as a test expression. If the test expression is False, the loop is terminated. For example, day<=3; tests whether control variable day is less than or equal to 3; if day is greater than 3, the loop is terminated.

3. An increment expression that increases or decreases the control variable. The increment can be in the positive direction (increment) or negative direction (decrement). The increment of the control variable can be achieved by using an increment operator such as ++. For each loop, the ++ will add 1 to the control variable. Similarly, the decrement operator, --, will subtract 1 from the control variable.

In addition, the counter variable can be int, double, or other type of variable. Its value can be positive, negative, or 0. To avoid round-off error, we recommend that you use integer type variables as counter variables. Figure 4.14 schematically displays the structure of a for loop. In this figure, observe that the first time through the loop, the initialization expression is executed. Other times through the loop the increment expression is executed in place of the initialization expression. Also, when the condition evaluates as False, the body of the loop is not executed and the loop is exited.

What we have described here is a typical loop in a way that we will commonly use it in this book and that you will commonly see in practice. However, be aware that the three loop expressions can be used in other ways. Later in this book, we use for loops in ways that differ somewhat from this description.

FIG. 4.14
Order of operation of a for loop. The first for loop in this lesson's program is used as an example.

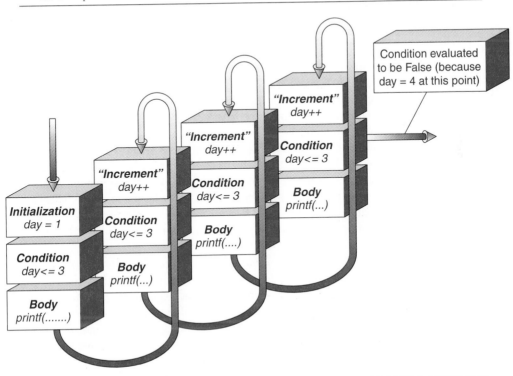

2. What is the general structure of a for loop? In general, the for loop takes the form of

```
for (loop_expressions)
    {
    for_loop_body
    }
```

where the *for_loop_body* consists of the statements between the two braces. The *for_loop_body* describes all the processes to be executed repeatedly in the loop. For example, the *for_loop_body*

```
{
 minutes = 60 * hour;
 printf("Hour = %2d, Minutes = %3d\n", hour, minutes);
}
```

contains an arithmetic statement and a printf statement. In this lesson, both statements will be executed three times, when hour is equal to 5, 4, and 3. As you begin to learn how to write loops, it can be helpful to clarify where your for loops start and end. You can sketch open brackets as we have done to indicate the starting and ending points. This is especially useful when you have many for loops in your program. One purpose of doing this is to make sure that none of the brackets cross. Like if control structures, we cannot have loop control structures that cross.

Note that the general structure of the for loop is different from the simplest structure. The general structure must be used when the for loop body contains more than one C statement. If the for loop body is to contain only one C statement, then the simplest for loop structure can be used.

As we saw with the if control structure, the braces create a *block of code*. To create a code block, simply enclose a group of statements within braces.

A conceptual illustration of the for loop used in this lesson's program is shown in Fig. 4.15. You should note the similarity in structure of this loop with the while loop in Fig. 4.12. If you compare these two, you will see the similarity of the placement of the test expression for both the while and for loops. However, in a for loop, the increment expression is forced to be the first expression executed each time through the loop (after the first time), whereas in a while loop, the increment expression can be embedded deeper in the loop body. Also, a for loop sometimes is easier to follow in a code than a while or a do-while loop because, in a for loop, all three expressions used in a looping structure are in one location.

3. Can we change the value of the counter variable in the body of a for loop? Yes, but the practice is not recommended. A counter variable is used to control the loop, so its value should be changed in the increment expression, not in the loop body. You easily may cause errors if you change the value of the counter variable in both the increment expression and the loop body.

For example, the following loop is "correct." However, it is an infinite loop and never ends. For each loop, the increment expression increases the counter variable, k, by 1, but the loop body always restores it back to 1:

FIG. 4.15
Conceptual illustration of for loop from this lesson's program. Note that hour is decremented each time through the statement block and the statement block is not executed when hour = 2.

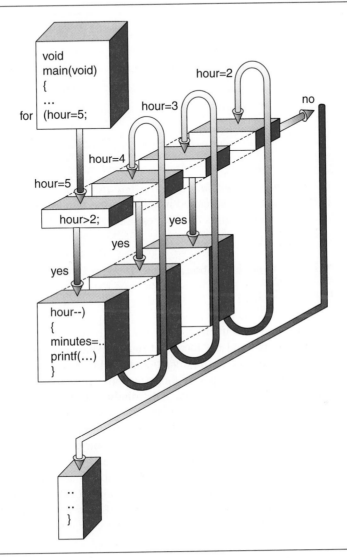

for (k=1;k<3;k++) k=1;

If you use this for loop in one of your programs, you will find that your program never completes executing. Try it. You can leave your computer on for days and your program will never finish. Not modifying the counter variable in the loop body will help you avoid creating infinite loops in your programs.

4. What are the differences between while loops and for loops? In general, while loops and for loops are similar in their order of execution. You can use a for loop to replace a while loop, but you must realize that the bodies of the loops must be different. The differences between the two follow:

Item	For loop	While loop
Initialization expression	Is one of the loop expressions	Must be given prior to the loop
Test expression	Is one of the loop expressions	Is one of the loop expressions
Increment expression	Is one of the loop expressions	Must be in the loop body
When number of iterations is known	Is very convenient and clear to use	Is less convenient and clear
When number of iterations is unknown	Is less convenient and clear	Is more convenient than for loop

5. How are for loops sometimes used when the number of iterations is unknown? It is dangerous but a so-called infinite loop to iterate indefinitely until it is terminated by a break statement is sometimes used. An example follows:

```
for(;;)
{
    printf("please type a number, type 0 to quit\n");
    scanf("%d",&number);
    if (number==0) break;
}
```

The initialization, test, and increment expressions in this for loop are blank, which means nothing will be tested and the loop will run forever. The loop will be terminated only when the user enters the number 0 and the break statement is executed. Be careful with the use of these types of loops in your programs. If you do not do them correctly, your program will execute indefinitely. We do not recommend this form, but we show it here because you may see it in code written by others.

Also, be aware that there are different schools of thought regarding using for loops that require break statements in the loop body. Some programmers feel that doing so violates the rules of structured programming and that it is acceptable only to break out of while or do-while loops. Others feel that it is acceptable to break out of for loops as well. In this book, we occasionally break out of for loops as we feel you may see it in practice. However, before you use a break statement in a for loop in your programs, you should check with your instructor or employer. In this book we will not use the so-called infinite loop form.

EXERCISES

1. True or false:
    a. In a for loop expression, the starting counter value must be smaller than the ending counter value.
    b. The three loop expressions used in for loops must be separated by commas.
    c. Some programmers do not break out of for loops.

d. While loops can be used to replace for loops without changing the loop body.

2. Find the error(s), if any, in each of these statements:
   a. `for day=1,3,1`
   b. `for (day=1,day<3,day++)`
   c. `for (day=10;day<=20;day++);`
   d. `for (day=10;day<5;day++)`
   e. `for (day=100;day<100;day--)`
   f. `for (day=10;day>100;day--)`
   g. `for (i=20;i>10;i--) i=i*3;`

3. Use a for loop and this formula to find the value of π. The result should at least be as accurate as 3.1416:

$$\frac{\pi}{2} = \frac{2}{1} \cdot \frac{2}{3} \cdot \frac{4}{3} \cdot \frac{4}{5} \cdot \frac{6}{5} \cdot \frac{6}{7} \cdots$$

4. Use a for loop to construct this conversion table:

```
inch       feet        meter

1          0.0833      0.0254
2          0.1667      0.0762
3          0.2500      0.0762
...
100        8.3333      2.5400
```

Solutions

1. a (false), b (false), c (true), d (false)
2. a. `for (day=1;day<=3;day++)`
   b. `for (day=1;day<3;day++)`
   c. No error, the do nothing statement, ;, will be executed ten times.
   d. No error, but its loop body will not be executed.
   e. No error, but its loop body will not be executed.
   f. No error, its loop body will be executed once.
   g. No error, but the loop is an infinite loop.

■ LESSON 4.11 NESTED FOR LOOPS

Topics
• Using the += type operator in an increment expression
• Nested for loops

The loops in this program are "nested." *Nested* means that there is a loop within a loop. A nested loop is executed, essentially, from the inside out. Look at the output for the values of i and j. Can you deduce how a nested loop is executed?

Source code

```
#include <stdio.h>
void main(void)
{
    int i, j, k, m=0;

    for (i=1; i<=5; i+=2)
        {
        for (j=1; j<=4; j++)
            {
            k = i+j;;
            printf("i=%3d, j=%3d, k=%3d\n", i, j, k);
            }
        m=k+i;
        }
}
```

Outer loop.

Inner loop.

Output

```
i= 1, j= 1, k= 2
i= 1, j= 2, k= 3
i= 1, j= 3, k= 4
i= 1, j= 4, k= 5
i= 3, j= 1, k= 4
i= 3, j= 2, k= 5
i= 3, j= 3, k= 6
i= 3, j= 4, k= 7
i= 5, j= 1, k= 6
i= 5, j= 2, k= 7
i= 5, j= 3, k= 8
i= 5, j= 4, k= 9
```

Explanation

1. What is the effect of the loop expression i+=2*?* In this lesson's outer for loop, it is an increment expression that increases the value of i by 2 for each loop. You also will find that not all of your loops involve addition as the increment expression. For instance, an equally valid expression is i*=2. What is used depends entirely on the problem being solved.

2. What is a nested for loop? A *nested for loop* has at least one loop within a loop. Each loop is like a layer and has its own counter variable, its own loop expression, and its own loop body. In a nested loop, for each value of the outermost counter variable, the complete inner loop will be executed once. This means that the inner loop will be executed more frequently than the outer loop. The example in this lesson has two counter variables, i and j, where i is the outer loop counter and j is the inner loop counter. The outer loop is executed three times, when i = 1, 3, and 5. For each i value, the j loop is executed four times. Since the j values in each j loop can be 1, 2, 3, and 4, the total number of times that the inner loop is executed is 3 * 4 or 12 times. A conceptual illustration of the nested for loop for this lesson's program is shown in Fig. 4.16. Observe from this figure how the value of j changes each time through the inner loop and how i does not change until many complete passes through the inner loop. This figure demonstrates how control flows from the outer loop to the inner loop and back again.

Note that, in a nested for loop, if either loop body contains more than one C statement, that loop body must be enclosed with a pair of braces. Arrowhead brackets for each loop body (such as those displayed in the source code) have been used to clarify which loop body belongs to which loop. The brackets for any inner loop must be within the brace pair for its outer loop. For all cases, brackets for different loops must not cross.

FIG. 4.16

Nested for loop for this lesson's program. The unlabeled numbers are the values of j. For simplicity, the test expressions and their proper locations are not shown.

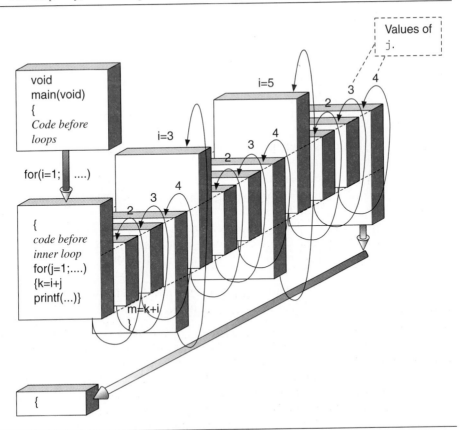

In general, the syntax of a nested for loop is

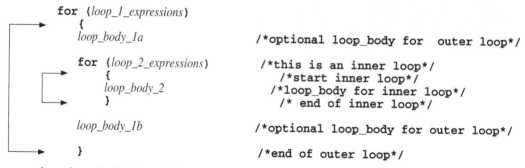

```
for (loop_1_expressions)
    {
    loop_body_1a                /*optional loop_body for  outer loop*/

        for (loop_2_expressions)    /*this is an inner loop*/
            {                       /*start inner loop*/
            loop_body_2             /*loop_body for inner loop*/
            }                       /* end of inner loop*/

    loop_body_1b                /*optional loop_body for outer loop*/

    }                           /*end of outer loop*/
```

where *loop_body_1a* and *loop_body_1b* form the loop body for the outer loop. They are optional and can be left blank. Similarly, the *loop_body_2* is the loop body for the inner loop.

3. How many nests can be used for nested for loops? ANSI C requires that a compiler allow at least 15 nesting levels of iteration control structures. Your compiler may allow even more. In any event, 15 levels almost certainly is more than you will ever need.

4. What nesting patterns for looping control structures are allowable and how can looping and decision making fit together? The pattern of nested loops can be anything, provided that the loops do not cross (meaning that the arrowhead brackets do not cross). The sketches in Fig. 4.17 illustrate some of the many possibilities. The way the loops are nested is determined primarily by the problem to be solved by the program.

Figure 4.18 illustrates looping and decision making together in one program. We have not shown the code that corresponds to this figure; however, from what you know at this point, you could create one that does. Of course, the actual problem to be solved dictates how control should flow. We will illustrate this in the application examples.

5. How can you make your code easy to read when using loops? Indentation commonly is used by programmers to make the code easy to read. Indentation is not necessary and has no effect on the code's performance; however, it is considered to be standard practice.

There is no absolute standard style of indentation. Commonly, three spaces or more is considered to be an indent. We recommend you use the following style at this point in your programming career:

```
for (...)
    {
    ......
    ......

    for (...)
        {
```

FIG. 4.17
Patterns of nested for loops

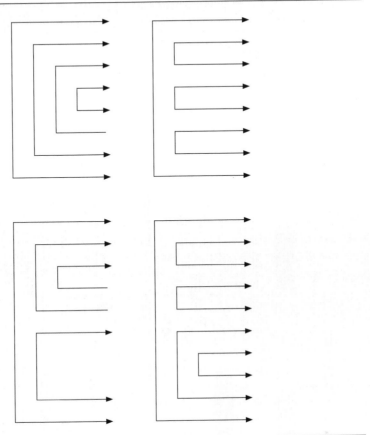

```
      . . . . . .
      . . . . . .
   }
 . . . . . .
 . . . . . .
 }
```

The advantage of the indentation is that, even in the absence of drawn arrows, one can discern the nesting of loops and program flow. In many instances in this text, we would like to have shown more indentation than is illustrated in the source codes; however, publishing restrictions preclude us from doing so.

6. How can we debug loops that do not work correctly? It is extremely help-ful to make a table of the values of each variable calculated in the loop and the loop

FIG. 4.18

Combination of iteration and selection structures. There are an infinite number of combinations. One goal in programming is to develop reliable and efficient combinations of these structures.

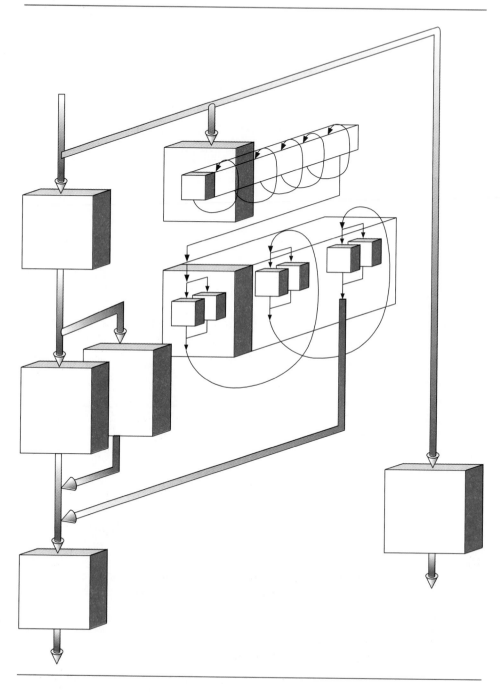

counters. Then, as you look at the code, you can trace the code and fill in the values of each of the variables and counters on each step. For instance, for this lesson's program, the table would be

i	j	k	m
1	1	2	0
1	2	3	0
1	3	4	0
1	4	5	6
3	1	4	6
3	2	5	6
3	3	6	6
3	4	7	10
5	1	6	10
5	2	7	10
5	3	8	10
5	4	9	14

Making a table like this also is helpful when you initially write the code. You usually can make a first attempt at writing a nested loop, then make a table and write in the values as shown here. You gain confidence that you have written the loops correctly if the values in the table are what you expect.

When debugging code that you have already written, you can write printf statements within the loops to print the variables and counters on each step. If you put printf statements both within and outside of the nests, you can see the variables change as they are printed out. These printf statements should be written so that a table similar to the one here is displayed.

Your compiler's debugging option (if it has one) is extremely useful in tracing loops. You can set markers causing the program execution to pause at various locations within the loops. During the pause you can check the values of variables and counters. If any of these have unexpected values, you can rewrite the code of the loops.

EXERCISES

1. True or false:
   a. In C, day*=5 may be used as a third expression in a for loop.
   b. In C, day=+5 is similar to day = day + 5.

2. Find the error(s), if any, in each of these statements:
   a. ```
      for (i=1;i<3;i++) i=i+100;
      ```
   b. ```
      for (i=1;i<3;i++)
      for (j=100;j>10;j/=5)  k=i+j;
      ```
   c. ```
      for (i=1;i<3;i++)   k=i+10;
      for (j=100;j>10;j/=5) k=i+j;
      ```
      ```

3. Hand calculate the final value of c in this program. Create a table of all of the values. After doing this, run the program to check your results:

```
void main(void)
{
   int a,b,c;
   for (a=1;a<3;a++)
      {
      c=a;
      for (b=1;b<3;b++) c+=a+b;
      c+=10;
      }
}
```

4. Write a program to calculate the mean, variance, and standard deviation of the first ten positive integers. The mean, *m*, variance, *v*, and standard deviation, *s*, are defined as follows:

$$m = \sum_{i=1}^{n} \frac{x_i}{n}$$

$$v = \frac{\sum_{i=1}^{n} (x_i - m)^2}{n}$$

$$s = \sqrt{v}$$

5. Run the following program and discuss the effect of using different types of variables as counter variables. Remember that it is recommended that you use only integers as your counter variables.

```
#include <stdio.h>
void main(void)
{
  int i, isum1=0, isum2=0, N=9999;
  float f, x, sum1=0.0, sum2=0.0;

  x=1.0/N;
  printf("x=%f\n",x);

  for (i=1;i<=N;i++) {sum1 +=x; isum1+=1;}
  printf("sum1=%f, isum1=%d\n",sum1, isum1);

  for (f=x;f<=1.0;f+=x) {sum2+=x; isum2+=1;}
  printf("sum2=%f, isum2=%d\n",sum2, isum2);
}
```

Solutions

1. a (true), b (false)
2. a. No error, but it is not recommended to modify the counter variable i in its loop body.
   b. No error. This is a nested loop. Division of integer by integer would cause the fractional part to be lost if the test expression had a smaller comparison value.
   c. No error. This is not a nested loop.

▩ APPLICATION PROGRAM 4.1 GIRDER INTERSECTION—IF-ELSE CONTROL STRUCTURE

Problem statement

Two long girders are to be placed on a bridge in a manner such that they will cross each other. Their placement can be represented by the typical slopes and intercepts we know from trigonometry. To properly place the connector bolts on these girders, we need to know the coordinates of the intersection point. Write a program capable of finding the intersection point of a pair of lines that represent the girders, assuming the girders are long enough to guarantee intersection if they are not placed parallel to each other. The input data is to consist of the slope and intercept for the pair. The data are to come from a file (INTSECT.DAT) that is to consist of two lines:

line 1	**m1**	**b1**
line 2	**m2**	**b2**

where
$m1$ = slope of the first line of the pair
$b1$ = intercept of the first line of the pair
$m2$ = slope of the second line of the pair
$b2$ = intercept of the second line of the pair

The output is to be the x and y coordinates of the intersection point and output is to the screen.

Solution

RELEVANT EQUATIONS
The general equation for a line in terms of x and y is

$$y = mx + b$$

where
m = slope
b = y intercept

If m_1, b_1, m_2, and b_2 are defined as previously, the equations of the two lines are

$$y = m_1x + b_1 \tag{4.1}$$
$$y = m_2x + b_2 \tag{4.2}$$

The intersection of the two lines can be found by solving the two equations simultaneously. By substitution

$$m_1x + b_1 = m_2x + b_2 \tag{4.3}$$
$$x = \frac{b_2 - b_1}{m_1 - m_2} \tag{4.4}$$
$$y = m_1x + b_1 \tag{4.5}$$

SPECIFIC EXAMPLE
Consider the following lines:

$$y = 2x - 3$$
$$y = 5x + 1$$

For these lines

$$m_1 = 2$$
$$b_1 = -3$$
$$m_2 = 5$$
$$b_2 = 1$$

From equations (4.4) and (4.5),

$$x = \frac{1 - (-3)}{2 - 5} = -1.333333$$
$$y = 2\,(1.333333) + (-3) = -5.666666$$

which are the coordinates of the intersection point.

ALGORITHM

The algorithm can be written by following the steps in the specific example. A basic algorithm is the following:

```
Read the values of slopes and intercepts
Calculate the values of x and y (intersection point)
Print the values of x and y
```

This algorithm, however, does not account for the condition where the two lines are parallel ($m_1 = m_2$), in which case there is no single intersection point. To account for this situation the algorithm is changed to the following:

```
Read the values of slopes and intercepts
If the lines are not parallel
     Calculate the values of x and y (intersection point)
     Print the values of x and y
```

The if control structure for this algorithm is shown in Fig. 4.19. From this algorithm, the source code can be written. Look at the program line by line. Compare the algorithm with the source code so that you can see all of the steps of the program.

SOURCE CODE

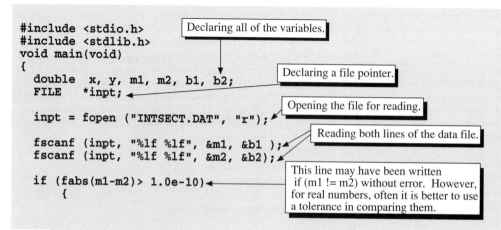

```
#include <stdio.h>                Declaring all of the variables.
#include <stdlib.h>
void main(void)
{
   double  x, y, m1, m2, b1, b2;        Declaring a file pointer.
   FILE    *inpt;
                                        Opening the file for reading.
   inpt = fopen ("INTSECT.DAT", "r");
                                        Reading both lines of the data file.
   fscanf (inpt, "%lf %lf", &m1, &b1 );
   fscanf (inpt, "%lf %lf", &m2, &b2);
                                        This line may have been written
   if (fabs(m1-m2)> 1.0e-10)           if (m1 != m2) without error. However,
      {                                for real numbers, often it is better to use
                                       a tolerance in comparing them.
```

```
x = (b2 - b1) / (m1 - m2);
y =  m1 * x +  b1;

printf ("\n\The slopes and intercepts are:\n\n"
        "m1  =  %lf \n"
        "b1  =  %lf \n"
        "m2  =  %lf \n"
        "b2  =  %lf\n",m1,b1,m2,b2);

printf ("\nThe intersection point is (%lf, %lf).", x,y);
}
fclose(inpt);
```

> Calculating the x and y of the intersection point according to the equations written.

> Closing the input file.

> Printing the results.

INPUT FILE INTSECT.DAT

```
2   -3
5    1
```

OUTPUT

```
The slopes and intercepts are
m1  =   2.000000
b1  =  -3.000000
m2  =   5.000000
b2  =   1.000000
The intersection point is (-1.333333, -5.666667).
```

COMMENTS

This program is very simple to write after the equations have been developed. You will find that, for many programs, a key aspect is the development of the equations.

FIG. 4.19
The if control structure for Applicaton Program 4.1

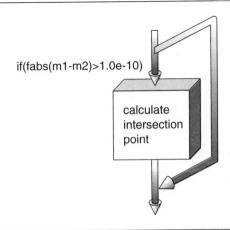

if(fabs(m1-m2)>1.0e-10)

calculate
intersection
point

Many times, you will have to develop those equations yourself, and this is considered to be part of the programming process.

Note that, in writing each equation, the variable of interest should appear alone on the left side of the equal sign. This allows for simple translation into C source code. Remember, in algebra it is acceptable to write an equation of the sort $m_1x+b_1 = y$ but in C this must be written `y=m1*x+b1`, because we are allowed to have only a single variable on the left side of an assignment statement.

For this program, it was necessary to account for the possibility of parallel lines. Parallel lines occur when $m_1 = m_2$. However, when we compare real numbers in a program in many cases it is better to consider a tolerance. In this program, we have selected $1.0e-10$ for the tolerance. Any lines whose slopes are different by less than that amount are regarded as parallel.

Using this method of comparing real numbers keeps the programmer aware of the number of significant digits being carried in the program. For instance, if we wanted to calculate whether two lasers being shot at the moon would intersect at the correct location, then this program may not work because the slopes of the lines representing the laser light may be closer than $1.0e-10$. We may need to make the variables long doubles and reduce the tolerance to $1.0e-20$ or less to solve the problem.

Remember, one reason we do not use the float type is that the decimal precision usually is only about six significant digits. Doubles usually have about 15 significant decimal digits and long doubles 19 significant decimal digits and so are much better suited for engineering calculations. See your compiler documentation for more details on this.

Also note that here we have used $m_1 - m_2$ to compare to the tolerance. We might have used something like $(m_1 - m_2)/m_1$ to compare to the tolerance. This might have been appropriate if both m_1 and m_2 were very small to begin with. For instance, if $m_1 = 1.0 \times 10^{20}$ and $m_2 = 0.5 \times 10^{20}$ then $m_1 - m_2 = 0.5 \times 10^{20}$. Although the lines are nearly parallel, m_1 has twice the slope of m_2. This may be important and we may want the tolerance to reflect this. Using $(m_1 - m_2)/m_1$ (which would evaluate to 0.5 in this case) to compare to the tolerance may help us recognize the differences between two very small slopes. However, it also creates the problem that, if $m_1 = 0$, the calculation causes the program to crash. Therefore, a check on m_1 equaling zero must be included.

These are not the only techniques to use in comparing real numbers. As you go further in your studies you will probably encounter others. We describe this dilemma, though, because it is an issue that arises frequently in engineering programming. To resolve it properly, you should be aware of the number of significant digits involved and the type of problem you are solving.

Modification exercises

1. Modify the program so that it prints an error message indicating that the lines are parallel if the two slopes are equal.

2. Modify the program so that, if both slopes are equal, it prompts the user to enter a new value for one of the slopes from the keyboard. Then it solves the problem.

3. Modify the program so that it can handle four pair of lines and, therefore, find four intersection points. You should put in a loop to do this.

4. Modify the program so that it can handle a variable (n) number of lines. The input data file would be

```
n
m₁ x₁
m₂ x₂
m₃ x₃
 ·
 ·
 ·
mₙ xₙ
```

◼ APPLICATION PROGRAM 4.2 AREA CALCULATION (1)—FOR LOOP

As illustrated, for loops can be used to repeat execution of C statements. As such, they can be used in programs that perform the same tasks as those done by many of the application programs in Chapter 3.

This application program uses a for loop to perform the same task as Application Program 3.1, which calculates the areas of four different right triangles as an example of using patterns to write programming statements.

Because the sample calculation and the algorithm have been developed in Application Program 3.1, we will not develop them again here. Please look at this portion of the program development in Application Program 3.1 and then read Application Program 4.2 and follow the steps through the for loop.

Compare both application programs. Closely follow the flow of Application Program 4.2, statement by statement. The flow is illustrated in Fig. 4.20. Use your calculator and the loop in the program to fill in this table.

i	area	horizleg	vertleg

Pay particular attention to the way the variables horizleg and vertleg are used. You can see that they are initialized before the for loop. Once in the for loop, the area is first calculated and then printed. Then new values of horizleg and vertleg are calculated. Note that these new values of horizleg and vertleg are retained and used to calculate the area the second time through the loop. The values of horizleg and

For loop for Application Program 4.2

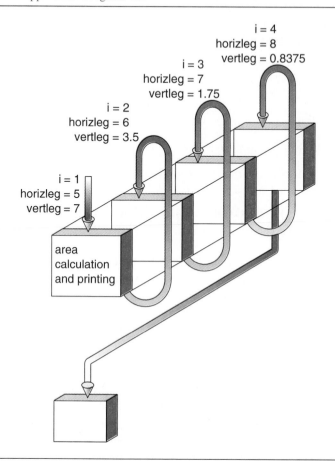

vertleg calculated during the second time through the loop are retained and used for calculating the area during the third time through the loop. And the area on the fourth time through the loop is calculated from the values of horizleg and vertleg calculated during the third time through the loop. Note that the values of horizleg and vertleg calculated during the fourth time through the loop are not used.

This program is much shorter than Application Program 3.1. In addition, unlike Application Program 3.1, it easily can be changed to calculate the areas of 50 triangles that follow the same pattern of leg lengths.

SOURCE CODE

```
#include <stdio.h>
void main(void)
{
    int     i;
    double horizleg, vertleg, area;

    horizleg = 5.0;        Initializing the variables.
    vertleg  = 7.0;

    for ( i=1; i<=4; i++ )        Using a for loop to calculate four areas.
        {
        area = 0.5 * horizleg * vertleg;
        printf ("Triangle area number %d = %6.2lf \n",i, area);

        horizleg += 1.0;        Modifying the leg lengths
        vertleg  /= 2.0;        each time through the loop.
        }
}
```

OUTPUT

```
Triangle area number 1 = 17.50
Triangle area number 2 = 10.50
Triangle area number 3 =  6.12
Triangle area number 4 =  3.50
```

COMMENTS

If you compare the output from this program to the output from Application Program 3.1, you can see that this program does not use the words First, Second, Third, and Fourth. Instead, it prints out the numbers 1, 2, 3, and 4. This is because the printf statement is contained within the for loop. Each time the statements in the for loop are executed, only variables can change. Therefore, we print out the counter variable, i, to indicate the triangle we are calculating.

In Chapter 7, we will work with character variables and see how to treat words as character strings.

Modification exercises

Modify the program to

1. Calculate the areas of 30 triangles following the same pattern. Note how small the values become. How can you print out at least four significant digits?

2. Calculate the areas of ten triangles with the vertical leg length doubling instead of halving each time through the loop.

3. Calculate the areas of 40 rectangles with the horizontal leg length being reduced by 5% each time through the loop and the vertical leg length increasing by 3%.

4. Calculate the areas of 20 trapezoids. Read in the height, bottom length, and top length. Invent your own variable names. Make this pattern: The bottom length is reduced by 3 percent each time through the loop, the top length is increased by 8 percent, and the height is increased by 2 percent.

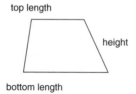

▨ APPLICATION PROGRAM 4.3 AREA CALCULATION (2)—FOR LOOP

Problem statement

Write Application Program 3.1 another way. Use the loop counting variable n to create new equations.

Solution

RELEVANT EQUATIONS

We can develop independent equations representing the values of the lengths of the legs of the triangles by looking again at the pattern exhibited by the lengths. Recall that the lengths of the legs are

Triangle number	Horizontal leg length	Vertical leg length
1	5	7
2	5 + 1 = 6	7/2 = 3.5
3	5 + 2 = 7	7/4 = 1.75
4	5 + 3 = 8	7/8 = 0.875

If the triangle number is n then the horizontal leg length is

$$\text{Horizontal leg length} = 5 + (n-1) \qquad (4.6)$$

And the vertical leg length is

$$\text{Vertical leg length} = 7/\, 2^{(n-1)} \qquad (4.7)$$

We can use these equations in our C program. Note that it is up to us to recognize the patterns and write the equations that represent the patterns. Recognizing patterns and writing the necessary equations to use in your programs are parts of programming in which (after reading this book) you will develop skill.

SPECIFIC EXAMPLE

Using equations (4.6) and (4.7), we get

n	Horizontal leg length	Vertical leg length	Area
1	5 + (1 − 1) = 5	$7/2^{(1-1)} = 7.0$	0.5 * 5 * 7 = 17.5
2	5 + (2 − 1) = 6	$7/2^{(2-1)} = 3.5$	0.5 * 6 * 3.5 = 10.5
3	5 + (3 − 1) = 7	$7/2^{(3-1)} = 1.75$	0.5 * 7 * 1.75 = 6.12
4	5 + (4 − 1) = 8	$7/2^{(4-1)} = 0.875$	0.5 * 8 * 0.875 = 3.50

ALGORITHM

From the specific example we can see that we can put equations (4.6) and (4.7) into a loop with n going from 1 to 4. Therefore, the algorithm becomes

```
Include files
Declare variables
Loop n from 1 to 4
        Calculate horizontal leg length using equation (4.6)
        Calculate vertical leg length using equation (4.7)
        Calculate area as 0.5*horizleg*vertleg
        Print results
```

SOURCE CODE

This source code follows the algorithm step-by-step.

```c
#include <math.h>
#include <stdio.h>
void main(void)
{
    int     n;
    double nminus1, horizleg, vertleg, area;

    for ( n = 1; n <=4; n++ )
        {
        nminus1 = n - 1;          Using the counter variable in the
        horizleg = 5.0 + nminus1;  calculation in the for loop body.
        vertleg  = 7.0 / pow(2,nminus1);
        area     = 0.5 * horizleg * vertleg;
        printf ("\tTriangle area number %d = %6.2lf\n",n, area);
        }
}
```

OUTPUT

```
Triangle area number 1 = 17.50
Triangle area number 2 = 10.50
Triangle area number 3 = 6.12
Triangle area number 4 = 3.50
```

COMMENTS

As you write more programs, you will find that often it is convenient to use the counter variable in calculating the values of other variables within a for loop. Just remember that you have to be a little clever in writing the equations that use the counter variable and cautious about mixed type arithmetic.

Notice that we created a new variable, nminus1. It was not absolutely necessary to create this variable. It was created only for convenience and to improve the look of the program. You will find that, as you write programs, you often will create variables for these two reasons. To make your programs as understandable as possible, we recommend that you choose names that are very descriptive of what the variables represent.

Notice also that, in the calculation of nminus1, the expression on the right side of the assignment statement is int type. By assigning this to a double variable (nminus1), a double is placed in the memory cell for nminus1. Thus, we have done no mixed type arithmetic in this code.

Also observe that we included math.h because we used the pow function. Remember this function. You likely will use it in your programs.

Modification exercises

Modify the program to

1. Calculate the result for ten triangles following the same pattern.

2. Calculate the result for four triangles, with each one having the vertical leg length shrink by one-third rather than one-half each time through the loop.

■ APPLICATION PROGRAM 4.4 TEMPERATURE UNIT CONVERSIONS—FOR LOOP

Problem statement

Write a program that creates a table of degrees Celsius with the corresponding degrees Fahrenheit. Begin at 0°C and proceed to 100°C in 20°C increments using a for loop. In other words, rewrite Application Program 3.2 using a for loop.

Solution

RELEVANT EQUATIONS

The equation developed in Application Program 3.2. is

$F = C*(9/5) + 32$

where F is Fahrenheit, in degrees, and C is Celsius, in degrees.

SPECIFIC EXAMPLE

A specific example was given in Application Program 3.2 so it will not be repeated here. We note, though, that we print out six values of degrees, at $C = 0$, 20, 40, 60, 80, and 100.

ALGORITHM

Because we want to print the results in the form of a table, we must print the headings of the table before we begin the loop. Therefore, the algorithm becomes

```
include files
declare variables
initialize Centigrade degrees
loop from 1 to 6
        calculate new value of Centigrade degrees
        calculate corresponding value of Fahrenheit degrees
        print values for table
```

The source code follows this algorithm step by step. Read the program to see how it is done. Once again, follow through the for loop carefully to understand how it is executed. What is the purpose of the variable i? Also, note the initial value given to the variable, degC. How does this initial value influence the order in which the variables are calculated in the for loop?

SOURCE CODE

```
#include <stdio.h>
void main(void)
{                              The counter for the for loop is a
    int    i;                  variable that must be declared.
    double degC, degF ;

    printf (" \n"
            "Table of Celsius and Fahrenheit degrees \n\n"
            "        Degrees                Degrees         \n"
            "        Celsius                Fahrenheit      \n");

    degC    = -20. ;           Initializing degC to be less than
                               the first value of interest.

    for ( i = 1; i<= 6; i++ )
        {                                  Incrementing the value of degC
        degC += 20.;                       each time through the loop.
        degF = degC * 9./5. +32.;
        printf ("\n%7.2lf          %7.2lf        ",
        degC, degF);
        }
}
```

OUTPUT

```
Table of Celsius and Fahrenheit degrees
Degrees     Degrees
Celsius     Fahrenheit
  0.00        32.00
 20.00        68.00
 40.00       104.00
 60.00       140.00
 80.00       176.00
100.00       212.00
```

COMMENTS

You can see that the variable i has no real function except to serve as a counter for the for loop. Since it is not used anywhere within the for loop it is called a *dummy variable.* You often will need to create dummy variables in your programs.

Note that a slightly different approach (from that of Application Program 4.1) is taken in this program for initializing the variables used in the for loop. First, the variable degC is set to −20. This value is below the first value of interest. Then the first statement in the for loop increments the value of degC to the first value of interest, 0.

An alternative to this approach is to have the statement degC=0; before the for loop and then increment degC after degF is calculated and after the printf statement within the for loop. This type of approach was used in Application Program 4.2 for calculating the triangle areas.

Modification exercises

Modify the program to

1. Create a table of values that go from *C = 0* to *C = 100* in increments of 1 degree.

2. Create a table with *F* in the left column and *F* incrementing by 5 degrees from −50 to +300.

3. Create a table including Rankine degrees.

APPLICATION PROGRAM 4.5 ENGINEERING ECONOMICS, INTEREST—NESTED FOR LOOPS

Problem statement

A major part of engineering economics is in managing money. Write a program capable of computing the value of the money in a bank account at the end of each month for a period of five years. The account will start with $5000 and have neither deposits nor withdrawals. The interest will be compounded monthly. The annual interest rate is to be input from the keyboard. The output will be to a file (MONEY.OUT) and consist of a table listing the year, month, and new principal (balance).

Solution

Relevant equations. If the annual interest rate is *i*, then the monthly interest rate is *i*/12. And the amount of interest for one month is

$$I = P_o\left(\frac{i}{12}\right) \qquad (4.8)$$

where $\quad I$ = interest for the month
$\qquad P_o$ = principal at the beginning of the month
$\qquad i$ = annual interest rate

The principal at the end of the month (which is the beginning principal for the succeeding month) is

$$P_f = P_o + I = P_o + P_o\left(\frac{i}{12}\right) = P_o\left(1 + \frac{i}{12}\right) \tag{4.9}$$

where P_f is the principal at the end of the month.

SPECIFIC CALCULATION

For a starting principal of $5000 and an annual interest rate of 6 percent, the principal at the end of the month, using equation (4.9), is

$$P_o = 5000$$
$$i = 0.06$$

$$P_f = 5000\left(1 + \frac{0.06}{12}\right) = 5025.00$$

This calculation can be done for four months of the first year:

$i \quad = 0.06$
$year = 1$

beginning of month 1 $\qquad P_o = 5000.0$

$$P_f = 5000\left(1 + \frac{0.06}{12}\right) = 5025.00$$

end of month 1

beginning of month 2 $\qquad P_o = 5025.0$

$$P_f = 5025\left(1 + \frac{0.06}{12}\right) = 5050.13$$

end of month 2

beginning of month 3 $\qquad P_o = 5050.13$

$$P_f = 5050.13\left(1 + \frac{0.06}{12}\right) = 5075.38$$

end of month 3

beginning of month 4 $\qquad P_o = 5075.38$

end of month 4 $\qquad P_f = 5075.38\left(1 + \frac{0.06}{12}\right) = 5100.75$

ALGORITHM

One can begin by writing an algorithm for the main portion of the program directly from this specific calculation. In this problem we illustrate various styles of writing an algorithm or pseudocode. Remember that the purpose of writing an algorithm is to make writing of the actual source code easier. There are no strict rules in writing algorithms. As you do more programming, you will find that you will develop your own style. Any style that facilitates writing the actual source code is acceptable.

General algorithm with no equations. This algorithm is based on the specific example.

```
read annual interest rate
open output file
initialize year

First month:
initialize principal at beginning of first month
calculate principal at end of first month
print year, month and principal

Second month:
set principal at beginning of second month = principal at end of first month
calculate principal at end of second month
print year, month and principal

Third month:
set principal at beginning of third month = principal at end of second month
calculate principal at end of third month
print year, month and principal

Fourth month:
set principal at beginning of fourth month = principal at end of third month
calculate principal at end of fourth month
print year, month and principal

etc. (continue for a total of 5 years)
```

Algorithm with equations. By writing an algorithm with equations, your algorithm more closely imitates the actual source code, and stepping from the algorithm to the source code becomes easier. This algorithm is

```
read annual interest rate, i
open output file

year = 1

Po = 5000
Pf = Po ( 1 + i/12 )
print year, month, Pf

Po = Pf
Pf = Po ( 1 + i/12 )
print year, month, Pf

Po = Pf
Pf = Po ( 1 + i/12 )
print year, month, Pf
```

```
Po = Pf
Pf = Po ( 1 + i/12 )
print year, month, Pf

Po = Pf
Pf = Po ( 1 + i/12 )
print year, month, pf


etc. (for 5 years)
```

Algorithm with equations and loops. You can use loops in your algorithm also. Trace the flow of the following algorithm to assure yourself that, indeed, it performs the same functions as the algorithm without a loop.

```
read annual interest rate, i
open output file


po = 5000
year = 1
loop on month = 1 to 12
      Pf = Po ( 1 + i/12 )
      print year, month, Pf
      Po = Pf


year = 2
loop on month = 1 to 12
      Pf = Po ( 1 + i/12 )
      print year, month, Pf
      Po = Pf


year = 3
loop on month = 1 to 12
      Pf = Po ( 1 + i/12 )
      print year, month, Pf
      Po = Pf


year = 4
loop on month = 1 to 12
      Pf = Po ( 1 + i/12 )
      print year, month, Pf
      Po = Pf


year = 5
loop on month = 1 to 12
      Pf = Po ( 1 + i/12 )
      print year, month, Pf
      Po = Pf
```

Algorithm with equations and nested loop. We can see the previous algorithm has repetition due to the calculations being done over five years. This means that this calculation also can be put into a loop, which results in a nested loop as follows:

```
read annual interest rate, i
open output file

Po = 5000
loop on year = 1 to 5
   loop on month = 1 to 12
            Pf = Po( 1 + i/12 )
            print year, month, Pf
            Po = Pf
```

Algorithm with equations, nested loop, and efficient use of variables. We can simplify the algorithm even further, because we know that assignment statements work differently from algebraic equations. For instance, we can use the statement:

$$P = P * (1 + i/12)$$

As described earlier, a statement of this type is a valid assignment statement but not a correct algebraic expression (unless $i = 0$). Recall that an assignment statement of this type accomplishes the following:

$$P_{(\text{new value})} = P_{(\text{old value})} \times (1 + i/12)$$

This is exactly the operation that we want to accomplish when we write:

```
Pf = Po * ( 1 + i/12 )
```

Therefore, we can use the assignment statement

```
P = P * ( 1 + i/12 );
```

and eliminate the need to use the assignment statement

```
Po = Pf;
```

Recall also that the statement

```
P *= ( 1 + i/12 );
```

accomplishes the same task as

```
P = P * ( 1 + i/12 );
```

Consider the following algorithm and follow the flow of the program, paying particular attention to the value of P:

```
read annual interest rate, i
open output file
P = 5000
```

```
loop on year = 1 to 5
    loop on month = 1 to 12
                P *= ( 1 + i/12 )
                print year, month, P
```

To check the loop, fill in the following table for a total of 15 months, using your calculator and looking only at the preceding algorithm to guide you. Do you feel confident that this works?

Year	Month	P

SOURCE CODE

This source code can be written using the preceding algorithm as a guide. Figure 4.21 shows the flow of the program.

```c
#include <stdio.h>
void main(void)
{
    int     month, year;
    double i, p;
    FILE *outpt;

    outpt = fopen ("MONEY.OUT", "wt");
    printf ("Enter the annual interest rate\n");

    scanf  ("%lf",&i);

    p = 5000.;

    for ( year = 1;year<=5;year++ )
        {
        for ( month = 1;month<=12;month++ )
            {
            p *= ( 1 + i/12 );
            fprintf (outpt, "\n Year    = %d "
                "\n Month   = %d "
                "\n Balance = %7.2lf \n\n",year,month,p);
            }
        }
    fclose(outpt);
}
```

Nested loop that corresponds to algorithm.

Output to screen

```
                Enter the annual interest rate
```
Keyboard input   `0.06`

Output file MONEY.OUT

```
Year     = 1
Month    = 1
Balance = 5025.00

Year     = 1
Month    = 2
Balance = 5050.12

Year     = 1
Month    = 3
Balance = 5075.38

Year     = 1
Month    = 4
Balance = 5100.75

Year     = 1
Month    = 5
Balance = 5126.26
```

and so forth for five years.

COMMENTS

This program was written to illustrate the use of nested for loops, and less attention was paid to creating neat, efficient-looking output. How would you change the program if you were going to make this the output table:

```
Five-year monthly bank balance
Annual Interest rate = 6.0%
   Year        Month        Balance
    1            1           5025.00
                 2           5050.13
                 3           5075.38
                 4           5100.75
                 5           5126.26

     etc. for 5 years
```

Modification exercises

Modify the program to

1. Compound the interest not once per month but once every two weeks.

2. Compound the interest once per week.

3. Carry out the calculation for ten years.

4. Begin with a principal of $10,000, an interest rate of 5 percent, and interest compounded once per week for eight years. Print the result only every six months.

5. Create the output table shown in the preceding Comments.

FIG. 4.21
Flow of Application Program 4.5

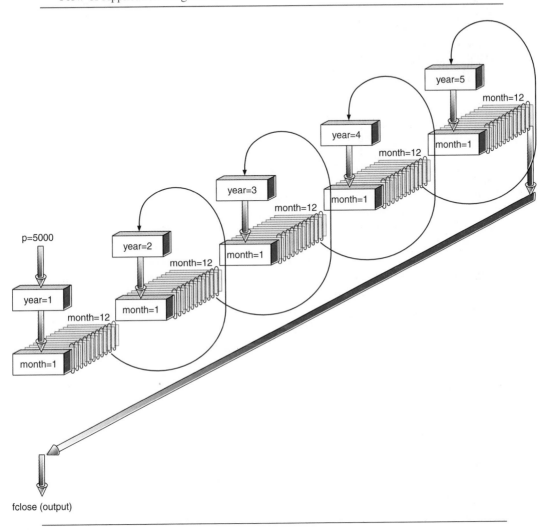

APPLICATION PROGRAM 4.6 MATHEMATICAL SERIES— FOR LOOP, DO-WHILE LOOP, IF-ELSE CONTROL STRUCTURE (NUMERICAL METHOD EXAMPLE)

Problem statement

Write a program capable of calculating $\tanh^{-1}(x)$ using an infinite series. Read the value of x from the keyboard and print the result to the screen.

Solution

RELEVANT EQUATIONS

As you solve more advanced problems in mathematics, science, engineering, or business, you may find you need to write programs capable of solving advanced equations, such as those that involve series representations. You often will find that a for loop or a while loop is ideally suited to simulating the series.

For instance the $\tanh^{-1}(x)$ function can be represented by the infinite series:

$$\tanh^{-1}(x) = x + \frac{x^3}{3} + \frac{x^5}{5} + \frac{x^7}{7} + \dots \qquad (4.10)$$

where $|x| < 1.0$. This infinite series can be written in summation form as follows:

$$\tanh^{-1}(x) = \sum_{n=1}^{n=\infty} \frac{x^{(2n-1)}}{(2n-1)} \qquad (4.11)$$

Recall that one characteristic of some (but not all) converging infinite series is that each successive term of the series decreases in value. In other words, the first term of the series is the largest. The second term of the series is smaller than the first but larger than the third. In general, the 50th term in this type of series is considerably smaller than the first term (this, of course, depends on the behavior of the particular series considered). The impact of this is that, if you were going to calculate a value from the series representation, you could take a finite number of terms (for instance, the first 50) and ignore the rest yet get a reliable value for use in practical applications. A question that arises, though, is, "How many terms are necessary to get a reliable value?"

The answer is, "When the values of the terms get small enough, they can be ignored." For instance, if you are interested in calculating the value of a number represented by a series to six decimal places, then when the value of the sum of the remaining terms becomes less than 0.000001, the remaining terms can be ignored. Knowledge of the convergence of the particular series to be calculated is necessary to guarantee a reliable result. However, from a practical standpoint, it may be possible to terminate a series when a particular term becomes substantially smaller than the desired accuracy. For example, if you are interested in calculating the value of a number represented by a series accurately to six decimal places and the series converges in the manner just described, then it might be possible to say that, when the value of a single term becomes smaller than 0.0000000000001, it is likely that the value of the number has reached the desired accuracy. (Please consult a mathematics text for further understanding of the convergence of infinite series.)

With the material in this chapter, you learned enough to write a program that can calculate a large number of terms in a series and decide whether the terms are small enough to be ignored.

SPECIFIC CALCULATION

Suppose we wanted to find $\tanh^{-1}(0.5)$ using the series representation in equation (4.10). Then,

$$x = 0.5$$

and equation (4.11) becomes

$$\tanh^{-1}(0.5) = \sum_{n=1}^{n=\infty} \frac{0.5^{(2n-1)}}{(2n-1)}$$

The first 14 terms of the series are

```
Term number                                     Term value
1             (0.5)¹   /( 1)                     = 0.500000000
2             (0.5)³   /( 3)                     = 0.041666667
3             (0.5)⁵   /( 5)                     = 0.006250000
4             (0.5)⁷   /( 7)                     = 0.001116071
5             (0.5)⁹   /( 9)                     = 0.000217014
6             (0.5)¹¹  /(11)                     = 0.000044389
7             (0.5)¹³  /(13)                     = 0.000009390
8             (0.5)¹⁵  /(15)                     = 0.000002035
9             (0.5)¹⁷  /(17)                     = 0.000000449
10            (0.5)¹⁹  /(19)                     = 0.000000100
11            (0.5)²¹  /(21)                     = 0.000000023
12            (0.5)²³  /(23)                     = 0.000000005
13            (0.5)²⁵  /(25)                     = 0.000000001
14            (0.5)²⁷  /(27)                     = 0.0000000003
                                      Σ         = 0.5493061443
```

If you have the $\tanh^{-1}$ function on your calculator, you can check this value. Unfortunately, the C compiler does not have this function built in (like it has sin, cos, tan, etc.). So, if you need to use the $\tanh^{-1}$ function in a program, you need to calculate from the series as just shown.

ALGORITHM

Clearly, an infinite number of terms cannot be used in the series. Generally, we have two choices:

1. Set a fixed number of terms to be used in the series.
2. Set a criteria by which the loop can be terminated.

If we set a fixed number of terms, we must make sure that we use a sufficiently large number to assure that we have a reliable result no matter what value of x is used.

If we set a criteria for terminating the loop (for instance, the value of a term being less than a particular value), then we must make sure that the criteria is sufficiently stringent (such as making the value that the term is checked against very small).

In either case, it is better to err on the side of selecting too many terms or making the check value too small. This may mean that the program is not efficient. However, reliability is more important than efficiency. The following algorithm is based on selecting a finite number of terms for the series (in this case, 500):

```
prompt user to input value of x
read value of x

for_loop with a counter going from 1 to 500
          calculate individual term in series
          add the term to the rest of the series
print result to the screen
```

This next algorithm is based on checking the absolute value of each term against a particular small value (in this case, 0.0000000001):

```
prompt the user to input a value of x
read value of x

do_while_loop (exit the loop when an individual term absolute
          value is less than 0.0000000001)
          calculate individual term in series
          add the term to the rest of the series
print result to the screen
```

Both algorithms suffer deficiencies. The first one may be inefficient, because it executes the loop 500 times no matter how quickly the series converges. The second one could execute the loop an infinite number of times if the series does not converge. In addition, neither algorithm checks to make sure that the value of x input by the user is between -1 and 1. An algorithm that utilizes a for loop and corrects the deficiencies follows:

```
prompt user to input value of x
read value of x

check to make sure that x is between -1 and 1
write message to the screen if x is not between -1 and 1
stop execution if x is not between -1 and 1
continue execution if x is between -1 and 1

for_loop with a counter going from 1 to 500
          calculate individual term in series
          add the term to the rest of the series
          exit loop if individual term absolute value is less than
          0.0000000001
print result to the screen
```

An algorithm that utilizes a do-while loop and corrects the deficiencies is listed follows:

```
prompt user to input value of x
read value of x

check to make sure that x is between -1 and 1
write message to the screen if x is not between -1 and 1
stop execution if x is not between -1 and 1
continue execution if x is between -1 and 1

do_while_loop (exit the loop when an individual term absolute value
          is less than 0.0000000001 or when the value of the
          counter is greater than 500)
          calculate individual term in series
          add the term to the rest of the series
          increment the counter

print result to the screen
```

SOURCE CODE

Source codes using both a for loop and a do-while loop follow. They are based on the last two algorithms and, therefore, check for the value of x between -1 and 1 and make sure that an infinite loop is not produced. The flow of these programs is shown in Fig. 4.22.

SOURCE CODE WITH FOR LOOP

```c
#include <stdlib.h>
#include <stdio.h>
#include <math.h>
void main(void)
{
    int    n;
    double  x, single_term, series_sum ;

    printf ("Input value of x (where -1< x <1 ) \n");
    scanf   ("%lf",&x);
    if (x<-1.   ||   x > 1.)
        {
        printf ("\n"
                "You have entered an invalid value of x.\n"
                "Program execution has halted\n");
        }
    else

        {
        series_sum = 0.0 ;
        for (n=1;n<=500; n++)
            {
            single_term = (pow(x,(2.*n-1.)))/(2.*n-1.);
            series_sum += single_term;
            if (fabs(single_term)< 0.0000000001) break;
            }
            printf ("x = %f       tanh-1(x) = %11.9lf",x,series_sum);
        }
}
```

If x is not between -1 and 1, no more calculations are performed.

The variable used to hold the sum (series_sum) must be initialized to 0 before the loop is begun.

The loop will execute 500 times unless it is exited early.

The loop will execute 500 times unless it is exited early.

The single_term is calculated from the counter, n.

The sum of the series terms is calculated using the compound assignment operator +=.

The loop is exited when the single_term is very small.

SOURCE CODE WITH DO-WHILE LOOP

```c
#include <stdlib.h>
#include <stdio.h>
#include <math.h>
void main(void)
{int     n;
 double  x, single_term, series_sum;
```

```
printf ("Input value of x (where -1<  x < 1 ) \n");
scanf  ("%lf",&x);

if (x<-1.  ||  x > 1.)
  {
   printf ("\n"
            "You have entered an invalid value of x.\n"
            "Program execution has halted\n");
  }
else
  {
     n=1;
     series_sum = 0.0;

     do
       {
          single_term = (pow(x,(2.*n-1.)))/(2.*n-1.);
          series_sum += single_term;
          n++;
       }
     while
       (fabs (single_term) > 0.0000000001  && n <=500 );
     printf ("x = %f       tanh-1(x) = %11.91f",x,series_sum);
  }
}
```

The conditional in this do-while loop checks both the single-term value and the number of iterations.

OUTPUT

Keyboard input
```
       Input value of x (where -1< x < 1 )
       0.5
       x = 0.500000
       tanh-1(x) = 0.549306144
```

COMMENTS

Either of these source codes is acceptable; however, the do-while loop is more structured because it has no break statement. We can also write the for loop with no break statement if we use a compound relational expression with a logical operator as a test expression as we did with the do-while loop. For instance, the for loop

```
single_term=1.0;
for (n=1; n<=500 &&(fabs(single_term)>0.0000000001); n++)
    {
    single_term=(pow(x,(2.*n-1.)))/(2.*n-1.);
    series_sum+=single_term;
    }
```

accomplishes the same task as the for loop shown in the original source code. Here, we do not need the break statement because we have used a more complex test expression. Notice that we have arbitrarily initialized `single_term` to be 1.0. This has no effect on the calculation of `series_sum`. It simply assures us of initially entering the for loop because it makes `fabs(single_term)` greater than 0.0000000001.

FIG. 4.22
Flow of Application Program 4.6

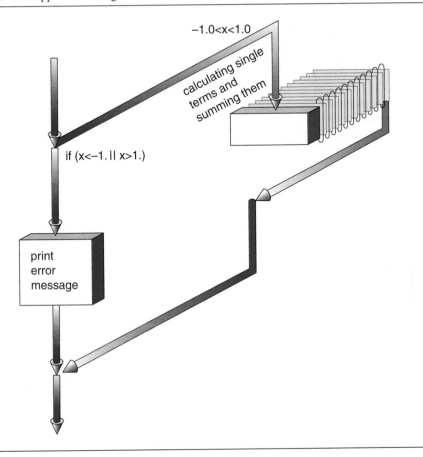

$-1.0 < x < 1.0$

calculating single terms and summing them

if (x<−1. || x>1.)

print
error
message

From this application example, you should observe how easy it is to develop a loop from an equation written in summation form (equation 4.11). In your assignments, if you express your equations in summation form whenever possible, you will find them very easy to program. Also, the include directives

```
#include<math.h>
#include<stdlib.h>
```

are needed for this program because the pow() and fabs() functions are used.

If you use your calculator to find the value of $\tanh^{-1}(0.50)$, you will see that it agrees with the value listed in the output. However, if you were to use float instead of double for the type of real variables in this program, the result would not agree with your calculator. The comparison follows for both 0.50 and 0.35:

x	$\tanh^{-1}(x)$ (from program using double)	$\tanh^{-1}(x)$ (from program using float instead of double)
0.50	0.549306144	0.549306273
0.35	0.365443754	0.365443736

This may seem like a minor difference, but because many calculations may be done with this number if it were to be used in other programs, this difference could have a significant effect on output results. Remember, use double or long double in your programs, not float.

Modification exercises

Modify the program to

1. Use float instead of double. Use the input values 0.1, 0.3, 0.7, and 0.9. Compare the printed results with results from your calculator. Are the program results satisfactory?

2. Calculate the natural log of 2 using the series $\ln 2 = 1 - 1/2 + 1/3 - 1/4 + 1/5 \ldots$ to eight significant digits.

3. Calculate $\pi^2/6 = 1/1^2 + 1/2^2 + 1/3^2 + 1/4^2 + \ldots$ Then calculate π to five significant digits.

4. Calculate $0.5 = [1/(1 * 3)] + [1/(3 * 5)] + [1/(5 * 7)] + [1/(7 * 9)] + [1/(9 * 11)] + \ldots$ to seven significant digits.

5. Calculate $0.75 = [1/(1 * 3)] + [1/(2 * 4)] + [1/(3 * 5)] + [1/(4 * 6)] + [1/(5 * 7)] + \ldots$ to seven significant digits.

6. Calculate $\pi^2/12 = 1/1^2 - 1/2^2 + 1/3^2 - 1/4^2 + \ldots$ Then calculate π to six significant digits.

APPLICATION PROGRAM 4.7 SOLVING A QUADRATIC EQUATION—IF-ELSE CONTROL STRUCTURE (NUMERICAL METHOD EXAMPLE)

Problem statement

Write a computer program capable of solving the quadratic equation

$$ax^2 + bx + c = 0$$

The input data is to consist of the values of a, b, and c and is to come from the keyboard. The output is to consist of the values of x and go to the screen.

Solution

RELEVANT EQUATIONS

The quadratic equation has two solutions:

$$x_1 = \frac{-b + \sqrt{b^2 - 4ac}}{2a} \qquad (4.12)$$

$$x_2 = \frac{-b - \sqrt{b^2 - 4ac}}{2a} \qquad (4.13)$$

Having been assigned to write a computer program, it is up to you do a thorough and correct job. Consider all the possibilities. In the case of the quadratic equation, no real solution may exist and your computer program must account for this possibility. If $b^2 - 4ac$ is positive, then equations (4.12) and (4.13) can be used directly to find the solutions x_1 and x_2. However, if $b^2 - 4ac$ is negative, the solutions become

$$x_1 = -\frac{b}{2a} + \frac{\sqrt{-(b^2 - 4ac)}}{2a} i \qquad (4.14)$$

$$x_2 = -\frac{b}{2a} - \frac{\sqrt{-(b^2 - 4ac)}}{2a} i \qquad (4.15)$$

where $i = \sqrt{-1}$.

SPECIFIC EXAMPLE

Consider the following equation:

$$2x^2 + 8x + 3 = 0$$

For this case:

$$a = 2$$
$$b = 8$$
$$c = 3$$

and $b^2 - 4ac = 40$, which is positive. The two solutions are

$$x_1 = \frac{-8 + \sqrt{8^2 - 4(2)(3)}}{2(2)} = -0.41886$$

$$x_2 = \frac{-8 - \sqrt{8^2 - 4(2)(3)}}{2(2)} = -3.58114$$

Consider also, the equation:

$$15x^2 - 2x + 3 = 0$$

For this case

$$a = 15$$
$$b = -2$$
$$c = 3$$

and $b^2 - 4ac = -176$, which is negative. The two solutions from equations (4.14) and (4.15) are

$$x_1 = \frac{-(-2)}{2(15)} + \frac{\sqrt{-((-2)^2 - 4(15)(3))}}{2(15)} i = -0.06667 + 0.44222i$$

$$x_2 = \frac{-(-2)}{2(15)} - \frac{\sqrt{-((-2)^2 - 4(15)(3))}}{2(15)} i = -0.06667 - 0.44222i$$

Just like your calculator, the computer indicates an error and stops executing when it tries to take the square root of a negative number. For your program to execute properly, it should reverse the negative number to the positive one and find the imaginary part, as indicated in equations (4.14) and (4.15).

ALGORITHM

Equations (4.12) through (4.15) have been written such that only a single variable appears on the left-hand side of the equations. This is a form that is useful because it fits the form of assignment statements in C code. As you write equations for programs, you should get your equations into this form so that you can easily write the source code.

The algorithm (including equations) and a check for taking the square roots of negative numbers is given below.

read the values of a, b, and c from the keyboard
compute the value of b²-4ac

if b²-4ac is positive then

$$x_1 = -\frac{b}{2a} + \frac{\sqrt{-(b^2 - 4ac)}}{2a}$$

and

$$x_2 = -\frac{b}{2a} - \frac{\sqrt{-(b^2 - 4ac)}}{2a}$$

print x_1 and x_2 to the screen.

if b²-4ac is negative then

$$x_1 = -\frac{b}{2a} + \frac{\sqrt{-(b^2 - 4ac)}}{2a} i$$

and

$$x_2 = -\frac{b}{2a} - \frac{\sqrt{-(b^2 - 4ac)}}{2a} i$$

print x_1 and x_2 to the screen.

SOURCE CODE

This source code below has been written from the preceding algorithm. The program flow is shown in Fig. 4.23.

```c
#include <math.h>
#include <stdio.h>
void main(void)
{
    double i, a, b, c, x1, x2, test, real, imag;

    printf("Enter the values of a, b, and c (each separated\n"
           "by a space) then press return\n\n");

    scanf("%lf %lf %lf", &a, &b, &c);

    test=b*b - 4*a*c;       Calculating the test value.

    if (test>=0)            Calculating the two values of x if
                            the test value is greater than 0.
    {
        x1 = (-b + sqrt(test))/(2*a);
        x2 = (-b - sqrt(test))/(2*a);
        printf(" Real result:\n x1=%10.5f\n x2=%10.5f\n\n",x1,x2);
    }

    else                    Calculating real and imaginary values if
                            test is less than 0.

    {
        real = -b/(2*a);
        imag = sqrt (-test) /(2*a);
        printf(" Imaginary result:\n"
               " x1=%10.5f %+10.5f i\n x2=%10.5f %+10.5f i\n",
               real, imag, real, -imag);
    }
}
```

OUTPUT

```
                    Enter the values of a, b, and c (each separated
                    by a space) then press return
Keyboard input      15 -2 3
                    Imaginary result:
                    x1=   0.06667 +   0.44222 i
                    x2=   0.06667 -   0.44222 i
```

COMMENTS

One can see that the quadratic equation itself never appears in the source code, only the *solution* to the quadratic equation. In general, you will find that you will need to solve your equation or equations before you can begin writing your algorithm or source code. This is considered part of the programming process and is integral in developing a reliable efficient program. If you solve your equations incorrectly, then your program will give incorrect results even though it is capable of executing without terminating abnormally.

FIG. 4.23
Flow of Application Program 4.7

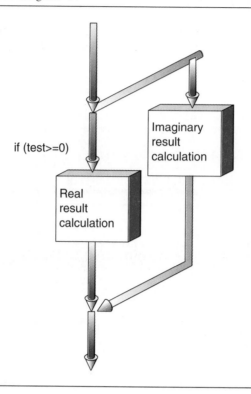

Your program also must be able to handle all possibilities. In Application Program 4.6, it was necessary to handle the case of a user accidentally entering a value of x less than -1 or greater than 1. In this program it was necessary to handle cases where the result is imaginary. Your responsibility as a programmer is to imagine all the possibilities and write a program to handle them.

Note that the variable test was used in the source code. This variable was used only for convenience and to simplify the look of the program. It was not necessary for this variable to be used. However, we recommend that you use variables for convenience and to simplify the look of your programs.

Note also that the include directive

#include <math.h>

is necessary for this program because the function sqrt is used.

Modification exercises

Modify the program to

1. Handle the input of five different equations. Have the user type in five lines when prompted. Each line should contain three coefficients.

2. Read the input from a data file.

3. Read five equations from a data file and print the results to a data file.

APPLICATION PROGRAM 4.8 ENGINEERING ECONOMICS, COMPUTING TAX—SWITCH AND IF-ELSE-IF CONTROL STRUCTURES, COMPOUND RELATIONAL EXPRESSIONS

Problem statement

If one day you run your own engineering consulting firm, you will find that getting your taxes correct is an important part of doing business. Write a program capable of computing the amount of tax you owe given the following tax table:

Tax bracket	Income	Marginal tax rate
1	0–$10,000	5%
2	$10,001–20,000	10%
3	$20,001–30,000	15%
4	$30,001–50,000	20%
5	$50,001–100,000	25%
6	greater than $100,000	30%

Input should be from the keyboard and consist of just one number, your income. Output should be to the screen and consist of your tax.

Solution

RELEVANT EQUATIONS

The total tax can be computed from the marginal tax rate by adding the tax from all of the lower tax brackets to the tax for the tax bracket that you are in. In general, the tax for each tax bracket as indicated from the table is

Tax bracket	Tax
1	$0.05\,(i)$
2	$t_1 + 0.1(i-10000)$
3	$t_1 + t_2 + 0.15(i-20000)$
4	$t_1 + t_2 + t_3 + 0.20(i-30000)$
5	$t_1 + t_2 + t_3 + t_4 + 025(i-50000)$
6	$t_1 + t_2 + t_3 + t_4 + t_5 + 0.30(i-100000)$

where $\qquad t_1$ = tax for bracket 1 = 0.05(10000) = 500
$\qquad\qquad\qquad t_2$ = tax for bracket 2 = 0.10(10000) = 1000
$\qquad\qquad\qquad t_3$ = tax for bracket 3 = 0.15(10000) = 1500
$\qquad\qquad\qquad t_4$ = tax for bracket 4 = 0.20(20000) = 4000
$\qquad\qquad\qquad t_5$ = tax for bracket 5 = 0.25(50000) = 12500
$\qquad\qquad\qquad i$ = your income

Therefore, the tax table becomes

Tax bracket	Tax
1	0.05 (i)
2	500 + 0.10(i − 10000)
3	1500 + 0.15(i − 20000)
4	3000 + 0.20(i − 30000)
5	7000 + 0.25(i − 50000)
6	19500 + 0.30(i − 100000)

SPECIFIC EXAMPLE

Determine the total tax for an income of $63,000. For this case, the tax bracket is 5; therefore, the total tax is

$$t = 7000 + 0.25(63{,}000 - 50{,}000)$$
$$t = \$10{,}250$$

ALGORITHM

In this case, decision making is involved in the specific example. The decision regards the tax bracket into which the income falls. The algorithm is

```
Print question to the screen asking for the income
Read keyboard input of income
Determine the tax bracket
Calculate the tax for that tax bracket
Print the tax to the screen
```

SOURCE CODE

We have written the source code for this example twice. The first time we illustrate using if statements with compound relational expressions and a switch statement. The second time we illustrate using if-else-if control structures. Read the codes. Compare and contrast them.

Source code using if statements with compound relational expressions and a switch statement. The source code has been written from the algorithm. Both if statements and a switch statement have been used. Examine the program and follow the flow.

```
#include <stdio.h>
void main(void)
{
  int tax_bracket;
  double i,income,tax;

  printf ("Enter your income:\n");
  scanf  ("%lf", &income);

  if (    0  <=income && income <= 10000)  tax_bracket = 1;
  if (10000 < income && income <= 20000)  tax_bracket = 2;
  if (20000 < income && income <= 30000)  tax_bracket = 3;
  if (30000 < income && income <= 50000)  tax_bracket = 4;
  if (50000 < income && income <=100000)  tax_bracket = 5;
  if (100000< income)                     tax_bracket = 6;

  i=income;

  switch (tax_bracket)
    {
    case 1: tax = 0.05 * i;
            break;
    case 2: tax = 500   + 0.10*(i-10000);
            break ;
    case 3: tax = 1500  + 0.15*(i-20000);
            break;
    case 4: tax = 3000  + 0.20*(i-30000);
            break;
    case 5: tax = 7000  + 0.25*(i-50000);
            break;
    case 6: tax = 19500 + 0.30*(i-100000);
            break;
    default:break;
    }

  printf ("\nYour tax bracket is: %2d \n"
          "Your tax is $ %-10.2lf\n", tax_bracket, tax);
}
```

Series of if statements to define the tax bracket.

The switch structure is used to determine the tax.

Source code with if-else-if control structure. (See Fig. 4.24 for program flow.)

```
#include <stdio.h>
void main(void)
{
  int tax_bracket;
  double i,income,tax;

  printf ("Enter your income:\n");
  scanf  ("%lf", &income);

  i=income;
```

```
if ( i <= 10000)
        {
        tax_bracket = 1;
        tax = 0.05 * i;
        }
```

> The if-else-if structure is used to determine both the tax bracket and the tax.

```
    else if (10000 < i && i <= 20000)
            {
            tax_bracket = 2;
            tax = 500 + 0.10 * (i-10000);
            }
    else if (20000 < i && i <= 30000)
            {
            tax_bracket = 3;
            tax = 1500 + 0.15 * (i-20000);
            }
```

> The form of this conditional represents the mathematical $20{,}000 < i \le 30{,}000$.

```
    else if (30000 < i && i <= 50000)
            {
            tax_bracket = 4;
            tax = 3000 + 0.20 * (i-30000);
            }

    else if (50000 < i && i <= 100000)
            {
            tax_bracket = 5;
            tax = 7000 + 0.25 * (i-50000);
            }

    else if (100000 < i)
            {
            tax_bracket = 6;
            tax = 19500 + 0.30 * (i-100000);
            }


    printf ("\nYour tax bracket is: %2d \n"
            "Your tax is $ %-10.21f\n", tax_bracket, tax);
}
```

OUTPUT

Keyboard input	Enter your income: 63000 Your tax bracket is: 5 Your tax is $10250.00

COMMENTS

The first source code is somewhat less efficient than the second because it involves more comparisons in the decision-making process. One way to assess the efficiencies of algorithms and code is to look at the number of comparisions. An algorithm that performs the same task with fewer comparisons often is more efficient.

FIG. 4.24
Flow of Application Program 4.8

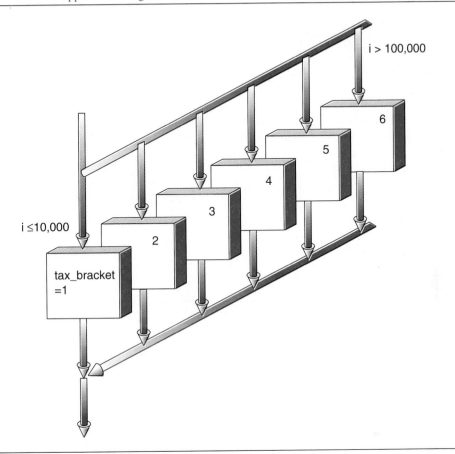

Note that the compound relational expressions within the if statements in this program have been written with a form that corresponds to the typical mathematical form. For instance, to express that the variable i is between 10,000 and 20,000, the mathematical form is

$$10{,}000 < i < 20{,}000$$

and an equivalent C relational expression is

```
(10000 < i && i < 20000)
```

This is not the only form that will work in C; however, we recommend this form because it corresponds to the mathematical form.

Modification exercises

Modify the programs to

1. Put a cap on the total amount of tax to be paid at $20,000. Run the program with an income of $300,000 to check your result.

2. Have the program accept three incomes from the keyboard and print the tax for each.

3. Have the program read ten incomes from a file and print the results to the screen for all ten incomes in the form of a table with two headings: Income and Tax.

4. Change the tax brackets to be:

   0–$5000
   $5000–30,000
   $30,000–60,000
   $60,000–100,000
   $100,000–300,000
   >$300,000

   Use the same percentages for the tax brackets as given in the program. How much less tax is paid on an income of $100,000 with these brackets compared to the ones in the program?

APPLICATION EXERCISES

4.1. Rewrite Application Exercise 3.1 using a for loop.

4.2. Rewrite Application Exercise 3.1 using a while loop.

4.3. Rewrite Application Exercise 3.1 using a do-while loop.

4.4. Rewrite Application Exercise 3.5 using for loops.

4.5. Rewrite Application Exercise 3.5 using while loops.

4.6. Rewrite Application Exercise 3.5 using do-while loops.

4.7. Rewrite Application Exercise 3.6 using a for loop.

4.8. Rewrite Application Exercise 3.6 using a while loop.

4.9. Rewrite Application Exercise 3.6 using a do-while loop.

4.10. Rewrite Application Exercise 3.8 using a do-while loop.

4.11. Rewrite Application Exercise 3.9 using a while loop.

4.12. Rewrite Application Exercise 3.10 using a for loop.

4.13. The equation of a parabola is

$$y = ax^2 + bx + c$$

Use a for loop and if statements in a program capable of finding the maximum or minimum of four different parabolas. As input, the program is to read the data consisting of a, b, and c for the four parabolas from a file called PARA.DAT, which consists of four lines:

line1	a_1	b_1	c_1	(coefficients for parabola 1)
line2	a_2	b_2	c_2	(coefficients for parabola 2)
line3	a_3	b_3	c_3	(coefficients for parabola 3)
line4	a_4	b_4	c_4	(coefficients for parabola 4)

An example data file is

```
1.1    -2    1
-1.3   1     15
-0.7   15    18.3
1.5    15.5  14.2
```

As output, print your result to file PARA.OUT in the following form:

Parabola number	Equation	(*x*, *y*) coordinates of minimum or maximum
1	$y = 1.1x*x - 2x \quad + 1$	(0.909,0.0909) min
2	$y = -1.3x*x + 1x \quad + 15$	—
3	$y = -0.7x*x + 15x + 18.3$	—
4	$y = 1.5x*x + 15.5x + 14.2$	—

4.14. The binomial theorem can be written as

$$(a + b)^n = a^n + na^{n-1}b + \frac{n(n - 1)}{2!} a^{n-2}b^2 + \frac{n(n - 1)(n - 2)}{3!} a^{n-3}b^3 + \ldots$$

Write a program for which $a = 1$ and $0 < b < 1$ that uses the binomial theorem to calculate $(a + b)^n$ accurately to eight decimal places. Your program also should calculate $(a + b)^n$ using the pow() function. Make sure that your program checks for b being between 0 and 1. (Note: This is a converging series.)

Input data from file BINO.DAT. The first line of the file consists of the number of values, m, of b and n that will be input. Then, m lines of data follow, each consisting of one value of b and one value of n. (Here, b is real, m and n are int.)

BINO.DAT

line 1	m	
line 2	b	n

```
line 3      b    n
line 4      b    n
...         m lines of b and n
...
...
line m+1    b    n
```

An example data file is

```
5
0.5  8
0.2  10
0.33 5
0.08 6
0.45 15
```

Show on the output

```
Binomial theorem and pow() output
a = 1          b = -----      n = -----

(a+b)^n                    (a+b)^n
From the                   From the
pow() function             binomial theorem
_____          _____
      ---                        ---
      ---                        ---
      ---                        ---
      ---                        ---
```

4.15. Write a program capable of using the month and day of a given date to calculate the number of days from January 1 that it represents. Make the program
capable of computing values for as many as 20 dates.

Data should be input from a data file called DAYS.DAT. The data file should
have the following format:

```
line 1      n                (number of dates to be computed)
line 2      month   day
line 3      month   day
line n+1 month      day      (all data are integers)
```

An example data file is

```
5
12   7
8    5
1    27
4    18
7    22
```

Print your results to file DAYS.OUT in the following format:

```
TABLE OF DATES AND DAYS FROM JANUARY 1
DATE                    DAYS FROM JAN. 1

December 7                    ...
August 5                      ...
January 27                    ...
April 18                      ...
July 22                       ...
```

Note that you must display the month in words and not numbers.

4.16. Write a program that computes the sum of all the negative integers in a list of integers.

As input, the program is to read the data from the data file SUMNEG.DAT. The data file has the following format:

line 1	**n**	*(number of integers in list)*
line 2	**int1**	
line 3	**int2**	
line 4	**int3**	
...		
...		
...		

Write the output to the screen in the following manner:

```
The sum of the negative integers is:
......
The list of integers is:
...
...
...
...
```

4.17. The grading structure for a class is the following:

```
90–100  A
80–89   B
70–79   C
60–69   D
< 60    F
```

Write a program that prints the grades for ten different numerical scores.

As input, the program is to read the ten scores (all integers) from the data file GRADE.DAT. The contents are

line 1 **score1 score2 score3 ....score10**

Print the results to file GRADE.OUT in the following table:

Numerical Score	Grade
...	...
...	...
...	...
...	...

4.18. Write a program that is capable of finding the 2 largest and 2 smallest integers in a list of 20 integers.

As input, the program is to read the data from the data file TWOMM.DAT. This data file consists of

```
line 1   int1   int2   int3   .... int10
line 2   int11  int12  int13  .... int20
```

Write the output to the screen in the form:

```
The two largest values in the list are:
   ...          ...
The two largest values in the list are:
   ...          ...
```

4.19. The constant percentage method of computing depreciation of an asset is based on the assumption that the depreciation charge at the end of each year is a fixed percentage of the book value of the asset at the beginning of the year. This assumption leads to the following relationship:

$$S = C (1 - d)^n$$

where

C = original cost of asset
d = depreciation rate per year
n = number of years
S = book value at the end of n years

Write a program to compute the number of years of useful life of an asset given the original cost, depreciation rate, and book value at the end of its useful life (called *scrap value*).

As input, the program is to read the data from the keyboard. Prompt the user to enter the data in the following way:

```
Enter the original cost and depreciation rate:
   ...          ...
Enter the book value at the end of the useful life of an asset:
   ...          ...
```

Print the result to the screen as follows:

```
The useful life of the asset is:    ...years
```

Functions

In Chapter 1, we discussed software engineering and described how functions are the modular components of the design of a C program. If we continue the analogy between building a structure and creating a C program, we will gain further insight into the value of functions in C programming.

For instance, suppose you were given the task of constructing a footbridge across a river. One way to do it would be to find a very large tree that is tall enough to span the river and wide enough to handle foot traffic (Fig. 5.1). You could chop down that tree, cut it so that it is just long enough to reach from one side of the river to the other, and then carve the tree so that it is flat and suitable to walk across. In doing this, you begin to realize that a large tree is difficult to work with because it is heavy and large, so you need very big equipment to maneuver it. Since there is only one tree, only a limited amount of work can be done on it at one time. Any mistake is difficult to correct. For instance, if one portion is gouged out too deeply, then to make it flat, the entire tree would need to be gouged out deeper. Thus, you could create a working bridge with just one tree, but you find that there are some major difficulties with it.

Another way to construct the bridge would be to cut down many small trees and make a number of boards from each of them. Then you could use bolts for connectors and assemble the bridge board by board. Because each tree is lighter and smaller than the one large tree, they are much easier to work with. Also, if a mistake is made on one of the boards, it can be replaced. Many people can work on this project at one time because it is straightforward to assign individual tasks to cutting many trees and creating many boards. The equipment involved is much smaller and therefore more readily available. However, unlike the one-tree bridge design, planning becomes very important in this many-board design. In other words, you must create detailed sketches of the bridge because you need to know how long each board needs to be before it is cut. Also, the connections are critical. It must be very clear which boards

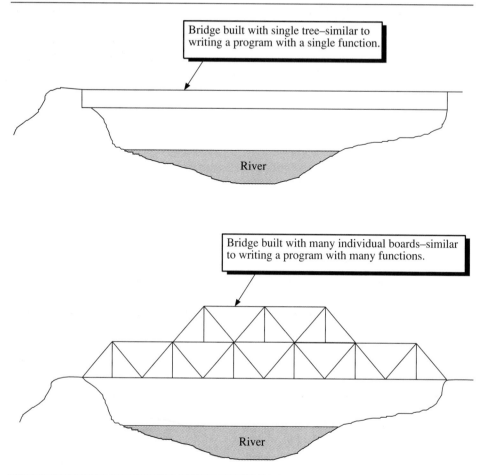

are connected together and how many connections there should be. The types of bolts, long or short, must be decided before they are ordered. In delegating tasks to many people, it is necessary to make sure that each worker understands the task and knows what the connections of each individual board should be. In addition, someone must be clearly in charge of organizing this and making sure that, after each individual board is constructed, everything fits together properly and the bridge created works properly. The result is that, again, a working bridge is created; however, the issues that must be overcome to efficiently create the bridge are quite different from those issues involved in creating a bridge with a single tree. If done correctly, though, the bridge design with many boards is superior because it can be done more quickly and cheaply than the single tree bridge.

The differences between these two methods of building a bridge are similar to the differences between creating a C program with just one function (`main`) and creating a program with many functions. To this point, you have written programs with just one function (other than library functions). In this chapter, we show you how to write functions other than `main`. In doing so, many new issues become important in your programming. First, planning becomes even more important. It is necessary to create a structure chart and data flow diagram and lay out exactly how the functions are to be connected together (meaning what information will be transferred into and out of the functions). As you learn how to write programs with many functions, you probably will become frustrated and want to write your program with no functions. Do not succumb to this temptation. Write your programs with functions, even if they are not completely necessary, so that you can learn how to overcome the issues raised by using functions.

Please read Table 5.1 to gain an understanding of the difference between writing a program with just one function and writing a program with many functions. In this table we continue the comparison between building a bridge and writing a program. After having read the table, you should understand why you need to learn how to write programs with functions.

In this text, we illustrate programs with functions in many cases. However, because sometimes we want to focus on issues different from those that are raised by using functions, we do not use functions in all the programs in which we could have. You will see that using functions sometimes adds length to your programs. This is fine in practice; however, here we have publishing constraints and have tried to keep the programs as short as possible.

To describe the details of writing and using functions, we broke them down into the types illustrated in Fig. 5.2. In this text, we already examined some library functions including printf, scanf, pow, and log. In this chapter, we focus on functions that are defined by you, the programmer.

■ LESSON 5.1 FUNCTIONS THAT DO NOT RETURN A VALUE

Topics
- Function prototypes
- Function calls
- Function definitions
- Transferring values to functions
- Types of functions

There are some similarities between functions and variables in C:

1. Both function names and variable names are considered to be identifiers; therefore, their names must conform to the rules for identifiers.
2. Functions (like variables) have types associated with them (such as int or double).
3. Like variables, functions and their type must be declared prior to their use in a program.

▨ **TABLE 5.1**
   Building bridges and programs

Issue	Bridge constructed with single tree	Program constructed with single function (main)	Bridge constructed with many boards	Program constructed with many functions
Connections	Because it is one piece, connections are not an issue.	Because it is a single function, transferring information between one function and another (that is, the connections between functions) is not an issue.	Connections are critical because, after the boards have been cut and the holes for the bolts have been drilled, they all must fit together properly.	The functions must be connected to each other. The information passed between functions (the connections) is critical. The information must be passed and received between functions correctly.
Number of workers	During cutting of the tree there can be only very few.	Only a few can work on a program constructed with a single function because a change in one portion of the program could necessitate change in another portion of the program.	Many workers can cut many trees and prepare boards.	Different individuals can work on many different functions independently, enabling more rapid creation of a program.
Planning	This simple design reduces the planning needed.	Even a program with a single function requires planning. However, without functions, the links between the functions need not be planned, and this reduces some of the planning.	The bridge must be thoroughly planned in advance. Drawings must be prepared. A time schedule is necessary to coordinate all of the materials and activities. All the parts must fit into the whole, and this can be done only with considerable planning.	The program must be thoroughly planned. Structure charts and data flow diagrams are needed. A time schedule is necessary to coordinate completion of functions on which other functions depend. All the functions must fit to create a whole program.
Equipment needs	Major heavy equipment is needed to handle a large tree. This equipment may not be readily available or reasonably priced.	For large programs, a very fast computer may be necessary, because it may take considerable time to compile the program each time it is modified unless a fast computer is used. A very fast computer such as a mainframe or supercomputer may not be readily available or reasonably priced.	The necessary equipment is much smaller than for the single tree design, because each individual tree is much smaller than the one large tree. However, greater numbers of pieces of equipment are needed. This equipment, though, is widely available and reasonably priced.	Even for very large programs, much of the work may be done on small computers because independently each individual function is small and easy to work with. Small computers are widely available and reasonably priced.

TABLE 5.1
Building bridges and programs

Issue	Bridge constructed with single tree	Program constructed with single function (main)	Bridge constructed with many boards	Program constructed with many functions
Supervision and organization	Supervision and organization are minimal because it is difficult to have multiple kinds of tasks done simultaneously.	Supervision and organization are minimal because few people are working on it.	A considerable amount of supervision and organization is necessary to coordinate the multiple kinds of tasks and various people involved.	A considerable amount of supervision and organization is necessary to coordinate all the different programmers.
Errors	To correct an error such as carving out too much in one location may mean having to recarve the entire tree.	To correct an error in one location may mean changing many different parts of the program.	An error in one board may mean simply replacing that board. Errors can be made affecting the entire bridge but in most cases they can be localized.	An error in one function may require correcting the error in just that one function. Errors can be made that cause the entire program to be modified; however, by using many functions, we hope to limit the scope of the errors.
Reusability of parts	There is only one bridge. It cannot be reused unless another location has exactly the same length and width requirements.	While it may be possible to copy some portions of the program and use it for other programs in most cases, modifications will be necessary before portions can be used for other programs.	The individual boards can be taken and used elsewhere in many different kinds of structures, not just bridges.	Individual functions can be copied and used in other programs that require similar tasks. For instance, a function that takes the log of a number can be reused in any program that requires the log of a number to be found.

The program for this lesson has two functions, function1 and function2, which are known as *void type functions*. Can you see which two lines in the program declare the functions to be type void? The locations of these declarations are important. Where are they located relative to main?

Recall that main is a function also. In previous lessons we learned that within main we can call other functions (like library functions printf, sin, and pow). To call those functions we used the function name followed by a left parenthesis then arguments and a right parenthesis. Look in the body of main here. Which two lines call function1 and function2?

FIG. 5.2
Categories of functions

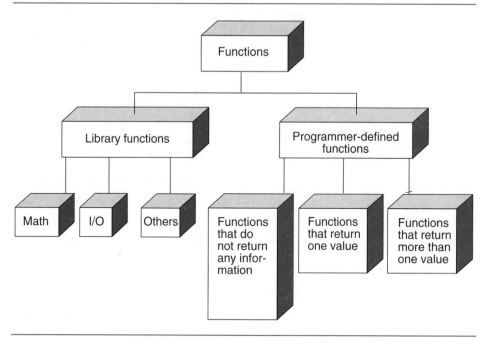

Note that the call to function1 contains no arguments. This matches the function declaration that has void enclosed in parentheses, meaning that no values are passed from main to function1. Note that the call to function2 contains two arguments. This fits its declaration, which also has two arguments.

Recall what you learned about writing the main function. The body of the function is enclosed in braces{}. In the body, the executable statements are written. In this program, find the bodies of function1 and function2. Observe that they are not located within the body of main but outside it.

The arguments for function2 are listed in three different lines. Find all three. Note that the arguments listed in the function calling statement are different from the arguments listed in the other two lines. However, the number of arguments (2) and the types of arguments (int and double) are the same for all three lines. The order of the arguments creates a correspondence between the variables in the function call and the variables in the function declaration. Determine the variable names in main that correspond to variables n and x in function2. By looking at the output from the program, can you see that the values of these variables in main have been transferred into the variables n and x in function2?

Look at the printf statements in the code and the output printed. Can you see the flow of the program? The variable m has been used in both main and function2. Does it have the same value in both locations (see the output)? What does this tell you about naming and declaring variables in different functions?

Source code

```
#include <stdio.h>

void function1(void);
void function2(int n, double x);
```
> Function prototypes indicate the number and types of arguments and the type of return value.

```
void main(void)
{
        int m;
        double y ;

        m=15;
        y=308.24;
        printf ("The value of m in main is m=%d\n\n",m);

        function1( );
        function2(m,y);
```
> Calls to function1 and function2. There are no arguments for function1. There are two arguments for function2.

```
        printf ("The value of m in main is still m=%d\n",m);
}
```

Definition of function1.

```
void function1(void)
{
```
> Function declarator or function header. The "voids" indicate that no values come into or go out of the function.

```
        printf("function1 is a void function that does not receive\n\
                \rvalues from main.\n\n");
}
```

Definition of function2.

```
void function2(int n, double x)
{
```
> Function declarator or function header. There are two arguments for the function. Both are values sent from main. The "void" indicates no value is returned.

```
        int k,m;
        double z;

        k=2*n+2;
        m=5*n+37;
        z=4.0*x-58.4;
```
> Assignment statements create new values from the values passed from main.

```
        printf("function2 is a void function that does receive\n\
                \rvalues from main.  The values received from main are:\n\
                \r\t n=%d \n\r\t x=%lf\n\n",n,x);

        printf("function2 creates three new variables, k, m and z\n\
                \rThese variables have the values:\n\
                \r\t l=%d \n\r\t m=%d \n\r\t z=%lf\n\n",k,m,z);
}
```

Output

```
The value of m in main is m=15

function1 is a void function that does not receive
values from main.

function2 is a void function that does receive
values from main.  The values received from main are:
        n=15
        x=308.240000

function2 creates three new variables, k, m and z
These variables have the values:
        k=32
        m=112
        z=1174.560000

The value of m in main is still m=15
```

Explanation

1. Briefly, how do we call (or invoke) a function? We write the function name followed by arguments enclosed in parentheses. For example, in this lesson's program, the line of code

`function2(m,y);`

calls `function2`.

2. Without going into detail, what does a function call do? It transfers program control to the function. In this lesson's program, after execution of the line

`function1();`

in `main`, control goes to `function1` and the next line of code executed is

```
printf("function1 is a void function that does not receive\n\
    \rvalues from main.\n\n");
```

This is illustrated schematically in Fig. 5.3.

3. What happens after the function has finished executing? As is also illustrated in Fig. 5.3, control goes back to the location at which the function was called. In this lesson's program, after `function1` has finished executing, control goes back to the location of the function call in `main`:

`function1();`

4. Is any information passed from the location of the function call to the function as control is transferred? Possibly, this depends on the type of the function and the way that it is implemented. We describe this process further later in the lesson.

FIG. 5.3
Direction of program flow when function1 is called

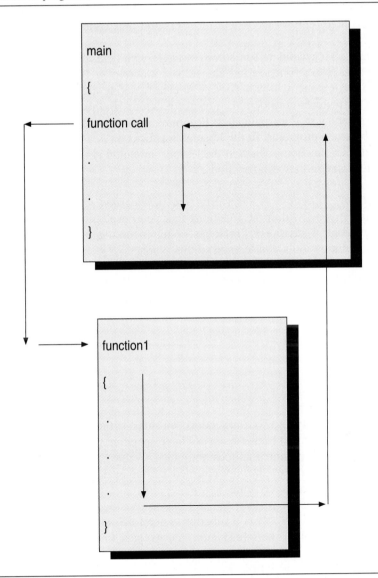

5. Is any information passed from the function back to the location from which it was called? Again, the answer is possibly. We describe functions that do return information back to the calling function in Lessons 5.2 and 5.3.

6. What is a major issue in writing programs with functions? A major issue with which a programmer must be concerned is correctly transferring information between the two functions involved. In other words, we pass to a function only the information that it needs and nothing more. A reason for this is that, if there is an error in a function, it will affect only the information to which the function has access. This helps control the extent of errors and thus contributes to writing error-free code.

Beginning programmers sometimes struggle with determining *what* information should be passed to a function and *how* to pass it correctly. To determine *what* information a function needs, it is necessary to clearly define the activities a function is to have. This requires considerable advance planning. In this book, we focus on the aspect of planning in the Application Programs at the end of the chapter, where we illustrate practical programs. In the lessons, we focus on *how* to correctly pass information between functions. Thus, in the lessons, attention is on the methods used for passing information between functions.

7. Within a function, is everything that we learned to this point still valid? Yes; in other words, within the functions that you develop, you will be using such things as if control structures, for loops, assignment statements, and many of the other programming tools you learned to this point.

8. What types of functions are there? The type of a function is the type of value that it *returns* to its calling function. Therefore, functions can be of type

```
int
long int
float
double
long double
char
void
```

and other valid data types. You may notice on this list the type void. We have not used data of this type, but to this point we have specified the function main to be of type void. A void type function does not return a value back to its calling function. Thus, main, in our program, has not returned a value back to its calling function (or calling unit), the operating system.

In this lesson we focus only on information *passing to* a function and not on information being *returned from* a function. Therefore, we use only void functions in this lesson and look at the information that we pass to them.

9. In this lesson's program, have we passed any information from main *to* function1? No, we can see this because the call to function1

function1();

has no variables enclosed in the parentheses. Thus, no variable values are passed from main to function1. We have done this because function1 needs no variable values from main. Its activities simply are to print a line of code with the statement

```
printf("function1 is a void function that does not receive\n\
     \rvalues from main.\n\n");
```

Because of this, it needs no more information, and therefore we choose to not send it any more information.

10. In this lesson's program, have we passed any information from main *to* function2? Yes, we can see this because the call to function2

```
function2(m,y);
```

has two variables enclosed in the parentheses, m and y. Thus, we are passing from main to function2 the *values* of the variables m and y. In main, at the location of the call to function2, the value of m is 15 and the value of y is 308.24. So, function2 receives the values 15 and 308.24.

11. Where does function2 *store these values?* Before we answer this, we must explain something about variable names. Each function has its own variable names, which can be confusing. As we write each function, we must choose the variable names we use in each of them. This could be difficult and cause you some confusion.

For instance, suppose you want to use the variable name x in one function to mean the *x* coordinate of a point, and in another function you want to use the variable x to mean the length of a side of a triangle. It is perfectly acceptable to use the variable name x in both functions, but when you look at your program you must remember that x means different things in the different functions! As you work with big programs you often will find that you will use the same variable names to mean different things in various functions, and you must keep these variables straight in your mind. Of course, you should be writing comments in your programs to help you avoid the confusion.

In function2 for this lesson's program, we created the variable names n and x. We did this in the line of code

```
void function2(int n, double x)
```

Because these variable names are given in parentheses after the type and name void function2, the values of 15 and 308.24 are stored in the memory cells reserved for variables n and x. We say more about this later in this lesson; however, the correspondence between the variables cause the values to be assigned to the variables in function2. This is illustrated as

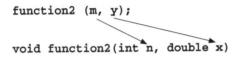

Because of the order of the variables listed in parentheses, the value of m in main (15) is assigned to the variable n in function2, and the value of y in main (308.24) is assigned to the variable x in function2.

Thus, we have successfully passed information from `main` to `function2`. Within `function2`, we can now work with the variables n and x and use the values of 15 and 308.24 to help us calculate other things of interest. In this program, we used n and x to calculate the values of k, m, and z with the statements

```
k=2*n+2;
m=5*n+37;
z=4.0*x-58.4;
```

12. In general, what do we need to define and make use of a programmer-defined function? We need three things:

1. a function prototype.
2. a function definition.
3. a function call.

13. What are these? First, a function prototype is a declaration, meaning essentially that it indicates that the function exists and probably will be used in the program. It usually is one line of code that gives the function's name, return type, and parameters. Function prototypes have the following form:

$$r\_type\ f\_name\ (arg\_type\ arg\_name,\ arg\_type\ arg\_name,\dots\dots);$$

where   *arg_type* is the argument type (int, double, or other)
         arg_name is the argument name (a valid identifier)
         f_name is the function name (a valid identifier)
         r_type is the value type returned by the function
                  (void if no value is returned, int, double, or other)
         . . . . represents more *arg_type arg_name* pairs separated by commas.

A primary purpose of a function prototype is to establish the type of arguments a function is to receive and the order in which it is to receive them. In other words, the order of the arguments in a call to a function must be in the same order as the argument list in the function prototype. Note the need for the semicolon at the end of the function prototype. For example, the prototype for function2 is

void function2 (int n, double x);

This indicates that function2 returns no value because it is of type void, and it has two arguments. The first argument is an integer and the second argument is a double. In this book, we give the argument names in our function prototypes. However, ANSI C does not require that they be given. This leads to an acceptable form of a function prototype to be

$$r\_type\ f\_name\ (arg\_type,\ arg\_type,\ arg\_type\ ,\dots);$$

where   . . . . represents more *arg_types* separated by commas.

So, an equally acceptable function prototype for function2 is

void function2 (int, double);

Second, a function definition includes the body of a function. It has the following form:

r_type f_name *(arg_type arg_name, arg_type arg_name,...)*
     {
      . . .
     function body - C declaration and statements
      . . .   *using the arg_names and other variables*
     }

where   *arg_name* is the name as it is used in the function body.

The programmer needs to develop the function body. Executable statements are given within the body of the function. In this lesson's program, `function1` simply calls printf in its body, whereas `function2` performs some mathematical calculations and calls printf. It is up to the programmer to write the function body to do the needed manipulations.

Note that no semicolon ends the line containing the function name and arguments. This line is called the *function declarator* or *function header.* Other than the semicolon, this line is the same as the first form that we gave for the function prototype. However, unlike in a function prototype, in the function header, the argument names *must* be included.

C considers a function definition to be an external definition, meaning that it is defined outside the body of any other function.

Third, a function call consists of a function name (which is an identifier) followed by an argument-expression list enclosed in parentheses. The form is

f_name *(exp, exp,...)*

where   *exp* is an expression that can be and commonly is a single variable or
          constant
      . . . represents more comma-separated expressions (the total number of
          expressions must equal the number of arguments in the function
          prototype)
     (exp, exp,...) is called the argument list

The argument list represents the values passed from the function in which the call exists to the function being called. For example, the call to `function2` in main is

`function2 (m,y);`

This causes the values of m and y (which are in the argument list) to be passed from `main` to `function2`.

14. What about the number, order, and type of parameters in the argument lists of a function call and its definition? These must match. This is sometimes referred to by the acronym *NOT* (number, order, type). In other words, if the function prototype and definition have three parameters then the function call must have three parameters. If the types are int, float, and double in the prototype and definition, then the types in the function call should be int, float, and double, respectively.

For example, for `function2` we have

*prototype*_____`void function2(int n,  double x);`

*function header*_____`void function2(int n,  double x)`

*function call*_____`function2 ( m, y);`

Note that there are two arguments for all three; the first is int and the second is double. With these three we have met the number, order, and type requirements.

When we call `function2` in main with the statement

function2(m,y);

we pass the values of m and y to the variables n and x in `function2`. The order of the variables in the argument list determines what is being passed into what. This is illustrated in Fig. 5.4.

Order not matching properly is a common source of error, especially for beginning programmers. The programmer establishes the order when he or she writes the function definition. You can use a style that you like, and sometimes a natural order is evident from the operation of the function. For instance, for the pow library function, the order is pow(x,y) which returns x^y, the first parameter is the base and the second parameter is the exponent. The programmer who wrote the pow function easily could have written the function definition in the reverse order. However, the order the programmer chose was natural and makes it easier for a user.

One way to avoid errors caused by order is to make your variable names very descriptive. For instance, if you had written the pow function as a programmer defined function, you could have used in your definition and prototype the names base and exponent (instead of x and y). When you later write a call in your program to that function, you can simply look at the prototype or definition to refresh your memory on the order of the parameters.

Another way to avoid errors caused by order is to use good commenting. Good commenting in your code, using a banner at the top of each function and clearly describing the meaning of each parameter, is helpful in reducing the likelihood of error caused by incorrect order.

Remember, if you have an incorrect order, your program may execute completely. Do not be fooled. This does not mean your answers are correct. If you find your answers are incorrect, you can check your parameters by inserting printf statements displaying the parameters just before the function call and at the very beginning of the function body. These printf statements should display values that you expect to see. If not, you may find that the order is incorrect.

FIG. 5.4

Functions for this lesson's program. Note that function1 neither receives nor returns values to main. Function2 receives the value of m and y from main and stores them in the locations for variables n and x.

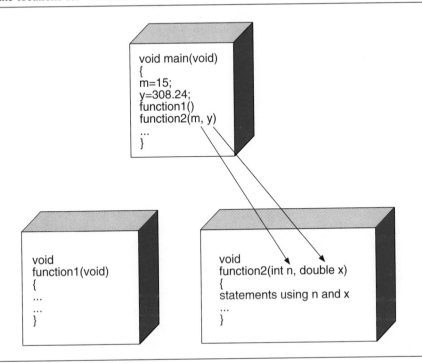

15. Can we say more about the relationship between main **and** function1**?** Because the prototype and function header for function1 show void in the argument list, the call to function1 simply transfers program control to function1 without passing any information. After function1 has finished executing, control passes back to main. Thus, a void function with void arguments simply transfers program control as illustrated in Fig. 5.3.

16. Is it necessary to have main as the first function defined? No, it is not necessary for the first function listed to be main. For the program in this lesson, it would be perfectly acceptable to write the program as

```
#include . . .
void function1(void)
void function2(int n, double x)
```

```
void function2(int n, double x)
{
        function body
}


void function1(void)
{
        function body
}


void main(void)
{
        function body
}
```

Traditionally, main is either the first function defined or the last.

17. Can a function prototype be written in a function body rather than outside the body of all functions? Yes; however, a function prototype given in a function body serves as a declaration only for the function in which it is prototyped. This means that the prototyped function can be called only from the body of the function where the prototype exists.

In contrast, a function prototype given outside the body of all functions declares the function for all functions. This means that the prototyped function can be used in any function in the program. Note that the function prototype must appear before the first function that calls it. More will be said on this topic when we talk about scope.

18. Is there any relationship between the variable m *in* main *and the variable* m *in* function2? No, as described earlier, variable names are declared for each function. As a result, we are allowed to use the same variable name in different functions without having any connection between those variables. We have done this with m in this lesson's program. So, m in main has no relationship to m in function2.

A primary value of functions and good modular design is that the only connections between functions occur through the argument list. Between main and function2, as is illustrated in Fig. 5.4, we created a connection between m in main and n in function2 and y in main and x in function2. So, in this program, m in main is related to n in function2, but there is no relationship between m in main and m in function2.

19. What would have happened if we had called function2 *from* main *with two integers in the parameter list instead of an integer and a double?* Because function2 needs to work with an integer and a double, the second integer would have been converted to a double. Converting an integer into a double usually does not cause errors. Therefore, even though the types do not match exactly, a successful execution can still take place. However, it is best to code your programs so that the types do match exactly.

20. *What would have happened if we had called* function2 *with two doubles in the parameter list instead of an integer and a double?* This type of situation is more likely to cause errors. This is because the first double would have been converted into an integer for function2, which usually causes a rounding of the value. For instance, consider the following short program. Look at the call to function2 and the function2 prototype. The call is with doubles even though function2 is prototyped to be ints. Look at the output. The values of a and b are truncated in being passed to i and j in function2 and printed to be only 12 and −2, losing a large number of digits.

The source code is

> The variables a and b are doubles. However, the protype states that they should be int.

```
#include <stdio.h>

void function2 (int i, int j);
void main(void)
{
        double a=12.34985, b=-2.5678;
        function2(a,b);
        printf ("a= %lf   b= %lf \n", a,b);
}

void function2 (int i, int j)
{
        printf ("i= %d   j= %d \n", i,j);
}
```

> When a and b are passed, copies are made and the fractional part of the double values is cut off from them. This type of loss of information may cause errors in your programs. Note that, because copies are made, the values of a and b remain unchanged.

The output is

```
i= 12   j= -2
a= 12.34985 b= -2.5678
```

Be careful, do not make this kind of error in calling your functions.

21. *We used the printf function for this program's lesson. Why is there no function prototype for it?* It only appears that there is no prototype for it. In reality, the prototype for the printf function appears in the file stdio.h that is given in the preprocessing directive #include <stdio.h>. If you want, you can use your editor to examine the contents of stdio.h. Although much of this file may look cryptic to you, you can find one line of the file that is the prototype for the printf function. This is a reason why we attach the file stdio.h to our program with the include directive.

EXERCISES

1. True or false:

   a. One reason for defining a function is to avoid writing the same group of C statements over and over again.

b.   A programmer defined function may be written before the main() function.

c.   A programmer defined function may be written after the main() function.

d.   A programmer defined function may be written within the main() function.

e.   A function body must be enclosed within a pair of braces.

f.   A programmer defined function must be called at least once; otherwise, you will get a warning message from the C compiler.

g.   In general, you write a programmer defined function if and only if no such function is in the C library. The reason is that the C library functions were written by professional programmers and have been used and tested many times. Therefore, they are more reliable, more efficient, and more portable than the functions you can write.

h.   If needed, keywords such as for, double, and while can be used as function names.

2.   Find the errors, if any, in the following function prototypes:

a.   `void (function1) void;`

b.   `void function2 (void)`

c.   `void function( n, x, a, b);`

d.   `void function1(int, double, float, long int, char);`

e.   `void function2(int n, double y, float, long int a, char);`

f.   `void function1(int, a, double, b, float,c);`

3.   Given the following function prototypes and variable declarations, find errors in the function calls:

```
void function1(void);
void function2(int n, double x);
void function3 (double, int, double, int);
void function4 (int a, int n, int b, int c);
void main(void)
{
        int a, b, c, d, e;
        double r, s, t, u, v;

 . . .

}
```

a.   `function1 (a, b);`

b.   `function2(a, b);`

c.   `function3(r,a,s,b);`

d.   `function4(a,b,c,d,e);`

e.   `function1( );`

f.   `function2(r, s);`

g. **`function3(r, a, r, a);`**
h. **`function4(r, s, t, u);`**

4. What will be the output from this program?

```
#include <stdio.h>
void function1(int a, double x);
void main(void)
{
        int a=1, b=2, c=3, d=4;
        double r=3.2, s=4.3, t=5.4, u=6.5;

        function1(a,b);
        function1(r,s);

}
void function1(int a, double x)
{
printf ("a= %d, x= %lf\n",a,x);
}
```


Solutions

1. a (true), b (true), c (true), d (false), e (true), f (false), g (true), h (false)
2. a. **`void function1 (void);`**
    b. **`void function2 (void);`**
    c. **`void function( int n, double x, int a, int b);`**
    d. No error, argument names are not required.
    e. No error but not a good form, argument names should be either all given or not given.
    f. **`void function1 ( int a, double b, float c);`**
3. a. **`function1( );`**
    b. No error detected by C, but the value of b will be transferred in double form.
    c. No error.
    d. **`function4(a,b,c,d);`**
    e. No error.
    f. No error detected by C, but the value of r will be transferred in integer form.
    g. No error, we can pass the same value to many arguments.
    h. No error detected by C, but the values of r, s, t, and u will be transferred in integer form.
4. **`a=1, x=2.000000`**
    `a=3, x=4.300000`

    Note that, because the types in the function call and function prototype do not match, b has been converted to 2.000000 and r has been converted to 3. This sort of conversion may cause errors in your programs.

■ LESSON 5.2 FUNCTIONS THAT RETURN JUST ONE VALUE

Topics
- Returning a value from a function
- Considerations about return value types

We already have learned how to make use of library functions that accept one value and return another. An example of one of these is the log() function from the program library. For instance if we write

```
y=log(x);
```

the log function accepts the value of x and returns the value of the natural logarithm of x. The assignment statement gives y the value of log(x).

You can utilize functions that you have defined in a similar manner. Suppose, for instance, that you needed to compute the factorial of an integer. You could check the function library and find out that C has no function that computes factorials like it does log, sin, tan, and others. Therefore, if you want to use a factorial in a calculation you need to write your own function. Then, you can call your function in the same way that you call a library function.

In this lesson's program, we have written a function named fact. Look at the code and find the function prototype for fact. From this, determine the return value type for fact. Why do you think we have chosen this particular type? Hint: It has to do with the size of the numbers when a factorial is calculated.

Recall that the factorial of *n* is $n * (n - 1) * (n - 2) * (n - 3) \ldots * (2) * (1)$. We can create a loop structure that performs this calculation. In the source code that follows we have put the loop into function fact. Look at the body of fact and the for loop that calculates the factorial. Note that the loop goes from i = m to i = 1. Trace the loop to assure yourself that the value of product indeed is equal to the factorial of m after the loop has fully executed. You can trace the loop by filling in the following table (as an example, try the case of m = 6). Before entering the loop, product = 1, m = 6. Then i goes from 6 to 1.

i	product
6	1 * 6 = 6
5	6 * 5 = 30
4	30 * 4 = 120
3	
2	
1	

Fill in these three blank squares by tracing the for loop in function fact of this lesson's source code.

Look again at the body of fact. Can you guess which statement returns control to the point where fact was called? What variable value is returned to this point? Of what type is this variable? Recall what you learned about calling library functions. Locate the call to function `fact` in main. Note that `fact` appears on the right side of an assignment statement. On the left side of the assignment statement is g. What type of variable is g? What can be deduced from knowing the types of g and `fact`?

This program actually computes 1/n!. Why did we choose %e as a format specification for displaying 1/g? The input value of n has been restricted to being less than or equal to 12. Why has this been done?

In all the programs previous to this one, we chose type void for main. For this program, what type is main? Recall that the calling unit for main is the operating system. What is returned to the operating system?

Source code

> Function prototype. Instead of void, double, or int, it uses unsigned long int as its return type. The function name is `fact`. It has one argument, m, that is of type int.

```
#include <stdio.h>
unsigned long int fact(int m);

int main(void)
{
```

> Instead of void, we have used the type int for main.

```
    int n;
    unsigned long int g;
    double one_over_nfactorial;
```

> Declaring the variable g as type unsigned long int.

```
    printf("This program calculates 1/nfactorial.\n\
            \rEnter a positive integer less than or equal to 12:\n ");
    scanf ("%d",&n);
    g=fact(n);
    one_over_nfactorial=1.0/g;

    printf("1/%d! = %e",n,one_over_nfactorial);
    return (0);
}
```

> This statement both calls `fact` and assigns the return value to g. Note that g and the return value for `fact` have the same type, unsigned long int. The value of n is passed from main to fact.

> We have returned an int type from main to the operating system.

```
unsigned long int fact(int m)
{
    int i;
    unsigned long int product;

    product = 1;
    for (i=m; i>=1; i--)
        {
        product*=i;
        }
    return(product);
}
```

> Function declarator. The variable m contains the value of n that was passed from main.

> Declaring the variable product as type unsigned long int. It serves as the variable returned by `fact`.

> Loop for calculating the factorial of m.

> Definition of function fact.

Output

| *Keyboard input* | ```
This program calculates 1/nfactorial.
Enter a positive integer less than or equal to 12:
9
1/9! = 2.755732e-06
``` |

Explanation

1. How do we define a function that returns a single value? In its prototype and declarator, a function that returns a value must have a type that reflects the type of value returned. For instance,

unsigned long int fact(int m);

declares the function `fact` to return a value that is of type unsigned long int.

In addition, a return statement must appear in the body of the function. For the function `fact` in this lesson's program, the value of product is the return value with the statement

return (product);

The form of the return statement is

return *expression;*

Many programmers put parentheses around the expression (which does not change the expression's value) to give a form of

return *(expression);*

C considers a return statement to be a jump statement because it causes execution to transfer unconditionally to another location. Another jump statement we have learned to this point is the break statement.

2. What information was passed from main *to* fact *and from* fact *to* main *in this lesson's program?* The value of n in `main` was passed to `fact` using the function call in the statement

g=fact(n);

With this statement, the value of n in `main` is passed to m in `fact`, as illustrated in Fig. 5.5. Also shown in this figure is the value of `product` in `fact` being passed to `main`. This occurs due to the statement

return(product);

3. To where does the value of the variable product *return?* The value of `product` returns to the location of the function call as illustrated in Fig. 5.5. For instance, if we were finding the factorial of 4 (which is 24), the value of `product` (24) is passed to the location of the function call in the statement
```

FIG. 5.5

This lesson's program. Note the transfer of information from main to fact (*n* in main corresponds to *m* in fact) and from fact to main (the return value, product).

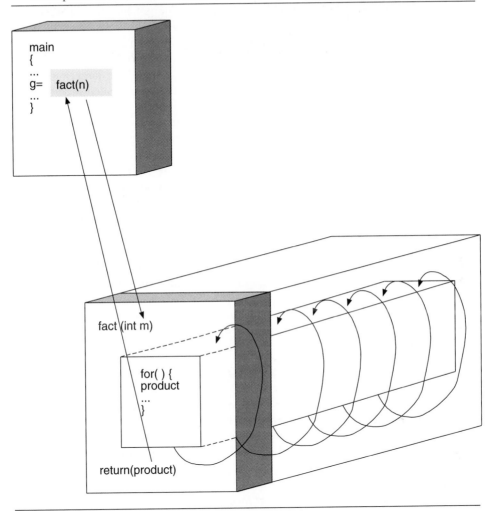

g=fact(n);

Thus, on execution of this assignment statement, the variable g has the value 24.

4. Why was the function fact *chosen to be unsigned long int, and why was the input value for* n *in this lesson's program restricted to being less than or equal to 12?* Numbers in a factorial calculation become big very quickly. For instance 13! is 6227020800. We have designated the data type for n to be an unsigned long int (also called unsigned long), which is required by ANSI C to be at least 4 bytes. Thus, unsigned long int may not be expected to handle a value greater than 4294967295 unless your compiler exceeds the ANSI C standard. Since 13! is greater than what ANSI C specifies for unsigned long int, we restricted n to be 12 or less. Should we want to calculate greater values, we would need to use double or long double. Using double or long double would allow us to calculate the factorials of larger numbers, but the calculated values may be only approximate due to the limited precision of the real variable types.

Note that library functions have similar restrictions on values received and returned. For instance in Lesson 3.5, we indicated the receiving and return types for a number of math functions. In your programs, you should make sure that you adhere to these types. Also, although not listed, there are restrictions on the arguments for some of the math functions. Consider the natural log function, $\log(x)$. You cannot use a value of *x* less than or equal to 0. If you call the log function with a value less than 0, your program will stop executing and an error (usually indicating something about the function domain) will occur. If this happens, it is up to you to determine why the value has become negative. If you have many calls to the log function in your program, you may find it difficult to locate the problem call. However, this is considered part of the debugging/programming process.

5. We learned in Lesson 5.1 that void functions do not return values. Can we use a return statement in a void function? Yes; however, we must not use an expression with it. In a void function, we can use a return statement with the following form

```
return;
```

6. If we do not include a return statement in a void function, how does control transfer back to the calling function? The ANSI C standard states that "reaching the } that terminates a function is equivalent to executing a return statement without an expression." Return statements are needed in void functions if there is to be more than one exit point.

7. Can a return statement be used at locations other than at the end of a function? Yes, a return statement can appear anywhere in the function body. In addition, more than one return statement can appear in a function. A common structure for returning a value is

```
if (expression)
          {
          return (a);
          }
else
          {
          return (b);
          }
```

8. Can the call to a void function appear on the right side of an assignment statement? No; because a void function does not return a value, a call to it cannot appear on the right side of an assignment statement.

9. We used the type int for `main`. *What does this do?* It means that we can return an int value to its calling function. The calling "function" for `main` is the operating system. We have returned the value 0 to the operating system, which the operating system interprets as meaning that the program has terminated normally. Operating systems are different, therefore, you should examine your operating system's documentation for information on sending other types of termination messages from main to the operating system.

EXERCISES

1. True or false:
   a. A correctly written int type function will return an int type value to the calling function.
   b. Only a void type function is allowed to have void as its argument.
   c. The arguments for an int type function need not conform to the number, order, and type agreement requirement.
   d. The call to an int or double type function must appear on the right side of an assignment statement.

2. Find the errors, if any, in the following function prototypes:
   a. `int (function1) void;`
   b. `double function2 (void);`
   c. `float function1( n, x, a, b);`
   d. `double function1(int, double, float, long int, char);`
   e. `int function2(int n, double y, float, long int a, char);`
   f. `double function1(int, a, double, b, float,c);`

3. Given the following function prototypes and variable declarations, find errors in the statements using the function calls:

```
double function1(void);
int function2(int n, double x);
double function3 (double, int, double, int);
double function4 (int a, int n, int b, int c);
void main(void)
{
        int a, b, c, d, e;
        double r, s, t, u, v;
        . . .
}
```

a. `a = function1 ( );`
b. `b = function2(a, b);`
c. `r = function3(r,a,s,b);`
d. `s = function4(a,b,c,d,e);`
e. `u = function1( );`
f. `c = d + function2(r, s);`
g. `t = s * function3(r, a, r, a);`
h. `a = v + function4(r, s, t, u);`

Solutions

1. a (true), b (false), c (false), d (false)
2. a. `int function1(void);`
    b. No error.
    c. `float function1( int n, double x, int a, int b);`
    d. No error.
    e. No error, but it is not good form.
    f. `double function1(int a, double b, float c);`
3. a. No error detected by C, but the double value returned by `function1` will be stored in an int.
    b. No error detected by C, but `b` will be transferred to `function2` in double form.
    c. No error.
    d. `s = function4(a, b, c, d);`
    e. No error.
    f. No error detected by C, but `r` will be transferred to `function2` as an int.
    g. No error.
    h. No error detected by C, but `r`, `s`, `t`, and `u` will be transferred to `function4` as int. Also, there are two doubles on the right side of the assignment statement but an int variable on the left side of the statement.

■ LESSON 5.3 SCOPE AND MECHANICS OF PASSING VALUES TO FUNCTIONS

Topics
- Scope
- Call by value
- Function with many arguments returning one value
- Multiple returns in a function
- Mechanics of passing values to functions

Now that we have introduced functions other than main and library functions, we have to be careful about where we put declarations in our programs. The program for this lesson is not considered to be good programming practice, but it illustrates the effect of having declarations outside the body of a function and the concept of scope.

Look at the source code and compare and contrast the declarations for m and n. Note that m has only one declaration and it appears outside the body of any function. However, observe that n has two declarations, one in main and one in function1. Note also that neither m nor n appears in the argument list for function1. Look at the first four lines of the output. What is the value of m in main and initially in function1? Why do you think they are equal? (Hint: The reason relates to the location of the declaration of m.)

Look at the line after the first printf statement in function1. Note that *m* has been given a new value in function1. Look at the output. What has this line in function1 done to the value of m in main? Why do you think m changed in main? (Hint: Again, it has to do with the location of the declaration of m.)

The values of n are different in main and function1. Is there any relationship between the two values of n?

The scope of a declaration is defined to be the region of the program in which that declaration is active. Two kinds of scope are function scope and file scope. A variable declaration that has function scope pertains only to the one function in which it is declared. A variable declaration that has file scope pertains to any function in the file. Would you say that the declaration for m has file scope or function scope? Consider also variable n in main. Does n in main have file scope or function scope?

Structured programming has developed over the years as an efficient method of programming with an intent on reducing errors. If you were working on a long program with a large number of people, each writing different modules, can you envision any difficulties in declaring variables outside the body of any function?

The variables a and e have correspondence (see the call to function1 in main and the definition of function1). The variable a in function1 initially is equal to e in main. In function1, a has been modified to be 1402. After returning to main, has the value of e become 1402? This reveals something about how C handles the passing of arguments to functions. Can you say what this tells us?

There is more than one return statement in function1. Trace the flow of the program and determine which return statement was executed. What is the value for i?

Source code

The variable n has function scope; therefore, n in main has no connection to n in function1.

Declaring the variable m outside any function definition gives the variable file scope. This means that any function can use m without it being passed through the parameter list. A variable with file scope also is generically called a *global variable*.

```c
#include <stdio.h>
int m = 12;
int function1 (int a, int b, int c, int d);

void main(void)
{
        int n = 30;
        int e,f,g,h,i;
        e=1;
        f=2;
        g=3;
        h=4;
        printf ("\n\n In main (before the call to function1): \n\
                \r m = %d\n\
                \r n = %d\n\
                \r e = %d\n\n",m,n,e );
```

Prototype for function1. It has four arguments, all of type int. It also returns an int.

The variable m is printed out. Note that, within main, its value does not need to be initialized because it is initialized in its declaration as a global variable.

```
              i=function1(e,f,g,h);
```
Call to function1. The values of e, f, g, and h are passed to function1. The return value is stored in i.

```
          printf ("After returning to main: \n");
          printf ("n = %d \n\
                    \r m = %d \n\
                    \r e = %d \n\
                    \r i = %d", n, m,e,i);
```
The variable m has been modified by function1. Because it is a global variable, the value of m also has changed in main and is printed here.

```
}
```
Definition of function1.

```
int function1 (int a, int b, int c, int d)
{
   int n = 400;
```
The variable n has function scope; therefore, n in main has no connection to n in function1.

```
          printf ("In function1:\n\
                    \r n = %d\n\
                    \r m = %d initially\n\
                    \r a = %d initially \n",n,m,a);

          m = 999;
```
The value of the global variable, m, is modified with this statement. Note that this variable has not been declared in function1. This is acceptable because its declaration is outside any function.

```
          if (a>=1)
                    {
                    a+=b+m+n;
                    printf ("m = %d after being modified\n\
                              \r a = %d after being modified\n\n",m,a) ;
                    return (a);
                    }
          else
                    {
                    c+=d+m+n;
                    return (c);
                    }
}
```
A function may have more than one return statement. However, only one is executed.

Output

```
In main (before the call to function1):
m = 12
n = 30
e = 1

In function1:
n = 400
m = 12 initially
a = 1 initially

m = 999 after being modified
a = 1402 after being modified

After returning to main:
n =30
m = 999
e = 1
i = 1402
```

Explanation

1. What is meant by scope? *Scope* refers to the region in which a declaration is active. In C, there are four kinds of scope: block, function, file, and prototype. Scope for an identifier is determined by the location of the identifier's declaration.

The variables n in this lesson's program are said to have function scope, which means that a value assigned to n is valid only within the function in which it was defined. For example, n was declared in main and was assigned the value of 30. This assignment pertains only to the function main because n was declared in main. The n that was declared in function1 has no relation to the n in main because its scope is only within the function function1.

On the other hand, the variable m has file scope because it was declared outside and before the body of any function. Its active region begins at its declaration point and extends to the end of the source code file. Note that m was not declared within the body of either of the functions yet was used by both main and function1. As a consequence, when m was modified in function1, the value of m in main also was modified. To follow good modular programming practice, this type of variable declaration is not recommended. Good modular programming style has values passing to functions through the parameter lists.

Also with file scope is the identifier function1. The prototype for function1 is outside the body of any function. Its active region extends from the point of declaration to the end of file.

Recall that a block of code begins with { and ends with }. An identifier declared within a block limits its active region to be within that block.

We earlier described how to define constants using the preprocessing directive #define. For example,

```
#define P 200
```

causes the preprocessor to change all instances of P to be 200 in the source code. These directives, whether placed within or outside a function body, have file scope. However, the file scope begins at the line in which the directive is placed and ends at the end of the file.

Parameters given names in function prototypes have function prototype scope. This means that, outside a one line function prototype, the identifier in that prototype is not considered to be declared. This is one reason why we said that the parameter names in a function prototype did not have to be exactly the same as the parameter names in the function definition. (However, the parameter types must be the same.)

Also considered to have scope are statement labels. Statement labels, such as those used in switch control structures, have function scope, meaning that a statement label cannot be duplicated within a function.

As you learn more advanced programming techniques and have the need, you will find that it is possible to expand the scope of an identifier to extend even to different files. We describe this in Chapter 8.

2. How does C handle the passing of arguments through an argument list? When a function is called, C allocates space within memory for the variables in the function's parameter list and the variables declared in the function body. It then copies the values of the expressions in the function call into the locations of the corresponding variables within the region allocated to the function. This concept is illustrated in Fig. 5.6 for the call to function1 in main for this lesson's program. Observe that, once copied, the variables within the function1 region can be changed without having any effect on the values stored in the region for main.

3. What happens in memory when the value a *is returned from function1 to main?* The following assignment statement is in main:

```
i=function1(e,f,g,h);
```

This means that the return value, a, from function1 is assigned to the value i in main. This is illustrated in Fig. 5.7. From Figs. 5.6 and 5.7, the distinct areas of memory for the variables of main and function1 can be seen clearly. An illustration of the passing of information for this lesson's program is shown in Fig. 5.8.

4. What are local and global variables? Local and global are not terms defined by the ANSI C standard but generic terms commonly used by programmers. Loosely speaking, *global variables* are variables with file scope and *local variables* are variables with function scope. For example, in this lesson's program, m is a global variable while n is a local variable. This is illustrated in concept in Fig. 5.8.

5. What are formal and actual parameters? These are other terms not defined by the ANSI C standard. Commonly, *formal parameters* refer to the parameters listed in the function prototype and function definition. *Actual parameters* are parameters in the parameter list of the function call.

6. What happens to the memory space allocated to the variables in function1 in this lesson's program after the execution of function1 has finished? For this program, the space is freed up. However, as you do more advanced programming, you will find that it is possible to specify the storage class of variables in their declaration so that they remain stored even after the function has finished.

If we choose to call function1 a second time, then the memory will be reserved again (possibly not at the same location depending on what else is occurring in memory). Thus, during execution of a program, each time a function is called, memory is reserved. Unless otherwise specified, when the function has finished executing, the memory is freed.

EXERCISES

1. True or false:
    a. In general, we should use global variables (i.e., variables with file scope) as often as we can.
    b. The argument types of a function must be the same as the function type.

FIG. 5.6

Actions taking place in memory when function1 is called in main. For this lesson's program, the two regions of memory may be adjacent to each other rather than separated as shown here.

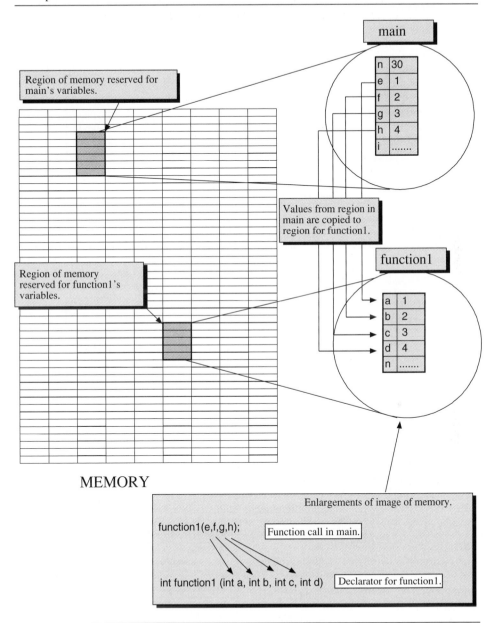

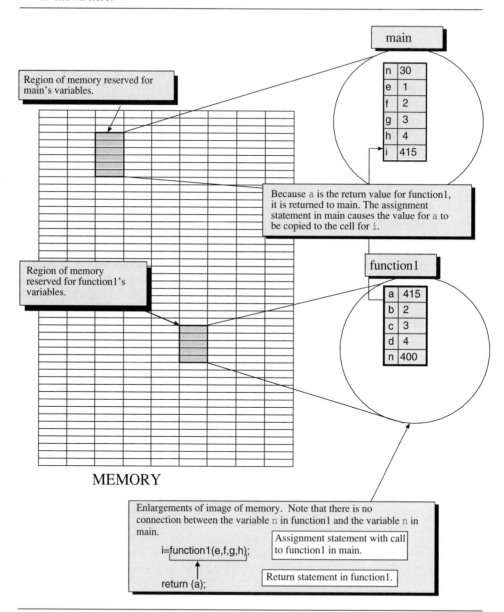

FIG. 5.7

Actions taking place in memory when the return statement in function1 and the assignment statement in main with the call to function1 are executed. For this lesson's program, the two regions of memory may be adjacent to each other rather than separated as shown here.

main

n	30
e	1
f	2
g	3
h	4
i	415

Region of memory reserved for main's variables.

Because a is the return value for function1, it is returned to main. The assignment statement in main causes the value for a to be copied to the cell for i.

Region of memory reserved for function1's variables.

function1

a	415
b	2
c	3
d	4
n	400

MEMORY

Enlargements of image of memory. Note that there is no connection between the variable n in function1 and the variable n in main.

i=function1(e,f,g,h);

Assignment statement with call to function1 in main.

return (a);

Return statement in function1.

FIG. 5.8

This lesson's program. Note that m is available for use in both main and function1 and that n in main is not related to n in function1.

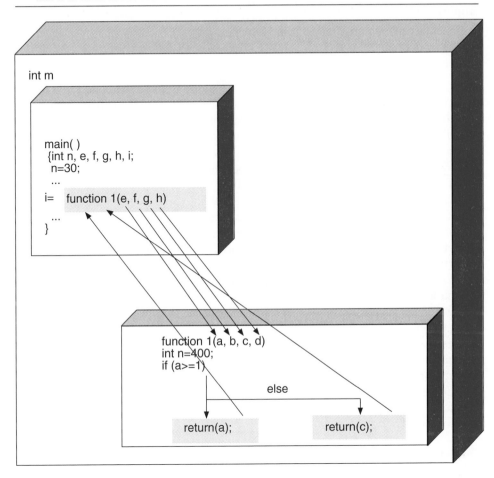

c. You may alter the value of a variable, n, by calling a function that uses n as its argument and changes its value within the function.

d. The memory reserved for the variables of other functions is the same as the memory reserved for the variables of main.

e. Memory is reserved when a function is called.

f. When variable values are passed to functions, a copy is made and put into the memory reserved for the functions' variables.

Solutions

1. a (false), b (false), c (false), d (false), e (true), f (true)

▨ LESSON 5.4 FUNCTIONS THAT "RETURN" MORE THAN ONE VALUE

Topics

- Functions that "return" more than one value
- The symbol &
- The symbol *

In this lesson, we describe the appearances of returning more than one value from a function. In the next lesson, we describe the mechanics of what is happening.

To this point, we have illustrated functions that return just one value using a return statement. Suppose, however, that you want to convey more than one piece of information from a function. C allows for this by using the parameter list both to send information to and receive information from a function.

Examine the source code. Find the calling statement for function1 in the body of main. What symbols are in the parameter list that you have not seen there before? Look at the prototype for function1. What symbols are in the prototype that you have not seen there before? Note that the fifth and sixth parameters in both the call and prototype have the odd symbols. Can you guess which variables are "returned" from function1 to main?

Look at the assignment statements in the body of function1. What symbols are on the left side of the assignment statements that we have not seen on the left side of assignment statements previously? Use what you have already learned about functions to determine the initial values of the variables a, b, r, and s in function1. What values should be computed for *c and *t? Look at the output and the last printf statement in main. Have the values of *c and *t from function1 been transferred to main? Through what variables in main have they been transferred?

Observe from function1's prototype the type of function that it is. What does this tell you about a function's type when all of the values are returned through the parameter list?

Source code

> Function prototype. The * indicates that to work with the value of the corresponding argument in the function call, we must use * in the function body.

```
#include <stdio.h>

void function1 (int a, int b, double r ,double s, int *c, double *t);

void main (void)
{
        int     i=5,    j=6,    k;
        double  x=10.6, y=22.3, z;

        printf (" i = %d \n\r j = %d \n\r x = %lf \n\r y = %lf \n\n",
                                i,j,x,y);
```

```
        function1 (i,j,x,y,&k,&z);

        printf (" k = %d \n\r z = %lf \n\n", k, z);

}

void function1 (int a,int b,double r,double s,int *c,double *t)
{
        *c = a+b;
        *t = r+s +(*c);

        printf (" *c = %d \n\r *t = %lf \n\n", *c, *t );

}
```

In the function call, we must use & for the last two arguments. These are the arguments that have * in the prototype.

In the function body, we must use * to work with the values.

Output

```
  i  = 5
  j  = 6
  x  = 10.600000
  y  = 22.300000

  *c = 11
  *t = 43.900000

  k  = 11
  z  = 43.900000
```

Explanation

1. In this lesson's program, which values have been transferred from main to function1? The values of i and j (which are 5 and 6, respectively) in main have been transferred to a and b in function1. Also, the values of x and y (which are 10.6 and 22.3, respectively) in main have been transferred to r and s in function1. This is illustrated in Fig. 5.9.

2. What variable values have been "returned" from function1 to main? The values of *c and *t (which are 11 and 32.9, respectively) in function1 have been transferred to k and z in main. This also is illustrated in Fig. 5.9.

3. In a function call, how can we specify that a variable will receive a value from the function? We can place the symbol & in front of a variable that we want to receive a value from the function. For function1 in this lesson's program, the fifth and sixth variables in the function call are k and z, both preceded by & and both variables receiving values from function1.

4. In a function prototype and function definition, how can we specify a variable whose value will be "returned" to the calling function? We use the symbol *

■ **FIG. 5.9**
Illustration of this lesson's program. This figure illustrates the appearance of passing of
values between main and function1.

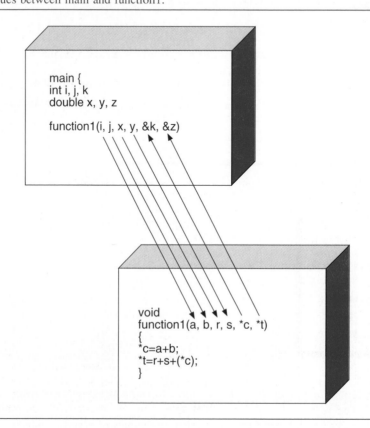

as the first symbol in a variable name whose value we want to "return" to the calling function. For function1 in this lesson's program, the fifth and sixth variables in
the parameter list are *c and *t; both begin with * and both variables sent values
from function1 to main.

*5. Within the body of a function, how do we handle the variables that begin
with *?* Handle them just like you would handle any other variables. However, you
can envision a problem, since the symbol * also is used as the multiplication operator. To avoid confusion, we recommend that you enclose these variables within
parentheses when they appear on the right side of assignment statements even
though parentheses may not always be necessary.

6. Why is function1 a void type function? Even though function1 returns parameters it is considered a void type function since nothing is returned with a return

statement. If a return statement were used in function1 to return a value (along with the parameter list), the function type would have to correspond to the value type returned with the return statement.

7. It was said earlier that C only copied values of the variables in the calling parameter list into the region of memory reserved for the function. If this is the case, how is the process of returning values through the parameter list accomplished? It is true that C only copies the values of the variables in the parameter list into the function's memory region. However, by using operators that we have not discussed previously, C can appear to pass values back to the calling function of a function. In other words, what we have described in this lesson is only what appears to be happening when we want to return many values from a function. We have described what appears to be happening because we can use appearances to write many useful and productive programs. Therefore, we recommend that you use this lesson as a guide to help you write your own functions that return more than one value. However, it also is necessary that you understand the actual mechanics of the way that C handles functions. In the next lesson, we discuss the mechanics in detail.

EXERCISES
A C program contains the following statements:

```
void plus(short a, long *b);

void main(void)
{
        short x=100;
        long y=9999;

        plus(x+200, &y);
. . . . . .
}
```

1. Determine whether each of the following statements is true or false:
    a.   The plus() function is of void type, therefore, it can never be used to return a value to the main()function.
    b.   The short integer variable a in the function plus(), can be used to transfer a value from plus() to main() function.
    c.   Without using a return statement, the plus() function can return a value to the main() function.
    d.   The symbol * in the plus function prototype is wrong, it should be &, not *.
    e.   The value of y can be modified by the plus() function.
    f.   The value of x can be modified by the plus() function.

2. Write a program that calls a void type function to find the maximum of three given integer numbers.

3. Write a program without calling a function to exchange the values of two long type integers.

4. Write a program that calls a function to exchange the values of two long type integers.

Solutions

1. a (false), b (false), c (true), d (false), e (true), f (false)

LESSON 5.5 MECHANICS OF "RETURNING" MORE THAN ONE VALUE FROM A FUNCTION—ADDRESSES AND POINTER VARIABLES

Topics
- Addresses
- Variables that store addresses, pointer variables
- "Address of" operator
- Indirection operator

The source code for this lesson is nearly identical to the source code for the previous lesson. The only difference is that we have added two printf statements to show other values.

C does the transferral of information from main to function1 in an indirect way. It uses the addresses of the locations where the variable values are stored to convey the information. Recall from Chapter 1 that we can store many different things in a memory location. We can store integers, real numbers, instructions, and addresses in memory cells. In other words, instead of storing integers such as 135 (converted to binary notation) in a variable's storage space, we can store an address. Also, remember that addresses often are written in hexadecimal notation. If you have forgotten hexadecimal notation, please refer to Chapter 1.

Variable addresses operate much the same way as home addresses. For instance, if you wanted to give a friend a birthday present and you knew his or her address, what would you do? Of course, you would take the present to your friend's address. C acts similarly. If it has a variable value and an address it takes the variable value to the address (if you command it to do so).

In this lesson's program, main has conveyed the addresses of two variables to function1. By looking at the call to function1, can you determine the names of the variables whose addresses are transferred from main to function1? What symbol in this parameter list do you think means "address of"?

Look at the second and third printf statements in main and the corresponding output from these statements. In the second printf statement, k and z are printed. In the third printf statement &k and &z are printed. What do you think the symbol & means? What type of notation is used in the display of &k and &z? What was the conversion specification in the string literal for &k and &z?

C operates by setting aside memory space (which we have described by cells, bytes, and bits) for the variables used in each function when it is called in a program. In other words, it creates a region of memory that is reserved for the variables in main and a separate region of memory for the variables in function1. For the process to work, the memory region for a function first must be created or reserved. To reserve the correct amount of space, C must know all the variables and variable types. C uses the declarations to learn the variables and their types. The variable type is important because the different variable types take up different amounts of space, or numbers of bytes, in memory (compare int and long int). Like an int, float, or double, a stored address takes a set number of bytes in memory. Given that an address is stored, it is also important to know the type of variable stored at that address indicated. Since main has conveyed addresses to function1, some of the memory cells for function1 must store addresses. Look at the prototype for function1. Can you guess which variables hold addresses? What symbol is used to indicate that an address is to be stored in that variable's memory cell? What types of variables do you think are stored in those addresses? For instance, the variable c in function1 holds an address. What type of variable do you think is located at the address that c has stored?

The next question becomes, "What do we do with addresses?" It makes little sense to divide one address by another. However, just like you would do if you had an address (which is to go to that address to do something with whoever was there), C has an operator that essentially says "go to the address and use what is inside." Look at the assignment statements in function1. Can you guess the symbol used as an operator that directs the computer to go to the address so that what is inside can be used? (Hint: It is the same symbol that indicates in declarations that a variable is to hold an address. Do not be confused, though. It is the same symbol, but in an assignment statement the symbol has a different purpose than it has when it is in a declaration.)

Look at the output. Note that the value contained in c is the same as the address of k and that the value contained in t is the same as the address of z. Look again at the call to function1 and the declarator for function1. Can you see, for instance, why the address of k and the value of c should be equal?

Source code

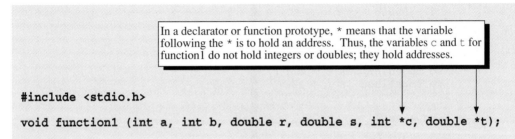

In a declarator or function prototype, * means that the variable following the * is to hold an address. Thus, the variables c and t for function1 do not hold integers or doubles; they hold addresses.

```
#include <stdio.h>

void function1 (int a, int b, double r, double s, int *c, double *t);
```

```
void main (void)
{
        int     i=5,j=6,k;
        double x=10.6,y=22.3,z;

        printf (" i = %d \n\r j = %d \n\r x = %lf\n\r y = %lf \n\n",
                                i,j,x,y) ;

        function1 (i,j,x,y,&k,&z);

        printf (" k = %d \n\r z = %lf \n\n", k, z) ;

        printf (" Address of k = %p\n\r Address of z = %p \n", &k, &z);

}

void function1 (int a,int b,double r,double s, int *c,double *t)
{
        *c = a+b;
        *t = r+s+(*c);

        printf (" *c = %d \n\r *t = %lf \n\n", *c, *t );
        printf (" Value contained in c = %p\n\
                \r Value contained in t = %p\n\n\n", c, t);
}
```

> Because & is the "address of" operator, a copy of the *addresses* of k and z are put into the memory cells reserved for c and t in the memory region reserved for the variables of function1.

> The * in a C statement has a different meaning than it has in the function declarator, prototype or in a declaration. In a C statement, * is the unary * operator (when it is not meant to mean multiplication) that means to go to the address stored in the variable following the * and get the value at that address or put a value at that address.

> The *values* of i, j, x, and y are copied into the memory cells reserved for a, b, r, and s in the region of memory for function1.

Output

```
i = 5
j = 6
x = 10.600000
y = 22.300000

*c = 11
*t = 43.900000

Value contained in c = FFF0
Value contained in t = FFD8

k = 11
z = 43.900000

Address of k = FFF0
Address of z = FFD8
```

Explanation

1. What is a conceptual image of the way variables are stored? As we described in Chapter 3, C conceptually creates a table of all of the variables in a function. In the table are the variable names, variable types, memory cell addresses, and variable values. For instance, for function main the following table was created:

Variable name	Variable type	Variable address	Variable value
i	int	FFF4	5
j	int	FFF2	6
k	int	FFF0	
x	double	FFE8	10.6
y	double	FFE0	22.3
z	double	FFD8	

Note that C is responsible for creating the addresses. You need not specify the addresses; they are taken care of automatically. C knows how much memory space must to be reserved for main by looking at the number and type of variables declared. The memory locations are not filled with the variable values until the variables have been initialized. In this table, we have shown the variable values for i, j, x, and y since they were initialized in the declarations. However, the variable values for k and z are not listed because they were initialized much later in the program. Their addresses, however, are listed because their addresses were set before values were put into the memory cells. Their memory cells simply wait to be filled.

For function1, the following table was created:

Variable name	Variable type	Variable address	Variable value
a	int	FFC0	5
b	int	FFC2	6
c	address to int	FFD4	FFF0
r	double	FFC4	10.6
s	double	FFCC	22.3
t	address to double	FFD6	FFD8

In this table, we show all the variable values because all these variables appear in the function declarator. All these values are copied from the parameter list in the call to function1. Note that the correspondence of a, b, r, and s to i, j, x, and y, respectively, give them their values. Also, note that the *value* of c is the *address* of k, and the *value* of t is the *address* of z. The transfer of information from main to function1 is shown in Fig. 5.10.

2. Where did the values for c *and* t *come from and why are they addresses rather than int or double?* The "address of" operator (&) was used to create these values, and the * symbol was used to indicate that an address rather than an int or double was to be placed in the variable value location. The variables c and t are known as *pointer variables* because they contain addresses.

The symbol & is an operator that means "address of." It is a unary prefix operator, meaning that it is placed before a single identifier. In the third printf statement

FIG. 5.10

The transfer of information for this lesson's program

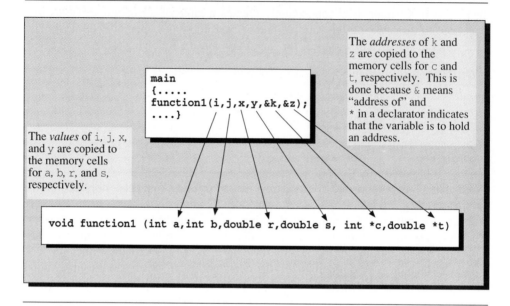

of this lesson's program, it causes the address of the variable k to be printed. In the call to function1, it creates the parameters "address of variable k" and "address of variable z" as the fifth and sixth parameters. These parameters are copied into the region for function1 at the fifth and sixth parameter locations.

In the declarator for the function, the symbol * appears in the fifth and sixth parameters. In this context, which is a declaration, the symbol * indicates that the variable following the * is to store an address. The declaration int *c indicates that c will store the address of an int, and double *t indicates that t will store the address of a double. Thus, the call to function1 (with &) and the declarator for function1 (with *) cause the addresses of variables k and z to be known to function1.

3. What does function1 do with those addresses? Since function1 knows the addresses, it knows where to store values that it wants to give k or z. Therefore, with an operator that says "go to the address," it can store the values it wants.

4. What operator says "go to the address"? The unary operator * (also called the *indirection operator*) used in the *body* of function1 causes the action of going to the address. Before we explain it in detail, just note that * has been used in the programs in this book for three different purposes. They are

1.   The binary multiplication operator; for example,
     ```
     d = e * f;
     ```

2. The declarator specifier indicating that an address is to be stored in a variable's memory cell; for example,

```
void function1 (int a, int b, double r,double s,int *c,double *t)
```

3. The unary operator indicating to go to the address; for example,

```
*c = a + b;
*t = r+s+(*c);
```

We point out these three different applications of * because it is important that you not confuse one application with another. Observe that, whenever * is used in a declaration, the usage corresponds to the second purpose. When * is used on the left side of an assignment statement, the usage corresponds to the third purpose. When * appears on the right side of an assignment statement, you will need to look at the expression closely and determine if it is a binary or unary operator to distinguish between the first and third purposes.

As a unary operator, * says to "go to the address contained in the memory cell of the identifier that follows." In other words, *c says "go to the address contained in c's memory cell." And *t says "go to the address contained in t's memory cell." For

```
*c = a + b;
```

the right side of this assignment statement causes the values stored in the memory cells of a and b to be added. The left side of the statement indicates go to the memory cell whose address is stored in variable c. As with all assignment statements, the = causes the value created on the right side to be stored in the location indicated by the left side. So, using the addresses shown in the previous table, the sum of a and b is put into the memory cell with the address FFF0. Note that this address also is the address of variable k in main. Therefore, this action causes the variable k in main to become the sum of a and b.

For

```
*t = r+s+(*c);
```

the right side of the assignment statement sums the values contained in the memory cells for r, s, and at the address stored in c's memory cell. The left side of the assignment statement indicates to go to the memory cell whose address is stored in variable t. The = causes the value created on the right side to be stored in the location indicated by the left side. Therefore, using the addresses shown in the table, the sum of r, s, and the contents of cell FFF0 is put into the memory cell with the address FFD8. Note that FFD8 also is the address of variable z in main. So, this action causes the variable z in main to become the sum of r, s, and FFF0's contents (which is the value of k).

*5. What is a visual image of the two address operators & and *?* We realize that the preceding words are confusing and difficult to follow. To work out problems with these operators, we believe a simple visual image will be more valuable. Visualize a table with variable names in the first column, followed by variable type, address, and value. The & operator can be visualized as an arrow from a variable's value to its address, as shown in Fig. 5.11. The & arrow in this figure indicates the operation &k. The unary * operator can be visualized as an arrow going from a variable's value to the address in its memory cell to the variable value at the address. The * arrow in this figure indicates the operation *t.

This figure illustrates that the & operator causes the address of a variable to be used, not the value of the variable. It also shows the path followed by using the unary * operator. It shows that the unary * operator causes the value of the variable whose address is stored to be used. Because the words describing the process are difficult to remember, we recommend you memorize the visual image. If you do this, you will be able to work with the & and unary * operators very effectively.

6. What are some examples of things we can and cannot do with these operators? Use the visual image described previously to understand the examples given here. For instance, if we had a statement in function1 that was

$$*t = 83.9;$$

FIG. 5.11

The operations of * and &. The action shown is *t. Since t has an address, *t indicates the value of z (because z's address is the value of t.) The other action shown is &k. The & is the "address of" operator. Therefore, &k is the address of k, which is FFF0.

Variable name	Variable type	Variable address		Variable value	
i	int	FFF4		5	
j	int	FFF2	&	6	
k	int	FFF0			main
x	double	FFE8		10.6	
y	double	FFE0		22.3	
z	double	FFD8			
a	int	FFC0		5	
b	int	FFC2	*	6	
c	address to int	FFD4		FFF0	function1
r	double	FFC4		5	
s	double	FFCC		6	
t	address to double	FFD6		FFD8	

its effect would be the same as the statement

$$z = 83.9;$$

in main. We cannot, however, use the statement (for example)

$$\&k = FDD2;$$

if we wanted for some reason to make sure that the variable k was stored in the memory cell whose address is FDD2, because we have no control over the address. We cannot know at the time we are writing our program that the memory cell FDD2 would be open. (There are other problems with this statement, not worth discussing at this time.) C determines the addresses itself, so, we cannot have & on the left side of an assignment statement.

We can use the & operator on any of the variables in this program (provided it is not on the left side of an assignment statement). However, we are not allowed to use the unary * operator on any variable. The unary * operator must be used only on variables declared with * in the declaration—that is, pointer variables. Thus, we cannot use

$$*y = 83.9;$$

because the variable y had no * in its declaration (and, so, has no address in its memory cell). We also cannot have the statement

$$t = 83.9;$$

because the variable t should hold only an address.

Also, for instance, we cannot write

$$t = FFC4;$$

because we do not know in advance which memory cells will be available (and other things). We can use the "address of" operator to handle the situation, however. For instance, a valid statement would be

$$t = \&r;$$

This would cause the address of r to be stored in the memory cell for the variable t. Notice that the variable r is a double and the declaration for t is double *t. It is important that the declarations match.

This description is not meant to be a complete discussion of pointer variables and the & and unary * operators. More will be said about them as we go further in this book.

7. What conversion specification can be used in the string literal of a printf statement to print addresses given by pointer variables? The conversion specification %p prints an address in a notation that is consistent with the addressing scheme used by your computer system. In many cases, this will be hexadecimal.

8. Will the same addresses be printed out when this lesson's program is run on a different computer system? Maybe and maybe not; however, the actual addresses are not important to the performance of this program. As long as the correspondence of the addresses for the different variables exists, the program will perform properly.

9. Is it possible to use a calling parameter as both an input parameter and an output parameter? Yes, all parameters in the function call with & preceding them indicate that the addresses of the variables are available to the function. This means that the value of the variable at the address can be used as input to the function. It also is possible that the function can change the value of the variable at that address. Thus, the value of the variable can serve as both input to the function and output from the function.

10. Where else have we encountered the & operator in this text? For using the scanf and fscanf functions, the & operator was used. For instance, to read an integer k from the keyboard, the statement would be

```
scanf ("%d", &k);
```

11. What does the & do in this statement? Again, the & is the "address of" operator. By using it in this statement, we pass to the scanf function the address that has been reserved for the variable k. Since k has not yet been initialized, no value is in that address; however, the address already has been designated to have k stored there when a value is indicated for it. Since the scanf function knows the address of k, it inserts the value read in from the keyboard into the memory cell with that address. It knows how many bytes the value is to occupy from the conversion specification %d.

12. It seems then that there are two ways to pass values from a function to the calling function; one uses the return statement and the other uses the addresses in the parameter list. If we wanted to, could we use both methods with the same function? Yes, you might like to return values both through the parameter list and the return statement. To do this, make sure that you do not declare the function to be type void in the prototype.

EXERCISES
1. True or false:
    a. A variable in a C program, regardless of its type, must have an address.
    b. Any type variable can be used to store the address of a variable.
    c. The address of a variable is expressed in hexadecimal form; therefore, only an integer type variable can be used to store the address of a variable.
    d. Any type pointer variable can be used to store the address of a double type variable.
    e. Pointer variables are more difficult to use than scalar variables because the programmer needs manually to find the addresses of variables to be stored in the pointer variables.

f.   The indirection operator * may appear only on the left side of an assignment statement.

g.   The indirection operator * may appear on both sides of an assignment statement.

h.   The address of operator & may appear only on the left side of an assignment statement.

i.   The address of operator & may appear on both sides of an assignment statement.

j.   For an int type variable, aa, its address, &aa, is a constant, not a variable.

2.   Write a program that calls a function to exchange the values of two long type integers; however, the prototype of the function may contain only one long type pointer variable.

Solutions

1.   a (true), b (false), c (false), d (false), e (false), f (false), g (true), h (false), i (false), j (true)

▧ APPLICATION PROGRAM 5.1 PASSING MANY VALUES TO A FUNCTION AND RETURNING ONE VALUE—INTEGRATION WITH THE TRAPEZOIDAL RULE (NUMERICAL METHOD EXAMPLE)

Problem statement

Write a program that can integrate a third order polynomial function using the trapezoidal rule. The input should come from the keyboard and consist of

1.   The number of trapezoids to be used
2.   The limits of integration
3.   The coefficients a, b, c, and d of the polynomial $ax^3 + bx^2 + cx + d$.

The output should be to the screen and consist of the area under the polynomial between the limits of integration given.

Solution

RELEVANT EQUATIONS

You may have learned the trapezoidal rule in your calculus class. If not, we review it here. The trapezoidal rule states that the area under a curve can be approximated by calculating the area of a number of trapezoids formed by connecting individual points on the curve. This is illustrated in Fig. 5.12.

This figure shows that, in some regions of the curve, a trapezoid is a good representation of the area. In other regions, a trapezoid is not particularly representative of the true area. Clearly, as more (and therefore narrower) trapezoids are used (between the same limits), the approximation improves. When using the trapezoidal rule, we decide the number of trapezoids to be represented. Since we are writing a computer program to do the calculations, there is little penalty for choosing a large number.

The area of the shaded trapezoid shown in Fig. 5.12 is

$$A = h\,(y_1 + y_2)/2$$

where A = area.

We first decide the number of trapezoids to be used, n. Then, with the given left and right limits of integration, the value of h is computed to be

$$h = (\text{right limit} - \text{left limit})/n$$

with h and the left limit we can get x_1 and x_2 for each trapezoid. Then, the corresponding y_1 and y_2 can be calculated from

$$y_1 = ax_1^3 + bx_1^2 + cx_1 + d$$
$$y_2 = ax_2^3 + bx_2^2 + cx_2 + d$$

With h, y_1, and y_2, we can get the area of each trapezoid. We then can sum the area of all trapezoids to get the total area between the limits.

SPECIFIC CALCULATION

Suppose we want to determine the area beneath the curve:

$$y = -2x^3 + 3x^2 + x - 4$$

FIG. 5.12

Trapezoid in trapezoidal rule. Note that the area of the trapezoid approximates the area beneath the curve between x_1 and x_2.

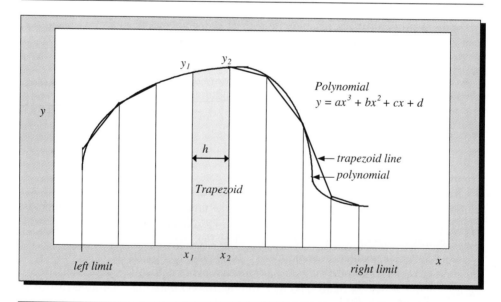

between $x = 1$ and $x = 5$. For illustration purposes, we will use four trapezoids to do this (for a real problem you would choose many more). This gives us a value of h:

$$h = (5 - 1)/4 = 1$$

For each trapezoid, we get the following. For trapezoid 1,

$$x_1 = 1$$
$$x_2 = 1 + 1 = 2$$

$$y_1 = -2(1)^3 + 3(1)^2 + (1) - 4 = -2$$
$$y_2 = -2(2)^3 + 3(2)^2 + (2) - 4 = -6$$

$$A = (-2 - 6)(1)/2 = -4$$

For trapezoid 2,

$$x_1 = x_2 \text{ from trapezoid } 1 = 2$$
$$x_2 = x_1 + h = 3$$

$$y_1 = -2(2)^3 + 3(2)^2 + (2) - 4 = -6$$
$$y_2 = -2(3)^3 + 3(3)^2 + (3) - 4 = -28$$

$$A = (-6 - 28)(1)/2 = -17$$
$$A_{tot} = -17 - 4 = -21$$

and so on (repeat the steps for each trapezoid). The following table of results summarizes the calculations for all four trapezoids:

Trapezoid	x_1	x_2	y_1	y_2	A
1	1	2	-2	-6	-4
2	2	3	-6	-28	-17
3	3	4	-28	-80	-54
4	4	5	-80	-174	-127

The total area is the sum of the values in the last column:

$$A_{tot} = -4 - 17 - 54 - 127 = -202$$

Note that the exact value of the integral is

$$A = \left[-\frac{x^4}{2} + x^3 + \frac{x^2}{2} - 4x \right]_1^5 = -192$$

Clearly, our trapezoid answer is approximate. We would do better with more trapezoids. For instance, with 100 trapezoids we get $A = -192.016000$, which is much closer to the correct result.

ALGORITHM

We choose to put the calculation of the area of each trapezoid into a function called `trapezoid`. This gives us the data flow diagram in Fig. 5.13 that illustrates that main passes the values of h, x_1, x_2, a, b, c, and d to function `trapezoid` and the area of a trapezoid is returned from the function to main.

This general algorithm can be written from the specific calculation and the data flow diagram.

```
prompt the user to enter the number of trapezoids, left limit, right limit
read number of trapezoids, left limit, right limit
prompt the user to enter the coefficients of the polynomial
read coefficients of the polynomial
calculate h
```

```
calculate x₁
calculate x₂
calculate y₁      Function trapezoid.      Repeat for each
calculate y₂                               trapezoid.
calculate A
```

```
accumulate A's from each trapezoidd
print the total A
```

▨ **FIG. 5.13**

Data flow diagram for integration using the trapezoidal rule program

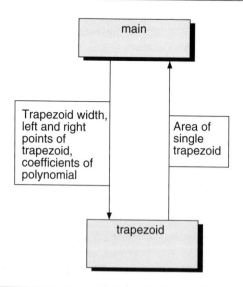

SOURCE CODE

This source code can be written from the algorithm and the data flow diagram:

```c
#include <stdio.h>
#include <math.h>

double trapezoid (double h, double x1, double x2,
                  double a, double b, double c, double d);
void main (void)
{
        double left_point, right_point, a, b, c, d, h, area;
        int n, i;

        printf ("Enter n, left point, right point\n");
        scanf ("%d %lf %lf", &n, &left_point, &right_point);
        printf ("a, b, c, d\n");
        scanf ("%lf %lf %lf %lf", &a, &b, &c, &d);

        area = 0.0;
        h = (right_point - left_point)/n ;
        for (i=1; i<=n; i++)
                {
                right_point = left_point + h ;
                area += trapezoid (h, left_point, right_point, a,b,c,d);
                left_point = right_point ;
                }

        printf ("Area = %lf \n", area);
}

double trapezoid (double h, double x1, double x2,
                  double a, double b, double c, double d)
{
        double  y1, y2, area_of_trapezoid;
        y1 = a*x1*x1*x1+b*x1*x1+c*x1+d;
        y2 = a*x2*x2*x2+b*x2*x2+c*x2+d ;
        area_of_trapezoid = (y1+y2)*h/2.;
        return (area_of_trapezoid) ;
}
```

Reading the input values.

Calculating the distance between points, h.

Looping over the number of intervals, *n*.

Advancing the points considered.

Calling trapezoid to calculate the area of a single trapezoid. The compound assignment operator causes the area to be the sum of the areas of all the individual trapezoids.

Function declarator. Because none of the variables has *, all of the values go in but do not go back.

Calculating the area of a single trapezoid.

Returning the area of a single trapezoid.

OUTPUT

	Enter n, left point, right point
Keyboard input	100 1 5
	a, b, c, d
Keyboard input	-2 3 1 -4
	Area = -192.016000

COMMENTS

This is a straightforward example of the use of a function that returns a single value. It illustrates a numerical technique of the sort commonly used in programs in engineering. While the trapezoidal rule is effective and direct, to get a reliable result, a large number of trapezoids may be needed, requiring a large number of computations. As you learn more about numerical methods, you will learn other techniques for evaluating integrals that require fewer computations to achieve the same reliability, being more efficient.

Modification exercises

1. Modify the program so that it can integrate a fourth order polynomial.

2. Modify the program so that it can integrate a fifth order polynomial.

3. Rewrite a portion of the program so that a function is used to accumulate the area. Give this function the name `area` and have it call the function trapezoid.

▨ APPLICATION PROGRAM 5.2 USING FUNCTIONS WITH COMPLEX LOOPS, WORKING WITH GRIDS, A LOGIC EXAMPLE

In developing engineering programs, a programmer often is faced with the task of working with a grid or a mesh. A grid may be used to analyze stresses in a plate or temperature distribution in a solid object, by subdividing the region of interest into smaller, more manageable parts. Then, each individual part can be evaluated.

The application program for this lesson is not a standard engineering example. We show it, however, to help you develop logic skills in dealing with grids and in writing loops.

Problem statement

A chessboard has eight rows and eight columns. If you are familiar with chess, then you know that a knight can move from its location (i, j) in the k direction one or two squares and in the m direction one or two squares. A knight in the center of the chessboard has eight possible moves as illustrated in Fig. 5.14 and listed in the table.

However, a knight located in the corner has fewer possible moves because of the edges of the chessboard, as illustrated in Fig. 5.15 and shown in the table.

For this application, write a program that can compute the number of possible moves for a knight located at any location on the chessboard. There is no input information. The output is to be to the screen in the form of a gridlike pattern, showing the chessboard with the number of possible moves as a numeral in each chessboard location.

Solution

With a grid problem, it is necessary first to number the grid locations. For the chessboard, we recommend using a two number sequence (i, j) to label the board. Fig. 5.16 shows some grid locations. The values of i and j increase from left to right and down to up, respectively. You should be able to deduce the values not shown from the ones shown.

RELEVANT EQUATIONS

If the initial knight location is (i, j), then we will call the new location $(i+k, j+m)$. A move is legal if

$$1 <= (i + k) <= 8 \qquad (5.1)$$

FIG. 5.14

Possible movement locations (shaded circles) of a knight (clear circle) in the center of a chessboard. Note the possible values of k and m.

k	m
-2	+1
-2	-1
-1	+2
-1	-2
+1	+2
+1	-2
+2	+1
+2	-1

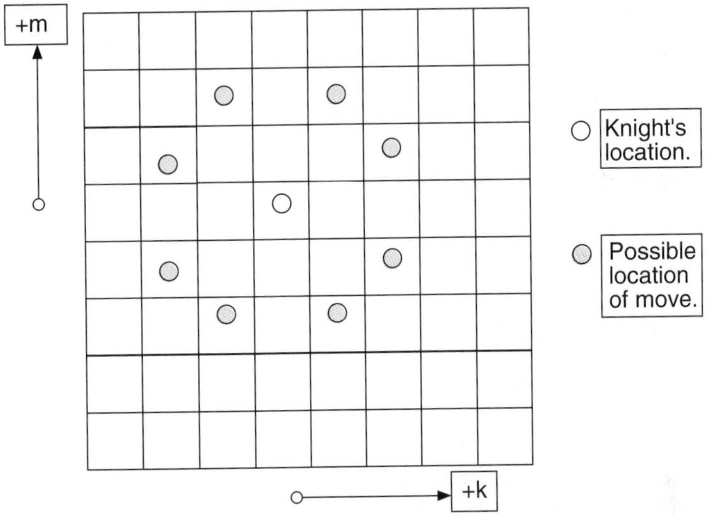

FIG. 5.15

Possible movement locations (shaded circles) of a knight (clear circle) located in the corner of a chessboard. Note the possible values of k and m for a knight in this location are fewer than those shown in Fig. 5.14.

k	m
-2	+1
-1	+2

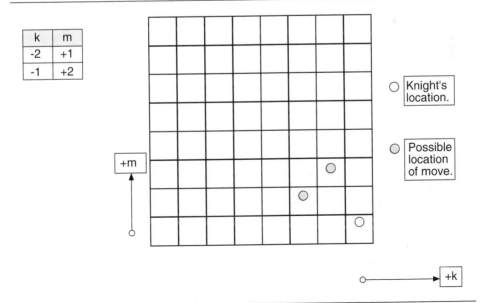

and

$$1 <= (j + m) <= 8 \qquad (5.2)$$

as seen from Fig. 5.16.

A list of all possible k and m are given in the table of Fig. 5.14. We must check each location (i, j) on the chessboard for all possible combinations of k and m to see if they are feasible. If a move is found to be feasible, then we will count it. All unfeasible moves simply will not be counted. Since there are 8 possible combinations of k and m and 64 (i, j) locations on the chessboard, we will check $8 * 64 = 512$ possible moves.

SPECIFIC CALCULATION

Consider the location $(7, 4)$ on the chessboard ($i = 7$ and $j = 4$). We list all the possible moves and check to see if the moves fall within the chessboard and not outside the edges. Using the values of k and m from the table in Fig. 5.14 and equations (5.1) and (5.2), they are

$(7 - 2, 4 + 1)$ ok
$(7 - 2, 4 - 1)$ ok
$(7 - 1, 4 + 2)$ ok
$(7 - 1, 4 - 2)$ ok
$(7 + 1, 4 + 2)$ ok
$(7 + 1, 4 - 2)$ ok

FIG. 5.16
Grid locations—the values of i and j are shown for just some of the squares

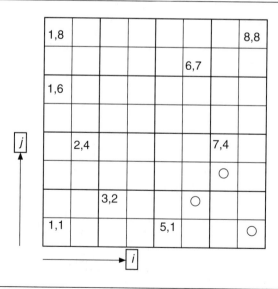

(7 + 2, 4 + 1)   not valid (9 is greater than 8 allowable, meaning the move goes past the edge)

(7 + 2, 4 − 1)   not valid (9 is greater than 8 allowable, meaning the move goes past the edge)

This location, therefore, has 6 valid moves.

To find a solution at each point on the board, we would need to do the preceding calculation for i from 1 to 8 and j from 1 to 8.

ALGORITHM AND CODING LOGIC

We can break our program into two distinct parts:

1. Creating the values of i and j that correspond to each point on the chessboard.
2. Checking a single point (i, j) for the number of permissible moves.

We create a function, check_point, to do the second part and we do the first part in main. This leads to the data flow diagram in Fig. 5.17.

Function check_point. To check an individual point, we must write code that does what is illustrated in the specific calculation. We realize that we must write a number of loops to accomplish this. We first focus on how we can create the combinations of k and m within a loop.

FIG. 5.17
Data flow diagram for knight moves program

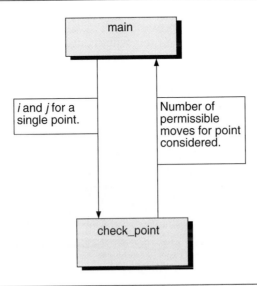

We note in the list of possible *k* (in the table of Fig. 5.14), *k* goes from -2 to 2 excluding 0. This implies a for loop of the sort

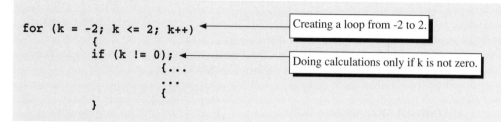

```
for (k = -2; k <= 2; k++)        Creating a loop from -2 to 2.
     {
     if (k != 0);                 Doing calculations only if k is not zero.
        {...
        ...
        {
     }
```

Note that, with this loop structure we create four iterations (for k $= -2, -1, 1,$ and 2). From the table of Fig. 5.14 we see that we need to create two values of *m* for each of these values of *k*.

We can calculate *m* from the value of *k* if we note that

$$|k| + |m| = 3$$

where |*k*| is absolute value of *k* and |*m*| is absolute value of *m*. Therefore,

$$|m| = 3 - |k|$$

which can be stated as

$$m = (3 - |k|)$$

or

$$m = -(3 - |k|)$$

So, we need to create a loop that executes twice within the loop that goes from -2 to 2 . The first time through the loop, $m = (3 - |k|)$, and the second time through the loop $m = -(3 - |k|)$. We can do this with the following:

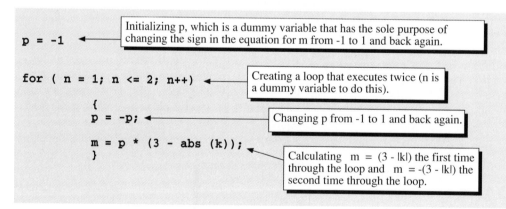

```
p = -1          Initializing p, which is a dummy variable that has the sole purpose of
                changing the sign in the equation for m from -1 to 1 and back again.

for ( n = 1; n <= 2; n++)        Creating a loop that executes twice (n is
                                 a dummy variable to do this).
     {
     p = -p;                      Changing p from -1 to 1 and back again.
     m = p * (3 - abs (k));       Calculating  m = (3 - |k|) the first time
     }                            through the loop and  m = -(3 - |k|) the
                                  second time through the loop.
```

Putting this into the loop for k gives

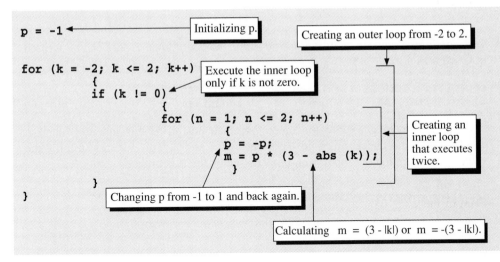

After developing loops that are somewhat complex, you should check that they create the values you want by making a table of the values and following the flow of the loop. For instance, for this loop we have this table:

Initialization of *p*	*k*	*n*	*p*	*m*
−1	−2	1	1	1
		2	−1	−1
	−1	1	1	2
		2	−1	−2
	0	End loop		
	1	1	1	2
		2	−1	−2
	2	1	1	1
		2	−1	−1

For this loop structure, we indeed have created the values of *k* and *m* that give us the combinations we want. Therefore, we can use this loop structure in our program. With a given combination of *i* and *j,* we can use each of these combinations of *k* and *m* to check whether we have a valid move for the knight. If we have a valid move, then we should add 1 to the count that keeps track of the number of valid moves.

If we call the counter icount, then we should add 1 to icount when a move is valid. This indicates that we need an if statement and a condition to evaluate. The condition was stated earlier to be

$$1 <= (i + k) <= 8$$

and

$$1 <= (j + m) <= 8$$

This conditional and the addition to `icount` if the conditional is true can be written in C form as

```
if ( 1 <= (i+k) && (i+k)   <=   8              &&
1   <=   (j+m)   &&   (j+m)   <=   8)       icount++;
```

This statement goes into the inner loop. We put all of this into the function check_point. Into check_point go the values of `i` and `j`. This function returns the value of `icount` for the input `i` and `j`. This gives the function:

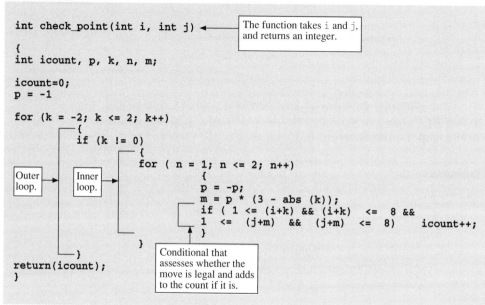

Function main. Within main, we need to make loops that create combinations of `i` and `j` that represent each point on the chessboard. This is an easier structure to create than what we did for function check_point. The following loop gives it to us:

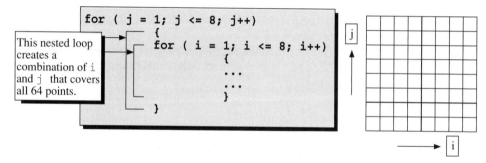

Using this nested loop, we create a combination of i and j for each square on the chessboard. This is shown in abbreviated form in the table that follows:

j	i
1	1 through 8
2	1 through 8
3	1 through 8
4	1 through 8
5	1 through 8
6	1 through 8
7	1 through 8
8	1 through 8

This loop indeed creates the values of i and j that cover the entire chessboard. For each combination of i and j, we then check all the possible combinations of k and m. To check k and m, we call the function check_point within the previous loop. This leads to the following for main:

```c
void main (void)
{
        int n, i, j, icount;
        printf ("Number of possible moves for a knight "
                "on a chess board \n\n");
        for (j = 1; j <= 8; j++)
                {
                for (i = 1; i <= 8; i++)
                        {
                        icount=check_point(i,j);
                        printf ("%5d", icount);
                        }
                printf ("\n");
                }
}
```

Function check_point returns the number of valid moves.

We print the number of valid moves.

To get the table correct, we advance a line after completing an entire inner loop.

Combining these two functions leads to the source code.

SOURCE CODE

```c
#include <stdio.h>
#include <math.h>
int check_point(int i, int j);
void main (void)
{
        int n, i, j, icount;
        printf ("Number of possible moves for a knight "
                "on a chess board \n\n");
        for (j = 1; j <= 8; j++)
                {
                for (i = 1; i <= 8; i++)
                        {
                        icount=check_point(i,j);
                        printf ("%5d", icount);
                        }
                printf ("\n");
                }
}
```

Calling check point to calculate the number of valid moves for each point.

Nested loop that creates i and j for each point.

The number of valid moves is assigned to icount.

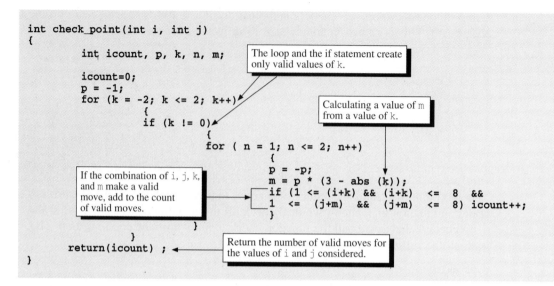

```
int check_point(int i, int j)
{
        int icount, p, k, n, m;              [The loop and the if statement create
                                              only valid values of k.]
        icount=0;
        p = -1;
        for (k = -2; k <= 2; k++)
                {
                if (k != 0)                  [Calculating a value of m
                        {                     from a value of k.]
                        for ( n = 1; n <= 2; n++)
                                {
                                p = -p;
                                m = p * (3 - abs (k));
        [If the combination of i, j, k,       if (1 <= (i+k) && (i+k)   <=  8  &&
        and m make a valid                     1  <=  (j+m)  &&  (j+m)   <=  8) icount++;
        move, add to the count                }
        of valid moves.]               }
                }
        return(icount) ;    [Return the number of valid moves for
}                            the values of i and j considered.]
```

OUTPUT

```
Number of possible moves for a knight on a chessboard:

2   3   4   4   4   4   3   2
3   4   6   6   6   6   4   3
4   6   8   8   8   8   6   4
4   6   8   8   8   8   6   4
4   6   8   8   8   8   6   4
4   6   8   8   8   8   6   4
3   4   6   6   6   6   4   3
2   3   4   4   4   4   3   2
```

COMMENTS

To keep this program reasonably simple and directly related to a chessboard, we have not used macros for such things as the number of squares in the *i* and *j* directions. However, the program would become more versatile had we done so, because it would be easy to modify the program to accommodate other size boards.

Also, function check_point was written for a knight. One could imagine a different function for each chess piece and create a program that could determine the number of options for any piece.

Modification exercises

1. Modify the program to handle a chessboard that is 10 by 10 rather than 8 by 8.

2. Modify the program to handle a chessboard that is 15 by 23 rather than 8 by 8.

3. Write a new function for the program that is capable of handling a pawn's moves.

4. Write a new function for the program that is capable of handling a rook's moves.

5. Write a new function of the program that is capable of handling a bishop's moves.

APPLICATION PROGRAM 5.3 MODULAR PROGRAM DESIGN—AREA OF PARALLELOGRAM, VOLUME OF PARALLELEPIPED (NUMERICAL METHOD EXAMPLE)

Now that we have covered functions, we can demonstrate modular program design. In practice, you probably would not make this particular program exactly as we have shown it. In this example, we are interested primarily in illustrating modular design concepts. The method is valuable, and we recommend that you get into the habit of writing programs in this manner. The experience will pay off when you begin to write much larger programs.

Problem statement

Write a program that can calculate the area of a parallelogram defined by two vectors and the volume of a parallelepiped defined by three vectors (Fig. 5.18). Input the **i, j,** and **k** components of each of the three vectors from the keyboard and print the result to the screen. Use a modular design and comment the program fully.

Solution

RELEVANT EQUATIONS

You may have learned in your math classes that the area of a parallelogram defined by the vectors

$$\mathbf{A} = a_1\mathbf{i} + a_2\mathbf{j} + a_3\mathbf{k}$$
$$\mathbf{B} = b_1\mathbf{i} + b_2\mathbf{j} + b_3\mathbf{k}$$

FIG. 5.18
Parallelogram formed by two vectors and parallelepiped formed by three vectors

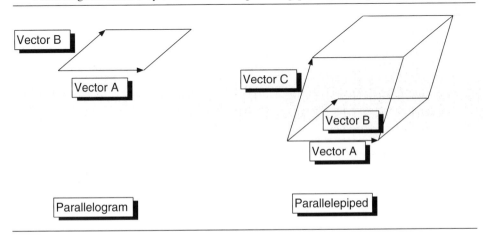

Vector B

Vector A

Vector C

Vector B

Vector A

Parallelogram

Parallelepiped

is the length of the vector created by the cross product of the two vectors. This is indicated in equation form as

$$Area = |\mathbf{A} \times \mathbf{B}|$$

where the $||$ symbol indicates the length of the vector.

Recall that the cross product of the two vectors is given by the following determinant:

$$\begin{vmatrix} \mathbf{i} & \mathbf{j} & \mathbf{k} \\ a_1 & a_2 & a_3 \\ b_1 & b_2 & b_3 \end{vmatrix} = (a_2 b_3 - a_3 b_2)\,\mathbf{i} - (a_1 b_3 - a_3 b_1)\,\mathbf{j} + (a_1 b_2 - a_2 b_1)\,\mathbf{k}$$

and that the length of a vector is given by the square root of the sum of the squares of the components. In equation form, this is the following. For a vector $\mathbf{G}$,

$$\mathbf{G} = g_1\mathbf{i} + g_2\mathbf{j} + g_3\mathbf{k}$$

The length of $\mathbf{G}$ is

$$|\mathbf{G}| = (g_1^2 + g_2^2 + g_3^2)^{0.5}$$

You also may have learned that the volume of the parallelepiped defined by the three vectors

$$\mathbf{A} = a_1\mathbf{i} + a_2\mathbf{j} + a_3\mathbf{k}$$
$$\mathbf{B} = b_1\mathbf{i} + b_2\mathbf{j} + b_3\mathbf{k}$$
$$\mathbf{C} = c_1\mathbf{i} + c_2\mathbf{j} + c_3\mathbf{k}$$

is

$$Volume = \text{abs } [\mathbf{A} \cdot (\,(\mathbf{B} \times \mathbf{C})]$$

where the dot product of two vectors $\mathbf{G}$ and $\mathbf{H}$:

$$\mathbf{G} = g_1\mathbf{i} + g_2\mathbf{j} + g_3\mathbf{k}$$
$$\mathbf{H} = h_1\mathbf{i} + h_2\mathbf{j} + h_3\mathbf{k}$$

is

$$\mathbf{G} \cdot \mathbf{H} = g_1 h_1 + g_2 h_2 + g_3 h_3$$

SPECIFIC CALCULATION

Consider the following vectors:

$$\mathbf{A} = 3\mathbf{i} + 2\mathbf{j} + 2\mathbf{k}$$
$$\mathbf{B} = 4\mathbf{i} + 3\mathbf{j} + 1\mathbf{k}$$
$$\mathbf{C} = 8\mathbf{i} + 2\mathbf{j} + 7\mathbf{k}$$

The area of the parallelogram is

$$\mathbf{A} \times \mathbf{B} = \begin{vmatrix} \mathbf{i} & \mathbf{j} & \mathbf{k} \\ 3 & 2 & 2 \\ 4 & 3 & 1 \end{vmatrix} = [(2 * 1) - (2 * 3)]\mathbf{i} - [(3 * 1) - (2 * 4)]\mathbf{j} + [(3 * 3) - (2 * 4)]\mathbf{k}$$

which gives

$$A \times B = -4i + 5j + k$$

The length of this vector is

$$|A \times B| = [(-4)^2 + (5)^2 + (1)^2]^{0.5} = 6.48074$$

Therefore, the area of the parallelogram is 6.48074.
To get the volume,

$$B \times C = \begin{vmatrix} i & j & k \\ 4 & 3 & 1 \\ 8 & 2 & 7 \end{vmatrix} = [(3 * 7) - (1 * 2)]i - [(4 * 7) - (1 * 8)]j + [(4 * 2) - (8 * 3)]k$$

which gives

$$B \times C = 19i - 20j - 16k$$
$$A \cdot (B \times C) = (3i + 2j + 2k) \cdot (19i - 20j - 16k)$$
$$= (3 * 19) + [2 * (-20)] + [2 * (-16)] = -15.0$$

The absolute value of -15.0 is 15.0. Therefore, the volume of the parallepiped is 15.0.

ALGORITHM

With the algorithm we include a structure chart and a data flow diagram for this modularly designed program.

From the hand calculation, the general algorithm can be seen to be

```
Read the components of the three vectors, A, B, and C.
Calculate the area defined by A and B by:
   Calculating the cross product of vectors A and B
   Calculating the magnitude of the result of the cross prod-
   uct (this is the area)
Calculate the volume defined by A, B, and C by
   Calculating the cross product of vectors B and C
   Calculating the dot product of A and B X C (the absolute
   value of this is the volume)
Print the results.
```

This algorithm has been deliberately written without a lot of detail to help develop a modular design for this program. From this algorithm we would get the following structure chart shown in Fig. 5.19. Note that main is used primarily to call other modules. A good modular design uses main not for specific calculations but for directing the modules that do specific calculations. Here, we have chosen to even read and print the results using modules. This is good practice and especially beneficial when large amounts of data are to be read.

You should observe the similarity of the algorithm to the structure chart. Note that both the area and volume calculations involve finding the cross product of two vectors. Therefore, both modules use the cross product module.

FIG. 5.19
Structure chart for area/volume program

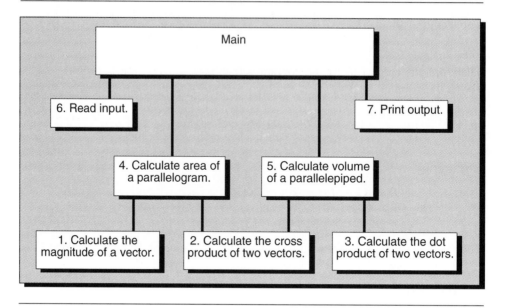

The next step of the process is to determine the information that passes into and out of each module. Table 5.2 lists the module, the required information, passed into and out of the function. We can put this information in a data flow diagram (Fig. 5.20).

With the structure chart and data flow diagram we define the purpose, input, and output of each module. We now can go forward with developing algorithms for each individual module. As you begin your programming career, you probably will not develop perfect structure charts and data flow diagrams on your first attempts. Do not be discouraged by this. Begin with what you believe is a good estimate of a structure chart and data flow diagram. Then, proceed with the development of the algorithm for each module. Thus, you will develop more insight into the problem. You always can go back to your structure chart and data flow diagrams and make adjustments. As you gain experience, you will get better at your first estimates of the structure charts and data flow diagrams and less repetition will be required.

Here are the algorithms for the functions. For the function `dot_product`,

```
Receive components of two vectors: g₁, g₂, g₃, h₁, h₂, h₃
Calculate dot product  G · H = g₁h₁ + g₂h₂ + g₃h₃
Return dot product
```

For the function `cross_product`,

```
Receive components of two vectors: g₁, g₂, g₃, h₁, h₂, h₃
Calculate cross product components
```

TABLE 5.2
Information flow

Module	Information passed into	Information passed out of
1. Vector magnitude	Three components of a vector (three values)	Length or magnitude of the vector (one value)
2. Cross product	Three components of two vectors for which the cross product is desired (six values)	Three components of the vector that is the cross product (three values)
3. Dot product	Three components of two vectors for which the dot product is desired (six values)	Dot product of the two vectors (one value)
4. Area of parallelogram	Three components of the two vectors which form the parallelogram (six values)	Area of the parallelogram (one value)
5. Volume of parallelepiped	Three components of the three vectors that form the parallelepiped (nine values)	Volume of the parallelepiped (one value)
6. Read input	Nothing	Three components of the three vectors that form the parallelogram and parallelepiped (nine values)
7. Print output	Three components of the three vectors that form the parallelogram and parallelepiped, area of parallelogram and volume of parallelepiped (11 values)	Nothing

$$k_1 = (g_2\, h_3 - g_3\, h_2)$$
$$k_2 = -(g_1\, h_3 - g_3\, h_1)$$
$$k_3 = g_1\, h_2 - g_2\, h_1$$

```
Return components of cross product k₁, k₂, k₃
```

Return components of cross product k_1, k_2, k_3

For the function `vector_magnitude`,

```
Receive components of vector g₁, g₂, g₃
Calculate magnitude = (g₁ g₁ + g₂ g₂ + g₃ g₃)^0.5
Return magnitude
```

Receive components of vector g_1, g_2, g_3
Calculate magnitude $= (g_1\, g_1 + g_2\, g_2 + g_3\, g_3)^{0.5}$
Return magnitude

For the function `area_parallelogram`,

```
Receive components of vectors g₁, g₂, g₃, h₁, h₂, h₃
Calculate cross product of G and H
Calculate magnitude of cross product
Return area of parallelogram
```

Receive components of vectors $g_1, g_2, g_3, h_1, h_2, h_3$
Calculate cross product of **G** and **H**
Calculate magnitude of cross product
Return area of parallelogram

FIG. 5.20

Data flow diagram for area/volume program

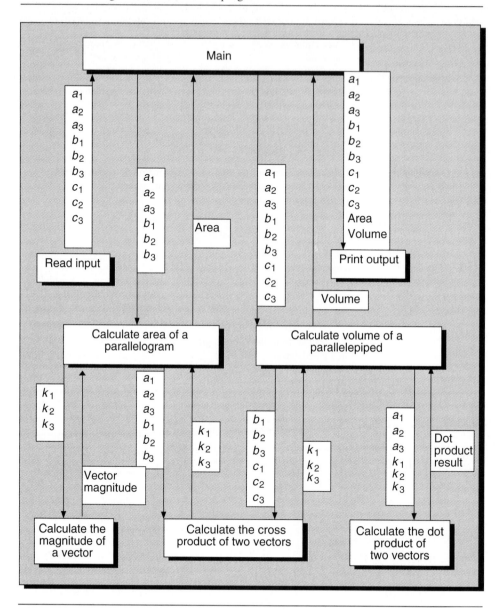

For the function `volume_parallelepiped`,

    Receive components of vectors g_1, g_2, g_3, h_1, h_2, h_3, k_1, k_2, k_3
    Calculate cross product of $\mathbf{H}$ and $\mathbf{K}$
    Calculate dot product of $\mathbf{G} \cdot (\mathbf{H} \times \mathbf{K})$

```
Take absolute value
Return volume of parallelepiped
```

For the function read_input,

```
Receive nothing
Print prompts to screen
Read components of three vectors
Return components of vectors
```

For the function print_output,

```
Receive components of three vectors, area of parallelogram,
  volume of parallelepiped
Print them to the screen
Return nothing
```

SOURCE CODE

From the structure chart, data flow diagram, and algorithms for each individual module, the following source code has been written. To understand this program, you should look at the algorithm for each function and compare it to the code for each function. Note how closely they correlate. Can you understand how writing an algorithm for each function can guide you to writing the code for the function?

Look at the data flow diagram and find each function that returns only one value. Note that each function on the data flow diagram that returns only one value is of type double. Look at the code for these functions and you will see a return statement that is used to return the value.

Look at the data flow diagram and find the functions that return more than one value. The functions that return more than one value must return the values through the parameter list. These functions are of type void. Note that they have pointer variables in their parameter lists. Compare the input and output for these functions on the data flow diagram to the parameter list for each of these functions. Can you see the correlation? Can you understand the value of preparing a data flow diagram, since it helps you write the parameter list for each function?

```
/*************************************************************************
 *************************************************************************
 **      APPLICATION PROGRAM TO CALCULATE
 **      AREA OF PARALLELOGRAM AND VOLUME OF
 **      PARALLELEPIPED DEFINED BY VECTORS A,B AND C
 **      AREA = LENGTH OF (A CROSS B)
 **      VOLUME = ABS VALUE OF (A DOT (B CROSS C))
 **
 **      PROGRAMMER NAMES         REVISION DATES
 **
 **
 *************************************************************************
 ************************************************************************/

#include <stdio.h>
#include <math.h>
```

```
/*******************************************
**         FUNCTION PROTOTYPES
*******************************************/

double dot_product (double g1, double g2, double g3,
                    double h1, double h2, double h3);

void cross_product (double g1, double g2, double g3,
                    double h1, double h2, double h3,
                    double *k1, double *k2, double *k3);

double vector_magnitude (double g1, double g2, double g3);

double area_parallelogram (double g1, double g2, double g3,
                           double h1, double h2, double h3);

double volume_parallelepiped (double g1, double g2, double g3,
                              double h1, double h2, double h3,
                              double k1, double k2, double k3);

void read_input (double *g1, double *g2, double *g3,
                 double *h1, double *h2, double *h3,
                 double *k1, double *k2, double *k3);

void print_output (double a1, double a2, double a3,
                   double b1, double b2, double b3,
                   double c1, double c2, double c3,
                   double area, double volume);

/*********************************************************************************
**         FUNCTION main
**              VARIABLES
**                    a1, a2, a3 = COMPONENTS OF VECTOR A
**                    b1, b2, b3 = COMPONENTS OF VECTOR B
**                    c1, c2, c3 = COMPONENTS OF VECTOR C
**                    area       = AREA OF PARALLELOGRAM DEFINED BY A AND B
**                    volume     = VOLUME OF PARALLELEPIPED DEFINED BY A,B AND C
*********************************************************************************/


void main (void)
{
        double a1, a2, a3, b1, b2, b3, c1, c2, c3, area, volume;
        read_input (&a1, &a2, &a3, &b1, &b2, &b3, &c1, &c2, &c3);
        area = area_parallelogram (a1,a2,a3,b1,b2,b3);
        volume = volume_parallelepiped (a1,a2,a3,b1,b2,b3,c1,c2,c3);
        print_output (a1,a2,a3,b1,b2,b3,c1,c2,c3,area,volume);
}

/***************************************************************
**       FUNCTION dot_product
**    CALCULATES THE DOT PRODUCT OF TWO VECTORS G AND H
**            VARIABLES - INPUT
**                  g1, g2, g3 = COMPONENTS OF VECTOR G
**                  h1, h2, h3 = COMPONENTS OF VECTOR H
**            RETURN
**                  SCALAR VALUE OF THE DOT PRODUCT OF G AND H
***************************************************************/


double dot_product (double g1, double g2, double g3,
                    double h1, double h2, double h3 )
{
        double g_dot_h ;
        g_dot_h = g1*h1+g2*h2+g3*h3;
        return (g_dot_h);
}
```

You should observe that the arguments listed in the function prototypes correspond to those shown in the data flow diagram.

Arguments listed with * are arguments that are "returned" to the calling function. Arguments without * are passed from the calling function to the function.

Body of main. Its purpose is to call other functions.

Because none of the arguments have *, their values come into the function but do not go back.

Equation for calculating the dot product.

```
/*****************************************************************
**      FUNCTION cross_product
**   CALCULATES THE CROSS PRODUCT OF TWO VECTORS G AND H
**        VARIABLES - INPUT
**            g1, g2, g3 = COMPONENTS OF VECTOR G
**            h1, h2, h3 = COMPONENTS OF VECTOR H
**        VARIABLES - OUTPUT
**            *k1, *k2, *k3 = COMPONENTS OF VECTOR THAT IS
**                            THE CROSS PRODUCT
*****************************************************************/

void   cross_product (double g1, double g2, double g3,
                      double h1, double h2, double h3,
                      double *k1, double *k2, double *k3)
{
        *k1 = g2*h3 - g3*h2;
        *k2 = -(g1*h3 - g3*h1);
        *k3 = g1*h2 - g2*h1;
}
```

The values calculated for *k1, *k2, and *k3 are returned to the functions that call cross_product.

Equations for calculating the cross product.

```
/*****************************************************************
**      FUNCTION vector_magnitude
**   CALCULATES THE MAGNITUDE OF A VECTOR G
**        VARIABLES - INPUT
**            g1, g2, g3 = COMPONENTS OF VECTOR G
**        RETURN
**            SCALAR VALUE OF THE MAGNITUDE OR LENGTH OF G
*****************************************************************/

double vector_magnitude (double g1, double g2, double g3)
{
        double mag;
        mag = sqrt (g1*g1+g2*g2+g3*g3);
        return (mag);
}
```

Because none of the arguments have *, their values come into the function but do not go back.

Equation for calculating the vector magnitude.

```
/*****************************************************************
**      FUNCTION area_parallelogram
**   CALCULATES THE AREA OF A PARALLELOGRAM DEFINED BY
**   TWO VECTORS G AND H
**   AREA = MAGNITUDE OF (G CROSS H)
**        VARIABLES - INPUT
**            g1, g2, g3 = COMPONENTS OF VECTOR G
**            h1, h2, h3 = COMPONENTS OF VECTOR H
**   RETURN
**      AREA OF THE PARALLELOGRAM
**   FUNCTION CALLS
**      CALLS cross_product TO CALCULATE G CROSS H
**      CALLS vector_magnitude TO GET MAGNITUDE OF (G CROSS H)
**
*****************************************************************/

double area_parallelogram (double g1, double g2, double g3,
                           double h1, double h2, double h3)
{
```

Because none of the arguments have *, their values come into the function but do not go back.

```
{
        double k1, k2, k3, area;
        cross_product (g1,g2,g3,h1,h2,h3,&k1,&k2,&k3 );
        area = vector_magnitude (k1,k2,k3);
        return (area);
}
```

The value of the area of the parallelogram is returned.

Calling function cross_product. The values of k1, k2, and k3 are modified by the function.

The new values of k1, k2, and k3 are used to call function vector magnitude.

```
/*****************************************************************
**      FUNCTION volume_parallelepiped
**   CALCULATES THE VOLUME OF A PARALLELEPIPED DEFINED BY
**   THREE VECTORS G, H AND K
**   VOLUME = ABS VALUE OF [G DOT (H CROSS K)]
**          VARIABLES - INPUT
**                  g1, g2, g3 = COMPONENTS OF VECTOR G
**                  h1, h2, h3 = COMPONENTS OF VECTOR H
**                  k1, k2, k3 = COMPONENTS OF VECTOR K
**                  d1, d2, d3 = COMPONENTS OF VECTOR D
**      RETURN
**         VOLUME OF THE PARALLELEPIPED
**   FUNCTION CALLS
**      CALLS cross_product TO CALCULATE D = H CROSS K
**      CALLS dot_product TO GET [G DOT D]
**
*****************************************************************/
```

```
double volume_parallelepiped (double g1, double g2, double g3,
                              double h1, double h2, double h3,
                              double k1, double k2, double k3)
{
        double d1,d2,d3,volume;
        cross_product (h1,h2,h3,k1,k2,k3,&d1,&d2,&d3);
        volume = fabs(dot_product (g1,g2,g3,d1,d2,d3));
        return (volume);
}
```

Because none of the arguments have *, their values come into the function but do not go back.

Calling function cross product. The values of d1, d2, and d3 are modified by the function.

The value of the volume of the parallelepiped is returned.

The new values of d1, d2, and d3 are used to call function dot_product.

```
/*****************************************************************
**        FUNCTION read_input
**   READS INPUT DATA FROM KEYBOARD
**          VARIABLES - OUTPUT
**               *g1, *g2, *g3 = COMPONENTS OF VECTOR G
**               *h1, *h2, *h3 = COMPONENTS OF VECTOR H
**               *k1, *k2, *k3 = COMPONENTS OF VECTOR K
*****************************************************************/
```

```
void read_input (double *g1, double *g2, double *g3,
                 double *h1, double *h2, double *h3,
                 double *k1, double *k2, double *k3)
{
        printf ("Enter vector A (a1 a2 a3): \n");
        scanf ("%lf %lf %lf", &*g1, &*g2, &*g3);
        printf ("Enter vector B (b1 b2 b3): \n");
        scanf ("%lf %lf %lf", &*h1, &*h2, &*h3);
        printf ("Enter vector C (c1 c2 c3): \n");
        scanf ("%lf %lf %lf", &*k1, &*k2, &*k3);
}
```

Because all of the arguments have *, their values are modified by the function read_input. In this case, they are initialized.

All of the arguments are initialized with scanf statements.

```
/*******************************************************************
**        FUNCTION print_output
**        PRINTS RESULTS TO SCREEN
**              VARIABLES
**                 a1, a2, a3 = COMPONENTS OF VECTOR A
**                 b1, b2, b3 = COMPONENTS OF VECTOR B
**                 c1, c2, c3 = COMPONENTS OF VECTOR C
**                 area       = area of parallelogram
**                 volume     = volume of parallelepiped
*******************************************************************/

void print_output (double a1, double a2, double a3,
                   double b1, double b2, double b3,
                   double c1, double c2, double c3,
                   double area, double volume)
{
        printf ("\n\n\rThe input vectors are:\n\n\
            \r   i    j     k \n\
            \r     %.2lf   %.2lf    %.2lf \n\
            \r     %.2lf   %.2lf    %.2lf \n\
            \r     %.2lf   %.2lf    %.2lf \n\n\n\n", a1,a2,a3,b1,b2,b3,c1,c2,c3);

        printf ("The results are:\n\n\
                \rArea of parallelogram (AxB)             = %lf \n\
                \rVolume of parallelepiped (abs(A dot (BxC)) = %lf\n\n",\
                area, volume);
}
```

Because none of the arguments have *, their values come into the function but do not go back. In this case, they do not need to go back because they simply are printed.

All of the arguments are printed.

Output

```
Enter vector A (a1 a2 a3):
3 2 2
Enter vector B (b1 b2 b3):
4 3 1
Enter vector C (c1 c2 c3):
8 2 7

The input vectors are:
             i       j       k
            3.00    2.00    2.00
            4.00    3.00    1.00
            8.00    2.00    7.00

The results are:

Area of parallelogram (AxB)             = 6.480736
Volume of parallelepiped (abs(A dot (BxC)) = 15.000000
```

Comments

It is worth commenting about some of the modules. Look at function read_input. Note that it is a void function because it returns many values (passed through the parameter list) rather than a single value. It receives nothing from main, so, all its parameters are pointer variables. Therefore, the variables used in the function have * in front of them. This results in the unusual looking scanf statement. Look at it. The scanf arguments in this statement are &*g1, &*g2, and so forth. If you follow the rules we developed in Lesson 5.4, you need not think much about this, and it makes some sense. Remember the actions followed by the operators & and unary *. Putting them together causes the action shown in Fig. 5.21. This figure shows the action &*g3, which causes a3's address (FFD8) to be transferred to the scanf function. This causes the scanf function to take the value read to be put into the memory location FFD8, which is the memory location for variable a3 (since g3 has a3's address). Another acceptable (and preferable) method to write the scanf statement simply is to use g3 with no &*. For example, scanf("%lf%lf,"g1,g2,g3); *is* acceptable because g1, g2, and g3 are pointer variables within function read_input. This means that they already hold addresses in their memory cells (Fig. 5.21). Note

FIG. 5.21

Illustration of the operation of &*. The action shown is &*g3. Since g3 has an address, *g3 indicates the value of a3 (because a3's address is the value of g3). The action of &*g3 follows the path from the value of a3 to the address of a3. Note that &*g3 is the same as g3.

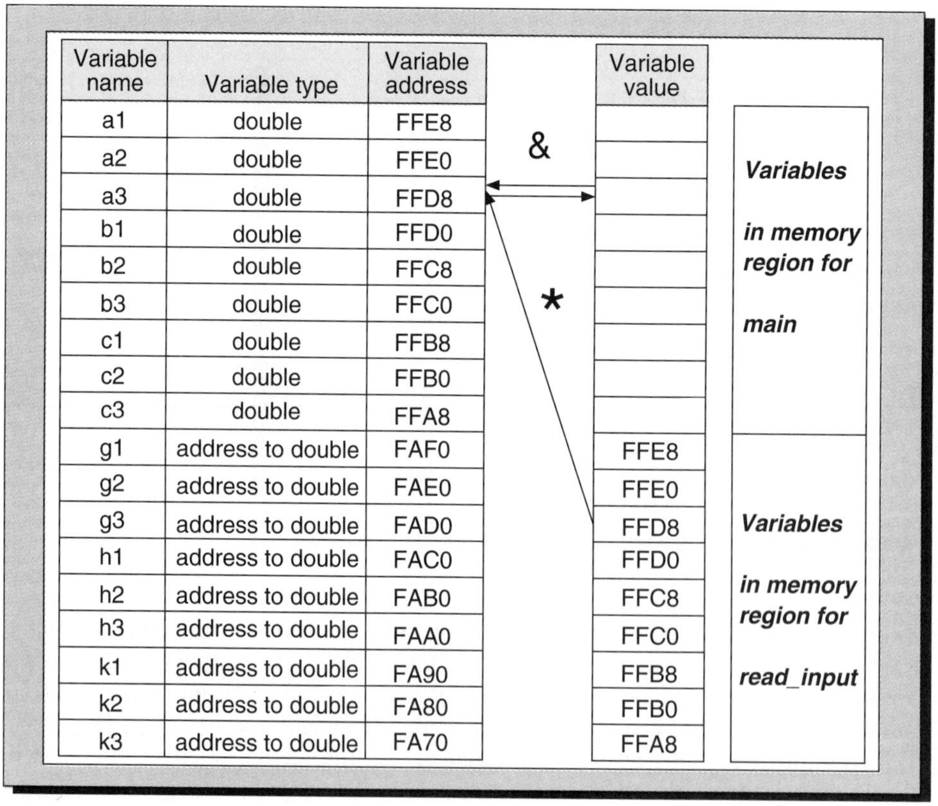

Variable name	Variable type	Variable address		Variable value	
a1	double	FFE8			
a2	double	FFE0	&		Variables
a3	double	FFD8			
b1	double	FFD0			in memory
b2	double	FFC8			region for
b3	double	FFC0	*		
c1	double	FFB8			main
c2	double	FFB0			
c3	double	FFA8			
g1	address to double	FAF0		FFE8	
g2	address to double	FAE0		FFE0	
g3	address to double	FAD0		FFD8	Variables
h1	address to double	FAC0		FFD0	
h2	address to double	FAB0		FFC8	in memory
h3	address to double	FAA0		FFC0	region for
k1	address to double	FA90		FFB8	
k2	address to double	FA80		FFB0	read_input
k3	address to double	FA70		FFA8	

that this seems to conflict with what we said in Chapter 3 where we indicated that & was required with the scanf arguments. In reality, the scanf statements are acceptable as long as the arguments represent addresses.

Observe that the functions cross_product, read_input, and print_output are void type functions. The functions cross_product and read_input are void type because they return multiple parameters through the parameter list. The function print_output is a void function because it returns nothing to main. The other functions, dot_product, vector_magnitude, area_parallelogram, and volume_parallelepiped are all type double because they return single values.

The function main serves primarily to call other functions. The main function in your modularly designed programs should do the same. When you are finished programming, you should be able to look at main and get an overview of what the program does without the details. If you find that you are putting a large number of calculations in main, then you should create a function or functions to perform the tasks. By looking at main for this program, you immediately get a general impression of the program's operations.

In this program, the comments were put at the beginning of each function enclosed in a banner of *s. One of the reasons for using the *s is to clearly distinguish comments from code. If commenting is not done properly, when you try to debug code, the comments can "get in the way" and make it difficult to follow the logic. We have found that using *s at the top and bottom of code blocks and two stars at the start of each line of comments works well for separating out comments and makes it easy to see the beginning of functions. This does not work very well, however, when it is necessary to write comments at the end of a line of code. Since most C code is written in lower case, writing comments in upper case also helps to make the comments stand out. However, there is no standard commenting scheme. We recommend that you follow your instructor's or employer's requirements in writing comments.

As you can see, making a modular design and including proper comments adds considerably to the length of a program. Do not let this bother you. The benefits of using modules and clear comments outweigh the disadvantages of added length. In this book, though, it is not efficient for us to use modular programming techniques in every example because we also are trying to teach you elements of the C language, such as looping and control structures. Therefore, we will show only a few modularly designed, fully commented programs. Because so many programs in this text are not written that way, please do not get the impression that we advocate not using a modular design or comments. We strongly endorse using both.

We devoted a considerable amount of space to this program because we want you to understand the concept of modular design. Clearly, a much simpler program could have been written to accomplish the same tasks. However, had the requirement been to work with a group of people to create the program, once the data flow diagram had been completed, each person in the group could have been assigned the task of writing a module. Then, each working module easily could have been put into a single program. This is the value of modular top-down design. As you become accustomed to it, you will be able to produce very powerful, reliable programs.

Modification exercises

Use the modules in this program to create programs that are capable of evaluating the following expressions (with **A, B, C,** and **D** being vectors expressed in the form **i, j,** and **k**):

1. $A \times (B \times C)$

2. $(A \times B) \cdot (C \times D)$

3. $(A \times B) \times (C \times D)$

4. $B(A \cdot C) - A(B \cdot C)$

5. $B(A \cdot C) - C(A \cdot B)$

▨ APPLICATION EXERCISES

1. The ancient Greek mathematician Euclid developed a method for finding the greatest common divisor of two integers, **A** and **B.** His method is
    a. If the remainder of **A/B** is 0, then **B** is the greatest common divisor.
    b. If it is not 0, then find the remainder of **A/B** and assign **B** to **A** and the remainder to **B.**
    c. Return to step a and repeat the process.

    Write a program that uses a function to perform this procedure. Display the two integers and the greatest common divisor.

2. Cost analysis is an important part of engineering. When you are in practice, you may be asked to write programs to determine the minimum cost for a number of different potential circumstances. Your programs can be used as decision-making tools for a project.
    Consider building an airport with the runway built on landfill. The contractor has two dump trucks, one with a capacity of 8 tons and the other with a capacity of 12 tons. The contractor uses the trucks to haul fill from a remote site to the airport location. The operating cost per trip for the 8 and 12 ton trucks is $14.57 and $16.26, respectively. One truck cannot make more than 60 percent of the total trips.
    Write a program that develops the minimum cost for a given number of tons. Prompt the user to enter the total number of tons. Display the number of trips required for each truck and the total cost. Use a modular design for this program.

3. The strength of an earthquake can be measured by its magnitude. In 1935, Charles F. Richter developed a scale, commonly known as the *Richter scale,* for determining an earthquake's magnitude. The amount of energy released in an earth-

quake, the length of an earthquake's fault rupture, and the number of worldwide earthquakes all have been correlated with magnitude. The following approximate equations relating these have been developed:

$$\log_{10} E \approx 11.8 + 1.5\, M$$
$$\log_{10} L \approx 1.02M - 5.77$$
$$\log_{10} N \approx 7.7 - 0.9M$$

where M = Richter magnitude ($0 < M < 8.2$)
        E = Energy released in ergs
        L = Length of the fault rupture in kilometers
        N = Number of worldwide earthquakes in a 100 year period

Variation from the equations can be $\pm$ 20 percent. Write a program that accepts values of E and L and computes the possible ranges of magnitude for this earthquake. Tell the user if the input data is totally incompatible. Determine the most likely numbers of earthquakes of these magnitudes.

4. This is a debugging problem: In this exercise you are to modify and debug a given code that almost works. The program is located in file Ae5_4 on the diskette with this book. Finding the root of a function has intrigued mathematicians for centuries. Unfortunately, most functions' roots cannot be found by direct analytical methods. Therefore, in many cases, numerical solutions are used. Most numerical methods use iteration to find the roots of a function. In essence, roots are found by trial and error, although other methods have been developed that work well for some functions.

    We may use the midpoint method to find the root of a function $y = f(x)$ within a given range $x_a <= x <= x_b$. This is illustrated in Fig. 5.22. The code that we give for this exercise is made for the function $y = 2 * x - 5$ and works only for functions of the type where the y value increases as x increases. It does not work for functions such as $y = -2 * x + 5$ (where y decreases as x increases) or $y = x * x - 5$ (where the values of y at x_a and x_b have the same sign).

    Do the following:

a. Modify the program so that it can find the root of $y = -2 * x + 5$ functions (test it with $y = -2 * x + 5$).

b. Modify the program so that it can find the root of $y = x * x - 5$ functions (test it with $y = x * x - 5$).

c. Modify the program so that it will tell the user that no root is in the given range for a given function (such as $y = x * x + 5$ for $-5 <= x <= 5$).

d. Modify the program so that the root is found and displayed accurately to within five decimal places.

Output from this program:

```
xa= -5.00   xb= 15.00   xc=  5.00----   ya=-15.00  yb= 25.00  yc=  5.00
xa= -5.00   xb=  5.00   xc=  0.00----   ya=-15.00  yb=  5.00  yc= -5.00
xa=  0.00   xb=  5.00   xc=  2.50----   ya= -5.00  yb=  5.00  yc=  0.00
The root is =  2.50
```

FIG. 5.22

Problem 4. Here $y_a<0$ and $y_b>0$. Given x_a and x_b, we take the mid-value x_c to look for the root. We would next look halfway between x_c and x_b for a root because $y_c<0$ and $y_b>0$.

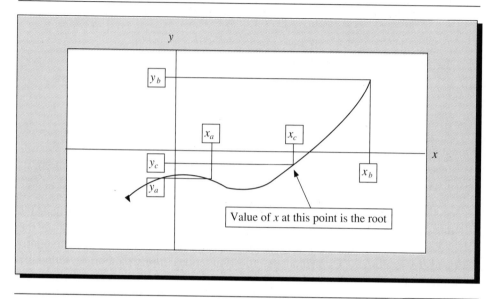

5. This is a debugging problem: In this exercise you are to modify and debug a given code that almost works. The program is located in file Ae5_5 on the diskette with this book. The program that follows shows how to find the largest divisor of an integer, N (other than the integer itself). The method tries to find a divisor one at a time. If N can be divided by one of the numbers 2, 3, 4, . . . , $N/2$, then the number is one of the divisors. There is a bug in the program so it works only for certain cases. For example, it can find the largest divisor of 84 as 42. However, it fails to find the divisor for 55. You are to correct the program so that it can find the largest divisor of any integer without error.

    Note that if an integer A can be divided by integer B, then B and A/B are divisors of A. Therefore, you can find two instead of one divisor at a time. In addition, you need to search the divisor B only from 2 to the square root of A (the largest possible value of B is when $B=A/B$; that is, when B is the square root of A). Use this method in the false block of the if statement in function find_max_divisor and complete this program.

Sample output

```
Please enter an integer
84
Please enter the method number (1 or 2)
1
```

```
Method1----- 2 is a divisor of 84
Method1----- 3 is a divisor of 84
Method1----- 4 is a divisor of 84
Method1----- 6 is a divisor of 84
Method1----- 7 is a divisor of 84
Method1----- 12 is a divisor of 84
Method1----- 14 is a divisor of 84
Method1----- 21 is a divisor of 84
Method1----- 28 is a divisor of 84
Method1----- 42 is a divisor of 84

Using method 1, the maximum divisor of 84 is 42
Please enter an integer
13
Please enter the method number (1 or 2)
1
Using method 1, the maximum divisor of 13 is 6
```

6*. A block is resting on a horizontal plane and is intended to be pulled with a force acting at an angle θ to the horizontal. The block weighs 30 kN and has a coefficient of friction of 0.2. Write a program that uses a function to calculate the magnitude of the force, F, needed to pull the block for θ = 5, 0, 10, 20, 30, 40, 50, 60, 70, and 80 degrees.

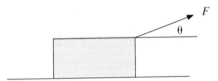

7*. For the same block as in problem 7, write a program that considers the effect of the plane being at an angle to the horizontal:

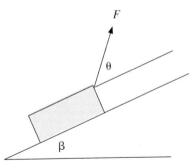

Solve the problem for the values of θ and β being 0, 10, 20, 30, 40, 50, 60, 70, and 80 degrees but with the sum of θ and β being less than 90 degrees.

8*. Solve problem 7 with the coefficient of friction being 0.1, 0.2, 0.3, and 0.4.

---

*Exercises beginning with an asterisk (*) require a basic engineering background or further conceptual information given by the instructor.

9*. Write a program that computes the forces at the supports, *A* and *B*, of the following beam for the distance *x* being 0, 0.25*L*, 0.5*L*, 0.75*L* and *L* and *F* being 100, 200, 300, 400, and 500 kN.

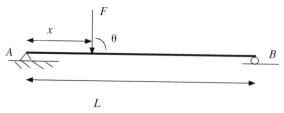

10*. Write a program that can solve problem 10 with θ being 0, 30, 60, and 90 degrees.

11*. Write a program to calculate the necessary forces in cables used to tow a car if it takes a total of 3 kN in the rolling direction of the car to get it moving with no net force component perpendicular to the rolling direction of the car. Consider the values of θ and β being 0, 10, 20, 30, 40, 50, 60, 70, and 80 degrees but with the sum of θ and β being less than 140 degrees.

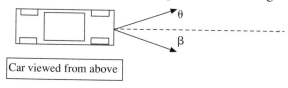

Car viewed from above

12*. Two vehicles collide and remain locked together after impact. Write a program that computes the direction (indicated by β) and speed (v_3) of these collided vehicles after contact takes place:

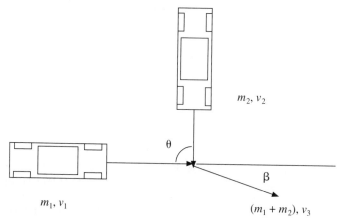

Use the following input data: m_1 = 1000 kg, m_2 = 3000 kg, v_1 = 30 m/sec, v_2 = 10, 20, 30, 40, 50, 60, 70, and 80 m/sec.

13*. Write a program for solving problem 13 but with the following variations: θ = 10, 20, 30, 40, 50, 60, 70, 80, and 90 degrees; m_1 = 1000, 2000, and 3000 kg, and m_2 = 2000, 4000, and 6000 kg.

14*. Write a program for solving problem 13 but with the following variations: $\theta = 10$, 20, 30, 40, 50, 60, 70, 80, and 90 degrees; $v_1 = 10$, 20, and 30 m/sec; and $v_2 = 20$, 40, and 60 m/sec.

15*. Write a program that can calculate the tensile force in a cable for holding down this submerged buoy:

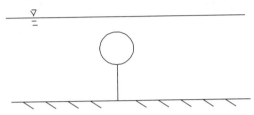

The buoy is filled with air. The water can be assumed to have a density of 9.8 kN/m^3. The program should compute the tensile force in the cable for a buoy radius of 1, 2, 3, 4, and 5 m.

16*. Write a program that can compute the cable tension for this partially submerged buoy:

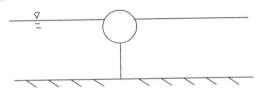

Consider the variations of the buoy being 0.25, 0.5, 0.75, and fully submerged. Consider both fresh water (density of 9.8 kN/m^3) and salt water (density of 10.05 kN/m^3). The program should compute the tensile force in the cable for a buoy radius of 1, 2, 3, 4, and 5 m.

17*. Write a program that can compute the current through each resistor for this circuit:

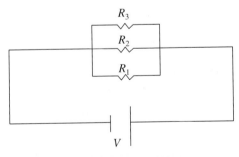

Take the voltage, V, equal to 30 volts. Use $R_1 = 10$, 20, and 30 ohms; $R_2 = 40$, 60, and 80 ohms; and $R_3 = 100$, 120, and 130 ohms.

18*. Repeat problem 17 with the number of resistors varying from one to five; with $R_4 = 50$, 70, and 90 ohms; and $R_5 = 140$, 160, and 180 ohms.

19. The flow of water through porous media is normally laminar (fluid flows in parallel layers without mixing) and therefore follows the Darcy (a French engineer) equation:

$$Q = kiA$$

where Q = the flow rate (volume/time, cm³/sec)
 k = the coefficient of permeability (volume/time/area, cm/sec)
 i = the hydraulic gradient (head loss per unit of length, no units,
     $i = H/L$)
 H = head loss (energy loss due to friction, cm)
 L = length of flow in the porous media (cm)
 A = the cross-sectional area through which flow occurs

A steel pipe that has an internal diameter of $D = 10$ cm and length of $L = 200$ cm is filled with sand; the permeability of sand is $k = 0.1$ cm/sec. The head loss is maintained at $H = 50$ cm. Write a program to calculate the rate of flow through the pipe. The input specifications are as follows. No external input from the keyboard or file. However, the program shall declare Q, k, i, H, L, A, and D as double type variables. The flow is calculated by calling a function named flow_rate() from the main function. Its prototype is

```
void flow_rate(double D, double L, double k, double H);
```

The output specifications are as follows. The output shall display the following data on the screen:

All raw data D, L, k, and H
Calculated values of i and k
Flow rate, Q

20. Redo problem 19 without using the function flow_rate.

21. During a rainstorm, water collected from a given area is guided to ditches or road culverts. To size the ditch or culvert, hydraulic engineers use what is called the *rational* method to calculate the peak rate of runoff water. The formula is

$$Q = CiA$$

where Q = peak rate of runoff (ft³/sec)
 C = weighted average runoff coefficient (no units)
 i = average precipitation intensity (in./hour)
 A = watershed (area of rainfall) tributary to the point of interest (acres)

Note that the value of C depends on the type of surface. In rural areas, the values of C are as follows:

Type of surface	Average runoff coefficient, C
Concrete	0.9
Bare earth	0.6
Cultivated fields	0.3
Forested areas	0.2

If the watershed area consists of different types of surfaces, C shall be calculated using the weighted average method. For example, if the area is 20 percent covered by concrete, 30 percent by bare earth, 50 percent by forest, the weighted average C is

$$C = [(0.9 * 20\%) + (0.6 * 30\%) + (0.2 * 50\%)] = 0.46$$

Write a program to determine the peak runoff in a given area based on the following. The input specifications are these. Call a function named read_data() to read an input file similar to this table (the third column is used for explanation only; it is not part of the input file):

6.9		*The first line only contains the rainfall intensity $i=6.9$ in./sec*
100	0.9	*The first column is the size of subarea in acres (100) that is covered by $C = 0.9$ type surface*
200	0.6	*The first column is the size of subarea in acres (200) that is covered by $C = 0.6$ type surface*
300	0.3	*The first column is the size of subarea in acres (300) that is covered by $C = 0.3$ type surface*
150	0.2	*The first column is the size of subarea in acres (150) that is covered by $C = 0.2$ type surface*

The prototype of read_data() is as follows:

```
double read_data(double *i, double *A);
```

The function shall

Calculate the total watershed area, A (A should be $= 100 + 200 + 300 + 150 = 750$ acres)

Calculate the weighted average runoff coefficient C and return this value to the main() function

In the main() function, calculate $Q = CiA$

The output specifications are these. The output shall display the following data on the screen:

Raw data i, and size of each subarea and its runoff coefficient

Size of total watershed area, A, and the weighted average runoff coefficient, C

Peak rate of runoff, Q

22. Redo problem 21; however, the prototype of read_data() is changed to

```
void read_data(double *i, double *A, double *C);
```

23. This table shows the coefficient of permeability, k, for various types of soils:

Soil type	Range of coefficient of permeability (cm/sec)
Clay	$1.0E - 10$ to $1.0E - 8$
Silt	$1.0E - 8$ to $1.0E - 4$
Sand	$1.0E - 4$ to 1.0
Gravel	1.0 to 100.0

Given the coefficient of permeability of a soil, write a program to determine its type. The input specifications are these: Call a function named soil_type() to

Read the coefficient of permeability from the keyboard
Find the range of the input k value
Determine the soil type

The function prototype is

```
void soil_type(void);
```

The output specifications are these: The output shall display the following data on the screen:

The coefficient of permeability entered by the user
The soil type

24. The power consumed by a heater can be calculated by the equation

$$p = vi$$

where
$$\begin{aligned} p &= \text{power (watts)} \\ v &= \text{voltage (volts)} \\ i &= \text{current (amperes)} \end{aligned}$$

Write a program to calculate the power consumed by various type of heaters. The input specifications are these: Read the voltage, v (first column), and the current, i (second column), from this file.

110	5.5
220	23.5
90	13.6
370	44.4

The output specifications are these: The output shall display the following data on the screen:

The input values of v and i
The power

For example, after reading the first line of input data, you should display the following on the screen:

```
Input voltage = 110.0 volts, current = 5.5 amperes
Power consumed by the heater = 605.0 watts
```

25. Redo problem 24, but display the resistance value, R ($R=v/i$, ohms) of each heater. The output for the first line of input data shall be

```
Input voltage = 110.0 volts, current = 5.5 amperes
Power consumed by the heater = 605.0 watts
Heater resistance = 20.0 ohms
```

Modification exercises

26. The program in file Ae5_26 on the diskette displays the $y = \sin(x)$ curve on the screen.

Modify the program so that the curve is displayed by calling a function named plot_curve() from the main() function. The prototype of the plot_curve() is

```
void plot_curve(void);
```

27. Modify the program so that the screen X axis is filled with character X, the Y axis is filled with character Y, and the vertical line where $y = 0$ is marked with the 0 characters.

28. Modify the program so that the plot_curve() function will call a function named xy_range() to find the ranges of x and y to be plotted. The prototype of the xy_range() is

```
void xy_range(double *xmin, double *xmax, double *ymin, double *ymax);
```

where xmin, xmax, ymin, and ymax are the ranges of x and y for the curve to be plotted. The values of xmin, xmax, ymin, and ymax are entered by the user from the keyboard.

29. Modify the program so that, in the function xy_range(), the user needs to enter the values of xmin and xmax only; after that, the xy_range() function will call a function named find_y_range() to find the value of ymin and ymax. The prototype of find_y_range() is

```
double find_y_range(double *ymin);
```

30. Modify the program so that the function can plot $y = x * \sin(x) - \tan(2 * x)$.

31. Modify the program so that the program can display both $y = x * \cos(x)$ and $y = \sin(x)$ without displaying the x and y values on the screen.

6

Numeric Arrays

A useful high-level programming language has features already built into it that can be used to simplify programming tasks. Arrays are a feature C has built into it that programmers can use.

An array is a *data structure* in C. It is a grouping of like-type data. In its simplest form, an array can represent a list of numbers; for instance, a list of the temperatures recorded every hour of a particular weather station for one year. In this case, all the numbers in the list represent temperatures and therefore are like-type data.

Another use of arrays might be in denoting the x and y coordinates of a number of data points. If there are 10,000 data points, then there are 10,000 x coordinates and 10,000 y coordinates. For this, we could create two arrays, one for x and another for y. In C, arrays are indicated with brackets containing a positive integer constant or expression following an identifier. For instance, the x coordinate of a data point might be represented as $x[129]$ or $x[4976]$ (anything from 0 to 9999 is acceptable in the brackets for an array of 10,000 points).

The reason why this is a convenient representation for like-type data is that we easily can write expressions for calculating and manipulating this data. The value enclosed in brackets is called a *subscript* or *index*. It can be used much as subscripts are used in algebraic expressions. For instance, to find the slope, m_1, of a line connecting two points indicated by the coordinates (x_1, y_1) and (x_2, y_2) we can use the following algebraic expression:

$$m_1 = (y_2 - y_1)/(x_2 - x_1)$$

With C's array structure, we can write this as an assignment statement:

```
m[1] = (y[2] - y[1])/(x[2] - x[1]);
```

348

Note the direct correspondence between the algebraic expression and the assignment statement. If we wanted to calculate the slope for many different pairs of points we could put the array in a loop, such as

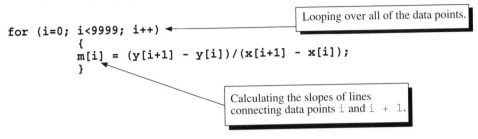

```
for (i=0; i<9999; i++)
        {
        m[i] = (y[i+1] - y[i])/(x[i+1] - x[i]);
        }
```

Looping over all of the data points.

Calculating the slopes of lines connecting data points i and i + 1.

At this point, you need not understand this loop completely, but let us just say that these few lines can be used to calculate all the slopes of the lines connecting all 10,000 data points! We cannot do a calculation of this type so simply without arrays. So, we find arrays to be very useful.

In this chapter, we introduce the concept of arrays and show how to work with them. The applications at the end of the chapter illustrate the many practical uses of arrays. With the material you learn in this chapter, you can write very useful and sophisticated programs.

LESSON 6.1 INTRODUCTION TO 1-D ARRAYS AND PRINTING ARRAY ELEMENTS

Topics

- Definition of arrays
- Characteristics of an array
- Using #define directive to define array size
- Printing array elements

Like ordinary variables, arrays must be declared before they can be used in a program. Values also can be assigned to array elements using assignment statements. This lesson's program declares two arrays, initializes some of the elements of the arrays, and prints them out.

In previous chapters, we guided you through the programs in detail. At this point, you should be familiar with how to look at a program and essentially read the code. Therefore, from this point forward, we simply list what you should observe about each program and the questions you should attempt to answer about the program before reading the explanation. To get the most out of this text, you should go through the lists and the source codes carefully, make the observations, and answer the questions.

What you should observe about this lesson's program.

1. In the first declaration, a[ ] is declared as an array that contains integers.
2. In the second declaration, b[ ] is declared as an array that contains doubles.
3. Both declarations use brackets [ ]. In the brackets are integers that declare the number of elements for which space is reserved in memory.
4. Only one pair of brackets is used for both a[ ] and b[ ]. This means that both are one-dimensional (1-D) arrays.
5. The a[ ] array is declared to have two elements.
6. The b[ ] array is declared to have ten elements.
7. The assignment statements have assigned values to all of the elements of the a[ ] array.
8. The assignment statements have assigned values to only two of the elements of the b[ ] array.
9. In the first two printf statements, we printed the values of all the array elements that have been assigned.
10. In the third printf statement, we printed the value of b[2], which previously had not been assigned a value. The output shows a meaningless value for b[2] because of this.
11. In the fourth printf statement, we printed the value of a[3], even though a[ ] has been declared to have only two values. The output shows a meaningless value for a[3].
12. No error of any kind has been indicated by the C compiler due to what is indicated in observations 10 and 11.

Questions you should attempt to answer before reading the explanation.

1. What is the advantage of using a constant macro in declaring the size of an array (or the number of elements for which space is reserved)?
2. Why has no error been indicated when we have printed a[3], which clearly is beyond the declared size of a[ ]?

Source code

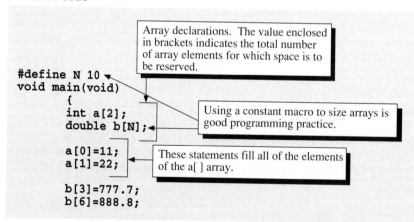

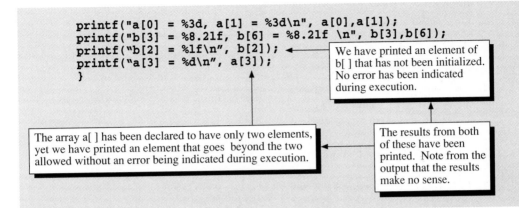

```
       printf("a[0] = %3d, a[1] = %3d\n", a[0],a[1]);
       printf("b[3] = %8.2lf, b[6] = %8.2lf \n", b[3],b[6]);
       printf("b[2] = %lf\n", b[2]);
       printf("a[3] = %d\n", a[3]);
       }
```

We have printed an element of b[ ] that has not been initialized. No error has been indicated during execution.

The array a[ ] has been declared to have only two elements, yet we have printed an element that goes beyond the two allowed without an error being indicated during execution.

The results from both of these have been printed. Note from the output that the results make no sense.

Output

```
a[0] =11, a[1] =22
b[3] = 777.70, b[6] = 888.80
b[2] = -33660644284456964.000000
a[3] = 373
```

Explanation

1. What is a one-dimensional array and how do we declare it? A one-dimensional array is a collection of the same type of variables (ANSI C calls an array an aggregate type variable) stored in contiguous and increasing memory locations. It is identified by its name, type, dimension, and number of elements. For example, as illustrated in Fig. 6.1, the statement

`int a[2];`

declares

> The name of the array is a
> The type of the *array elements* is int.
> The dimension is *1* (it has only one pair of brackets [ ]).
> The number of elements or size is *2* (meaning that memory for two elements is reserved).

In general, the syntax for a 1-D array declaration is

element_type   array_name   [number_of_elements];

where *element_type* specifies the type of the array's elements, such as int, float, double, or any other valid C data type, except void and function types; *array_name* is the name of the array; and *number_of_elements* specifies the maximum number of elements that can be stored in the array. This must have a value equal to a positive integer.

■ **FIG. 6.1**
  Array declaration

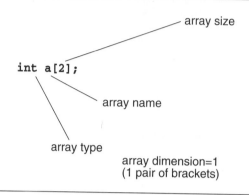

array size

`int a[2];`

array name

array type

array dimension=1
(1 pair of brackets)

2. How do we name an array? Array names are classified as identifiers. Therefore, the name of an array must follow the rules used for naming identifiers. Array names are given in their declarations. For instance, these are five valid array names:

```
c[15],    f23_rack[60]
a[ 10], fruit[300 ], country[100]
```

Invalid array names also are illegal identifiers and may begin with numbers or include illegal symbols, such as these three illegal array names:

```
12abd[15 ], y4#[30 ], io*jk[70 ]
```

3. How do we distinguish between a 1-D array and 2- or 3-D arrays? A one-dimensional array has only one pair of brackets following its name. A two-dimensional array has two pairs of brackets, and a 3-D array has three pairs of brackets, such as in this declaration:

```
int        b[2] [4],           c[6][9][5];
           2-D array           3-D array
```

For the next few lessons we will restrict our discussion to 1-D arrays.

4. How do we set the length of a 1-D array? It is up to the programmer to set the length (also called *size* or *number of elements*) of an array, based on the needs of that particular problem. The length of a 1-D array is determined by the result of the expression enclosed in brackets in the declaration. The length must be an integer greater than 0. The length of a 1-D array can be defined explicitly, such as:

```
int a[2], c[20], g[100];
```

Another common way to size arrays is to use a constant defined in a preprocessor directive, such as in this lesson's program, which uses:

```
#define N 10
double b[N];
```

Since integer expressions are allowed, the following declarations also would be valid (used together with #define N 10):

```
int   c[100+N], d[50*N];
```

The following are invalid numbers of elements:

```
int    c[-25], b[32.5]
```

These have an illegal number of elements because they are not positive integers.

Note that you may declare arrays and single variables in the same line. For instance,

```
double b[N], f, g, h;
```

is legal. The C compiler knows that b is an array because it is followed by brackets and that the others are single variables because they are not.

5. How can we calculate the amount of memory reserved for a one-dimensional array? By multiplying the number of elements expressed in the declaration and the amount of memory for each element, we can determine the amount of memory reserved by an array declaration. For instance, if an int occupies 2 bytes, then the int array a[ ] that has a length of 2 occupies $2 \times 2 = 4$ bytes. Similarly, the double array b[ ] with a length of 10 (with type double occupying 8 bytes each) occupies $10 \times 8 = 80$ bytes.

6. What is the advantage of using a constant defined in a preprocessor directive to size arrays? Using a constant symbol to size an array will make your program more flexible. For example, if you have 100 1-D arrays that all have the size of 9, that is, a1[9], a2[9], . . . , a100[9]; and you want to change the size from 9 to 25, you have to change 9 to 25 a total of 100 times. The effort involved in making this change and the likelihood of error is reduced by using a constant symbol for the subscripts; for example, a1[N], a2[N], . . . , a100[N] with N defined in a preprocessor directive. To increase the size from 9 to 25, all that is necessary is to change the value of N from 9 to 25 in the define directive.

7. What is the first index (also called subscript) of an array in C? In C, by default, the first index or subscript is 0.

8. What are the implications of this? This usually leads to confusion and error by students and even some experienced programmers who forget that C begins at 0 and not 1 (as some other languages do). For instance, in this lesson's program we declared

```
int a[2];
```

This means that there are two elements of the array a[ ]. Since C begins at 0, the two elements that are used in the program body are a[0], and a[1] with the lines

```
a[0] = 11;
a[1] = 22;
```

Observe that we did not use a[2] in the program body! This is because a[2] in the declaration means something completely different from a[2] in the program body. In the program body, a[2] means the third element of the array a[ ]. Since only two elements are declared for a[ ], we should not use a[2] in the program body.

9. If a[2] were used in the program body would C have indicated an error? No! C does not check to see if you try to access an array outside of its range. If a[2] were specified in the program, C actually would go to the memory location after a[1] and extract a value from that location. With this value, the program may execute completely and give answers (which very likely will be wrong).

In programs you write, it is up to you to recognize that your answers are wrong and find the source of the errors. With many arrays in your programs and many complicated control structures, it may be very difficult to find which array at which location causes the array size to be exceeded. If nothing meaningful is in that memory location, you very likely will get a run-time error, which may also be difficult to trace. Therefore, when writing your programs be very careful to not exceed the size of your arrays. Remember your arrays begin at 0 and end at one less than the number of elements.

For illustration only, in this lesson's program we printed out a[3], which very clearly is outside the range specified in the declaration for a[ ]. However, neither during compilation nor execution was any error indicated! (Try it yourself.) The integer 373 (which has absolutely no meaning in the context of this program) was found in the memory cell indicated by a[3]. We chose to simply print this value using the last printf statement in the program; however, we easily could have performed calculations with a[3] being on the right side of an assignment statement or even stored a new value in a[3]! Because it is outside the range specified in the declaration, doing anything with a[3] probably would have caused chaos in our program. This is because the value stored in a[3] is unpredictable (if you run this program on your computer, you probably will get something other than 373), and if we tried to store something into a[3], we probably would overwrite a value in another variable's memory cell that is supposed to be there. In summary, be careful. If you find that you are getting nonsensical answers from your programs, check whether you are exceeding the specified range for any of your arrays.

10. How do we print an array element using printf? We treat an array element like we treat a single variable. For instance, the statements

```
printf("a[0] = %3d, a[1] = %3d\n", a[0],a[1]);
printf("b[3] = %8.2f, b[6] = %8.2f \n", b[3],b[6]);
```

cause the values of a[0], a[1], b[3], and b[6] to be printed.

Note that the array type determines the format to be used to display its elements. For example, to display an int type array element, we can use the %d format; to display a float type, we can use the %f, %e, and %E type formats.

11. What types of variables can we use for array indices? Only integer type variables such as int and char with their modifiers signed, unsigned, short, and long can be used as index variables.

12. In this program, we assigned values for b[3] and b[6]. However, b[ ] has been declared to have ten elements. What is stored in the other elements of the b[ ] array? Nothing meaningful; since we have not assigned any values to the other elements of the b[ ] array, no meaningful values are stored in the memory locations for these elements. The statement

```
printf("b[2] = %d\n", b[2]);
```

has printed the value of b[2]. The value of $-33660644284456964.000000$ simply is representative of the miscellaneous bits currently in the memory location reserved for b[2]. Should you run this program, you probably would get a different value for b[2]. Therefore, before you use any array elements in your programs you must initialize them. Note that this program ran with no errors being indicated during compilation or execution; however, the result for b[2] was nonsensical. So, if you find that you are getting nonsensical answers from your programs, check whether you have initialized all the elements of your arrays. The next lesson illustrates various ways of initializing array elements.

EXERCISES

1. True or false:
    a. All elements of a given array have the same data type.
    b. All elements of a given array are placed randomly in computer memory.
    c. All elements of a given array may be displayed using formats with different field widths.
    d. The subscript of the first element of a 1-D C array is 1.
    e. A 1-D array has 99 elements; its size is 100.

2. Find the error(s), if any, in each of these statements:
    a. `int a, b(2);`
    b. `float a23b[99], 1xy[66];`
    c. `void city[36], town[45];`
    d. `double temperature[-100];`
    e. `long phone[200];`
    f. The first and last array elements in the array just defined are phone[1] and phone[200].

3. Find errors in this program:

```
#define (N=2)
void main(void)
float a[N],b;
a[1]=N;
N=99;
a[2]=N;
}
```

Solutions

1. a (true), b (false), c (true), d (false), e (false)
2. a. `int a, b[2];`

b. **`float a23b[99], xy1[66];`**
c. An array cannot be of the void type.
d. **`double temperature[100];`** *subscript must be > 0*
e. No error.
f. The first and the last array elements are `phone[0]` and `phone[199]`.

▨ LESSON 6.2 ARRAY INITIALIZATION

Topics

- Methods for initializing array elements in declarations
- Initializing array elements using scanf
- Initializing array elements using assignment statements in loops

In the previous lesson, we saw the importance of assigning values to the array elements used in our programs. We assigned values one element at a time. The first time that we assign a value to an array element (or to a simple variable) is called *initializing* the element or variable.

In Chapter 3, we illustrated initializing the values of simple variables in their declarations. We did this by writing an assignment type statement within the declaration. We can do something similar for arrays. That is, we can initialize arrays in declarations. However, with arrays, the situation is somewhat different, because arrays have many elements and, if we want to initialize the values of many elements in a declaration, we need to list many values in the declaration. In C, we do this using braces { } to enclose the values of the array elements we are initializing.

What you should observe about this lesson's program.

1. The array, a[ ], is declared to have three elements. Two of the elements are initialized in the declaration as indicated by what is enclosed in the braces { }.
2. The array b[ ] is declared without the size explicitly stated; however, three elements are initialized in the declaration.
3. The first printf statement prints a[0 to 2] and b[0 to 2]. The value printed for a[2] is 0, even though it has not been explicitly initialized.
4. Arrays x[ ] and y[ ] are declared but not initialized in the declaration.
5. The scanf statement reads values for x[0] and x[1] from the keyboard.
6. The variable i is used as the index for the for loop and as the subscript for y[ ].
7. The for loop initializes values for all ten elements of the y[ ] array.

Questions you should attempt to answer before reading the explanation.

1. How many elements does b[ ] have?
2. How many times is the for loop executed?
3. What does the for loop initialize the y[ ] values to be?

Source code

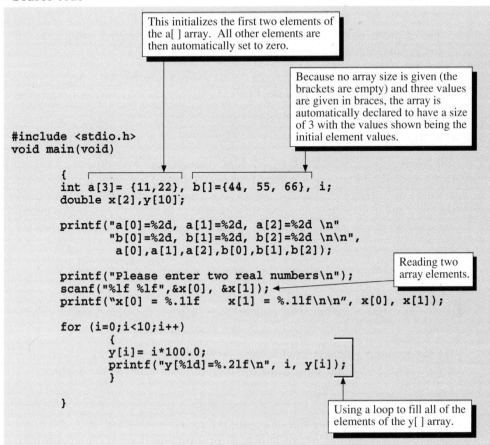

This initializes the first two elements of the a[ ] array. All other elements are then automatically set to zero.

Because no array size is given (the brackets are empty) and three values are given in braces, the array is automatically declared to have a size of 3 with the values shown being the initial element values.

```c
#include <stdio.h>
void main(void)

    {
    int a[3]= {11,22}, b[]={44, 55, 66}, i;
    double x[2],y[10];

    printf("a[0]=%2d, a[1]=%2d, a[2]=%2d \n"
           "b[0]=%2d, b[1]=%2d, b[2]=%2d \n\n",
             a[0],a[1],a[2],b[0],b[1],b[2]);

    printf("Please enter two real numbers\n");
    scanf("%lf %lf",&x[0], &x[1]);
    printf("x[0] = %.1lf    x[1] = %.1lf\n\n", x[0], x[1]);

    for (i=0;i<10;i++)
          {
          y[i]= i*100.0;
          printf("y[%1d]=%.2lf\n", i, y[i]);
          }

    }
```

Reading two array elements.

Using a loop to fill all of the elements of the y[ ] array.

Output

Keyboard input

```
a[0]=11, a[1]=22, a[2]=0
b[0]=44, b[1]=55, b[2]=66

Please enter two real numbers
77.0 88.0
[0] = 77.0    x[1] = 88.0

y[0]=0.00
y[1]=100.00
y[2]=200.00
y[3]=300.00
y[4]=400.00
y[5]=500.00
y[6]=600.00
y[7]=700.00
y[8]=800.00
y[9]=900.00
```

Explanation

1. What are two ways to initialize the elements of a 1-D array in a declaration? The elements of a 1-D array can be initialized in a declaration using either of the following two methods (Fig. 6.2):

1. Declare the array, including the number of elements in brackets, and immediately list values of at least some of the array elements enclosed in braces. For example, the declaration

   `int a[3]={11,22};`

   initializes a[0] = 11 and a[1] = 22. Note, though, that with this statement we have explicitly initialized only the first two elements of the array a[ ], which has a length of 3. However, because we use this particular method of initialization, the third element, a[2], is automatically initialized to be 0. In general, to initialize some elements and to set the rest of the elements to 0, we use the following form:

   type   name   [number_of_elements]={value_0,   value_1, ... value_n};

   where n is less than or equal to one less than the *number_of_elements*. All values of the array elements with indices from [n + 1] through [*number_of_elements − 1*] will be initialized to 0 by this declaration.

2. Declare the array without including the number of elements in brackets and immediately list values of the array elements in braces. For example, the declaration

   `int b[]={44, 55, 66};`

   automatically declares the size of b[ ] to be three elements (b[0], b[1], and b[2]), because three values are listed in braces. In this case, if we try to access b[3], we would be exceeding the range of b[ ]. In general, the form of this type of declaration is

   type   name   [ ]={value_0,   value_1, ... value_n};

   This form initializes all elements of the array. The number of elements in the array is *n* + 1.

   Two advantages of this type of declaration and initialization are that it is unnecessary to count the number of array elements and, should more elements be added to the array during program modification, it is unnecessary to change the number of elements value (since it is not included in the declaration). A disadvantage is that it is not readily apparent to a programmer how many elements are in the array. This can lead to out-of-range errors. Therefore, in this text, when initializing arrays in a declaration, we will use primarily the first method described.

FIG. 6.2
Initializing an array in its declaration

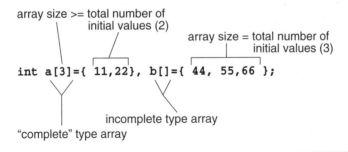

2. *How else can we initialize an array?* We can declare the array first, then use a read function, such as scanf(), to initialize it. For example, the statements

```
double x[2];
scanf("%lf%lf",&x[0], &x[1]);
```

initialize the array elements x[0] and x[1] with the values entered by the user. Since a double type array element is like a scalar double type variable, we use a %lf format and place an & in front of the array element.

We can declare the array first, then use an assignment statement to initialize the array elements one by one. For example, the statements

```
double y[10];
for (i=0; i<10, i++)
     {
     y[i] = i*100.0;
     }
```

initialize array elements y[0] through y[9] to be 0.00 through 900.00 in increments of 100.00. Note that, in this for loop, we have gone from 0 to 9 only (because when $i = 10$, the conditional $i < 10$ is false). Another way to write this for loop is

```
for (i=0; i<=9, i++)
     {
     y[i] = i*100.0;
     }
```

Check with your instructor for the preferred method for writing these loops. Some programmers prefer the second method because it clearly shows that only 0–9 are valid array indices. Despite this shortcoming, for most cases, we use the first method for writing for loops. However, we recommend that you not forget that the largest valid array index is one less than the declared size.

3. In this lesson's program, would the statement

```
y = 100.;
```

have initialized all the elements of array y[ ] to be 100? No; in C, we must initialize each element individually. Also, if we want to modify an entire array, we must modify it element by element.

4. Would the declaration

```
int b[2]={44, 55, 66};
```

have caused an error in C? Yes, because we stated that *b* has only two elements but we listed three. Although C will not detect out-of-range errors with arrays, it will detect this error.

EXERCISES

1. True or false:
   a. There is more than one method for initializing a 1-D array.
   b. An incomplete array can be initialized using the scanf() function.
   c. The number of initial values of a given array must be less than or equal to the array size.

2. Find error(s), if any, in each of these statements:
   a. `int a={11,22},b[33];`
   b. `float c[3]={11,22,33,44};`
   c. `double d(4)=(11,22,33,44);`
   d. `d[4]={11 22 33 44};`
   e. `int a[3]/11,22,33/;`

3. The number of cars crossing a bridge from Monday through Sunday in a given week are 986, 818, 638, 763, 992, 534, and 683. Use these numbers to initialize an array and write a program to generate the following output. The daily average number of cars crossing the bridge is 773. The maximum number of cars crossing the bridge in a day is 992, which occurs on Friday.

Solutions

1. a (true), b (false), c (true)
2. a. `int a[2]={11,22},b[33];`
   b. `float c[4]={11,22,33,44};`
      or
      `float c[ ]={11,22,33,44};`
   c. `double d[4]={11,22,33,44};`
   d. Must show array data type.
   e. `int a[3]={11,22,33};`

▓ LESSON 6.3 BASIC ARRAY INPUT/OUTPUT

Topic
- The value of EOF
- More about scanf and fscanf
- Reading a file with an unknown number of array elements
- Using arrays in arithmetic statements
- Writing data to a file using arrays

For many practical problems, arrays have very large numbers of elements. Therefore, most commonly, array values from external sources are not input from the keyboard but read from a data file. Also, to print an array to the screen often is not practical because, with a large number of values, more than one screen is required and one cannot do further manipulations with values displayed on the screen. Therefore, output to a file is very common. In this lesson, we illustrate input from and output to a file.

Obviously, many math type functions return a value. For instance, in the assignment statement $y = \log(x)$, the log function returns a value that can be used on the right side of an assignment statement. We know that the log function is not a void type function since it returns a value. However, for other functions we need to know (or find out from references) whether a function returns a value before we use it on the right side of an assignment statement. The fscanf function is such a function. If we look it up, we find that the fscanf function returns an int value. The question becomes, "What does the int value mean?" In this lesson's program, we use the fscanf function on the right side of an assignment statement.

We previously discussed constant macros and noted that the convention used (but not required) is that they be written in all capital letters. We did not mention, though, that there are constant macros defined in header files. If we attach a header file, we can use the constant macros defined in it. A commonly used macro defined in stdio.h is EOF. In this lesson's program, we illustrate its value.

From what you have learned to this point, you would be very comfortable with a loop that begins with

```
while ( log(x)  !=  1.0 )
```

In this program, we use a similar loop beginning but with the fscanf function instead of log and EOF instead of 1.0 of the form

```
while ( fscanf(....)   !=   EOF)
```

We can interpret this to mean that, while the fscanf function does not return EOF, the rest of the loop is to be executed. Stated another way, fscanf is repeatedly called until it returns EOF. We can tell you that EOF stands for "end of file" and that fscanf returns the value of EOF (set by the operating system in most cases to be -1) when fscanf attempts to read past the end of data in a file. With this background information, read this lesson's source code and make the following observations.

What you should observe about this lesson's program.
1. Three arrays are declared.
2. The first elements of x[ ] and y[ ] are read using the first fscanf statement.
3. The first fscanf statement is on the right side of an assignment statement, which means that fscanf returns a value. The value returned from fscanf is an integer assigned to the variable k. The value of k is printed in the output.
4. EOF is used in the program but is not declared in the variable list. Its value is printed using the second fprintf statement.
5. The conditional in the while loop compares the value returned by fscanf to EOF. The array index increases each time through the loop.
6. The for loop loops over the number of elements in the array. Within the for loop, the array elements are manipulated and printed.

Questions you should attempt to answer before reading the explanation.
1. Where do you think EOF comes from?
2. How many times will the while loop execute?
3. What is the value of i after the while loop has executed?

Source code

```
#include <stdio.h>
#include <math.h>
void main(void)
{
        int i, j, k, num_elem;
        double x[20], y[20], z[20];
        FILE *infile, *outfile;

        infile = fopen ("C6_3.IN","r");
        outfile= fopen ("C6_3.OUT","w");

        k = fscanf(infile,"%lf %lf",&x[0],&y[0]);
        fprintf (outfile, "k = %d\n",k);

        fprintf (outfile,"Value of EOF = %d\n", EOF);

        i = 1;

        while ( fscanf(infile,"%lf %lf",&x[i],&y[i])   !=   EOF)   i++;

        num_elem = i;

        fprintf(outfile,"   x[i]     y[i]     z[i]\n");
```

> Declaring all of the arrays to be size 20.

> The fscanf function can be used on the right side of an assignment statement because it returns an integer equal to the number of values read.

> EOF is a constant macro defined in stdio.h. The fscanf function returns EOF when it attempts to read past the end of a file.

> This loop executes repeatedly until fscanf returns EOF. The value of *i* increments each time through the loop and, thus, counts the number of array elements.

```
        for (j=0;j<num_elem;j++)
           {
           z[j]=sqrt(x[j]*x[j]+y[j]*y[j]);
           fprintf(outfile,"%7.1f %7.1f %7.1f\n",x[j],y[j],z[j]);
           }

        fclose(infile);
        fclose(outfile);
}
```

> We work with arrays element by element so we often create a loop over the number of elements.

> Each element of z[ ] is calculated from the corresponding elements of x[ ] and y[ ].

Input file C6_3.IN

```
3.0    4.0
6.0    8.0
9.0    12.0
```

Output file C6_3.OUT

```
k = 2
Value of EOF = -1
    x[i]       y[i]      z[i]
     3.0        4.0       5.0
     6.0        8.0      10.0
     9.0       12.0      15.0
```

Explanation

1. How do we use fscanf() to read array data from a file? Reading array data from a file using fscanf() is similar to reading scalar data from a file (Fig. 6.3). After opening the input file, we use the fscanf() function to read the data, one by one, using array elements. Note that the & symbol must precede each array element. For example, the expression

```
fscanf(infile,"%lf%lf",&x[0],&y[0]);
```

reads two double data from a file that has a file pointer `infile`. The data are saved in the first elements of arrays x and y, x[0] and y[0].

2. Does the function fscanf( ) return a value, meaning can it be used on the right side of an assignment statement? Yes, although to this point in the book we have not made use of this particular feature of the function. The function fscanf is set up within C to return an int value after it has read or attempted to read from the specified input file. It returns an integer value equal to the number of items it successfully read on being called. In this lesson's program, the line

```
k = fscanf (infile, "%lf%lf", &x[0], &y[0]);
```

FIG. 6.3
Using array to read data from a file

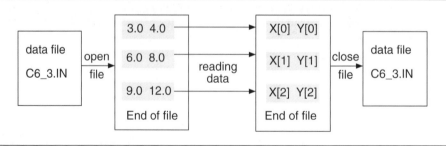

caused fscanf to read two items x[0] and y[0] and, therefore, returns the integer value of 2. Therefore, k = 2 after execution of the assignment statement, and this is the value printed for k.

3. What is EOF? EOF is a constant type macro defined in the header file stdio.h. Observe that, like the constant macros we used in programs, EOF is all capital letters (standing for end of file). ANSI C requires only that EOF be equal to a negative integer; however, in most cases it is equal to −1 as we have shown in this lesson. This fprintf statement was used in the program to print EOF:

fprintf (outfile,"Value of EOF = %d\n", EOF);

Note that, even though we have not defined EOF in our program, we can use it as if it were defined because it is defined in stdio.h, which is attached to the top of our source code by the preprocessor. We will find as we continue programming other header files define other constant macros that we may wish to use in our programs.

Using macros such as EOF can make your code confusing for those who are unfamiliar with the macros you are using, because it is impossible to see what the macros are meant to be by looking at your code alone. Therefore, we recommend that you write comments at the beginning of your programs, in the location where macro definitions normally are located, indicating macros you are using, what they mean, and from where they come. This will reduce the confusion. If you are looking at someone else's code that is not well commented, be aware that identifiers written in all capital letters very likely may be macros defined in header files.

4. How is EOF *used in the standard C library?* A number of functions use EOF, most notably fscanf. The function fscanf returns EOF if an input failure occurs before it is able to successfully read the first value it attempts to read. In other words, if fscanf encounters the end of a file before it is able to read anything, it returns to its calling location the value of EOF (which in most cases is −1).

5. How can we use EOF *in our programs?* EOF is especially useful when we want to read data from a file into an array and do not know how many values are in the data file. For instance, in this lesson's program, the arrays x[ ] and y[ ] have been given a size of 20 elements. This means that these arrays may have as many as 20 elements but may have fewer. In input file C6_3.IN only three elements of these arrays are given. We may not know this in advance and therefore can use a while loop and the fact that fscanf returns EOF when it encounters the end of the file to read the data. The loop

```
while ( fscanf(infile,"%lf %lf",&x[i],&y[i])   !=   EOF)   i++;
```

causes fscanf to read the file C6_3.IN until it encounters the end of the file. Each time it reads two values (once each of x[ ] and y[ ]), it increments the value of the variable i. In our case, this line causes fscanf to read x[1], y[1], x[2], and y[2]. Since it increments i after reading the values of x[ ] and y[ ], the value of i after reading x[2] and y[2] is 3. When fscanf tries to read x[3] and y[3], it fails because it encounters the end of the file. Therefore, fscanf returns EOF and the loop terminates. At the end of executing this loop, all of the values of x[ ] and y[ ] have been read and the value of i is 3. Note that the value of i is important to us because it indicates how many values of x[ ] and y[ ] were in the file. Therefore, we save this value of i in the variable num_elem and use num_elem in the terminating condition for other loops that we write regarding the arrays x[ ] and y[ ].

6. How do we use arrays in arithmetic expressions? We must use arrays element by element, and we frequently use arrays in loops. For example, the loop

```
for (j=0; j<num_elem; j++)
        {
        z[j]=sqrt(x[j]*x[j]+y[j]*y[j]);
        }
```

first calculates

```
z[0]=sqrt(x[0]*x[0]+y[0]*y[0]);
```

The second time through the loop,

```
z[1]=sqrt(x[1]*x[1]+y[1]*y[1]);
```

is calculated. This process is repeated until num_elem values of z[ ] have been calculated. The algebraic form of this expression is

$$z_1 = \sqrt{x_1^2 + y_1^2}$$

Note that we cannot write the following line of C code:

```
z=sqrt( x*x + y*y);
```

to accomplish the same task as the for loop. We must work with arrays element by element. So, when working with arrays, we will find that much of our effort will be in writing the loops to do the proper manipulations.

7. How do we use fprintf to write array data to a file? Writing array data to a file using fprintf is similar to writing scalar variable data to a file using fprintf. After opening the file, we use the fprintf() function to write the data, one by one, using array elements. We usually do this using a loop. For example, the loop

```
for (j=0; j<num_elem; j++)
        {
        fprintf(outfile,"%7.1f%7.1f%7.1f\n",x[i],y[i],z[i]);
        }
```

writes three double type data stored in array elements x[i], y[i], and z[i] to a file that has a file pointer `outfile`.

EXERCISES

1. Based on

    ```
    int A[3]={1,2,3}, B[3]={4,5,6}, C[3]={1,2};
    FILE *in;
    ```

    determine whether the following statements are true or false:
    a.   The statement

        ```
        C = A+B;
        ```

        will add the data stored in arrays A and B and assign the result to array C.
    b.   The statement

        ```
        C[0]= A[1]+B[2];
        ```

        is incorrect because 0, 1, and 2 are different subscript numbers.
    c.   The statement

        ```
        fscanf(in, "%d %d %d",&A);
        ```

        can be used to read three int type integers from a file and store them in array A.

2. Find error(s), if any, in each of these statements (assume i[5] is an int type array, f[6] is a float type array, and in is a file pointer):
    a.   `fscanf(in,"%d %d",i[2],i[4];`
    b.   `fscanf(in,"%d %d",&i(2), &i(4));`
    c.   `fscanf("in,%f %d",&i[2],&f[2]);`
    d.   `fscanf(in,"%d %f",&i[5],&f[6]);`
    e.   `fprintf(in,"%d %d",i[2],i[4];`
    f.   `fprintf(in,"%d %d",&i(2), &i(4));`
    g.   `fprintf("in,%f %d",&i[2],&f[2]);`
    h.   `fprintf(in,"%d %f",&i[5],&f[6]);`

3. Use arrays to read the following input file

    ```
    1   30.0
    2   45.0
    3   60.0
    4   90.0
    ```

and then process the data and generate the following file:

```
N     X(degree)     cos(X)
1     30.0          0.8667
2     45.0          0.7071
3     60.0          0.5000
4     90.0          0.0000
```

4. A surveyor's notebook contains a figure like Fig. 6.4. Find the area between the straight fence and the riverbank, assume all units are in meters and all lines between the riverbank and the fence line are perpendicular to the fence line. Use the trapezoidal rule as described previously, two double type arrays, fscanf, and fprintf functions to solve the problem.

FIG. 6.4
Problem 4

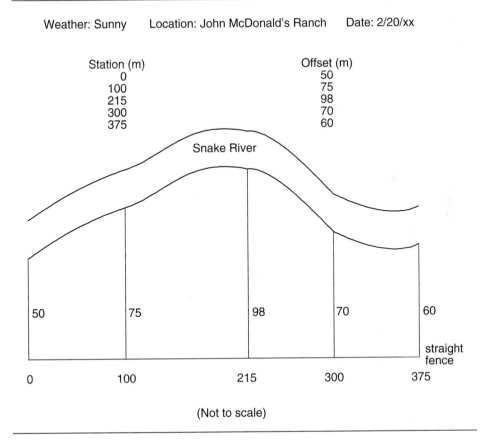

Weather: Sunny   Location: John McDonald's Ranch   Date: 2/20/xx

Station (m)	Offset (m)
0	50
100	75
215	98
300	70
375	60

Snake River

(Not to scale)

Solutions

1. a (false), b (false), c (false)
2. a. `fscanf(in,"%d %d",&i[2], &i[4]);`
   b. `fscanf(in,"%d %d",&i[2], &i[4]);`
   c. `fscanf(in,"%d %f",&i[2],&f[2]);`
   d. Array elements i[5] and f[6] do not exist.
   e. `fprintf(in,"%d %d",i[2],i[4]);`
   f. `fprintf(in,"%d %d",i[2],i[4]);`
   g. `fprintf(in,"%d %f",i[2],f[2]);`
   h. Array elements i[5] and f[6] do not exist.

▨ LESSON 6.4 MULTIDIMENSIONAL ARRAYS

Topics

- Concept of multidimensional arrays
- Comparing one- and multidimensional arrays
- Initializing multidimensional arrays
- Printing multidimensional arrays

In your algebra courses you encountered the use of matrices in calculations. For instance, you found that you could express the following algebraic equations:

$$3x + 4y + 8z = 15$$
$$2x - 3y + 9z = 8$$
$$4x + 7y - 6z = 5$$

in the form

$$\begin{bmatrix} 3 & 4 & 8 \\ 2 & -3 & 9 \\ 4 & 7 & -6 \end{bmatrix} \begin{Bmatrix} x \\ y \\ z \end{Bmatrix} = \begin{Bmatrix} 15 \\ 8 \\ 5 \end{Bmatrix}$$

With a slight modification, this form of expressing equations is particularly useful. Instead of representing the variables x, y, and z, we prefer to use x_0, x_1, and x_2 as follows:

$$\begin{bmatrix} 3 & 4 & 8 \\ 2 & -3 & 9 \\ 4 & 7 & -6 \end{bmatrix} \begin{Bmatrix} x_0 \\ x_1 \\ x_2 \end{Bmatrix} = \begin{Bmatrix} 15 \\ 8 \\ 5 \end{Bmatrix}$$

There are two advantages of this form:

1. As matrices get bigger it is simple to include more variables, even up to x_{100} or x_{1000}.
2. Using subscripts on our variables leads us to use arrays to store the values of those variables.

We could write a computer program to solve this particular set of equations; however, a computer program of this sort would have limited usefulness because it is unlikely that we would need to solve this particular set of equations very frequently. It would be more useful if we could write a computer program to solve a set of three equations with any values in the coefficient matrix and any values in the right-hand side vector. To do this, in our computer program, we need to treat the values in the matrix and vector as variables, because we will be allowed to change them as we try to solve different problems. The traditional naming of variables in the coefficient matrix and the right-hand side vector is the following:

$$\begin{bmatrix} a_{00} & a_{01} & a_{02} \\ a_{10} & a_{11} & a_{12} \\ a_{20} & a_{21} & a_{22} \end{bmatrix} \begin{Bmatrix} x_0 \\ x_1 \\ x_2 \end{Bmatrix} = \begin{Bmatrix} b_0 \\ b_1 \\ b_2 \end{Bmatrix}$$

If we call the first row, row 0, and the first column, column 0, we see that the double subscript naming scheme for a is $a_{\text{row column}}$. (Note: sometimes the first row is row 1 and the first column is column 1.)

In the application examples, we work a problem with matrices, so at this point we will not go through all of the details. For now, we just want you to realize that a very common usage of double subscripts is matrix coefficients. Just as single subscripts led us to use one-dimensional arrays, double subscripts lead us to use two-dimensional arrays. In other words, the variable a_{02} would be represented in C as a[0][2]. We will find that this representation in C will make it very simple for us to write programs in matrix arithmetic.

C allows us to use arrays with even more dimensions. Suppose that we are going to start collecting daily rainfall data at a particular airport rain gauge in the year 2000 and continue collecting it for a period of ten years (years 2000–2009). At the end of each day, we could obtain the amount of rain (in cm) collected by the rain gauge. After the first year, for instance, we may be interested in performing calculations on this data, determining the average rainfall per day, the month in which the rainfall is heaviest, and other such useful information. We realize that we are going to want to do this computation periodically over the ten year period. To write a computer program to do these calculations, we would need to store the rainfall data in an array (all of the data being like-type data, rainfall).

For this, a one-dimensional array, rainfall[ ] is not particularly convenient. This is because we would need $10 \times 365 = 3650$ array elements (one for each day), and there is no apparent correspondence between the array index and the date. For instance, it is not obvious to us the date corresponding to array element rainfall[827]. Instead, it is easier for us to use a three-dimensional array, rainfall[ ][ ][ ]. With this three-dimensional array we could use the year as the first index, the month as the second index, and the day as the third index. Thus, rainfall[6][11][23], would represent the rainfall on November 23, 2006. Also, rainfall[6][11][1] to rainfall[6][11][30] represents the 30 elements of the array rainfall[ ][ ][ ] that contain rainfall data for the month of November 2006. For instance, the loop

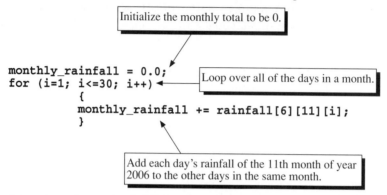

determines the total rainfall for the month of November 2006. You need not understand the details of this loop at this time; however, we want you to realize the convenience that multidimensional arrays provide in writing programs.

It is difficult to conceptually envision arrays of greater dimension than two. In many cases it is not worth the effort to try to develop a physical picture. For instance, for our array, rainfall[ ][ ][ ], it was not necessary to create a mental image of a three-dimensional array, because the division of year/month/day is natural. In fact, we easily could create a four-dimensional array of rainfall by simply breaking the data down into hours instead of just days giving us rainfall[year][month][day][hour]. A five-dimensional array could take us down to the minute. All these are natural divisions with which you can work without developing a complex mental image. In other words, you need not struggle trying to develop a mental picture of a five-dimensional array. When you create large-dimensional arrays in programs, your divisions should be natural. If you make your divisions natural, you will find that it is straightforward to work with multidimensional arrays.

In summary, because of the correspondence between the way that engineers and scientists typically group information, we find that multidimensional arrays are a very convenient and useful form for storing and manipulating data.

What you should observe about this lesson's program.

1. The array b[ ][ ] is declared to be a two-dimensional array. Each element of the array is an integer.
2. The elements of the b[ ][ ] array are initialized in the declaration.
3. The first for loop is a nested for loop used to print all of the values of the b[ ] [ ] array.
4. In the output, the b[ ][ ] array is displayed as it would appear in a matrix. In the printf statement, the subscripts are in the order "row column" and all of the values for one row are printed, then a new line is created and the next row is printed.
5. The array rainfall[ ][ ][ ] is declared to be a three-dimensional array of integers.
6. The "month" index for rainfall[ ][ ][ ] is declared to have a size of 13 not 12.
7. The "day" index for rainfall[ ][ ][ ] is declared to have a size of 32 not 31.

8. The 31 days (all of the days for one month) of rainfall data are read in the second for loop.

9. In the third for loop, the 31 days of rainfall data are printed. The if statement in the loop makes the output appear like a calendar, by creating a new line each seven days.

A question you should attempt to answer before reading the explanation. Why have we sized the rainfall[ ][ ][ ] array to have 13 months and 32 days?

Source code

> Both b[ ][ ] and rainfall[ ][ ][ ] are multi-dimensional arrays, b[ ][ ] is a two-dimensional array because it has two sets of brackets, rainfall[ ][ ][ ] is a three-dimensional array because it has three sets of brackets. The array b[ ][ ] has 2 x 3 = 6 elements, rainfall[ ][ ][ ] has 10 x 13 x 32 = 4160 elements.

```
#include <stdio.h>
void main(void)
{
        int i, j, year, month, day;
        int b[2][3]={51,52,53,54,55,56};
        int rainfall[10][13][32];
        FILE *infile;

        infile = fopen ("L6_4.DAT","r");

        for (i=0;i<2;i++)
                {
                for (j=0;j<3;j++)
                        {
                        printf("b[%1d][%1d]= %5d   ", i,j,b[i][j]);
                        }
                printf("\n");
                }

        fscanf (infile,"%d %d",&year,&month) ;
        for (day=1; day<=31; day++)
                {
                fscanf(infile, "%d",&rainfall[year][month][day]);
                }

        printf ("Rainfall for Year = %d, Month = %d\n\n",year, month);
        for (day=1; day<=31; day++)
                {
                printf("%d ",rainfall[year][month][day]);
                if (day==7 || day==14 || day==21 || day==28) printf("\n");
                }
}
```

> The maximum values of the loop control variables correspond to the declared array size, b[2][3].

> This printf statement simply creates a new line. It is needed only to make the output look the way that a two-dimensional array typically is written. It is placed between the two nested loops.

> A nested loop structure often is used with multidimensional arrays. Here, each element of the b[ ][ ] array is printed.

> This loop reads one month of rainfall data. The loop loops over 31 days.

> This loop prints one month of rainfall data. The loop loops over 31 days.

> Array output must look neat. This if structure creates a new line each seven increments of the variable day. This makes the output look like a calendar.

Input file L6_4.DAT

```
4 12
0 2 0 29  0  1 2
3 0 7 22 11 12 6
0 3 4  2  8  7 5
7 6 0  4  9  7 8
1 9 8
```

Screen output

b[0][0]= 51	b[0][1]= 52	b[0][2]= 53
b[1][0]= 54	b[1][1]= 55	b[1][2]= 56

```
Rainfall for Year = 4, Month = 12
0 2 0 29  0  1 2
3 0 7 22 11 12 6
0 3 4  2  8  7 5
7 6 0  4  9  7 8
1 9 8
```

Explanation

1. What is a multidimensional array? Like a one-dimensional array, a multidimensional array is a collection of the same type of data stored in contiguous and increasing memory locations. However, for convenience and clarity, a multidimensional array often is not thought of as a long list of values but as a table or matrix (for a 2-D array as shown in Fig. 6.5) or more complex images (such as a block for 3-D arrays as is shown in Fig. 6.6). Although these images may be good visual tools for working with multidimensional arrays, C actually stores the values in adjacent memory cells so that an image of a long list of values is equally valid. In this text we will work frequently with 2-D arrays and mostly use the matrix image to describe them.

FIG. 6.5

Conceptual image of the two-dimensional array, b[][], for this lesson's program

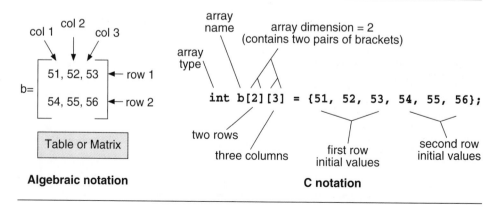

Algebraic notation   C notation

FIG. 6.6
Image of 3-D array of size [2][3][4]

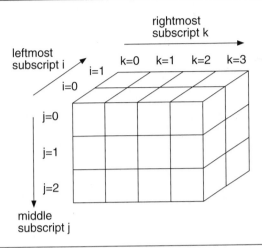

2. How do we declare a multidimensional array? It is similar to declaring a one-dimensional array. For example,

```
int b[2][3];
```

declares

> The name of the array to be b.
> The type of the array elements to be int.
> The dimension to be 2 (it has two pairs of brackets [ ]).
> The number of elements or size to be 2 × 3 = 6 (the product of the numbers in brackets).

3. Can we initialize a multidimensional array when we declare it? Yes; in C, similar to a 1-D array, a multidimensional array can be initialized directly in the declaration statement. Using this method, for a 2-D array, the array elements are initialized by row. For example, array b[ ][ ] with six elements is initialized in the declaration statement

```
int b[2][3]={51,52,53,54,55,56};
```

which initializes the elements to be

```
b[0][0]= 51    b[0][1]= 52    b[0][2]= 53
b[1][0]= 54    b[1][1]= 55    b[1][2]= 56
```

Note again that C begins its subscripts at 0. Also in this list, the rightmost subscript is incremented first.

4. How else can we initialize arrays in declarations? We can use braces to separate rows in two-dimensional arrays. We have not shown the following declaration in this lesson's program, but we could have declared a two-dimensional array, c[ ]. For instance, the declaration and initialization

```
int c[4][3]={{1,   2,  3},
             {4,   5,  6},
             {7,   8,  9},
             {10,11,12}};
```

has each row contained in a separate set of braces. The advantage of this display is that the subscripts for each array element become more obvious. In addition, if we leave some values out of each row, we implicitly initialize them to 0. For instance,

```
int c[4][3]={{1,   2},
             {4,   5,  6},
             {7},
             {10,11,12}};
```

initializes c[0][2], c[2][1], and c[2][2] to be 0.

We also can declare the far left dimension size (number of rows for a 2-D array) implicitly. For instance, the declaration

```
int c[ ][3]={{1,   2,  3},
             {4,   5,  6},
             {7,   8,  9},
             {10,11,12}};
```

implicitly declares the number of rows to be 4.

5. How many dimensions can we use? ANSI C does not explicitly specify how many dimensions a compiler must be capable of supporting; however, most compilers can handle 7 dimensions and some can handle 12 or more.

Although you may have use for a large number of dimensions at some time in your programming career, you probably will find that it is unusual to need more than three dimensions. Therefore, in this text, we will focus on arrays with fewer dimensions. Should you need more dimensions, it is straightforward to apply the concepts for fewer-dimensional arrays to those with more dimensions.

6. How are data from multidimensional arrays stored in memory? In computer memory, the elements of all multidimensional arrays are stored contiguously in increasing memory locations, essentially in a single list (shown as a two column list to save space). In C, the order of storage is that the first element stored has 0 in all its subscripts. The second element stored has all of its subscripts 0 except the far right, which has a value of 1. For instance, for a 3-D array a[ ][ ][ ] (with a declared size of a[2][3][4], which has a total of $2 \times 3 \times 4 = 24$ elements), the array elements are stored in this order

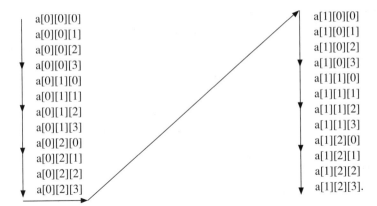

As you can see, the far right subscript increments first. The other subscripts increment in order from right to left.

For an array element, a[x][y][z] of an array declared with a size a[I][J][K], C uses the following type of formula to locate the array element's position in the list and therefore its position in memory:

$$\text{sequence location} = x \times (J \times K) + y \times (K) + z + 1$$

For example, for the previous 3-D array, a[ ][ ][ ], we have $I = 2$, $J = 3$, and $K = 4$. The sequence location for element a[0][1][2] ($x = 0$, $y = 1$, $z = 2$) is

$$\text{sequence location} = 0 \times (3 \times 4) + 1 \times (4) + 2 + 1 = 7$$

Therefore, in computer memory, a [0][1][2] is the seventh element stored (as illustrated in the list).

Note that in this formula the variable I never appears. In fact, one can develop similar formulas for greater numbers of dimensions (we leave this to you to do as an exercise) and it would be found that the size of the far left dimension is never needed to find the location in memory of an array element, whereas all the other sizes are required. Knowing this is significant because it plays a role in transferring multidimensional arrays to functions. When transferring multidimensional arrays to functions, we must pass along enough information to the function for it to be able to calculate the location of each array element in memory. Therefore, we must pass to the function the sizes of all of the dimensions except the far left one. Should we choose, we can transfer the far left one as well; however, it is not required. We discuss this in more detail when we cover arrays and functions.

Another conceptual image used to represent 3-D arrays is given in Fig. 6.6. Although this image does not represent computer memory well, it is an easily remembered mental image.

7. How do we determine the amount of memory occupied by a multidimensional array? The amount of memory occupied by a multidimensional array is determined by the cumulative product of its size in all dimensions. For example, the size of the int array a[2][3][4] is 2 × 3 × 4 = 24. It contains 24 elements and occupies 24 × 2 = 48 bytes memory (using 2 bytes for each integer).

8. How do we use a loop to print all the elements of a multidimensional array?
Usually, we use a nested for loop to print the elements of a multidimensional array. For a two-dimensional array, we use a two-deep nested loop with one variable controlling one subscript and another variable controlling the other. For example, for the array b[2][3] from this lesson's program, the following nested loop prints all of the elements in the form of a matrix:

```
for (i=0;i<2;i++)
    {
    for (j=0;j<3;j++)
        {
        printf("b[%1d][%1d]= %5d   ", i,j,b[i][j]);
        }
    printf("\n");
    }
```

Note that the outer loop loops over the number of rows and the inner loop loops over the number of columns. The inner loop (using j as the controlling variable) prints all the columns of one row and the elements appear to be in one row in the output (see the output for this lesson's program). On leaving this loop, we have a single printf statement (contained within the outer loop) with the sole purpose of creating a new line. After creating a new line, control goes back to the inner loop to print another row. You should trace the flow of this loop (use a table of i and j values) to make sure that you understand it. You will use this form frequently in printing the results of your array calculations.

9. In this lesson's program, why have we chosen the size of rainfall[ ][ ][ ] to be rainfall[10][13][32] rather than rainfall[10][12][31]? Since there are only 12 months and at most 31 days in a month, we have sized our array to be larger than is needed. We did this for conceptual reasons because C begins its subscripting at 0. Had we sized rainfall for only 12 months, then the month of December would be represented by the subscript 11. Also, had we sized rainfall for only 31 days, then the 22nd day of the month would be represented by the subscript 21. It can be envisioned that this could cause some difficulties at one point in the programming process and could lead to bugs in a program. So, we have chosen to avoid this potential problem by simply adding one to the size of these dimensions.

10. What is the disadvantage of doing this? The disadvantage is that we require the program to reserve more memory space than it needs. For small programs, this is not an issue and you are better off having a programming strategy that reduces the likelihood of error than one that saves insignificant amounts of memory. However, for larger programs, it may be important for you to reduce memory consumption. If this is the case then you must adjust your programming to accommodate the fact that C begins at 0.

11. In this lesson's program, how did we read the rainfall data from a file?
We began by reading the first line of the file that contains the year and month for
which the rainfall data pertains with the line:

```
fscanf (infile,"%d %d",&year,&month);
```

We then used the values of year and month read in from this file to partially fill the
rainfall array with the for loop:

```
for (day=1; day<=31; day++)          Loop over each day in a month.
        {
        fscanf(infile, "%d",&rainfall[year][month][day]);
        }
```

Using the previously read values for year and month,
read each day's rainfall value.

Note that, in executing this loop, the values of the first two subscripts for rainfall
(indicating the year and month) do not change. Only the last subscript (day) changes.
In this manner, we have filled the portion of the array reserved for that year and
month's days.

12. How did we print the rainfall data to the screen? We began with a short
heading for the array giving the year and month for the data. For simplicity, we did
not make it very elaborate. This printf statement was used:

```
printf ("Rainfall for Year = %d, Month = %d\n\n",year, month);
```

Then, with the following for loop,

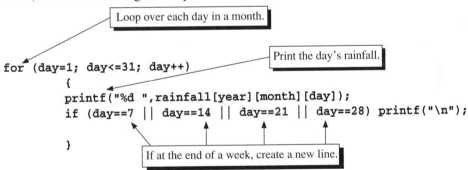

we printed each day's rainfall data to the screen. Note, that to make the output look
like a calendar, we needed to create a new line at the end of each week. The if statement with the compound conditional in the preceding loop creates a new line each
seven days. While this may not be perfect for the month being considered, it illustrates some of the concepts of printing arrays.

**13. Why did we not need an if statement in the reading for loop with fscanf to
advance to the new line?** The function fscanf automatically skips to the next line

when it is looking for the next non-white-space character. Therefore, no statement is needed to get fscanf to advance to the next line.

EXERCISES

1. Based on the declarations:

    ```
    int a[3][1]={1,2,3}, b[3], c[3][2], d[2][3];
    ```

    determine whether each of the following statements is true or false:
    a.   Array c[3][2] contains 3 + 2 = 5 elements.
    b.   The 1-D array b[12] can be used to store all data saved in the 2-D array a[3][4].
    c.   Array d[2][3] contains six elements: d[0], d[1], d[2], d[3], d[4], and d[5].
    d.   Array c[3][2] contains six elements: c[1][0], c[1][1], c[1][2], c[0][0], c[0][1], and c[0][2].
    e.   Array c[3][2] contains three 1-D subarrays while array d[2][3] contains only two 1-D subarrays.

2. Find the error(s), if any, in each of these declarations:
    a.   `int a[2][0];`
    b.   `float a23b[99][77], 1xy[66][77];`
    c.   `double city[36][34], town(12)(34);`
    d.   `int a(2,3)={11,22,33,44};`

3. Use a nested for loop to print array a[3][2] and a nested while loop to print array b[2][3] defined here and print both in matrix form:

    ```
    int a[3][2]={11,22,33,44,55,66}, b[2][3]={111,222,333,444,555,666};
    ```

4. Array a[3][4][2] is initialized as follows:

    ```
    int a[3][4][2] = {1,2,3,4,5,6,7,8,9,10,11,12,
                      13,14,15,16,17,18,19,20,21,22,23,24};
    ```

    What are the values of a[1][1][1], a[2][1][1], and a[2][2][1]?

5. Array x[2][3][4][5] is initialized as follows:

    ```
    int x[2][3][4][5] = {1,2,3, ... through 120};
    ```

    What are the values of x[1][2][3][4], x[0][1][3][1], and x[1][0][0][4]?

Solutions

1. a (false), b (true), c (false), d (false), e (true)
2. a.   The minimum subscript number is 1, not 0.
    b.   `float a23b[99][77], xy1[66][77];`
    c.   `long city[36][34], town[12][34];`
    d.   `int a[2][3]={11,22,33,44};`

■ LESSON 6.5 FUNCTIONS AND 1-D ARRAYS

Topics
- Passing individual array elements to functions
- Passing entire arrays to functions
- Passing entire arrays to functions with a restriction

In developing modularly designed programs, we need to write functions that pass and receive arrays. Recall that, when we passed simple variables, we had only two choices: We could pass a copy of the *value* of the simple variable or we could pass a copy of the *address* of the variable. In passing a copy of the value, we were able to change only the value of the copy and not the value of the original variable being passed. In passing the address, we allowed the function to modify the original variable's value. In this lesson's program, we use three functions—function1, function2, and function3—to illustrate how values and address of array elements are passed.

What you should observe about this lesson's program.
1. The first argument in function1's prototype is a pointer variable and the first argument in the call to function1 is the address of an array element.
2. The first argument in function2's prototype has brackets, indicating an array, and the first argument in the call to function2 is the name of an array but without following brackets.
3. The first argument in function3's prototype has the keyword const before double b[ ] and the first argument in the call to function3 is the name of an array but with no following brackets.
4. The second argument in function1's prototype is an ordinary integer variable and the second argument in the call to function1 is a single array element.
5. The second argument in function2's and function3's prototypes is an ordinary integer variable representing the number of elements in the array represented by the first argument.
6. The for loop in main prints all the elements of the c[ ] array.
7. In the body of function1, the variable *d modifies the value of the first argument passed to function1.
8. In the body of function2, the array b[ ] is modified in the for loop. The loop loops over the number of elements in the b[ ] array. Also, c[ ] in main is modified by this action.
9. The for loop in function3 sums all of the elements of the b[ ] array. The actions in this function *do not* modify the values of c[ ] in main.

Questions you should attempt to answer before reading the explanation.
1. What does the keyword const mean in the function3 prototype?
2. How can we transfer access to an entire array to a function?

Source code

> Pointer variable for holding an address. This indicates that the call to the function should have an address as the first argument.

```
#include <stdio.h>
void function1 (int *d, int e);
void function2 (double b[ ],int num_elem);
double function3 (const double b[ ], int num_elem);
```

> The const means that, even though function3 receives the address of the first element of a 1-D array, it cannot use that address to modify the contents of the array.

> Variables with brackets in the prototypes. These also indicate that the calls to the functions should have addresses as the first arguments.

```
void main(void)
        {
        int i, a[10]={0,1,2,3,4,5,6,7,8,9};
        double x, c[5]={2.,4.,6.,8.,10};

        function1(&a[5], a[8]);

        function2(c,5);

        x = function3(c,5);

        printf("\na[5]=%d\n",a[5]);
        printf("c[ ]=");
        for (i=0; i<5 ; i++) printf("%.1lf",c[i]);
        printf("\nx = %.1lf",x);
        }
void function1(int *d, int e)
        {
        *d = 100+e;
        }

void function2 (double b[ ], int num_elem)
        {
        int i;
        for  (i=0; i<num_elem; i++) b[i]*=10.;
        }

double function3 (const double b[ ], int num_elem)
```

> This argument represents the address of the sixth element of the a[ ] array.

> These arguments, c without brackets, represent the address of the first element of the c[ ] array.

> In this function, we work with the address that it receives using pointer notation (meaning using the unary * operator).

> In this function we use array notation (meaning brackets) to modify the elements of the array whose address was passed.

> This loop modifies each element of the array whose address was passed.

> For a 1-D array, the brackets should be left empty in the function prototype and declarator.

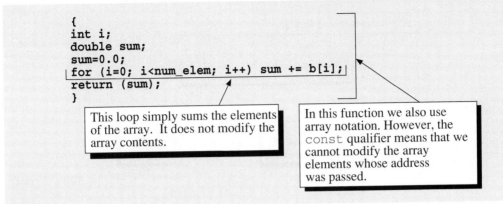

```
{
int i;
double sum;
sum=0.0;
for (i=0; i<num_elem; i++) sum += b[i];
return (sum);
}
```

This loop simply sums the elements of the array. It does not modify the array contents.

In this function we also use array notation. However, the const qualifier means that we cannot modify the array elements whose address was passed.

Output

```
a[5]=108
c[ ]=20.0 40.0 60.0 80.0 100.0
x = 300.0
```

Explanation

1. How do we write a function call and prototype to pass a single array element to a function? We treat a single array element like a simple variable. If we intend to change the value of the array element in the function, we use the "address of" operator before the array element in the function call. If we intend to pass the element without having it changed in the function, we simply put the array element in the parameter list. For instance, in this lesson's program, the call

```
function1(&a[5], a[8]);
```

passes a[5] and a[8]. Because this call causes the address of a[5] to be copied to the memory region of function1, the array element a[5] can be (and is) changed by function1. In contrast, the value (not the address) of a[8] is copied to the memory region of function1, and thus, while function1 can use the value of a[8] to perform calculations, it cannot change what is stored in the memory location for a[8].

The corresponding function prototype is similar to what it would be if a single variable's address and another variable's value were to be passed. When receiving an address, we must use a pointer variable (indicated in the declaration by *). When receiving a value, we use a simple variable. For this lesson's program, the prototype for function1 is:

```
void function1 (int *d, int e);
```

Therefore, we pass the address of a[5] into the pointer variable d and pass the value of a[8] into the variable e, as illustrated in Fig. 6.7.

FIG. 6.7

Passing array elements to a function

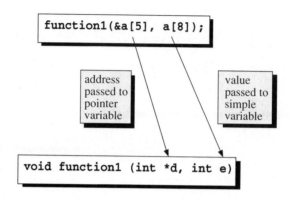

Within the function body, we use these two variables in the assignment statement:

*d = 100+e;

On return from the function, the value of a[5] is 100 plus the value of a[8], which is 108, and the value of a[8] remains unchanged.

2. How do we write a function call and prototype to pass the ability to access an entire one-dimensional array to a function? We could do it element by element, but this is not done in practice. Instead, it is much simpler to pass the address of the first element of the array. With the address of the first element (and the array type—int or double, for instance—to indicate the number of bytes per element), C can internally calculate the address of any element in the array. However, we do not need to use the "address of" operator to pass the address. This is because, in C, the address of the first element of an array is indicated by the array name with no brackets following it. In other words, for array c[ ] in this lesson's program the following two are equivalent:

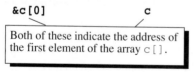

So, in the function call

function2 (c,5);

we are passing the address of the first element of the array c[] (along with the constant 5) to function2 (see Fig. 6.8). This then gives function2 the ability to modify the array c[ ].

The prototype for the function must indicate that it is receiving an address. To this point we have used * in the declarator to indicate that an address is to be received; however, C allows a second method that is more commonly used with

FIG. 6.8

Passing the ability to access all array elements to a function

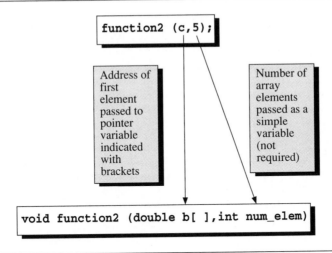

numeric arrays, using brackets. For instance, for function2, the following two pro-totypes are equivalent:

```
void function2 (double *b, int num_elem);
```

```
void function2 (double b[ ],int num_elem);
```

Observe that the only difference between these two is that the first uses *b and the second uses b[ ]. In this text, for numeric arrays, we will use the second method because it more clearly shows that we are working with arrays. If you use * instead of [ ] in the prototype, in the body of the function you still can use brackets to per-form array manipulations. In other words, the body of the function for both proto-types can be the same as shown in this lesson's source code. We see later in this book that we also can use pointer notation (that is, *) to access array elements if we choose.

3. Within a function that has received an array's address, how do we work with the array? We treat the array in a manner very similar to an array in the main func-tion. We must be cognizant of the number of elements that the array contains because C does not check whether we are trying to access an element that goes beyond the array's declared size. One way to do this is to transfer the number of ele-ments in an array through its parameter list. This number must be a separate vari-able in the list. For instance,

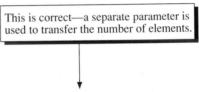

```
void function2 (double b[ ],int num_elem)
```

Note that it is not meaningful to put the number of elements in the brackets adjacent to the array name. In other words, we cannot transfer the information that function2 is to receive five elements by writing the decalarator as

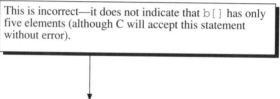

This is incorrect—it does not indicate that b[ ] has only five elements (although C will accept this statement without error).

```
void function2 (double b[5])
```

In this lesson's program, within the body of function2, we work with the b[ ] array element by element, as we have in previous examples. The loop that follows multiplies each element by 10, storing each result in the array itself:

```
for (i=0; i<num_elem; i++) b[i]*=10.;
```

After executing this function, we have modified our original array. Note that we have printed out the array in main. We can see from the output that array c[ ] in main indeed has been modified to be 10 times the original. By transferring the address of the first element of the array c[ ], we have given function2 the ability to modify c[ ].

   4. What if we want a function to have access to all of the values of an array but not the ability to modify the array values? This is a common occurrence because we often may want simply to use an array's elements in a calculation and keep the array intact. We can assure that an array is not modified in a function by using the const qualifier in the declarator. For instance, for function3 in this lesson's program, we have the following call and declarator:

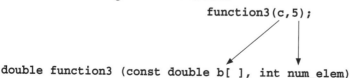

```
function3(c,5);
```

```
double function3 (const double b[ ], int num_elem)
```

By using the const qualifier, C will indicate an error if the function attempts to modify the b[ ] array. While this qualifier is not required, we recommend that you use it for functions that need to have access to an array without modifying it. In this lesson's program, function3 simply determines the sum of the values of the b[ ] array. The body of the function is

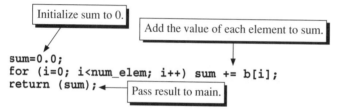

Initialize sum to 0.

Add the value of each element to sum.

```
sum=0.0;
for (i=0; i<num_elem; i++) sum += b[i];
return (sum);
```

Pass result to main.

Note that we have chosen a return statement to pass sum back to the main function. Therefore, in main, the function call appears on the right side of an assignment statement.

EXERCISES

1. True or false:
    a. In a function call, using one of the parameters as an array name not followed by brackets passes the entire array to the function by copying each element into the memory region for the function.
    b. If we want to pass a single array element, we need not use & in the parameter list of the function call.
    c. It is not possible in C to give a function access to all of an array's elements without giving the function the ability to modify those elements.
    d. A pointer variable is used to store the address of a variable.
    e. The name of an array, not followed by brackets, indicates the address of the first element of the array.
    f. If we want to transfer the number of elements in an array to a function, we should use a separate parameter in the parameter list to do so.

2. The following table shows the length, width, and thickness of a stack of steel plates:

Length (ft.)	Width (ft.)	Thickness (in.)
12.0	6.3	2.2
13.0	7.4	3.3
14.0	8.5	4.4
15.0	9.6	5.5

Use three 1-D arrays to store the length, width, and thickness information; then convert the units from foot or inch to meters, calculate the weights of the steel plates, and save the results in a 1-D array (assume the unit weight of steel is 7800 kg/m^3).

Solutions

1. a (false), b (false), c (false), d (true), e (true), f (true)

LESSON 6.6 FUNCTIONS AND 2-D ARRAYS

Topics
- Passing 2-D arrays to functions
- Declaring the size of a 2-D array
- Reading an entire 2-D array from a file
- Calling a function with a 2-D array
- Using a 2-D array in a function

This lesson's program is broken up into four sections:

1. Reading a 2-D array from a file
2. Calling a function with a 2-D array
3. Printing a 2-D array
4. Performing operations on a 2-D array

In working with arrays, it is necessary to remember that, in the array declaration, we specify the maximum possible number of elements not the actual number of elements to be occupied. Typically, we initialize the array elements by reading them in from a file.

With one-dimensional arrays, it was relatively simple to count the number of elements actually being read in and set the loop control variables that work with the arrays to match the actual number of elements. With two-dimensional arrays, it also is possible to count the total number of elements being read in, but to work with the arrays in loops, it is necessary to know how many rows and columns there are. One way to make this distinction is to have the number of rows and columns as input data. In this lesson's program, we read from a file the number of rows and columns of a 2-D array and the elements of the array. The array is modified in a function and printed to the screen.

What you should observe about this lesson's program.

1. Two constant macros are defined representing the maximum number of rows and columns of the array.
2. In the prototype for function1, the third argument is a two-dimensional array with the first set of brackets empty and the second set with a constant macro.
3. The input file has first the number of rows and columns of the array and then the values of the array elements.
4. The first fscanf statement reads the number of rows and columns for the two-dimensional array.
5. The first nested for loop in main reads the elements of the array from a file.
6. In calling function1, the actual number of rows and columns (not the maximum) are the first two arguments.
7. The second nested loop in main prints the array elements.
8. The nested for loop in function1 modifies the values of the array elements.

Questions you should attempt to answer before reading the explanation.

1. Why are the number of rows and columns needed for working with 2-D arrays?
2. Why is only one set of brackets filled for the b[ ][ ] array in the prototype for function1?

Source code

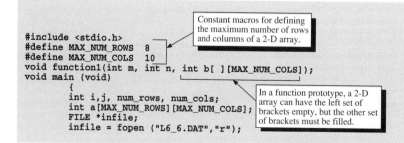

```
                                    Constant macros for defining
                                    the maximum number of rows
                                    and columns of a 2-D array.
#include <stdio.h>
#define MAX_NUM_ROWS   8
#define MAX_NUM_COLS   10
void function1(int m, int n, int b[ ][MAX_NUM_COLS]);
void main (void)
    {                                   In a function prototype, a 2-D
        int i,j, num_rows, num_cols;    array can have the left set of
        int a[MAX_NUM_ROWS][MAX_NUM_COLS]; brackets empty, but the other set
        FILE *infile;                   of brackets must be filled.
        infile = fopen ("L6_6.DAT","r");
```

```
/*************************************************************
'*      SECTION 1- READING A 2-D ARRAY FROM A FILE
'*************************************************************
        fscanf (infile,"%d %d", &num_rows, &num_cols) ;
        for (i=0; i<num_rows; i++)
                {
                for (j=0; j<num_cols; j++)
                        {
                        fscanf(infile,"%d ", &a[i][j]);
                        }
                }
/*************************************************************
'* SECTION 2 - CALLING A FUNCTION WITH A 2-D ARRAY
'*************************************************************/


        function1(num_rows, num_cols, a);



/*************************************************************
'* SECTION 3 - PRINTING A 2-D ARRAY
'*************************************************************/
        for (i=0; i<num_rows; i++)
                {
                for (j=0; j<num_cols; j++)
                        {
                        printf("%d ", a[i][j]);
                        }
                printf("\n");
                }

/*************************************************************
'* SECTION 4 - FUNCTION THAT PERFORMS OPERATIONS ON A 2-D ARRAY
'*************************************************************/
void function1 (int m, int n, int b[ ][MAX_NUM_COLS])
        {
        int i,j;

        for (i=0; i<m; i++)
                {
                for (j=0; j<n; j++)
                        {
                        b[i][j] += 100;
                        }
                }
        }
```

> The actual (not maximum) number of rows and columns are read from the file and then used to control the loop to read the values of the array elements.

> Elements of a 2-D array are read one at a time, so a nested loop is required.

> Calling function1. It is advisable to pass the number of rows and number of columns along with the array address to a function.

> Elements of a 2-D array are printed one at a time, so a nested loop is required.

> In a function declarator, a 2-D array can have the left set of brackets empty, but the other set of brackets must be filled.

> Within a function, we also work with a 2-D array element by element.

Input file L6_6.DAT

```
3 4
1 2 3 4
2 4 6 8
3 5 7 9
```

Output

```
101 102 103 104
102 104 106 108
103 105 107 109
```

Explanation

1. Why have we declared the size of the two-dimensional array a[ ][ ]with the constant macros MAX_NUM_ROWS and MAX_NUM_COLS? We used constant macros to declare the size of our array because we want to be able to easily change the maximum number of rows and columns that the array can handle. Remember, in C, memory is reserved for an array based on its declared size. If we are working with a series of problems that require only small arrays, then we can reduce our memory requirements by using small values for our constant macros. However, if we want to use the same program to handle large arrays, then we easily can change the values of the constant macros, recompile the program, and use it. If we did not put our array sizes in constant macros, it may be very tedious and error causing to change the array sizes everywhere they are used within a program. This is especially true when working with large numbers of functions and arrays. Therefore, we recommend you use constant macros to size your arrays.

2. Conceptually, in memory what does array a[ ][ ] from this lesson's program look like? The data file gives array a[ ][ ] as having three rows and four columns, meaning that it has a total of 12 elements. However, the declaration for a[ ][ ] is for eight rows and ten columns, meaning that a total of 80 elements is reserved for a[ ][ ]. Because C stores even multidimensional arrays in a linear type of fashion, we have the line of values shown in Fig. 6.9 (shown in four columns simply to save space).

3. What is another way to envision the array a[ ][ ]? At times, you will find it convenient to envision a two-dimensional array in matrix form, with the occupied rows and columns in the upper left corner and the rest of the array containing meaningless digits (represented by *):

```
1 2 3 4 * * * * * *
2 4 6 8 * * * * * *
3 5 7 9 * * * * * *
* * * * * * * * * *
* * * * * * * * * *
* * * * * * * * * *
* * * * * * * * * *
* * * * * * * * * *
```

Note that we have shown the entire declared array with eight rows and ten columns but only part of it filled.

4. For this lesson's program, why did we read the number of rows and columns of our array, a[ ][ ], from the data file? By reading the number of rows and columns, we can reduce the number of program operations. For arrays, the declared size gives the total number of elements that can be used by an array but does not indicate how many we actually use for a particular problem. By reading in the number of rows and columns, we give the program the information needed to limit the number of array elements on which to operate.

FIG. 6.9

The two-dimensional array a[][] for this lesson's program, as stored conceptually in memory. The symbol * represents a memory location that has not been filled but may contain meaningless bits.

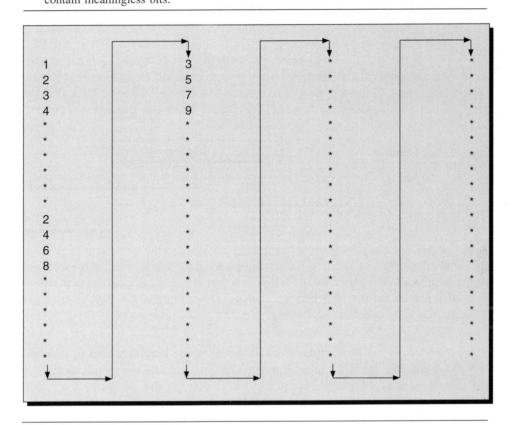

5. Are there other ways to have a program determine the actual numbers of rows and columns filled for two-dimensional arrays? Yes; we saw that, for one-dimensional arrays, it was possible for a program to use fscanf to count the number of array elements until EOF was encountered. While this works well for one-dimensional arrays, it is more difficult to implement for two-dimensional arrays, because we can get only the total number of elements and not the numbers of rows and columns.

Such things as *sentinel values* (values that are distinct from and cannot be confused with the other data values) can be used in a data file to indicate, for instance, the end of a row of data. Similar to how we handled EOF, a program can be used to count the number of array elements until the sentinel value is encountered, giving the program the number of columns. The number of times that the sentinel value is encountered gives the number of rows.

Another way would be to put only the number of rows and the values of the array elements in the input file. By counting the number of values until EOF is encountered, the number of columns can be calculated.

The method you use to determine or indicate the number of rows or columns can be developed based on the requirements of your problem.

6. How do we read a two-dimensional array from a data file? Similar to how we print a two-dimensional array, we use a two-deep nested for loop to read a two-dimensional array. Using the number of rows and number of columns, the following loop reads the elements of the array a[ ][ ] that are written in the data file pointed to by infile:

```
for (i=0; i<num_rows; i++)  ◄──── Loop over the number of rows.
       {
       for (j=0; j<num_cols; j++)  ◄──── Loop over the number of columns.
              {
              fscanf(infile,"%d ", &a[i][j]);  ◄── Read each
              }                                     individual element.
       }
```

Note that the outer loop loops over the number of rows and the inner loop loops over the number of columns and that we read just a single element with each execution of the fscanf call. Also, because the fscanf function automatically advances to a new line in its search for the next non-white-space value, there is no need to tell it to advance to a new line after a complete execution of the inner loop.

7. How do we use a multidimensional array in the parameter list of a function call? Because the array name without following brackets means the address of the first element of the array, we can use this directly in the parameter list of the function call. In this lesson's program, the call

```
function1(num_rows, num_cols, a);
```

calls function1 with the two-dimensional array a[ ][ ].

8. How should the array be given in the parameter list of the function prototype? Recall that, for a one-dimensional array, we indicated an array by using one pair of brackets following an identifier in the function prototype. For two-dimensional arrays, we use two pairs of brackets following an identifier in the function prototype. However, unlike the situation with a one-dimensional array for which the brackets were empty, we need to include the declared size of the second subscript (maximum number of columns) in the brackets. For instance, the prototype for function1 in this lesson's program is

```
void function1(int m, int n, int b[ ][MAX_NUM_COLS]);
```

The transfer of information is illustrated in Fig. 6.10.

FIG. 6.10
Passing the ability to access all array elements to a function

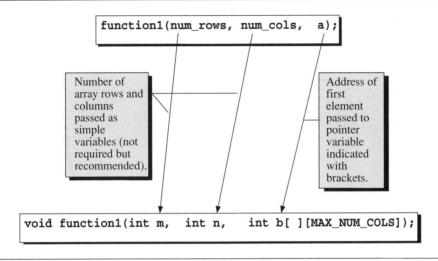

9. Why do we need to include the maximum number of columns in the second set of brackets of the array parameter in the prototype? Recall the formula used to calculate the location of an element in a multidimensional array given in Lesson 6.4. We can modify this formula to represent the location of an element of a two-dimensional array.

For an array element, a[x][y] of an array declared with a size a[MAX_NUM_ROWS][MAX_NUM_COLS], C uses a formula of the following type to locate an array element's position in the list and therefore its position in memory:

$$\text{sequence location} = (x \times \text{MAX\_NUM\_COLS}) + y + 1$$

For example, for this lesson's 2-D array a[ ][ ], we have MAX_NUM_COLS = 10. The sequence location for element a[1][2] ($x = 1$, $y = 2$) is

$$\text{sequence location} = (1 \times 10) + 2 + 1 = 13$$

From the input file it can be seen that a[1][2] = 6, which is the 13th element in the list according to Fig. 6.9.

Note that, in this formula and the calculation, MAX_NUM_COLS does appear but MAX_NUM_ROWS does not. So, to locate an array element's position, only MAX_NUM_COLS needs to be known by a function. This is accomplished in C by inclusion of it in the second pair of brackets for the array in the function prototype and function declarator.

10. If we were using a three- or four-dimensional array, what would be included in the function prototype? Except for the first pair of brackets, all of the brackets should be filled with their declared sizes. For instance, for a four-dimensional array declared as

```
#define I 10
#define J 5
#define K 8
#define L 3
int a[I][J][K][L];
```

a prototype for a function, function1, would be of the form:

```
void function1 (...int a[ ][J][K][L]...);
```

11. Must the function prototype and the function declarator be absolutely the same? No; in the function prototype, it is not necessary to give the names of the variables in the parameter list, whereas in the function declarator, the names are required. For instance, for this lesson's program, the function prototype

```
void function1(int m, int n, int b[ ][MAX_NUM_COLS]);
```

could have been

```
void function1(int, int, int [ ][MAX_NUM_COLS]);
```

because C ignores the variable names in the function prototype. We point this out primarily because you may see this form in other programs and it is perfectly acceptable. However, we recommend that you include variable names for ease of reading (you can see that, at first glance, the prototype without variable names is more difficult to read) and simplicity. For instance, if you make changes in your function definition and need to make corresponding changes in your function prototype, it is very straightforward to just copy the function declarator to the location of the function prototype and add a semicolon at the end. However, should your instructor or employer prefer not putting variable names in function prototypes, do not put them there.

EXERCISES

1. True or false:
    a.   C allows you to use arrays directly as function formal arguments.
    b.   C allows you to use arrays directly as function actual arguments.
    c.   C allows you to return arrays from a function to the main () function.
    d.   A pointer variable can be used to store the address of an array element.
    e.   We may use the address of an array as a function actual argument.
    f.   We may use the address of an array as a function formal argument.
    g.   We may use a pointer variable as a function formal argument.

2.   The table below shows the length, width, and thickness of a stack of steel plates:

Length (ft.)	Width (ft.)	Thickness (in.)
12.0	6.3	2.2
13.0	7.4	3.3
14.0	8.5	4.4
15.0	9.6	5.5

Within a function use a 2-D array to store the length, width, and thickness information, then convert the units from foot or inch to meter, calculate the weight of the steel plates and save the results in a 1-D array (assume the unit weight of a steel plate is 7800 kg/m^3).

Solutions

1.   a (false), b (false), c (false), d (true), e (true), f (false), g (true)

▦   LESSON 6.7 BUBBLE SORT, EXCHANGE MAXIMUM SORT, AND EXCHANGE MINIMUM SORT*

Topics

- Bubble sort
- Exchange maximum sort
- Exchange minimum sort
- Swapping values

A fundamental operation on arrays is sorting them; that is, arranging the elements such that they are in a specified order, usually from minimum value to maximum value.

For instance, suppose we have array b[ ] that has following elements: b[0] = 34, b[1] = 23, b[2] = 64, b[3] = 39, b[4] = 84, b[5] = 91, b[6] = 73. This array is considered to be sorted if we rearrange the values of the array elements to be b[0] = 23, b[1] = 34, b[2] = 39, b[3] = 64, b[4] = 73, b[5] = 84, b[6] = 91, because the values of the array elements increase with increasing subscript values.

To do this rearrangement, you can see that we needed to do a considerable amount of swapping of array element values. For instance, we swapped the values of 23 and 34 between elements b[0] and b[1]. The goal of writing sorting algorithms is to do this type of swapping very efficiently. We do not describe all the possible sorting methods or even all of the issues raised in writing sorting algorithms. In fact, entire college courses are devoted to simply the topic of sorting. You may take such a course later in your educational career. We lack the space here to fully cover sorting. However, in this lesson, we introduce you to the most basic sorting techniques, the bubble sort and two exchange sorts. In Chapter 8, we illustrate the quicksort.

For this lesson, you need not read the program in advance. In it, we sort the array {33, 44, 11, 22} using the three methods. First read the explanation and refer back to the program to understand how it operates.

---

*We recommend that you read Application Programs 6.1, 6.2, and 6.3 prior to reading this lesson.

Source code

```
#include <math.h>
#include <stdio.h>

#define START 0
#define END   4
#define SIZE  10

void main(void)
{
    int i, j, k, b[SIZE], c[SIZE], d[SIZE];
    int temp, max, wheremax=END-1, min, wheremin=START;
    int a[END]={33,44,11,22};

/*************************************************
**      INITIALIZE ARRAYS b, c, AND d
*************************************************/

    for (i=START; i<END ;i++) b[i]=c[i]=d[i]=a[i];

/*************************************************
**      BUBBLE SORT
*************************************************/
for (i=START;i <END;i++)
    {
    for (j=START; j <END-i-1;j++)
        {
        if (b[j] > b[j+1] )
            {
            temp=b[j+1];
            b[j+1]=b[j] ;
            b[j]=temp;
            }
        }
    }

/*************************************************
**    EXCHANGE MAXIMUM SORT
*************************************************/
 for (i=END-1;i>=START;i--)
    {
    max=c[i];
    for (j=i;j>=START;j--)
        {
        if (max <=c[j])
            {
            max=c[j];
            wheremax=j;
            }
        }
    c[wheremax]=c[i];
    c[i]=max;
    }
```

Beginning and ending of portion of array to be sorted.

Maximum size of array to be sorted.

Initializing the a[ ] array.

For this example, all of the arrays are the same as the a[ ] array.

Each time through this loop is one round for the bubble sort.

For the first round, this loop executes over the entire range (START to END). On the second round, one element is properly placed so this loop executes one fewer than the entire range. On the third round through the loop, two elements are properly placed so the loop executes two fewer than the entire range.

These three statements swap the values of b[j] and b[j+1].

The swap is performed if the lower subscripted array element is greater than the higher subscripted array element next to it.

Each time through this loop is one round for the exchange maximum sort.

At the beginning of the round, the maximum is taken to be the array element at the largest subscript used in the round.

For the first round, this loop executes over the entire range (START to END). On the second round, one element is properly placed so this loop executes one fewer than the entire range. On the third round through the loop, two elements are properly placed so the loop executes two fewer than the entire range.

If max is less than or equal to the element we are considering, then we have a new max and we store the subscript for it in wheremax.

At the end of the round, we swap the highest subscript element that has not already been set with max.

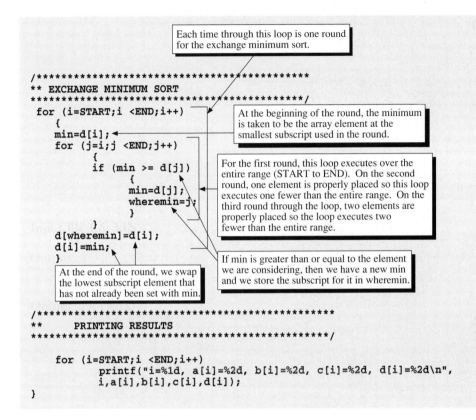

Each time through this loop is one round for the exchange minimum sort.

```
/***********************************************
** EXCHANGE MINIMUM SORT
************************************************/
  for (i=START;i <END;i++)
    {
    min=d[i];
    for (j=i;j <END;j++)
        {
        if (min >= d[j])
            {
            min=d[j];
            wheremin=j;
            }
        }
    d[wheremin]=d[i];
    d[i]=min;
    }

/***************************************************
**        PRINTING RESULTS
****************************************************/
    for (i=START;i <END;i++)
        printf("i=%1d, a[i]=%2d, b[i]=%2d, c[i]=%2d, d[i]=%2d\n",
        i,a[i],b[i],c[i],d[i]);
}
```

At the beginning of the round, the minimum is taken to be the array element at the smallest subscript used in the round.

For the first round, this loop executes over the entire range (START to END). On the second round, one element is properly placed so this loop executes one fewer than the entire range. On the third round through the loop, two elements are properly placed so the loop executes two fewer than the entire range.

If min is greater than or equal to the element we are considering, then we have a new min and we store the subscript for it in wheremin.

At the end of the round, we swap the lowest subscript element that has not already been set with min.

Output

```
i=0,  a[i]=33,  b[i]=11,  c[i]=11,  d[i]=11
i=1,  a[i]=44,  b[i]=22,  c[i]=22,  d[i]=22
i=2,  a[i]=11,  b[i]=33,  c[i]=33,  d[i]=33
i=3,  a[i]=22,  b[i]=44,  c[i]=44,  d[i]=44
```

Explanation

1. What concept underlies a bubble sort? We want to sort the 1-D array, $b[4] = \{33, 44, 11, 22\}$, in ascending order (see Fig. 6.11). We first compare $b[0]$ and $b[1]$. If $b[0] > b[1]$, then we swap the values of $b[0]$ and $b[1]$; otherwise, we do no rearrangement. Similarly, we perform the same actions for $b[1]$ and $b[2]$. If $b[1] > b[2]$, then we swap the values of $b[1]$ and $b[2]$. We continue this with the last pair, $b[2]$ and $b[3]$. This concludes the first round.

During the first round, the largest element, 44, bubbles down (meaning that, step by step, it moves into an array element of greater subscript) and stops in the last element of the array. Then, we use the same procedure (second round) to process only the top three elements (since the fourth element properly has the largest value). This

FIG. 6.11
Concept of bubble sort

FIRST ROUND

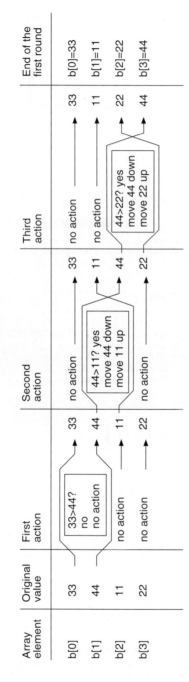

SECOND ROUND

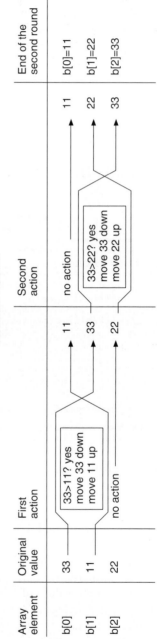

causes the second largest element, 33, to bubble down to the third element. In the third round, the third largest element, 22, bubbles into the second element. We are now finished because the last value is automatically in the correct top spot. In general, we need one fewer rounds than number of array elements.

We can trace the value of the indices in the loops. Look at the nested for loop in the source code. Tracing the loop we get

i	j	$j + 1$	Action	Comment
0	0	1	Larger of b[0] and b[1] put into b[1]	
	1	2	Larger of b[1] and b[2] put into b[2]	
	2	3	Larger of b[2] and b[3] put into b[3]	b[3] is the largest of the four values
1	0	1	Larger of b[0] and b[1] put into b[1]	
	1	2	Larger of b[1] and b[2] put into b[2]	b[2] is the largest of the three values
2	0	1	Larger of b[0] and b[1] put into b[1]	b[1] is the larger of the two values
3			No action	b[0] is automatically the smallest

2. What actions typically are involved in swapping the values of array elements? Because two actions cannot take place simultaneously, we must introduce a temporary storage value to help with the swap. In this lesson's program, we use the temporary storage variable called `temp`. For instance, we would use the following steps to swap the values of b[1] and b[2] (for the case of b[1] = 7 and b[2] = 9):

`temp=b[1];`   *Copy the value of b[1] into temp. (temp=7)*
`b[1]=b[2] ;`   *Copy the value of b[2] into b[1]. (b[1]=9)*
`b[2]=temp;`   *Copy the value of temp into b[2] which causes the original value of b[1] to be put into b[2].(b[2]=7)*

The result of these steps is that b[1] = 9 and b[2] = 7 and the values indeed have been swapped. Remember that, in your programs, when you want to swap values, you need to introduce a temporary storage value.

3. What concept underlies an exchange maximum sort? In this program, we performed the following steps for the exchange maximum sort (see Fig. 6.12):

1. Assign the last array element to the variable `max`.
2. One by one, compare `max` with the rest of the array elements; if `max` is smaller than any array element, then replace `max` with this element and use the variable `wheremax` to remember the location (array index) of the new maximum. Continue the process until the maximum value in the array is found.
3. Put the found maximum value at the last element of the array. This completes the first round.
4. Since the largest value of the group is found and stored in the last element, eliminate the last element from the search group.
5. Repeat steps 2, 3, and 4 until the array is in ascending order.

FIG 6.12
The concept of exchange maximum sort

FIRST ROUND

i	j	Array element	Original value	First action	Second action	Third action	Fourth action	End of Round 1	
0	0	c[0]	33				max<33? no action	Found: max=44 wheremax=1 i=3 c[i]=c[3]=22 do: c[wheremax]=c[i] → c[1]=22 c[i]=max → c[3]=44	c[0]=33
1	1	c[1]	44			max<44? yes max=44 wheremax=j=1			c[1]=22
2	2	c[2]	11		max<11? no action				c[2]=11
3	3	c[3]	22	max=22 i=3					c[3]=44

SECOND ROUND

i	j	Array element	Original value	First action	Second action	Third action	Fourth action	End of Round 2	
0	0	c[0]	33			max<33? yes max=33 wheremax=j=0		Found: max=33 wheremax=0 i=2 c[i]=c[2]=11 do: c[wheremax]=c[i] → c[0]=11 c[i]=max → c[2]=33	c[0]=11
1	1	c[1]	22		max<22? yes max=22 wheremax=j=1				c[1]=22
2	2	c[2]	11	max=11 i=2					c[2]=33

Final result   c[0]=11
c[1]=22
c[2]=33
c[3]=44

We can trace the indices and actions of the loop. Look at the nested for loop in the source code. Tracing the loop we get

i	j	Action	Comments
3	3	max = c[3]	For our data,
	2	if c[2] > max, max = c[2], wheremax = 2	max = 44, wheremax = 1, 44 moves
	1	if c[1] > max, max = c[1], wheremax = 1	to c[3] and the c[3] value (22)
	0	if c[0] > max, max = c[0], wheremax = 0	moves into c[1]
2	2	max = c[2]	For our data, max = 33,
	1	if c[1] > max, max = c[1], wheremax = 1	wheremax = 0, 33 moves to c[2]
	0	if c[0] > max, max = c[0], wheremax = 0	and the c[2] value (11) moves to c[0]
1	1	if c[1] > max, max = c[1], wheremax = 1	For our data, max = 22,
	0	if c[0] > max, max = c[0], wheremax = 0	wheremax = 1, no further movement needed
0		No action	c[0] is automatically the smallest

A detailed trace of the nested loop is given in Table 6.1. Follow this table and the source code to enhance your understanding of loops and arrays.

4. What is the concept of an exchange minimum sort? This sort is similar to the exchange maximum sort. However, instead of finding the maximum value, this method finds the minimum value and puts it in the first location. The process is then repeated until the array is in ascending order.

For this lesson's program, you can trace the loops to give

i	j	Action	Comments
0	0	min = c[0]	For our data,
	1	if c[1] < min, min = c[1], wheremin = 1	min = 11, wheremin = 2, 11 moves
	2	if c[2] < min, min = c[2], wheremin = 2	to c[0] and the c[0] value (33)
	3	if c[3] < min, min = c[3], wheremin = 3	moves into c[2]
1	1	min = c[1]	For our data, at this point, min = 22,
	2	if c[2] < min, min = c[2], wheremin = 2	wheremin = 3, 22 moves to c[1]
	3	if c[3] < min, min = c[3], wheremin = 3	and the c[1] value (44) moves to c[3]
2	2	if c[2] < min, min = c[2], wheremin = 2	For our data, at this point, min = 33,
	3	if c[3] < min, min = c[3], wheremin = 3	wheremin = 2, no further movement needed
3		No action	c[3] is automatically the largest

Fill in the blanks of Table 6.2. Doing so will help you understand loops and array manipulations.

EXERCISES

1. A 1-D array has these ten elements:

4.4   3.3   2.2   5.5   1.1   6.6   7.7   10.0   9.9   8.8

Use the three methods you learned from this lesson to sort the array in descending order.

TABLE 6.1

Detailed trace of nested loop for the exchange maximum sort. Use the source code and this table to understand each step.

Array initially is c[] = {33, 44, 11, 22}

			Inner loop—purpose is to find max and wheremax						Outer loop—purpose is to swap c[wheremax] with c[i]				
i	c[i]	max (max = c[i];)	j	c[j]	max	max≤c[j]?	max (max = c[j];)	wheremax (wheremax = j;)	i	c[i]	wheremax	c[wheremax] (c[wheremax] = c[i];)	max	c[j] (c[j] = max;)
3	22	22	3	22	22	yes	22	3						
			2	11	22	no	no action	no action						
			1	44	22	yes	44	1						
			0	33	44	no	no action	no action	3	22	1	22→c[1] = 22	44	44→c[3] = 44

Array before second round is c[] = {33, 22, 11, 44}, (note: 44 and 22 have been swapped; 44 is correctly the last element of the array)

2	11	11	2	11	11	yes	11	2						
			1	22	11	yes	22	1						
			0	33	22	yes	33	0	2	11	0	11→c[0] = 11	33	33→c[2] = 33

Array before third round is c[] = {11, 22, 33, 44}, (note: 11 and 33 have been swapped; 33 is correctly the second-to-last element of the array)

1	22	22	1	22	22	yes	22	1						
			0	11	22	no	no action	no action	1	22	1	22→c[1] = 22	22	22→c[1] = 22

Array before fourth round is c[] = {11, 22, 33, 44}, (note: no change from previous because 22 was already correctly placed)

0	11	11	0	11	11	yes	11	0	0	11	0	11→c[0] = 11	11	11→c[0] = 11

Final array c[] = {11, 22, 33, 44}, (note: no change from previous because 11 was already correctly placed)

Blank table for tracing nested loop of the exchange minimum sort. Follow the source code line by line and fill in this table.

Array initially is d[ ] = {33, 44, 11, 22}

Outer loop—purpose is to swap d[wheremin] with d[i]

Inner loop—purpose is to find min and wheremin

i	d[i]	min (min = d[i];)	j	d[j]	min	min>d[j]?	min (min = d[j];)	wheremin (wheremin = j;)	i	d[i]	wheremin	c[wheremin] (d[wheremin]=d[i];)	min	d[i] (d[i]=min;)

2. A 2-D array has these 20 elements:

```
3    33    333    3333
5    55    555    5555
1    11    111    1111
4    44    444    4444
2    22    222    2222
```

Use the three methods to write a program to do the following:

a. Sort the array so that it will look like

```
5    55    555    5555
4    44    444    4444
3    33    333    3333
2    22    222    2222
1    11    111    1111
```

b. Sort the array so that it will look like

```
1111    111    11    1
2222    222    22    2
3333    333    33    3
4444    444    44    4
5555    555    55    5
```

APPLICATION EXAMPLE 6.1 LINEAR INTERPOLATION— EVALUATING VOLTAGE MEASUREMENT DATA (NUMERICAL METHOD EXAMPLE)

Problem statement

Voltage measurements have been made for an electronic circuit at various times. However, after collecting the data, it is desired to know the voltage at a time that is different from the times at which the data were collected. Write a program that will read the collected voltage data, the corresponding times, and the time at which the voltage is desired and compute the desired voltage based on a linear interpolation using the nearest measured values. Have the program read the time and voltage measurement data from a file, the desired time from the keyboard, and print the interpolated voltage to the screen.

Solution

RELEVANT EQUATIONS

Interpolating linearly between two points, (x_1, y_1) and (x_2, y_2), essentially means drawing a straight line connecting the points and obtaining a y value on that line for a given x value that is between the x values of the two endpoints. Figure 6.13 illustrates this.

The slope of the line, m, is:

$$m = \frac{(y_2 - y_1)}{(x_2 - x_1)} \tag{6.1}$$

FIG. 6.13
Interpolating linearly between two points

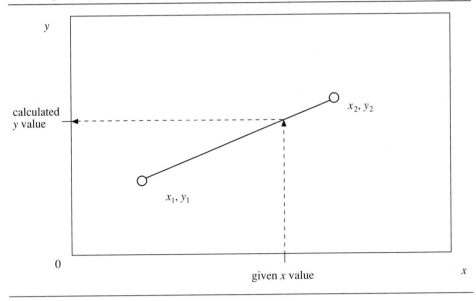

The x distance from x_1 to the given x value is $(x_{\text{value}} - x_1)$. So, the calculated y value is

$$y_{\text{value}} = \frac{(y_2 - y_1)}{(x_2 - x_1)} (x_{\text{value}} - x_1) + y_1 \qquad (6.2)$$

If we have a set of n data points as shown in Fig. 6.14 and are interested in interpolating between points $(i - 1)$ and i, then the equation becomes

$$y_{\text{value}} = \frac{(y_i - y_{i-1})}{(x_i - x_{i-1})} (x_{\text{value}} - x_{i-1}) + y_{i-1} \qquad (6.3)$$

We will use this equation in our source code.

SPECIFIC EXAMPLE
Suppose we have the measured voltages and times that follow and want to estimate the voltage at 25 milliseconds.

Data point	Time (milliseconds)	Voltage (millivolts)
0	0	23
1	2	78
2	3	89
3	6	−12
4	8	0
5	19	90
6	29	18
7	34	−23
8	37	76
9	45	98

FIG. 6.14
Location of *n* data points

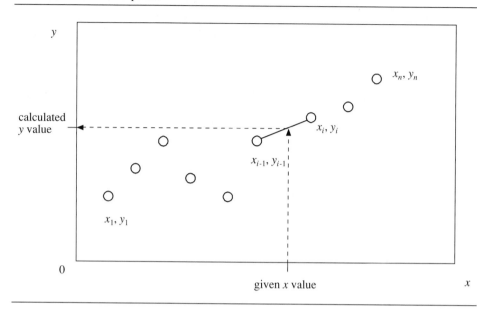

At 25 milliseconds, we realize that we are between data points 5 and 6. Using equation (6.3) with $(i - 1) = 5$ and $i = 6$, we have

$$y_{\text{value}} = \frac{(y_6 - y_5)}{(x_6 - x_5)} \left(x_{\text{value}} - x_5 \right) + y_5 = \frac{(18 - 90)}{(29 - 19)} (25 - 19) + 90 = 46.8$$

Therefore, our estimate of the voltage at 25 milliseconds is 46.8 millivolts.

ALGORITHM

In performing the specific example, we followed these steps:

1. Read all of the data point x and y coordinates.
2. Read the x value for which the y value is to be calculated.
3. Determine the data point numbers that are immediately to the left and right of the x value specified.
4. Interpolate using equation (6.3).
5. Print the results.

We can generate the source code from this algorithm by breaking down each task and writing each step somewhat individually. Steps 2 and 5 are straightforward. Next we describe steps 1, 3, and 4.

Step 1. Read all of the data point x and y coordinates. For instance, using one-dimensional arrays of type double, if we call the time x[ ] and the voltage y[ ], the following loop reads ten data points from the data file specified by infile:

```
for (i=0; i<10; i++)     fscanf (infile, "%lf %lf", &x[i],&y[i]);
```

Note that using C's array data structure, the first data point has an array index of 0, so the preceding loop goes from 0 to 9 (because 9 is the last integer that is less than 10).

Step 3. Determine the data point numbers that are immediately to the left and right of the x value specified. The following loop determines the data point number immediately to the right of the given x value.

```
i=0;
while (x_value > x[i]) i++;
```

After completing this loop, the value of i is equal to the index of the x[ ] array value immediately to the right of the given x_value. To verify this, we tabulate the values for each step of the loop for this specific example (with x_value = 25):

i	x[i]	x_value $> x$ [i] ?	i
0	0	yes	1
1	2	yes	2
2	3	yes	3
3	6	yes	4
4	8	yes	5
5	19	yes	6
6	29	no	

Thus, (i - 1) = 5 and i = 6 at the end of the loop. These two indices give the values of x[ ] to the right and left of the x_value. With these indices, the values of x[ ] and y[ ] for equation (6.3) can be obtained.

Step 4. Interpolate using equation (6.3). The following line of code applies equation (6.3) using the indices i and i - 1.

```
y_value = ((x_value-x[i-1])/(x[i]-x[i-1]))*(y[i]-y[i-1])+y[i-1];
```

SOURCE CODE

The source code has been developed from the algorithm. Look at the declared size of x[ ] and y[ ]. Why has a constant type macro been used for this? Where else in the program has this constant macro been used?

```
#include <stdio.h>
#define NUM_PTS 10
void main(void)
{
        double x[NUM_PTS], y[NUM_PTS], x_value, y_value;
        int i;                          Reading the data from the input file.
        FILE *infile;
        infile = fopen ("interp.dat","r");

        for (i=0; i<NUM_PTS; i++)     fscanf (infile, "%lf %lf", &x[i],&y[i]);
        printf ("Enter a value of time between %.2lf and %.2lf\n",x[0],x[NUM_PTS-1]);
        scanf ("%lf", &x_value);
```

```
i=0;                                    ┌─────────────────────────────────────┐
while (x_value > x[i]) i++;             │ Finding the index for the x[ ] array element
y_value = ((x_value-x[i-1])/(x[i]-x[i-1]))*(y[i]-y[i-1])+y[i-1];
printf("The value of the voltage at time = %.2lf is %.2lf", x_value, y_value);
}
```

Finding the index for the x[ ] array element to the right of the interpolation *x* value.

Interpolation equation (6.3).

INPUT DATA FILE

0	23
2	78
3	89
6	−12
8	0
19	90
29	18
34	−23
37	76
45	98

OUTPUT

Keyboard input

```
Enter a value of time between 0.00 and 45.00
25
The value of the voltage at time = 25.00 is 46.80
```

COMMENTS

The constant type macro NUM_PTS has been used to declare the size of x[ ] and y[ ] to facilitate a future change in the number of data points collected. (We will see in our modular design example for this chapter an even more versatile way of sizing arrays.) The macro NUM_PTS has been used in defining the loop used to read the input data file and in printing out the rightmost value in the input data prompt of x[ ]. Should we want to change the number of points input, we need to change only the number 10 in the macro definition. Using macros to declare the sizes of arrays is good programming practice.

In reading the input with the fscanf statement, it is necessary to use the conversion specification %lf, because both x[ ] and y[ ] are declared to be doubles. Unlike printf, when using scanf or fscanf, the conversion specification of %f will not work for double type variables. (Try changing this program to use %f and you will see that it does not work.) Remember, for scanf or fscanf, use %lf for doubles.

Modification exercises

1. Replace the while loop in the program with a do-while loop.
2. Replace the while loop in the program with a for loop.
3. Modify the program to handle 100 data points instead of 10.
4. Make the program print out the coordinates of the two nearest data points to (x_value, y_value).
5. Create a modular design for this program. Make two functions—one for input and the other for calculations and output.

■ APPLICATION EXAMPLE 6.2 MEAN AND MEDIAN OF MEASURED WAVE HEIGHTS (NUMERICAL METHOD EXAMPLE)

Problem statement

As part of a study to evaluate the reasons why a particular beach is eroding quickly, a number of wave height measurements have been made. To calculate the movement of the sand, it is necessary to determine an average wave height from the measurements. Two different types of averages can be taken, the mean and the median. Write a program that can calculate the mean and median values of measured wave heights. Read the input wave heights from a file and print the results to the screen.

Solution

RELEVANT EQUATIONS

We define the following terms:

$$x_i = i\text{th value in a list of numbers}$$
$$n = \text{number of } x \text{ values in list}$$

Using these terms we have the definition of the mean:

$$\bar{x} = \frac{\sum_{i=1}^{n} x_i}{n} \tag{6.4}$$

where $\bar{x}$ = mean of the set of n values. In other words, the mean is the sum of all the values in a list divided by the number of values.

The median of a set of n data points commonly is described as the value in a list that has an equal number of values greater than and less than the median value. For example, for the five values 10, 13, 24, 9, 1, the median value is 10 because two values (13 and 24) are greater than 10 and two values(1 and 9) are less than 10.

This definition is not quite accurate because it does not account for the possibility of having like values. For instance, for the five values 9, 10, 10, 13, 24, the median is 10. Here, it can be seen that the median is the value for which both the number of values less than or equal to it and greater than or equal to it is greater than half of the total number of values. In this case, the number of values less than or equal to 10 is 3 (values 9, 10, and 10) and the number of values greater than or equal to 10 is 4 (values 10, 10, 13, and 24). Since both 3 and 4 are greater than half of the number of values (5/2 = 2.5), 10 is the median.

We can write these conditions in equation form:

$$n_{\text{lower}} = \text{number of values less than or equal to } x_i$$
$$n_{\text{higher}} = \text{number of values greater than or equal to } x_i$$

if n_{lower} is greater than $n/2$ and n_{higher} is greater than $n/2$, then

$$\hat{x} = x_i$$

where $\hat{x}$ = median value.

Note that, in this example, we consider only an odd number of values in a list, because the definition of median becomes less clear for an even number of values. Another approach in finding the median is to first sort the values and then select the value at the center of the sort as the median. Because we discuss sorting in other examples, we will not use sorting here to find the median.

SPECIFIC EXAMPLE

Once per day over a period of approximately one month (29 days), wave heights have been measured. The following values have been found for each day (measurements in centimeters):

67 87 56 34 85 98 56 67 87 90 45 42 31 97 58 78 12 16 22 42 83 95 53 27 49 85 58 79 79

Using equation (6.4), the mean of these is found to be the sum of the values (1778) divided by the number of them (29):

$$\bar{x} = 1778/29 = 61.3$$

Therefore, the mean wave height is 61.3 cm.

To find the median with the set of equations that we developed, we take each value and compare it to all of the others to get the number of lower and higher values. For instance, if we take the first value in the list, 67, and compare it to all of the other values, we find that 17 values are less than or equal to 67 and 14 values are greater than or equal to 67. Therefore, 67 is not the median value because the total number of values (29) divided by 2 is 14.5, and both 17 and 14 are no greater than 14.5. Next, we list the values, in order, and the number of values above and below that value:

Value x_i	Number of values $<= x_i$	Number of values $>= x_i$
67	17	14
87	25	6
56	13	18
34	6	24
85	23	8
98	29	1
56	13	18
67	17	14
87	25	6
90	26	4
45	9	21
42	8	23
31	5	25
97	28	2
58	15	16

Clearly, this is not a complete list. We stopped the list at 58 because the number of values $<= 58$ and the number of values $>= 58$ both are greater than 14.5 (being 15 and 16, respectively). So, we need to go no further. We have found the median. It is 58.

ALGORITHM

It can be seen from the specific example that we follow these steps. For input,

1. Open the data file.
2. Read the input data values of wave heights from the data file.

For mean,

1. Sum all of the values in the list.
2. Divide the sum by the number of values in the list.

For median,

1. Compare a value to the other values, and
    a. Count the number of values less than or equal to the compared value.
    b. Count the number of values greater than or equal to the compared value.
2. If both 1a and 1b are greater than $n/2$, then we have found the median, and we stop.
3. If either 1a or 1b are not greater than $n/2$, then we repeat step 1 with the next value on the list.

For the results, we print the mean and median to the screen.

SOURCE CODE

We develop the source code for the mean and the median step by step from the algorithm. We do not develop the source code for the input and results in detail because we have previously described the techniques used for doing this. This code, however, is shown in the final source code.

Mean calculation.

1. Sum all of the values in the list. The following loop sums all of the values of x[ ], where `num_pts` is the number of x[ ] values:

```
sum = 0.0;
for (i=0; i<num_pts; i++) sum += x[i];
```

Observe the importance of making sum = 0.0 before the loop. This assures that only the values of the array elements are summed by `sum+=x[i]`.

2. Divide the sum by the number of values in the list. The following code does this calculation. Note that typically `num_pts` will be specified as an integer whereas `sum` and `mean` are doubles. This leads to mixed arithmetic, of which you should be careful. Although in this case we could use an implicit conversion (which is done automatically by C) to a double type, we choose to explicitly make `num_pts` as a double using a cast operator to illustrate the use of the cast operator. As you become more comfortable with mixed type arithmetic in a situation like this, it can be omitted.

```
mean = sum/ (double)num_pts;
```

Median calculation.

1. Compare a value to the other values, and

    Count the number of values less than or equal to the compared value.
    Count the number of values greater than or equal to the compared value.

Using x[j] as the value being compared to all the others, the loop below performs these operations:

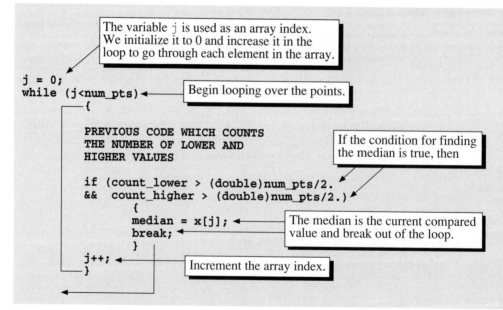

```
count_higher = 0;
count_lower = 0;
for (i=0; i<num_pts; i++)
        {
        if (x[j] <= x[i]) count_higher++;

        if (x[j] >= x[i]) count_lower++;

        }
```

Initializing the counter variables to be 0. This must be done before the beginning of the loop.

Looping over all of the points.

If the compared value is less than or equal to another value, add 1 to the count_higher variable.

If the compared value is greater than or equal to another value, add 1 to the count_lower variable.

We must use each value in the list as the compared value x[j], so we put the above code into a loop over all of the data ponts. However, we do not necessarily want to loop over all of the data points, because we may encounter the median as one of the first points checked. Therefore, we can break out of the loop early. To illustrate different types of loops and the break statement, we show two different methods for writing this control structure. First we show a while loop with a break statement and then we show a do-while loop with no break statement.

   For the while loop, the condition for breaking out of the loop (as described in the algorithm section) is that both count_lower and count_higher are greater than num_pts/2. We have not previously used a break statement in anything other than a switch body; however, a break statement also is allowed in a loop body. A break statement in a loop body causes control to transfer out of the nearest enclosing loop as follows:

The variable j is used as an array index. We initialize it to 0 and increase it in the loop to go through each element in the array.

```
j = 0;
while (j<num_pts)
        {

        PREVIOUS CODE WHICH COUNTS
        THE NUMBER OF LOWER AND
        HIGHER VALUES

        if (count_lower > (double)num_pts/2.
        &&  count_higher > (double)num_pts/2.)
                {
                median = x[j];
                break;
                }
        j++;
        }
```

Begin looping over the points.

If the condition for finding the median is true, then

The median is the current compared value and break out of the loop.

Increment the array index.

By breaking out of the loop early we have saved computation and found the median value. However, if possible, it is considered to be better programming practice to avoid the use of break statements in looping structures. This can be done by putting the break condition within the loop's controlling condition. For instance, another way to look at this problem is to realize that looping should continue if either `count_higher` or `count_lower` is less than `num_pts/2`. Using a do-while loop gives the following code:

```
j=-1;  ◄──── Initialize the array index variable.
do
{
        j++;  ◄──────────── Increment the array index variable
                            each time through the loop.

        PREVIOUS CODE WHICH COUNTS
        THE NUMBER OF LOWER AND        Continue looping while either count
        HIGHER VALUES                  _lower or count_higher is <=
                                       num_pts/2, otherwise, stop. Also, j
                                       must be less than num_pts.

}       while (j<num_pts &&
        (count_lower <=((double)(num_pts)/2.) ||
        count_higher <=((double)(num_pts)/2.)));
median = x[j];
```

The median is the compared value at the time that looping has stopped.

Note that the structure of the do-while loop is cleaner than that of the while loop because it does not use the break statement. It is possible to write a while loop for this case without using a break statement. How would you do it?

The complete source code (which incorporates the above described code with the do-while loop) follows:

```
#include <stdio.h>
#define MAX_NUM_PTS 100

void main (void)
{
        int x[MAX_NUM_PTS], num_pts, i, j, count_lower, count_higher, median;
        double sum, mean;
        FILE *infile;

        infile = fopen ("average.dat", "r");
        fscanf (infile, "%d", &num_pts);
        for (i=0; i<num_pts; i++) fscanf (infile, "%d", &x[i]);

/***********************************************************
**   CALCULATION OF THE MEAN
***********************************************************/

        sum = 0.0;
        for (i=0; i<num_pts; i++) sum += x[i];
```

```
        mean = sum/(double)num_pts;
/************************************************************
**   CALCULATION OF THE MEDIAN
************************************************************/

        j=-1;
        do
          {
          j++;
          count_lower = 0;
          count_higher = 0;
          for (i=0; i<num_pts; i++)
                  {
                  if (x[j] <= x[i]) count_higher++;
                  if (x[j] >= x[i]) count_lower++;
                  }
          }         while (j<num_pts && (count_lower <=((double)(num_pts)/2.)
                          ||count_higher <=((double)(num_pts)/2.)));
        median = x[j];

        printf ("The mean of the values is: %.3lf \n"
                "The median value is:     %d \n", mean, median);
}
```

INPUT FILE AVERAGE.DAT

29
67 87 56 34 85 98 56 67 87 90 45 42 31 97 58 78 12 16 22 42 83 95
53 27 49 85 58 79 79

OUTPUT

```
The mean of the values is: 61.310
The median value is:        58
```

COMMENTS

In this program, we defined the maximum number of points as 100 and read in the actual number of data points as the first item in the data file. Should we want to analyze more than 100 points, we would need to change this value.

We would like to comment here about developing efficient code. Because we are very interested in developing efficient code, we are interested in assessing the efficiency of our algorithms. Part of assessing the efficiency of an algorithm that involves comparisons is evaluating how many comparisons are made in executing the algorithm. Determining the number of comparisons is not necessarily straightforward because different situations cause different numbers of comparisons to be made. For instance, for our algorithm to evaluate the median of a list of n numbers, we see that if the median value is the first value in our list (just by chance) we will make only n comparisons (because just one pass through the list gives us the median).

However, should the median be the last value in the list (again by chance) we would make n comparisons for each of the n values; that is, n^2 comparisons to perform a median evaluation. If we had 1000 values in our list, this would mean we would make $1000^2 = 1$ million comparisons. You can see that, for this particular algorithm, the number of comparisons can be quite great. Therefore, developing a more efficient algorithm may be quite beneficial. We will not develop one here; however, we want to make you aware that a part of engineering and computer science involves the search for efficient algorithms. You very well may take courses later in your educational career that focus on algorithm development.

Modification exercises

1. Replace the do-while loop with a while loop that needs no break statement.
2. Make x[ ] an array of doubles rather than integers.
3. Modify the program to handle 12 lists of waveheight data (one for each month in a year) in the input file. The input data file would be

    $n1$
    $h_1 \, h_2 \, h_3 \ldots h_{n1}$
    h_2
    $h_1 \, h_2 \, h_3 \ldots h_{n3}$
    .
    .
    .
    .
    $n12$
    $h_1 \, h_2 \ldots . h_{n12}$

4. Remove the cast operators. Does the program still work properly? Why or why not?
5. Create a modular design for this program. Make three functions—one for input, one for the mean, and one for the median.

APPLICATION EXAMPLE 6.3 MATRIX-VECTOR MULTIPLICATION (NUMERICAL METHOD EXAMPLE)

Problem statement

Write a program that multiplies a matrix, a[ ][ ], and a vector, x[ ], giving a vector, b[ ], result. Read the input data from a file that contains the data in the form that looks like a matrix and vector; for example,

$$a_{11} \, a_{12} \, a_{13} \, a_{14} \, a_{15} \, a_{16} \, a_{17} \, x_1$$
$$a_{21} \, a_{22} \, a_{23} \, a_{24} \, a_{25} \, a_{26} \, a_{27} \, x_2$$
$$a_{31} \, a_{32} \, a_{33} \, a_{34} \, a_{35} \, a_{36} \, a_{37} \, x_3$$
$$x_4$$
$$x_5$$
$$x_6$$
$$x_7$$

where the subscripts for the elements of the matrix are $a_{row\ column}$. Print the result to the screen. Write the program to handle the case of the number of rows being less than the number of columns.

Solution

First, assemble the relevant equations. Here, we briefly review multiplying a matrix and vector. For the given matrix and vector, the following equations give the components of the b[ ] vector:

$$b_1 = a_{11}x_1 + a_{12}x_2 + a_{13}x_3 + a_{14}x_4 + a_{15}x_5 + a_{16}x_6 + a_{17}x_7$$
$$b_2 = a_{21}x_1 + a_{22}x_2 + a_{23}x_3 + a_{24}x_4 + a_{25}x_5 + a_{26}x_6 + a_{27}x_7$$
$$b_3 = a_{31}x_1 + a_{32}x_2 + a_{33}x_3 + a_{34}x_4 + a_{35}x_5 + a_{36}x_6 + a_{37}x_7$$

We choose to write these equations in summation form because we found in Chapter 4 that, once written in this form, the equations easily are translated into code with loops. For this case,

$$b_1 = \sum_{j=1}^{7} a_{1j}x_j$$

$$b_2 = \sum_{j=1}^{7} a_{2j}x_j$$

$$b_3 = \sum_{j=1}^{7} a_{3j}x_j$$

In general, we find that if we have n columns in our matrix, any value b_i can be represented as

$$b_i = \sum_{j=1}^{n} a_{ij}x_j \tag{6.5}$$

If we have m rows in our matrix, then we will have m values of b. Therefore, we should compute b_i from $i = 1$ to $i = m$.

SPECIFIC EXAMPLE

We evaluate the product [**a**][**x**] = [**b**], of the following matrix [**a**] (with 4 rows and 5 columns) and vector [**x**] (with 5 elements).

$$\begin{bmatrix} 2 & 4 & 5 & 3 & 6 \\ 9 & 8 & 4 & 1 & 4 \\ 0 & 9 & 1 & 3 & 9 \\ 9 & 8 & 2 & 4 & 1 \end{bmatrix} \begin{bmatrix} 2 \\ 5 \\ 2 \\ 5 \\ 1 \end{bmatrix}$$

We get

$$[b] = \begin{matrix} 2 \times 2 + 4 \times 5 + 5 \times 2 + 3 \times 5 + 6 \times 1 = \\ 9 \times 2 + 8 \times 5 + 4 \times 2 + 1 \times 5 + 4 \times 1 = \\ 0 \times 2 + 9 \times 5 + 1 \times 2 + 3 \times 5 + 9 \times 1 = \\ 9 \times 2 + 8 \times 5 + 2 \times 2 + 4 \times 5 + 1 \times 1 = \end{matrix} \begin{bmatrix} 55 \\ 75 \\ 71 \\ 83 \end{bmatrix}$$

ALGORITHM

The algorithm is straightforward:

1. Read the number of rows and columns of the matrix.
2. Read the input matrix and vector.
3. Calculate an element of the **b** vector by applying equation (6.5).
4. Repeat step 3 for all elements of the **b** vector.
5. Print the **b** vector result.

SOURCE CODE

Reading input data. It is worth discussing reading the input data as the loops are not simple. The input file takes the shape of the matrix and vector to be multiplied, where the vector has a number of elements equal to the number of columns of the matrix. For instance, the following is a sample data file with the first line being the number of rows and columns in the matrix and the other lines being the matrix and vector:

```
3 8
2 4 6 4 3 6 8 9 4
9 8 7 6 5 8 9 6 3
8 7 6 4 1 0 2 8 6
                3
                4
                6
                0
                3
```

For `infile` representing the input data file and `num_rows` and `num_cols` representing the number of rows and columns of the matrix, the line

`fscanf(infile,"%d %d",&num_rows,&num_cols);`

reads the first line of the data file. We can establish the loops using the variables `num_rows` and `num_cols` read with this line. Using the subscript `i` to represent the column and `j` to represent the row, the nested loop that follows reads the matrix and part of the vector from the data file. (Note: Because C begins at 0 not 1 for its subscripts, we do not follow exactly the subscript numbering shown at the beginning of this application example. Everything is shifted by 1.)

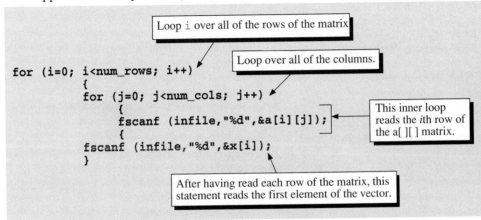

With this code, we will have read lines 2, 3, and 4 of the sample data file. We now need to read lines 5, 6, 7, 8, and 9, which contain only elements of the vector. Again, because C begins at 0 for its subscripts, we do not begin the loop for reading x[ ] at num_rows+1, we begin at num_rows. The next loop reads the rest of the x[ ] values:

```
for (i=num_rows; i<num_cols; i++) fscanf (infile,"%d",&x[i]);
```

Remember that fscanf skips over white space when reading numeric data, so shifting the column vector many spaces to the right has no significance.

Calculate the b vector. We use equation (6.5) to create the loop for calculating each element of the vector b[ ]. Compare the following code to this equation to understand how the loop is created:

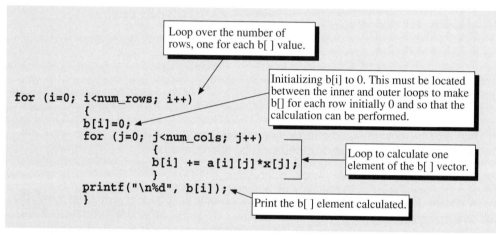

Note that within this loop we also have printed the value of each element of the b[ ] vector to the screen.

The entire source code follows:

```
#include <stdio.h>
#define MAX_NUM_ROWS 20
#define MAX_NUM_COLS 20

void main (void)
{
        int a[MAX_NUM_ROWS][MAX_NUM_COLS], x[MAX_NUM_COLS];
        int b[MAX_NUM_ROWS];
        int i, j, num_rows, num_cols;
        FILE *infile;

        infile=fopen("matvect.dat","r");
        fscanf(infile,"%d %d",&num_rows,&num_cols);
```

```
for (i=0; i<num_rows; i++)
        {
        for (j=0; j<num_cols; j++)
                {
                fscanf (infile,"%d",&a[i][j]);
                }
        fscanf (infile,"%d",&x[i]);
        }
for (i=num_rows; i<num_cols; i++) fscanf (infile,"%d",&x[i]);

printf("\nb vector");

for (i=0; i<num_rows; i++)
        {
        b[i]=0;
        for (j=0; j<num_cols; j++)
                {
                b[i] += a[i][j]*x[j];
                }
        printf("\n%d", b[i]);
        }
}
```

INPUT DATA FILE

```
4 5
2 4 5 3 6 2
9 8 4 1 4 5
0 9 1 3 9 2
9 8 2 4 1 5
            1
```

OUTPUT

```
b vector

55
75
71
83
```

COMMENT

You will see that, as you work with arrays in programming, much of your effort will be in manipulating the subscripts of the components of the arrays.

We also can see that it is relatively straightforward to take equations written in summation form and create loops that perform the calculations. The loops simply manipulate the subscripts used in the summation form. Because of the ease of conversion from equation to code, we recommend you write your equations in summation form whenever possible.

Modification exercises

1. Make the matrix and both vectors doubles instead of integers.
2. Modify the program to handle the situation of the number of rows being greater than the number of columns. Why will the current program not work in this situation?
3. Create a modular design for this program. Make two functions—one for input and the other for calculations and output.

▨ APPLICATION EXAMPLE 6.4 BEST FIT LINE— LINEAR REGRESSION, MODULAR DESIGN (NUMERICAL METHOD EXAMPLE)

Problem statement

For a particular stretch of highway it is believed that there is a correlation between the vehicle density (number of vehicles per 100 m) on the highway and the number of accidents that occur. From casual observation, the number of accidents has been found to increase with an increase in vehicle density up to a certain point. However, once the vehicle density exceeds a certain value, the average vehicle speed is reduced due to congestion, thereby reducing the number of accidents. To predict accident rates and as an aid to produce an improved highway design, we wish to develop equations relating the vehicle density to the number of accidents from observed data.

Our goal is to create two straight lines that represent a best fit through the data, one that rises until it reaches the vehicle density at the peak number of accidents and another that decreases from this point. Figure 6.15 illustrates an example of data collected (represented by the circles) and the best fit lines through the data.

The problem is to write a program that can create a best fit straight line through a set of 14 (x, y) data points. The points represent observed data with x being the vehicle density and y being the number of accidents. The program should print out a total of five values:

1. Slope of line 1 (m_1)
2. Intercept of line 1 (b_1)
3. Slope of line 2 (m_2)
4. Intercept of line 2 (b_2)
5. Vehicle density at the peak number of accidents

Solution

RELEVANT EQUATIONS

We will not go through the derivation (because it goes beyond the scope of this book) but just list the equations representing the slope and intercept of a best fit line. We use the following quantities:

$$n = \text{number of points}$$

$$c = \sum_{i=1}^{n} x_i$$

FIG. 6.15

Number of accidents related to the vehicle density. Straight lines represent best fit through the data.

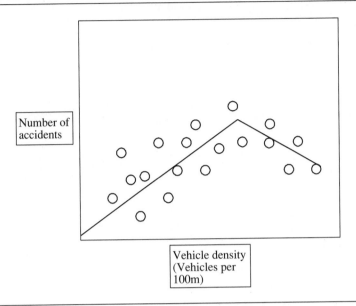

$$d = \sum_{i=1}^{n} y_i$$

$$e = \sum_{i=1}^{n} x_i^2$$

$$f = \sum_{i=1}^{n} x_i y_i$$

Note that c is simply the sum of all the x values of the data points and d is the sum of all the y values of the data points. The variable e is the sum of the squares of the x values and f is the sum of the xy products of the data points. Therefore, these values, despite looking somewhat complicated, can be calculated relatively simply.

The slope (m) and intercept (b) of the best fit line are

$$m = \frac{nf - cd}{ne - c^2}$$

$$b = \frac{de - cf}{ne - c^2}$$

(6.7)

SPECIFIC EXAMPLE

We will perform the calculations for the following data:

Data point	Vehicle density (x)	Number of accidents (y)
1	1.4	3
2	2.0	6
3	2.3	4
4	4.5	7
5	6.2	10
6	6.7	15
7	7.0	11
8	8.5	18
9	9.0	13
10	12.7	17
11	13.1	15
12	17.7	16
13	18.5	11
14	20.3	5

We note that the peak accident count is 18 (at point number 8). So, we use points 1–8 to create one line and points 8–14 to create another line. For the first line,

$$c = \text{sum of the } x \text{ values} = 1.4 + 2.0 + \ldots +7.0 + 8.5 = 38.6$$
$$d = \text{sum of the } y \text{ values} = 3 + 6 + \ldots + 11 + 18 = 74$$

To get the value of e, we square each of the x values and sum them:

$$e = 1.4^2 + 2.0^2 + \ldots + 7.0^2 + 8.5^2 = 236.08$$

To get the value of f, we get the product of each x, y pair and sum them:

$$f = 1.4(3) + 2.0(6) + \ldots 7.0(11) + 8.5(18) = 449.4$$

For the second line,

$$c = 8.5 + 9.0 + \ldots + 18.5 + 20.3 = 99.8$$
$$d = 18 + 13 + \ldots + 11 + 5 = 95$$
$$e = 8.5^2 + 9.0^2 + \ldots + 18.5^2 + 20.3^2 = 1553.78$$
$$f = 8.5(18) + 9.0(13) + \ldots + 18.5(11) + 20.3(5) = 1270.6$$

We can put these values into the equation (6.7) for m and b, giving

$$m_1 = \frac{8(449.4) - 38.6(74)}{8(236.08) - (38.6)^2} = 1.853$$

$$b_1 = \frac{74(236.08) - 38.6(449.4)}{8(236.08) - (38.6)^2} = 0.3087$$

$$m_2 = \frac{7(1270.6) - 99.8(95)}{7(1553.78) - (99.8)^2} = -0.6403$$

$$b_2 = \frac{95(1553.78) - 99.8(1270.6)}{7(1553.78) - (99.8)^2} = 22.70$$

Therefore, the equations of the best fit lines ($y = mx + b$) are

Line 1 $y = 1.853x + 0.3087$ applies from $x = 1.4$ to $x = 8.5$
Line 2 $y = -0.6403x + 22.70$ applies from $x = 8.5$ to $x = 20.3$

ALGORITHM
The algorithm can be developed from the specific example. The basic steps are

1. Read the input data.
2. Find the maximum value of accidents (x_{max}).
3. Find the best fit line for for $x < x_{max}$.
4. Find the best fit line for for $x > x_{max}$.
5. Print the results.

Because this is our modular design example for this chapter, we will develop a function for each of the steps. Even though it is a bit of an overkill to do this, we do it to illustrate the use of arrays in a modular design. We begin by developing a structure chart/data flow diagram.

To develop a data flow diagram, we must determine the information that is to go into and out of each function. This is not a simple step. With experience, this will get easier. Make your best first attempt at this and develop your functions based on that attempt. As you do the actual coding, you can modify your data flow diagram. For this program, we envision that the information being transferred between main and each function is

Function	Information going from main to the function	Information going from the function to main
readinput	Nothing	The vehicle density (x[ ]), the number of accidents (y[ ]), and the number of data points
findmax	The number of accidents (y[ ]), and the number of data points	The maximum number of accidents and the data point number corresponding to the maximum number of accidents
bestfit	The vehicle density (x[ ]), the number of accidents (y[ ]), the left and right bounding data point numbers of the data points to be considered	The slope and intercept of the best fit line
printresult	The slope and intercept of the first line, the slope and intercept of the second line, the maximum number of accidents, and the vehicle density at the maximum number of accidents	Nothing

The corresponding data flow diagram is shown in Fig. 6.16.
We develop the algorithm and source code for each function individually.

FIG. 6.16
Data flow diagram for accident/vehicle density program

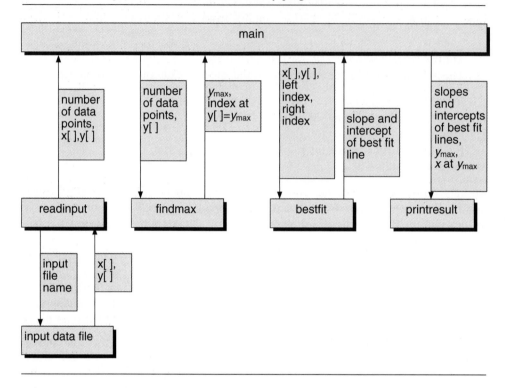

Function readinput. This function reads from the data file. In the data file are only the *x* and *y* values of the data points and not the number of data points. Example data file:

```
1.4     3
2.0     6
2.3     4
4.5     7
6.2     10
6.7     15
7.0     11
8.5     18
9.0     13
12.7    17
13.1    15
17.7    16
18.5    11
20.3    5
```

This function must be capable of counting the number of data points. Recall that the function fscanf returns a value just like other functions and that the constant macro EOF (meaning "end of file") is defined in stdio.h. The following loop reads the input file until the end of file character is read. The *x* and *y* coordinates of the data points are put into the vectors r[ ] and s[ ]. At the same time, this loop counts the number of data points using the variable i and puts this value into the variable `*num_pts`:

```
i=0;
while ( (fscanf(infile,"%lf%lf",&r[i],&s[i])) != EOF ) i++;
*num_pts=i;
```

The function call and the function header are as follows:

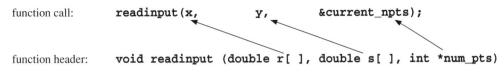

function call: **readinput(x,   y,   &current_npts);**

function header: **void readinput (double r[ ], double s[ ], int *num_pts)**

With these argument lists, the values in r[ ], s[ ], and `*num_pts` are transferred to x[ ], y[ ], and current_npts in main. Note that, in the function call, because x and y are one-dimensional arrays in main, we simply use the array name without & preceding it or [ ] following it.

Function findmax. With the function findmax, we want to find the maximum *y* value (maximum number of accidents). Therefore, we need to allow the array y to be accessed by findmax by sending the address of the first element of the y[ ] array to findmax. Also helpful would be to send the number of y values, current_npts. The function findmax then should return the maximum value, y_max, and the array index at this value i_at_y_max. The following function call and header do this:

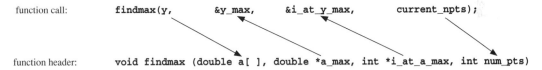

function call: **findmax(y,   &y_max,   &i_at_y_max,   current_npts);**

function header: **void findmax (double a[ ], double *a_max, int *i_at_a_max, int num_pts)**

Here, in findmax, we used the one-dimensional array a[] to contain the values of the number of accidents. Note that, in the function call, single variables received from findmax are preceded by & and single variables sent to findmax are not. Also, in the function header, single variables sent to main are preceded by *, others are not. Therefore, within function findmax, we must use *a_max and *i_at_a_max, not a_max and i_at_a_max.

The following if control structure compares the value of one item in the array, a[i], to the currently found maximum value, *a_max. If the value of a[i] is greater than the current value of *a_max, then we have a new value of *a_max

(the new value is a[i]). We also save the value of the index using the variable
`*i_at_a_max`:

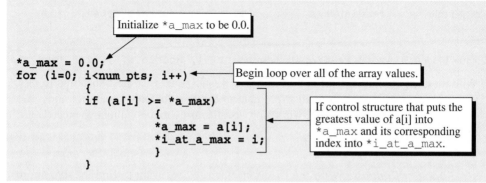

```
if (a[i] >= *a_max)          If one array value is greater than the max.
    {
    *a_max = a[i];           Set the maximum to that array value.
    *i_at_a_max = i;
    }                        Save the array index for that array value.
```

By initializing `*a_max` as 0 and putting the if control structure into a loop from 1 to
the number of data points, `num_pts`, we can get the maximum value of the array
a[ ] and save it using the variable `*a_max` as follows:

```
                              Initialize *a_max to be 0.0.

*a_max = 0.0;
for (i=0; i<num_pts; i++)     Begin loop over all of the array values.
    {
    if (a[i] >= *a_max)
        {                     If control structure that puts the
        *a_max = a[i];        greatest value of a[i] into
        *i_at_a_max = i;      *a_max and its corresponding
        }                     index into *i_at_a_max.

    }
```

This code is used in the function `findmax`. After executing this loop we have saved
the maximum value and the index at which the maximum value is located.

   Function bestfit. The function bestfit uses the equations described earlier to
develop the slope and intercept of the best fit line through a set of data points. Since
these equations were written in summation form, the for loop is especially easy to
write. For instance, the equation for c

$$c = \sum_{i=1}^{n} x_i$$

that sums all of the value of x is easily duplicated with the for loop:

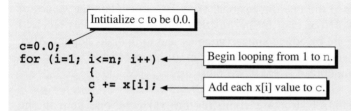

```
                  Intitialize c to be 0.0.

c=0.0;
for (i=1; i<=n; i++)       Begin looping from 1 to n.
    {
    c += x[i];            Add each x[i] value to c.
    }
```

Given the straightforward conversion from summation type equations to code in the
form of loops, we recommend that you write your equations in summation form
prior to developing your code whenever possible.

We can put all of the summation equations into code in the form of a for loop. Compare the code that follows to the equations to observe how easily equations of this type are coded:

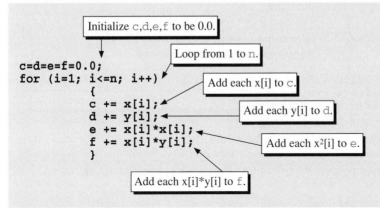

```
c=d=e=f=0.0;
for (i=1; i<=n; i++)
        {
        c += x[i];
        d += y[i];
        e += x[i]*x[i];
        f += x[i]*y[i];
        }
```

Initialize c,d,e,f to be 0.0.

Loop from 1 to n.

Add each x[i] to c.

Add each y[i] to d.

Add each $x^2[i]$ to e.

Add each x[i]*y[i] to f.

We need to call function bestfit twice from main: once to fit the line for the first part of the data (from the first data point to the peak point) and again to fit the second part of the data (from the peak point to the last data point). So, when we create the loops we do not go from 1 to n, we go from a specified first point to a specified last point.

Since in C the first array index is not 1 but 0, we begin at 0 for the first part of the data and end at the index i_at_y_max (which corresponds to the peak point). In addition we pass the data point arrays x[ ] and y[ ] and return to main the slope (*m*) and intercept (*b*). The function call (and corresponding header) for the first time for bestfit is

function call   `bestfit(x,      y,       0,    i_at_y_max,    &m1,    &b1);`

function header   `void bestfit (double x[ ], double y[ ], int i_left, int i_right,  double *m, double *b)`

For the second call to bestfit, we change the first and last points for the loops (note that since C starts at 0 and not 1, it ends at num_pts-1 not num_pts). So, we call bestfit with arguments different from the arguments on the first call:

function call   `bestfit(x,      y,    i_at_y_max,    num_pts -1    &m1,    &b1);`

function header   `void bestfit (double x[ ], double y[ ], int i_left, int i_right,  double *m, double *b)`

Given the variables listed in the function header (i_left and i_right), the loop for c, d, e, and f becomes

```
c=d=e=f=0.0;
for (i=i_left; i<=i_right; i++)
        {
        c += x[i];
        d += y[i];
        e += x[i]*x[i];
        f += x[i]*y[i];
        }
```

Loop from i_left to i_right, which are the upper and lower indices of the values to be summed (as passed to the function from its call).

The slope *m, intercept, *b, and number of data points, n, are calculated based on equations (6.7); when looking at this code do not confuse the unary * operator on the left side of the assignment statements with the multiplication operator * on the right side of the assignment statements:

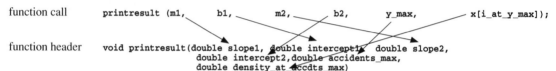

Calculating n, the number of points.

```
n = i_right - i_left +1;
*m = (n*f - c*d)/(n*e - c*c);
*b = (d*e - c*f)/(n*e - c*c);
```

Calculating *m and *b from equations (6.7).

Function printresult. This normally would not be a function, because it is much too simple; however, we show it as a function to demonstrate the creation of a function for the sole purpose of printing the results and to illustrate the transfer of information.

The function call and header for this function are

function call   printresult (m1,   b1,   m2,   b2,   y_max,   x[i_at_y_max]);

function header   void printresult(double slope1, double intercept1, double slope2,
                              double intercept2,double accidents_max,
                              double density_at_accdts_max)

Here, observe that a single value of the array x[ ], x[i_at_y_max], is passed to a single variable, density_at_accidents_max. Also, all of the information is transferred to printresult and none is returned back to main.

The single line

```
printf ("The first line is y = %.3lfx + %.3lf \n\
        \rThe second line is y = %.3lfx + %.3lf \n\n\
        \rThe maximum number of accidents is %.1lf \n\
        \rat a vehicle density of %.2lf\n",
        slope1, intercept1, slope2, intercept2,
        accidents_max, density_at_accdts_max);
```

is the entire body of the printresult function.

SOURCE CODE

This source code is commented and combines all of the functions described previously.

```
/*****************************************************************************
*******************************************************************************
**      APPLICATION PROGRAM TO CALCULATE
**      THE RELATIONSHIP BETWEEN THE VEHICLE DENSITY AND NUMBER
**      OF ACCIDENTS OF A STRETCH OF HIGHWAY.
**      TWO LINES ARE USED TO FIT THE DATA.  ONE THAT GOES FROM 0 TO
**      THE PEAK NUMBER OF ACCIDENTS AND THE OTHER FROM THE PEAK TO
**      THE LAST DATA POINT.
**
**      PROGRAMMER NAMES    REVISION DATES
**
**
*******************************************************************************
*****************************************************************************/
```

```
#include <stdio.h>
#define MAX_NUM_PTS 50
/*************************************************************************
**      FUNCTION PROTOTYPES
*************************************************************************/

void readinput (double r[], double s[], int *num_pts);
void printresult(double slope1, double intercept1,
                                double slope2, double intercept2,
                                double accidents_max,
                                double density_at_accdts_max);
void findmax (double a[], double *a_max, int *i_at_a_max, int num_pts);
void bestfit (double x[], double y[], int i_left, int i_right,
                                double *m, double *b);
/*************************************************************************
**      FUNCTION main
**      CALLS FUNCTIONS TO:
**      READ THE DATA POINTS OF VEHICLE DENSITY AND NUMBER OF ACCIDENTS
**      FIND THE MAXIMUM NUMBER OF ACCIDENTS
**      FIT A STRAIGHT LINE THROUGH THE FIRST PART OF THE DATA (UP TO THE PEAK
**           NUMBER OF ACCIDENTS)
**      FIT A STRAIGHT LINE THROUGH THE SECOND PART OF THE DATA (FROM THE PEAK
**           NUMBER OF ACCIDENTS TO THE LAST POINT)
**      PRINT THE SLOPES AND INTERCEPTS OF THE TWO LINES
*************************************************************************/


void main(void)
{
        double x[MAX_NUM_PTS], y[MAX_NUM_PTS], m1, m2, b1, b2;
        double y_max;
        int  i_at_y_max, num_pts;

        readinput(x,y,&num_pts);
        findmax(y,&y_max,&i_at_y_max,num_pts);
        bestfit(x,y,0,i_at_y_max,&m1,&b1);
        bestfit(x,y,i_at_y_max,num_pts-1,&m2,&b2);
        printresult (m1,b1,m2,b2,y_max,x[i_at_y_max]);
}


/*************************************************************************
**      FUNCTION readinput
**      READS THE INPUT DATA POINTS AND COUNTS THE NUMBER OF POINTS
**           VARIABLES - INPUT
**           none
**           VARIABLES - RETURNED THROUGH PARAMETER LIST
**              r[ ] = x values of data points = vehicle density
**              s[ ] = y values of data points = number of accidents
**              *numpts = number of data points
*************************************************************************/
void readinput (double r[], double s[],int *num_pts)
{
        int i;
        FILE *infile;
        infile = fopen ("bestline.dat","r");
        i=0;
        while ((fscanf(infile,"%lf %lf",&r[i],&s[i]))!= EOF) i++;
        *num_pts = i;
}

/*************************************************************************
**      FUNCTION printresult
**      PRINTS THE SLOPES AND INTERCEPTS OF THE BEST FIT LINES.
**      DISPLAYS THE MAXIMUM NUMBER OF ACCIDENTS AND
**      THE VEHICLE DENSITY AT THE MAXIMUM NUMBER OF ACCIDENTS
**           VARIABLES - INPUT
**           slope1, intercept1 = slope and intercept of first line (from 0 to peak accidents)
**           slope2, intercept2 = slope and intercept of second line (from peak accidents to last pt.
**           accidents_max = maximum number of accidents in data set
**           density_at_accdts_max = vehicle density at maximum number of accidents
**           VARIABLES - RETURNED
**           none
*************************************************************************/
```

```c
void printresult(double slope1, double intercept1,
                                double slope2, double intercept2,
                                double accidents_max,
                                double density_at_accdts_max)
{
        printf ("The first line is y = %.3lfx + %.3lf \n\
                        \rThe second line is y = %.3lfx + %.3lf \n\n\
                        \rThe maximum number of accidents is %.1lf \n\
                        \rat a vehicle density of %.2lf\n",
                        slope1, intercept1, slope2, intercept2,
                        accidents_max, density_at_accdts_max);
}

/*************************************************************************/
**      FUNCTION findmax
**      FINDS THE MAXIMUM VALUE IN A 1-D ARRAY.
**          VARIABLES - INPUT
**          a[ ] = array containing number of accidents
**          num_pts = number of points in the array
**          VARIABLES - RETURNED THROUGH THE PARAMETER LIST
**          *a_max = maximum number of accidents
**          *i_at_a_max = array index at maximum number of accidents
/*************************************************************************/

void findmax ( double a[], double *a_max, int *i_at_a_max, int num_pts)

{
        int i;
        *a_max = 0.0;
        for (i=0; i<num_pts; i++)
                {
                if (a[i] >= *a_max)
                        {
                        *a_max = a[i];
                        *i_at_a_max = i;
                        }
                }
}

/*************************************************************************
**      FUNCTION bestfit
**      FINDS A BEST FIT STRAIGHT LINE THROUGH A SET OF X, Y DATA POINTS
**          VARIABLES - INPUT
**          x[ ] = array containing vehicle density
**          y[ ] = array containing number of accidents
**          i_left, i_right = array indices, all x-y values of data points from i_left
**                              through i_right are included in the best fit analysis
**          VARIABLES - RETURNED THROUGH THE PARAMETER LIST
**          *m = slope of the best fit line
**          *b = intercept of best fit line
**          OTHER VARIABLES
**          c,d,e,f = intermediate values for best fit calculation - see accompanying documents
**          n = number of data points used in best fit evaluation
/*************************************************************************/
void bestfit (double x[], double y[], int i_left, int i_right,
                                double*m, double *b)
{
        double c, d, e, f, n;
        int i;
        c=d=e=f=0.0;

        for (i=i_left; i<=i_right; i++)
                {
                c += x[i];
                d += y[i];
                e += x[i]*x[i];
                f += x[i]*y[i];
                }

        n = i_right - i_left+1;
        *m = (n*f - c*d)/(n*e - c*c);
        *b = (d*e - c*f)/(n*e - c*c);

}
```

```
The first line is y = 1.853x + 0.309
The second line is y = -0.640x + 22.701

The maximum number of accidents is 18.0
at a vehicle density of 8.50
```

COMMENTS

Note that we have used the constant macro MAX_NUM_PTS to set the size of the arrays x[] and y[]. In this program we set it to be 50. Should we want to use more than 50 points, we need to change only this one line of the program.

Note also that our method for reading in the data, using while (fscanf ( )!= EOF), is very versatile in that it counts the actual number of data points in the file. Thus, if MAX_NUM_PTS were set at 1 million while only 50 data points were in our data file, our program would go through the loops only 50 times not 1 million and save a considerable number of computations. This method of counting the number of data points works well for files that contain only data points and EOF as an indicator of the end of the data point information. Should this not be the case, you may need to develop other methods for determining the end of the data point information in order to count your data points.

Although a modular design is not needed for this simple program, the example again illustrates that once the data flow diagram has been established, each of the functions can be developed independently. Thus, if you were working in a team to develop this program, each of you could develop a function. After each function was debugged and operational, all of them could be put together to give a final working program.

Note the simplicity of function main. It consists only of calls to other functions. In a program with a good modular design, main primarily consists of function calls.

Modification exercises

1. Modify the program to request a user to type a vehicle density and have the program print the corresponding number of accidents.
2. Modify the program to request a user to type a number of accidents and have the program print the corresponding vehicle density (two values).
3. Modify the program to accept data from five different roads and print the results for each.

APPLICATION EXAMPLE 6.5 SEARCHING AND FILE COMPRESSION

Problem statement

Write a program that compresses a file by creating a new file that can be used to generate it. The new file should be smaller than the original file. Check the result by making the program capable of using the new file to re-create the original file.

Solution

One method for compressing or encoding a file is to recognize repeated digits and replace the runs of repetition with a number representing the repetition length. For instance, the line

`000000000000000111111111111111111110000011111111111111111111111111111111`

can be represented with the line

15 20 5 33

because there are 15 zeros, 20 ones, 5 zeros, and 33 ones. It must be understood in advance that only zeros and ones appear and that zeros are the first to appear in a line. The new line of code is shorter and requires less memory than the original line. Thus, storing the new line is more efficient than storing the original. To re-create the original line from the new line, one need only carry out a reverse implementation. For instance, an encoded line

19 23 8 17

is decoded to be

`0000000000000000000111111111111111111111111000000000111111111111111111`

For this application example, we will work with a file containing 30 lines with 60 digits in each line as shown in Fig. 6.17.

This is a type of bitmap with a treelike shape (traced by the ones). Our goal will be to encode these lines to create a new, shorter file. Then we will decode the new file to re-create the bitmap.

By counting the numbers of zeros and ones in this file, we get an encoded file being as shown in Fig. 6.18. You can see that this file is considerably smaller than the original file. We can use this file to generate the original file by simply printing zeros and ones the number of times listed for each line.

ALGORITHM

The following algorithm was used for each line in the specific example for creating the new file:

1. Search the original file for the first value in a line that is different from the previous one. (In other words, find the first one after a number of zeros.)
2. Count the number of digits passed to get to this value.
3. Restart the count.
4. Repeat steps 1–3.

FIG. 6.17
Bitmap type sequence of zeros and ones that forms a treelike figure. This figure serves as the input file for this application program.

```
000000000000000000111111111111111000000000000000000000000000
000000000000001111111111111111111111000000000000000000000000
000000000000011111111111111111111111111000000000000000000000
000000000000111111111111100000011111111000000000000000000000
000000000000111111111111000000001111111100000000000000000000
000000000000111111111110000000000111111100000000000000000000
000000000000111111111110000000000111111110000000000000000000
000000000001111111111110000000000111111110000000000000000000
000000000001111111111110000000000111111110000000000000000000
000000000001111111111110000000000111111111000000000000000000
000000000011111111111110000000000111111111000000000000000000
000000000011111111111110000000000111111111000000000000000000
000000000011111111111110000000000111111111000000000000000000
000000000011111111111110000000000111111111000000000000000000
000000000001111111111110000000000111111111000000000000000000
000000000000111111111111100000011111111100000000000000000000
000000000000011111111111111111111111111000000000000000000000
000000000000001111111111111111111111000000000000000000000000
000000000000000000111111111111110000000000000000000000000000
000000000000000000011111111111110000000000000000000000000000
000000000000000000001111111111100000000000000000000000000000
000000000000000000000111111110000000000000000000000000000000
000000000000000000000011111110000000000000000000000000000000
000000000000000000000011111110000000000000000000000000000000
000000000000000000000011111110000000000000000000000000000000
000000000000000000000011111110000000000000000000000000000000
000000000000000000000011111110000000000000000000000000000000
000000000000000000000011111110000000000000000000000000000000
000000000000000000000011111110000000000000000000000000000000
000000000000000000000000000000000000000000000000000000000000
```

This process is illustrated in Fig. 6.19. We take the array as being x[j][i], where *j* represents the row number and *i* the column number. We move from the left until we encounter the first one. In doing so, our array index, *i*, is equal to that at the last zero (12), and the count is also equal to 12 since 12 zeros are present.

FIG. 6.18

Compressed file representing type of bit map shown in Fig. 6.17

```
18 15 27
14 22 24
13 26 21
12 13 6 9 20
12 12 8 9 19
12 11 10 8 19
12 11 10 9 18
11 12 10 9 18
11 12 10 9 18
11 12 10 10 17
10 13 10 10 17
10 13 10 10 17
10 13 10 10 17
10 13 10 10 17
11 12 10 10 17
12 13 6 9 20
13 26 21
14 22 24
18 15 27
19 13 28
20 11 29
21 9 30
22 7 31
22 7 31
22 7 31
22 7 31
22 7 31
22 7 31
22 7 31
60
```

To decode a compressed file, we follow the steps:

1. Set the first value to be printed as zero.
2. Read the count.
3. Print the value to the screen count times.
4. Toggle the value to be one from zero or zero from one.
5. Sum all of the counts in a line.
6. Repeat steps 2–5 until count = 60 (the number of digits in one line).

FIG. 6.19
Search through a line with the loop shown

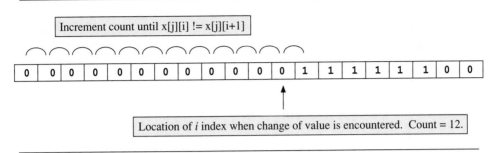

SOURCE CODE

The following code sequence performs the steps listed in the algorithm for compressing the file.

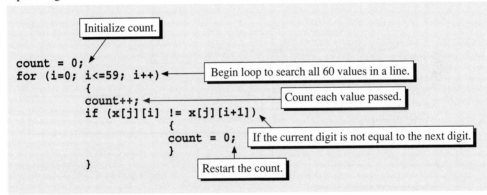

We need to add four things to this sequence:

1. Search each line.
2. Account for being at the end of a line (i being 59).
3. Print the count to a file and to the screen before it gets reset.
4. Print a blank line after completing the search of one line.

The following code includes these added features:

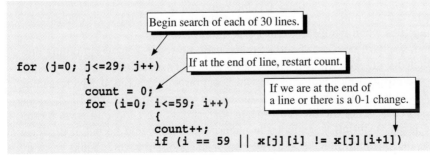

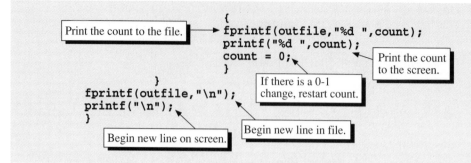

We also need to decode a compressed file. If we call the value to be printed a (where a is either zero or one), the following code performs the steps listed in the algorithm.

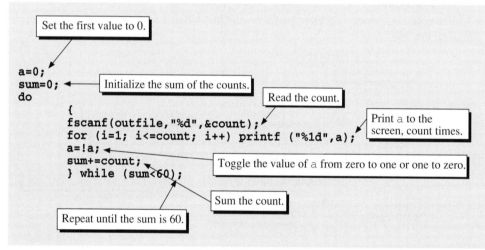

In writing this program, a few issues arise regarding reading from and writing to files. First, we need to both write to and read from the compressed file. We write to the file the counts from each line and then we read the counts to print the original file. Previous to this, we only opened a file for either reading or writing but not both. To open a file for both reading and writing we use the mode r+. The line

```
outfile = fopen("output.out","r+");
```

allows both reading from and writing to file output.out. We will discuss other modes for opening files when we discuss file I/O in more detail.

Second, we have not described this previously; however, it is worth noting now that associated with each file is a file position indicator (sometimes called a *file off-set*). The file position indicator indicates the location within a file at which reading or writing is to begin. In other words, at the end of writing to a file, the file position indicator is at the end of the file. This means that, should we choose to write further in the file using an fprintf statement, the material printed would be put at the location of the file position indicator—at the end of the file. However, should we want to read from the file after having written to it, we need to reset the file position indicator to the beginning of the file. If we were to try to read from the file without having reset the file position indicator, we would encounter the end of the file without having read anything. Therefore, we need to add a line to the previous code that resets the file position indicator. The C library function for doing this is rewind. It has the following form:

rewind (*file pointer*)**;**

· If we open the file with the following statement

outfile = fopen("output.out","r+");

then the line

rewind (outfile);

resets the file position indicator to be at the beginning of the file.

For instance, after the first ten lines have been written, the file position indicator is located as shown here:

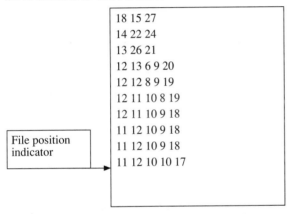

Rewinding this causes the file position indicator to be located at the beginning of the file:

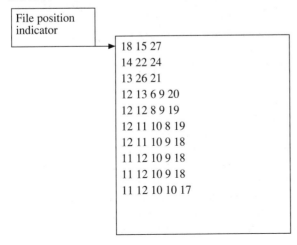

Adding the rewind and a loop over all 30 lines of the file, we get

```
rewind(outfile);          Reset the file position indicator.
for (j=1; j<=30; j++)
        {
        a=0;              Loop over all of the lines.
        sum=0;
        do
                {
                fscanf(outfile,"%d",&count);
                for (i=1; i<=count; i++) printf ("%1d",a);
                a=!a;
                sum+=count;
                } while (sum<60);      Previous code that prints
        printf("\n");                   a decompressed file.
        }
        Begin new line on screen.
```

Third, to get the entire process going we need to read from the bitmap type file. This file has ones and zeros but no spaces. Using fscanf with a %d conversion specification is not possible because we want each one and zero to occupy a different element of our array, and the conversion specification %d looks for a space separator to determine the end of each integer. Therefore, we need to use a %1d conversion specification. With this conversion specification, fscanf takes one digit at a time and does not need a space separating each digit. The code to read the input file is

```
for (j=0; j<30; j++)           Loop over all of the lines.
        {
        for (i=0; i<60; i++)           Loop over all of the digits in one line.
                {
                fscanf(infile,"%1d",&x[j][i]);
                }
        }
        Read one digit at a time with %1d not %d.
```

We have taken the blocks of code and put them in a final program. In addition we have added the following with which at this point you should be familiar:

1. We defined the constant macros SIZE1 and SIZE2 as 30 and 60, respectively. These macros are used in place of all references to the line length and number of lines in the program. Should we encounter a bitmap type file of a different size, we can easily modify our program to accommodate it.
2. We printed the input file to the screen before doing any operations on it.

```c
#include<stdio.h>
#define SIZE1 30
#define SIZE2 60
void main (void)
{
        int x[SIZE1][SIZE2];
        int i,j,a,count,sum;
        FILE *infile, *outfile;
        infile = fopen ("input.dat","r");
        outfile = fopen("output.out","r+");

/*********************************************
**              READ INPUT FILE
*********************************************/

        for (i=0; i<=SIZE1-1; i++)
                {
                for (j=0; j<=SIZE2-1; j++)
                        {
                        fscanf(infile,"%1d",&x[i][j]);
                        }
                }

/*********************************************
**              PRINT INPUT FILE
*********************************************/

        for (i=0; i<=SIZE1-1; i++)
                {
                for (j=0; j<=SIZE2-1; j++)
                        {
                        printf("%1d",x[i][j]);
                        }
                printf("\n");
                }

/*********************************************
**                 COMPRESS FILE
*********************************************/

        count = 0;
        printf("\n\n\n");
        for (j=0; j<=SIZE1-1; j++)
                {
                for (i=0; i<=SIZE2-1; i++)
                        {
                        count++;
                        if (i == SIZE2-1 || x[j][i] != x[j][i+1])
                                {
                                fprintf(outfile,"%d ",count);
                                printf("%d ",count);
                                count = 0;
                                }
                        }
                fprintf(outfile,"\n");
                printf("\n");
                }
```

```
/**********************************************
**   EXPAND FILE
**********************************************/

        rewind(outfile);
        for (j=1; j<=SIZE1; j++)
                {
                a=0;
                sum=0;
                do
                        {
                        fscanf(outfile,"%d",&count);
                        for (i=1; i<=count; i++) printf ("%1d",a);
                        a=!a;
                        sum+=count;
                        } while (sum<SIZE2);
                printf("\n");
                }

}
```

INPUT FILE
Identical to Fig. 6.17.

OUTPUT FILE
Identical to Fig. 6.18.

COMMENT
We indeed have created a file that is considerably smaller than the original file, and it can be used to create the original file should we want to do so.

Modification exercises

1. Modify the program and input file to handle bitmaps of size 20 by 40. Is it easy to do?
2. Create a modular design for this program. Make four functions—one for reading the input file, one for printing the input file, one for compressing the file, and one for expanding the file.

APPLICATION EXERCISES

1. The number of million gallons of sewage that are disposed of each day for a major city is measured continuously for about a month. The records saved in a file, EX6_1.DAT, follow:

   123, 134, 122, 128, 116, 96, 83, 144, 143, 156, 128, 138, 121,
   129, 117, 96, 87, 148, 149, 151, 129, 138, 127, 126, 115, 94,
   83, 142

   Write a program to calculate the frequency distribution using an interval of 10 million gallons per day. The input specification is to use the array sewage_amt[100] to read the number of millions of gallons from file EX6_1.DAT. The output specification is to display the following data on the screen:

Day no.	Millions of gallons
1	123
2	134
3	122
. . .	. . .

Sewage per day	Frequency of occurrence
81–90	3
91–100	3
101–111	0
. . .	

2. Modify the program so that the input is read by calling a function named get_data().

3. Modify the program so that the output is displayed by calling a function display() from the get_data() function. The prototype of the display() function is

   ```
   void display (int *mil_gal, int array_size);
   ```

   where mil_gal is a pointer to be used to pass the information in the array sewage_amt[100] and the array_size is the total number of records.

4. Construction engineers use concrete to build high-rise buildings. Since concrete nearly is incapable of resisting tension loads, reinforcing steel, known as *rebar*, is embedded in concrete to resist tension. The bar is available in a number of sizes. The following table shows the ASTM (American Society for Testing and Materials) standard reinforcing bars size, weight, and diameter:

Size	Weight (lb./ft.)	Diameter (in.)
2	0.167	0.250
3	0.376	0.375
4	0.668	0.500
5	1.043	0.625
6	1.502	0.750
7	2.044	0.875
8	2.670	1.000
9	3.400	1.128
10	4.303	1.270
11	5.313	1.410
14	7.650	1.693
18	13.600	2.257

The next table shows the type and length of rebar used for the basement of a parking garage:

Size	Length (ft.)
4	5000.0
10	2000.0
14	1200.0
18	900.0

Write a program to calculate the total weight of the rebar used.
The input specifications are these:

Use a 2-D array named bar_data[20][3] to read the ASTM standard reinforcing bars size, weight, and diameter.
Use a 2-D array named bar_used[10][2] to read the bar used and length for
the basement.

The output specification is to display the following data on the screen:

Size	Diameter (in.)	Length (ft.)	Weight (lb.)
4	0.500	5000.0	3340.0
10	1.270	2000.0	. . .
14	1.693	1200.0	. . .
18	2.257	900.0	. . .
Total		. . .	. . .

5. Modify the program so that the output is generated by calling a function named
output() from the main() function. The prototype of the output() function must
contain at least the following two formal parameters:

**void output (double *input_bar_data, double *input_bar_used,
...)**

6. Modify the program to convert the output to metric units. Use the following conversions:

$$1 \text{ in} = 2.54 \text{ cm}$$
$$1 \text{ ft} = 12.0 \text{ inch}$$
$$1 \text{ lb} = 0.454 \text{ kg}$$
$$1 \text{ m} = 100 \text{ cm}$$

The output specifications are these: The output shall display the following data
on the screen:

Size	Diameter (cm)	Length (m)	Weight (kg)
4	1.27	1524.0	1516.4
10	. . .	. . .	. . .
14	. . .	. . .	. . .
18	. . .	. . .	. . .
Total		. . .	. . .

7. Write a program that can calculate the sum of three equally sized matrices, [A]
+ [B] + [C].

The input specifications are these:

Read the input from a file with the first line of the file being the number of
rows and columns of each matrix.
The rest of the file has the elements of the matrices.

The output specification is to print the results to a file.

8. Write a program that can multiply two 6 × 6 matrices. The input specification is to read the matrices line by line from a file. The ouput specification is to print the output matrix to the screen.

9. Write a program that can multiply an $n \times m$ and $m \times n$ matrix together.

The input specifications are these:

> Read n and m from the file.
> Read the two matrices from a file.

The ouput specification is to print the results to a file.

10. Write a program that can solve a 3 × 3 set of equations such as

$$3x_1 + 4x_2 - 5x_3 = 2$$
$$-2x_1 + 6x_2 - 12x_3 = -8$$
$$6x_1 - 3x_2 + 2x_3 = 5$$

Write the algorithm by solving these equations by hand and using the following steps. The input specifications are these:

> Read the coefficients of x_1, x_2, and x_3 from a file.
> Read the right-hand side of the equations from the same file.

The output specification is to print the solutions to the equations to the screen.

11. A 1-D array has these ten elements.

    4.4   3.3   2.2   5.5   1.1   6.6   7.7   10.0   9.9   8.8

Write a bubble sort program to sort the array in descending order.

12. A 2-D array has these 20 elements:

    3   33   333   3333
    5   55   555   5555
    1   11   111   1111
    4   44   444   4444
    2   22   222   2222

Write an exchange maximum program to do the following:
a.   Sort the array so that it will look like this:

        5   55   555   5555
        4   44   444   4444
        3   33   333   3333
        2   22   222   2222
        1   11   111   1111

b.   Sort the array so that it will look like this:

```
1111    111    11   1
2222    222    22   2
3333    333    33   3
4444    444    44   4
5555    555    55   5
```

13. Use the following method to calculate the determinant of a 2-D n 3 n integer array. Assume n 5 3 and the array a[3][3] is as follows:

```
11 12 13
21 22 23
31 32 33
```

The determinant $D = (11 \times 22 \times 33) + (13 \times 21 \times 32) + (12 \times 23 \times 31) - (13 \times 22 \times 31) - (11 \times 23 \times 32) - (12 \times 21 \times 33) = 0.$

The input specifications are these:

Read array a[3][3] as shown.
Use a for loop to generate eight 2-D arrays b[n][n] for which to find the determinants, where $n = 2, 3, 4, 5, 6, 7,$ and 8.

The value stored in array element b[i][j] = IJ (the syntax IJ means placing J on the right side of I), where

$$I = i + 1 \quad \text{and} \quad J = j + 1$$

For example, when $n = 8$, B[0][0] = 11, B[0][7] = 18, B[7][0] = 81, B[7][7] = 88.

The output specification is to display the following data on the screen:

Array a[3][3] and its determinants.
Arrays b[n][n] and their determinants.

Modification exercises

14. The program in file Ae6_14 on the diskette displays the $y = \sin(x)$ curve on the screen.
    Modify the program so it will plot the curve of the x[50] and y[50] data as follows instead of plotting the $y = \sin(x)$ curve:

```
x[i]     y[i]
0.0      0.4
0.2      0.5
1.5      0.9
2.3      0.7
2.8      0.2
3.3     -0.3
4.4     -0.8
6.1     -0.9
```

15. Modify the program so that the main() function calls a function named xy_range(). The function will find the values of xmin, xmax, ymin, and ymax based on the data just given.

Exercises involving random numbers

The rand() function is a standard C function that returns a pseudo-random integer in the range 0 to RAND_MAX (which is a macro built into C), where ANSI C requires RAND_MAX to be at least 32767.

Because we often want a random number to be within a specified range, we frequently use the mod operator with the function, rand(). Recall that the mod (%) operator computes the integer remainder of a division operation. For instance, for an integer n the value of n%25 is a number from 0 to 24. Similarly, n%50 is a number from 0 to 49. If we wanted to create a number in the range 50 to 99, then (n%50) + 50 would do it. Thus, we find the mod operator handy for creating integers within a specified range. If we wanted to simulate the roll of a single die, where the result is an integer from 1 to 6, an action such as (n%6) + 1 would work. If n is a random integer, then we could simulate a random roll. The following pair of statements (with roll being an integer representing the value of the roll) would simulate, somewhat, a roll of the die:

```
n=rand();
roll=(n%6)+1;
```

We say that these statements only somewhat simulate the roll of the die because the rand function does not produce a true random number. The theory of random number generation is complex, so we will not go through the details here; however, rand returns only a pseudo-random number. To more closely generate a true random number, we should work with the functions rand and srand together as they have been designed in C to complement each other (this makes them unusual functions in the C library).

The function srand automatically "seeds" the function rand by giving rand a value to begin its calculation. The functions rand and srand are linked together through a global variable that basically is invisible to a programmer. The function srand modifies this global variable and rand uses the global variable in calculating its pseudo-random number. If the value of the global variable is the same each time rand is called, then the number returned by rand is the same. However, if the global variable is different each time rand is called, then unrelated and nearly random numbers are returned by rand. Therefore, the challenge in getting rand to work randomly is to get the global variable to be different on each call to rand. Even more challenging is to make the global variable different each time the program is run.

C also has a time function that can be called to give a number that cannot be guaranteed to be different each time the program is run but will likely give such a number. The function time returns the single integer time of day in seconds by reading the computer's clock. Unless you execute a program at precisely the same time every day, the function time returns a different value each time you run the program. Because of this feature, the time function is well suited for use in creating a different integer each time the program is run and can be used with function srand to create a new global variable for rand, which means rand then returns a nearly random number. We will not discuss the details, but if we call the time function with the macro NULL, we get the desired results. Thus,

```
srand(time(NULL));
```

creates a new global variable each time srand is called. We then can call rand with the statements

```
n=rand();
roll=(n%6)+1;
```

and more realistically simulate a roll of the die.

The functions rand and srand require stdlib.h. The time function uses the header file time.h.

Similarly, if we wanted to simulate the number of a playing card being dealt, we would want a random number from 1 to 13. Using the integer variable card the statements

```
srand(time(NULL));
n=rand();
card=(n%13)+1;
```

produce such a random number. However, a playing card has a suit (clubs, spades, hearts, or diamonds) as well. The suits can be generated using a random number from 1 to 4. However, to properly simulate dealing a deck of cards, it is necessary to make sure that none of the cards is dealt twice. In engineering, generating random numbers is useful for performing what are called Monte Carlo methods of analysis. We will not describe them here; however, you may encounter them in some of your more advanced classes. With this background, you should be capable of writing the programs that follow.

16. Write a program that asks a user to roll a single die twice to get a sum value of 7. If the sum is 11, the user loses. If the sum is neither 7 nor 11, the user neither wins nor loses, meaning that there is no decision.

17. Modify the program to allow the user to collect points. Give the user 100 points to begin and keep track of how many points the user has. Allow the user to choose how many points he or she can win or lose for each decision. The amount of points that the user chooses may not exceed the amount of points that the user currently has. Give the user the option to stop at any time.

18. Write a program to simulate the card game 21 for a single player. Give the user two cards to begin. Ask the user if he or she would like another card, up to a total of five. A user who exceeds 21 loses. A user whose total is greater than or equal to 17 wins. If the user has five cards with the total being less than or equal to 21, the user wins. Otherwise no decision is given. Make sure that no card is dealt twice.

19. Modify the program of problem 18 by allowing the user to collect points as described in problem 17.

Strings and Pointers

$\mathbf{I}$n this chapter we again work with arrays. However, instead of examining arrays that contain numerical values in each array element, as we did in Chapter 6, we will look at arrays that contain a single character in each element. When the last element of these arrays is the null character (\0), these character arrays are known as *strings.*

It made some sense to work with numerical arrays element by element. However, we will see that it is often more convenient to work with strings by using the address of the first element of the string, so we will find that pointers become very important. In this chapter, therefore, we discuss using pointer variables both for transferring the address of the beginning of a string and for manipulating the individual characters of a string.

To completely understand string and pointer operations, we need to understand how memory is arranged and how it can be manipulated. At the end of this chapter, we discuss how memory can be reserved during program execution in a manner that fits the needs of the particular problem being analyzed. Lesson 3.9 was devoted to single characters. If you skipped this lesson, or have forgotten how to work with single characters, read Lesson 3.9 now.

LESSON 7.1 DECLARING, INITIALIZING, AND PRINTING STRINGS AND UNDERSTANDING MEMORY ARRANGEMENT

Topics

- Character arrays
- Initializing single characters
- Initializing strings
- Printing strings
- Memory arrangement

In this lesson, we describe strings, which are arrays of characters. We declare three character arrays and a single character variable. Using printf, fprintf, and the functions that we introduce in this lesson (putc, fputc, puts, and fputs), we illustrate how strings are printed. This program has two sections. The first illustrates initializing a single character and string, and the second illustrates printing a single character and a string.

What you should observe about this lesson's program. For section 1,
1. The keyword char is used to declare both single characters, aa, and arrays of characters, bb[ ], cc[ ], and dd[ ]. The size of each array is given in the declarations.
2. The single character aa is initialized in an assignment statement to be 'g' with single quotes.
3. The character array bb[ ] is initialized with assignment statements character by character with single quotes.
4. The last character of the bb[ ] array is \0.
5. The character array cc[ ] is initialized using the function strcpy.
6. The second argument in the first strcpy statement is a sentence enclosed in double quotes.
7. The two arguments for the second strcpy statement are the names of two arrays (without brackets).

For section 2,
1. The single character aa is printed with putchar, putc, and fputc.
2. The string stored in the array bb[ ] is printed with puts, fputs, printf, and fprintf.
3. The conversion specification for bb[ ] in the printf and fprintf statements is %s.
4. From the output, the strings stored in cc[ ] and dd[ ] can be seen to be identical.
5. The last printf statement prints the addresses of the first element of the arrays using the %p format. The output shows that hexadecimal notation is used.

Questions you should attempt to answer before reading the explanation.
1. How many characters are stored in the memory reserved for the bb[ ] array?
2. What is the last character in memory for all strings?

Source code

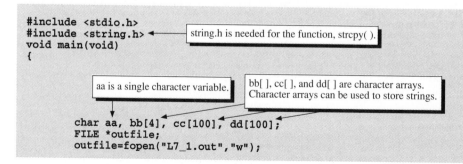

```
#include <stdio.h>
#include <string.h>          string.h is needed for the function, strcpy( ).
void main(void)
{
                   aa is a single character variable.    bb[ ], cc[ ], and dd[ ] are character arrays.
                                                         Character arrays can be used to store strings.

        char aa, bb[4], cc[100], dd[100];
        FILE *outfile;
        outfile=fopen("L7_1.out","w");
```

We use single quotes to enclose a single character. We store a single character in each element of the bb[ ] array.

```
printf ("********** Section 1 - Initializing ************\n");
aa='g';
bb[0]='C';
bb[1]='a';
bb[2]='t';
bb[3]='\0';
```

The last element of the bb[ ] array is \0. This is called the null character and is the last character of all strings. Thus, bb[ ] stores the string Cat.

An array name without following brackets represents the address of the first memory cell reserved for the array. Here, cc represents the address of the cc[0] memory cell. The function strcpy uses only addresses as arguments.

The function strcpy( ) copies strings into reserved memory cells. A total of 100 cells are reserved for cc with the declaration: char cc[100]. This call to strcpy fills the first 56 of those cells with the string literal used as the second argument. Note that double quotes (not single quotes) are used to enclose the string literal. We cannot use an assignment statement such as cc="This is a string constant"; to do this. Assignment statements, in general, do not work with strings.

A string literal written in a program body is stored in memory also. Because it is not a variable, it is not stored in memory near the variables. A string literal in a program body actually represents the address where the string literal is stored. Therefore, strcpy, which accepts only addresses as arguments, accepts a string literal as its second argument.

```
strcpy(cc,"This is a string constant, also called a string literal.");
strcpy(dd,cc);
```

Here, we copy the string stored at the address indicated by cc into the memory cells at the address indicated by dd.

```
printf ("\n********** Section 2 - Printing **********\n");
putchar(aa);
putc(aa, outfile);
fputc(aa, outfile);
```

A single character can be printed to the screen using putchar or to a file using putc or fputc.

```
puts(bb);
fputs(bb,outfile);
printf("%s\n",bb);
fprintf(outfile,"%s\n",bb);
```

puts prints a string to the screen.

fputs prints a string to a file.

printf can be used to print a string to the screen and fprintf can be used to print a string to a file. The %s conversion specification is for strings. Note that bb (which is an address) corresponds to %s.

```
puts(cc);
puts(dd);
```

We see from the output that cc and dd represent identical strings. We copied the string at cc to dd using strcpy.

```
printf("addresses aa=%p, bb=%p, cc=%p, dd=%p\n",&aa,bb,cc,dd);
```

%p is used as a conversion specification for addresses.

```
}
```

Output to the screen

```
********* Section 1 - Initializing ************
********* Section 2 - Printing *********
gCat
Cat
This is a string constant, also called a string literal.
This is a string constant, also called a string literal.
addresses aa=FFF5, bb=FFF0, cc=FF8C, dd=FF28
```

Output to the file

ggCatCat

Explanation

1. What is a string? A string is an array of characters including the terminating null (\0) character. For instance, we have stored a string in the bb[ ] array character by character in this lesson's program with the statements:

```
bb[0]='C';
bb[1]='a';
bb[2]='t';
bb[3]='\0'
```

Recall that the elements of an array are stored in contiguous and increasing memory locations. As such, a string consists of character codes (usually either ASCII or EBCDIC) stored in contiguous memory cells. The last memory cell contains the escape sequence, \0, which is treated as a single character. The character, \0, is called the *null character* and should not be confused with the null pointer, which we describe later. The null character, written as \0, is stored in memory as 1 byte with all of the bits set to 0.

When more than one character is written, we use double quotes to surround the characters. When this is done, C automatically adds the terminating null character when it is stored in memory. Therefore, it is incorrect for a programmer to write a number of characters enclosed in double quotes with the last character being \0. For instance, in this lesson's program, we have written the string:

"This is a string constant, also called a string literal."

in one of our statements. Although it does not end in \0, C recognizes this to be a string and adds \0 after the period when it is stored in memory. As a result, the character array declaration must be large enough to include the \0 character, even though it is not shown! For this string an array size of at least 57 is needed (one element for each character plus one for \0).

Note that what is stored in the variable aa is a single character, not a string. A string is stored in a character array, whereas a single character is stored in a character variable.

2. When a string of characters enclosed in double quotes is written in a program body, how does C treat it? C treats it like an address. C actually works with the address of the first memory cell at which the string is stored. We later look at the addresses used for the string constants in more detail. For the moment, you need to remember only that, where a string constant is written in a program body, C sees an address. Wherever we use string constants as arguments to functions, for instance, we also can use addresses. These addresses often will be represented by the names of arrays without brackets.

3. How do we initialize strings without doing it character by character? Initializing strings character by character with assignment statements is tedious. Fortunately, C has library functions that make it easy for us to initialize strings. For instance, the function strcpy, whose prototype is in the file string.h (which must be included), copies a string that begins at one address into the memory cells beginning at another address. For instance, the statement

```
strcpy(cc,"This is a string constant, also called a string literal.");
```

causes the string indicated by the address "This is a string constant, also called a string literal." to be copied to the memory cells indicated by the array name without brackets, cc (which also represents an address). Note that cc[ ] has a declared size of 100, which is greater than the 57 that is needed to store the string.

Also in this program, we have used the statement

```
strcpy(dd,cc);
```

Here, clearly, both arguments represent addresses, as both are the names of arrays without brackets. This statement causes the string in the memory cells beginning at the address cc to be copied into the memory cells beginning at the address dd.

4. Could we have used an assignment statement of the sort:

```
cc="This is a string constant, also called a string literal.";
```

to store a string in the cc[ ] array? No, this is a common error. We cannot use this assignment statement because C sees an address for the string literal given on the right side of the assignment statement. Therefore, such a statement causes C to try to store an address in the location indicated by cc. But the cc[ ] array is declared to store characters not addresses, and therefore the assignment statement will not work. Remember, when working with strings, most of the time you use string functions, not assignment statements. Assignment statements work well with numeric data but not very well with character information.

5. What functions can we use to print a single character? We previously showed how the functions putchar and printf printed single characters to the screen and will not describe them here.

To print a character to a file, C has the functions putc and fputc. In use they essentially are the same. For example, the statements

```
putc(a,outfile);
fputc(a,outfile);
```

cause the character represented by the variable a to be printed to the file pointed to by the file pointer, outfile (meaning file L7_1.OUT for this program).

The general form for these is

```
putc  (character, file_pointer);
fputc (character, file_pointer);
```

We also can use the fprintf function with the %c conversion specification to print single characters to a file. Since fprintf works like printf we will not describe it further.

6. What functions can we use to print a string to the screen, and how do they work? We can use puts and printf to print a string to the screen. For example, the statements

```
puts(bb);
printf("%s\n",bb);
```

cause the string at the address indicated by the identifier bb to be printed to the screen.

The function puts prints element after element of the bb[ ] array to the screen until it encounters the null character (which it does not print), at which point it prints a newline character (meaning it advances the cursor to a new line) even though it has not been explicitly told to do so. Because of the way puts operates, it is important that a null character be present at the end of the characters to be printed.

The general form for puts is

```
                    puts (address);
```

where *address* often is the address of the first element of an array.

For printf, the conversion specification for a string is %s. With this conversion specification, printf expects an address or a pointer to a string to be given. For instance, in the printf statement shown the %s in the format string is used together with the address indicated by bb to print the string in the bb[ ] array. From the output we can see that the entire bb array has been printed, because the %s conversion specification causes printf to print a string up to the terminating null character. However, unlike puts, printf does not print a newline character unless \n is included explicitly in the printf statement.

7. How can we print a string to a file? We can use either the fputs or fprintf functions. For instance, each of the two statements

```
fputs(bb,outfile);
fprintf(outfile, "%s\n", bb);
```

causes the bb[ ] char array to be printed to the file pointed to by outfile.

The general form for fputs is

```
           fputs (address, file_pointer);
```

which causes the string indicated by *address* to be printed to the file indicated by *file_pointer.* Unlike puts, fput does *not* automatically print a newline character where it encounters a null character in a string. Observe from the output of this lesson's program that puts has printed the string "Cat" on two different lines whereas fputs has printed "Cat" twice on the same line.

Again, we will not describe fprintf in detail as it operates much the same as printf.

8. If we were to write

```
char aa, ee[2];
aa='g';
strcpy(ee, "g");
```

would the contents of the memory for aa and the contents of the memory for ee[ ] be the same? No, and this illustrates a fundamental characteristic of strings. The following represents what would be in the memory cells:

Memory for character variable aa	g

Memory for character array ee[ ]	g	\0

Clearly, the array form has the terminating null character. Because of this difference, we cannot treat aa as a string.

9. How will we illustrate character arrays in our table of variables? We will show one-dimensional character arrays in the three-dimensional format shown in Fig. 7.1. Observe that the address of the first element of the array is given in the Address column. The contents of the first element of the array are given in the Value column. The contents of the other elements are illustrated by the three-dimensional part of the figure. These are indicated as being behind the first element. For short strings, we show all the characters, as is shown with Cat\0. With long strings, we show simply the first character and maybe a few others. Remaining characters will be shown as a dashed line behind the first character.

FIG. 7.1
One-dimensional character arrays from this lesson's program in a three-dimensional format

Variable name	Type	Address	Value
aa	char	FFF5	g
bb	array of char	FFF0	C
cc	array of char	FF8C	T
dd	array of char	FF28	T

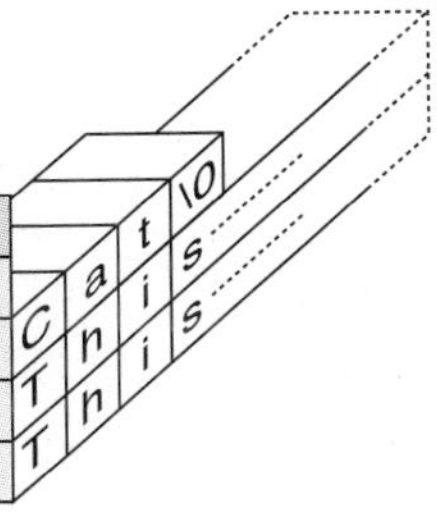

10. With the addresses printed out, can we get an understanding of where in memory the strings for this lesson's program were stored? Yes; we note that, in increasing order, the memory locations and corresponding string names are

FF28   FF8C   FFF0   FFF5
  dd   cc   bb   aa

We can convert these hexadecimal values to decimal values and work with the decimal values to see how many bytes lie between the addresses indicated. Converting these to decimal gives

65320   65420   65520   65525
  dd   cc   bb   aa

Between dd and cc we have $65420 - 65320 = 100$ bytes, between cc and bb we have $65520 - 65420 = 100$ bytes, and between bb and aa we have $65525 - 65520 = 5$ bytes. The arrays cc[ ] and dd[ ] were declared to have 100 elements, bb[ ] to have 4 elements, and aa to have just 1 element. The addresses illustrate that the memory for all of them is closely packed. Note that bb[ ] had only four elements declared, but five bytes separate bb[ ] and aa. The exact memory addresses are said to be *implementation dependent,* meaning a different C compiler may have only four bytes between bb[ ] and aa. We will not concern ourselves with this particular detail in this text. For this particular compilation and execution of this program, we show the order of memory occupied (in linear rather than tabular fashion) in Fig. 7.2.

With this illustration of memory, you can see that, if we attempt to write a string into dd that covers more than 100 elements, we will extend into cc's memory region. In fact, because C does not check whether or not boundaries are exceeded, it indeed will write over a portion of cc[ ] without indicating an error even has occurred! Therefore, it is up to the programmer to make sure that this does not happen. If you accidentally exceed the bounds of an array, you will cause unexpected results and have a difficult time finding your error.

FIG. 7.2
Arrangement of memory for this lesson's program

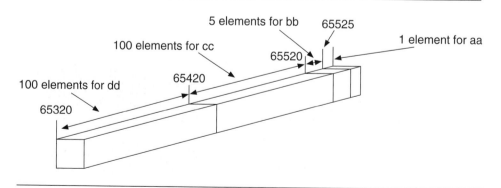

11. When is this memory arrangement done? When the program is compiled, C uses the declarations to set how memory cells for variables and arrays will be arranged (meaning which arrays will be next to which arrays and variables) and the size of the block of memory needed for all the declared data structures. Because this is done during compilation, we must make sure that our arrays are sized to the maximum that may be needed in any possible problem. In other words, we cannot enlarge the declared size of our arrays during execution to accommodate extra data without using advanced techniques, which we describe later in this chapter.

Remember that the addresses of the variables have been set during compilation (not the actual addresses but what can be regarded as relative addresses, meaning where in relation to other variables a variable's cell exists). We cannot move these set memory cells around during execution. If we try to do this we will crash our program.

12. What happens when the program executes? At this time, the program reserves the block of memory in the arrangement that the C compiler has already determined. In a modularly designed program, a block of memory for the local variables in a function is not reserved until the function is called. After a function has been called, the memory for the declared variables and arrays is reserved and stays reserved until the function has completed execution. Then, the memory is released (unless specified to do otherwise, which is an advanced technique that we cover in Chapter 8) so that the next function being called can use this memory.

After the memory has been reserved, we can change the *values* in the memory cells. However, we cannot change an *address* of a particular variable or shrink, swell, or rearrange the memory blocks. For example, for this lesson's program, we cannot assign a new address for the dd array. In other words, dd is set to be at address FF28. If we attempt to change the address, our program will crash or not compile. We mention this because you accidentally may try to do this. Especially when working with character arrays (strings), it is easy to forget that the name of the array without brackets means an address and write an assignment statement with the name of the string on the left side. For instance, in this lesson's program, we cannot write

```
dd=cc;
```

or

```
dd="A sample string";
```

because dd is an address and cannot appear on the left side of an assignment statement (also said that dd is not an l-value, *l* meaning left). An assignment statement of this type tries to get us to change an address, which we cannot do. Remember, if we want to copy the string stored in cc[ ] to the memory cells for dd[ ], we must use strcpy(dd,cc).

We will see that, in working with strings, we must keep in mind what represents addresses and use these only in their proper locations. As you go through this chapter, note the many different ways that addresses are represented.

13. Can we initialize character arrays in the declaration as we did with numeric arrays? Yes; however, we will not illustrate this until near the end of the chapter. The reason for delay is to avoid confusion because what appear to be illegal assignment statements are legal when used in a declaration. At this point, if you see character arrays that are initialized in their declaration, be aware that things that appear to be illegal are being done, but because they are done within a declaration they are allowed.

14. Why would we want to shrink, swell, or rearrange memory blocks during execution? We might want to do this to make more efficient use of memory. For instance, if we make declarations in a program to accommodate the largest possible problem, we may find that some computers will not be able to accommodate it due to a lack of memory. If this is the case, we would not be able to run our program on these systems without making changes to the declarations and recompiling it. This is not practical for commercial software.

A better way to handle the problem is to make the memory set during compilation be reasonably small, so that all modern computers can handle it. Then, if large problems are to be run, have the program reserve other memory when it becomes necessary (for instance, when a large array needs to be used) *during execution.* In doing this, on computers with only small amounts of memory, we could simply print an error message saying that the computer needs more memory to execute this large problem. No error message would be printed for small problems run on small systems or large problems run on large systems, and so we have been able to write one program that can be used on systems of many sizes. Later in this chapter, we illustrate how to do this.

15. Can you give us a hint about some of the details to do this? Yes; although we cannot assign the *addresses* of our variables during execution (because they are done at compilation), we can assign *values* to our variables. This becomes important to us, because we can use pointer variables. Remember that the value of a pointer variable is an address. To help us get more memory during execution (meaning just enough memory for our needs and no more), we will use a C function to tell us an address that has enough free memory for our needs. We can assign this address as a value for a pointer variable. Once this is done, we can store information in this new area of memory.

16. Does every ANSI C compatible compiler arrange memory the same way? No; if you run this lesson's program you will likely get a different memory arrangement from what we have shown. Throughout this chapter, we illustrate how memory is arranged for various programs. Be aware that, should you run these programs, the actual addressing that you get will likely differ from what we have shown. However, the concepts we illustrate still apply to your programs.

17. Can we summarize the C character/string printing functions we learned?
Yes; here is a list.

putchar and printf with %c	print a character to the screen
putc, fputc, and fprintf with %c	print a character to a file
puts and printf with %s	print a string to the screen
fputs and fprintf with %s	print a string to a file

EXERCISES

1. Given these declarations

   ```
   char aa, bb[10], cc[15], dd[15];
   ```

   find errors, if any, in the following statements:
   a. `strcpy(bb,aa);`
   b. `strcpy(bb,"This is 23");`
   c. `puts(aa);`
   d. `fputs(dd);`
   e. `strcpy(cc,'Many words');`

2. Find the errors in this program:

   ```
   #include <stdio.h>
   void main(void)
   {
           char dd[20], pp[30], rr[5];

           dd[6]="D";
           pp="Panda",
           strcpy("abcd",rr);
           printf("%s, %s,%s\n",dd,pp,rr[1]);
   }
   ```

3. Use the printf() and putchar() functions alternatively to display the following output on the screen:

   ```
   12345678901234567890123456*654321098765432109876543211
   A                          *                          A
   BB                         *                         BB
   CCC                        *                        CCC
   DDDD                       *                       DDDD
   ...                        *                        ...
   XXXXXXXXXXXXXXXXXXXXXXXXX   *   XXXXXXXXXXXXXXXXXXXXXXXXX
   YYYYYYYYYYYYYYYYYYYYYYYYY * YYYYYYYYYYYYYYYYYYYYYYYYY
   ZZZZZZZZZZZZZZZZZZZZZZZZZ*ZZZZZZZZZZZZZZZZZZZZZZZZZ
   12345678901234567890123456*654321098765432109876543211
   ```

4. Redo problem 3, but use the puts() function only.

5. Redo problem 3, but use the fprintf() and fputc() functions alternatively to save the output to a file named EX7_1_5.OUT.

6. Redo problem 3, but use the fputs() function only to save the output to a file named EX7_1_6.OUT.

Solutions

1. a. **`strcpy(bb,cc)`**   (We cannot use a single character as an argument to strcpy.)

    b. **`strcpy(bb,"This is 2");`**   (The string "This is 23" has ten characters showing and the size of bb is declared to be 10. However, there is no room for \0 that C must add.)

    c. **`putchar(aa);`**

    d. **`fputs(dd,outfile);`**   *where outfile is a file pointer.*

    e. **`strcpy(cc,"Many words");`**

LESSON 7.2 DETERMINING INFORMATION ABOUT STRINGS AND CHARACTERS AND USING PRINTF

Topics

*   %s conversion specification
*   String functions for determining information about strings
*   Single character functions for determining information about characters

In this lesson we learn more about using the %s conversion specification in printing strings (section 1) and how we can determine information about strings within our programs (section 2). This program prints strings using printf, finds the length of a string, and performs other string manipulations.

What you should observe about this lesson's program.

1. The header file ctype.h is included.
2. The many_lines[ ] array has \n included in the string and is initialized with strcpy.
3. In section 1, long_string[ ] is printed using a number of different %s type specifications. The double brackets [[ ]] help illustrate the different spacings caused by the different %s specifications (see the output).
4. In section 1, the last printf statement illustrates how three string constants can be printed using just one statement.
5. In section 2, the strlen( ) function is called with an array name as its argument. The return value from this function is assigned to the variable occupied, which is printed in the output.
6. In section 2, the sizeof operator (which looks like a function) is used on both an array name (many_lines) and a type (char).

7. The BUFSIZ macro is used in section 2. Its value is assigned to the integer variable `buffer_size` and is printed in the output.
8. The isdigit and tolower functions are used in section 2.

Questions you should attempt to answer before reading the explanation.
1. What is the value of the variable `occupied`, and what does it mean?
2. What does the sizeof operator do?
3. What do the isdigit and tolower functions do?

Source code

```
#include <stdio.h>
#include <string.h>
#include <ctype.h>
void main(void)
{
        char cc, long_string[50], many_lines[70];
        int occupied, reserved, buffer_size;



        printf ("****** Section 1 - conversion specifications ********\n");
        strcpy(long_string,"This is a complete sentence." );
        strcpy(many_lines,"This sentence \ncovers two lines.");
        printf ("[[%s]]\n",long_string);
        printf ("[[%40s]]\n",long_string);
        printf ("[[%-40s]]\n",long_string);
        printf ("[[%40.10s]]\n\n",long_string);



        printf("%s%s\n%s","This is"," one method for printing","string constants.\n");



        printf("\n****** Section 2 - finding information about strings **\n");
        occupied = (int)strlen(many_lines);
        reserved = sizeof(many_lines)/sizeof(char);



        buffer_size = BUFSIZ;
        printf ("occupied = %d \nreserved = %d\n\nOur buffer size is %d\n",
                        occupied, reserved, buffer_size);


        if(isdigit(many_lines[0])) putchar(many_lines[0]);
        cc=tolower(many_lines[0]);
        putchar(cc);
        putchar('\n');
        puts(many_lines);
}
```

Annotations:
- String functions require the file string.h.
- Functions for checking or changing case or kind of character require ctype.h.
- Escape sequences can be put into strings.
- The %s conversion specification can be used to print strings in a formatted manner.
- The first comma in the printf statement separates the format string from the other string literals.
- The function strlen uses an address as an argument and determines number of characters in the string at that address.
- The sizeof operator determines the number of bytes declared for its argument.
- BUFSIZ is a constant macro defined in stdio.h.
- The functions isdigit and tolower operate on single characters.

Output

```
****** Section 1 - conversion specifications ********
[[This is a complete sentence.]]
[[                      This is a complete sentence.]]
[[This is a complete sentence.                      ]]
[[                                     This is a ]]
This is one method for printing
string constants.

****** Section 2 - finding information about strings **
occupied = 32
reserved = 70

Our buffer size is 512
t
This sentence
covers two lines.
```

Explanation

1. In this lesson's program, what do the conversion specifications %s, %40s, %-40s, and %40.10s mean? These conversion specifications work like those used for the integer type d. The form is

$$\% \; flag \;\; field\_width \, . \, precision \quad \mathbf{s}$$

where *flag, field_width, the decimal point,* and *precision* are optional. As before, if the specified *field_width* is less than what is needed, C expands the width to accommodate the value to be printed. We have the following specifications and results:

Command and Explanation	Result
printf ("%s\n",long_string); Since no field width is given, the field width is expanded to exactly the size of the string.	[[This is a complete sentence.]]
printf ("%40s\n",long_string); Field width is 40. With no flag, result is right justified	[[This is a complete sentence.]]
printf ("%-40s\n",long_string); Field width is 40. Flag causes left justification	[[This is a complete sentence.]]
printf("%40.10s\n\n",long_string); Field width is 40. Precision is 10 meaning that only the first ten characters are printed. No flag means right justification	[[This is a]]

2. What is a method for using printf to print string constants? It results in an odd looking printf statement, but the following form, for example, can be used:

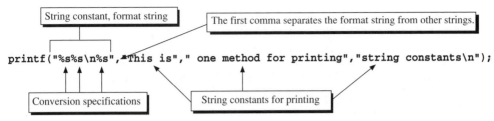

In this statement, we have the first string constant (or string literal), which is the format string, and three string constants to be printed. Within the format string are three conversion specifications as indicated. They correspond to each other in their respective orders. This causes the first %s to be the format specification for "This is" and the second %s to be the format specification for the string " one method for printing". The third %s corresponds to "string constants\n". Although we are accustomed to seeing variables to be printed in our printf statements, in fact we can have string constants. Note that the string before the first comma is the format string. Remember, a string constant used in a program body actually represents an address, and the strings after the first comma are ordinary string constants. C essentially replaces string constants in the program body with the address in memory of the string constant. So, even though we write a string constant, we actually are indicating an address, and with an address, %s is the correct conversion specification.

Be aware that we can have a number of format strings before the first comma. For instance, we could have written

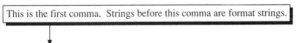

```
printf("%s%s" "\n%s","This is"," one method for printing","string constants\n");
```
After the first comma are the string constants to be printed. The printed output

```
This is one method for printing
string constants.
```

results from this statement. Note that the \n in the format string causes the last string literal to be printed on the line following the line on which the other two are printed.

3. What does the function strlen do? The function strlen is used to determine the number of characters actually stored in a particular character array. It does this by using the address of the first element of the array to begin its search for the null character. It counts the number of memory cells until it encounters that character. The form to use strlen is

strlen(address);

where the *address* tells strlen where to begin its search. The function returns the number of bytes or memory cells up to but not including the null character in the string. For all practical purposes the return value is an integer; however, to emphasize that it is being stored in an integer variable, it is prudent to use the (int) cast operator on the return value. For example, the statement:

```
occupied = (int)strlen(many_lines);
```

assigns to the variable occupied the value of the number of characters in the string many_lines (excluding the null character).

4. What use does strlen have? It allows us to know exactly how much memory is needed for a particular string. We can use this to optimize our memory usage. In addition, it can help us avoid inadvertently overwriting memory by going beyond the bounds that we want to write.

5. What does the sizeof operator do? The sizeof operator computes the number of bytes reserved for storage of a specified variable or variable type. That it is an operator rather than a function is of little importance to us, except to note that we need not always put parentheses around the argument. In this text we will use parentheses in all cases to avoid potential problems from not using them.

The form for using sizeof is

sizeof(*name*)

where *name* can be a variable name or type. For instance, two legitimate applications of sizeof are

```
sizeof(int);
sizeof(many_lines);
```

In the first of these, sizeof returns the number of bytes used by the system to represent an int. In the second, sizeof returns the number of bytes reserved for the string many_lines. Note that, in this case, sizeof returns a value different from the value returned by strlen. This is because sizeof returns the number of bytes *reserved* for many_lines, which is 100, whereas strlen returns the number of characters that actually are *stored,* which is 34.

6. In this lesson's program we divided sizeof(many_lines) by sizeof(char). Why? This allows us to determine the number of element places reserved for the array. If we want to know the number of elements reserved for an array, it is not sufficient to know only the number of bytes reserved. We must divide that number of bytes by the number of bytes used for the array's element type, because ANSI C does not explicitly set the number of bytes to be used for each type except char. For instance, one C implementation may use two bytes for an integer while another may use four bytes. To keep our code portable, we divide by the number of bytes in the element type. By definition, C gives a char one byte. Therefore, sizeof(char) is always equal to 1 and we really did not need to divide by sizeof(char) in this program. However, we did so to emphasize that you should divide by the sizeof the element type in your programs when you want to determine the number of elements reserved for an array.

7. In the expression sizeof(many_lines), the array name, many_lines, does not have brackets with it. Does this mean that sizeof actually is working with an address? No, this is one time in C (and we will not see any others in this book) where an array name without brackets does *not* mean an address. So, remember this one exception. When using the sizeof operator, an array name without brackets does not mean an address, it is simply an identifier. The operator sizeof finds the allocated storage associated with that identifier.

8. What is BUFSIZ? As you can guess, BUFSIZ is a constant macro representing the size of the input buffer. It is defined in stdio.h and represents the size for your C implementation. ANSI C requires that it be at least 256 bytes. A common size is 512. It gives us the maximum number of characters that can be transferred through our standard input stream, stdin. It is useful because we know that no input string transferred in this manner can exceed that size. Therefore, we can use the macro BUFSIZ to size char arrays to be read from stdin (which represents the keyboard, in most cases). Later in this chapter, when we read in strings from the keyboard, we see that we cannot read more than 512 at a time for our C implementation.

9. Both strcpy and strlen used addresses as input arguments. Do most string functions use addresses rather than variable names as input arguments? Yes, for instance, unlike a math library function, that takes a variable *name* as an input argument, a string function usually takes an *address* of the first element of a string or character array as an input argument. So, in using string functions, we must be cognizant of how to refer to addresses in C, of how much memory is associated with a particular address, and of the way memory is being manipulated in our programs.

10. What does the function isdigit do? This function works with a single character, not a string. This function determines whether or not the character used as the argument is 0, 1, 2, 3, 4, 5, 6, 7, 8, or 9. If it is not, the function returns 0. If it is, the function returns a nonzero integer. For example, in this lesson's program, isdigit was used in the line

```
if (isdigit (many_lines[0]))  puts(many_lines);
```

The argument, many_lines[0], is the character *T*. So the function isdigit returned the value 0, which the if statement used as indicating False, causing the puts statement to not be executed.

11. What use may we have for isdigit in our programs? We find isdigit useful in helping us evaluate the content of strings. For instance, we may be searching in a string for a number, such as a measurement or price. With isdigit, we can exclude the characters and focus on the portion of the string that contains numbers.

It also is helpful for checking input data. For instance, if we prompt the user to enter a digit from the keyboard, we can use isdigit to check that indeed a digit has been typed. If something other than a digit has been typed, an error message can be displayed and the user can be prompted to reenter the data.

12. What does the function tolower do? This function also works with a single character, not a string. If the argument for tolower is an uppercase letter, this function returns the lowercase version of this letter. It does not modify the argument itself. For instance, in this lesson's program, we have used the following statements

```
cc = tolower(many_lines[0]);
putchar(cc);
puts(many_lines);
```

Because many_lines[0] is the uppercase letter *T*, tolower(many_lines[0]) returns the lowercase letter *t*. The assignment statement assigns *t* to the variable cc. This is verified in the output with the statement, putchar(cc);. However, because the function tolower does not modify the argument many_lines[0], the first character of many_lines[ ] remains the uppercase letter *T*. This is verified in the output with the statement puts(many_lines);.

13. What is one way that we can print a new line using puts? If a string that we are printing has the new line character (\n), then a new line will be begun. The newline character will not be printed. Given the string:

```
"This sentence \ncovers two lines.",
```

the following is printed with puts

```
This sentence
covers two lines.
```

Other escape sequences, such as \t, embedded in strings perform in a manner similar.

14. Are there any other character functions in the C library? Yes, Table 7.1 lists the character functions available in the C library whose prototypes are in the file ctype.h. We will not show the use of these functions; however, we believe the descriptions in the table give you sufficient information to use them in your programs. Note that each of the functions takes a single character as an argument, indicated as int because C treats characters as integers. The sketch at the bottom of the table illustrates the naming of C's character groups.

EXERCISES
1. Find errors, if any, in the following statements:
   a. ```
      int b;
      b=strlen(double);
      ```
   b. ```
      int d;
      d=strlen("1234567890");
      ```
   c. ```
      long f;
      f=strlen('q');
      ```
   d. ```
      char g[20];
      int h;
      strcpy(gg,"1234567890");
      h=strlen(g[ ]);
      ```
   e. ```
      char aa[30];
      int bb;
      strcpy(aa, "APPLE");
      bb=sizeof(aa[]);
      ```
      ```

TABLE 7.1
Character functions in C library

Function	Operation
isalnum (int)	Returns a nonzero integer if the argument is any lower- or uppercase letter or 0–9 digit. Otherwise, it returns the integer 0.
isalpha (int)	Returns a nonzero integer if the argument is any lower- or uppercase letter. Otherwise, it returns the integer 0.
iscntrl (int)	Returns a nonzero integer if the argument is any control character (being new line ('\n'), horizontal tab ('\t'), carriage return ('\r'), form feed ('\f'), vertical tab ('\v'), backspace ('\b'), or alert ('\a')). Otherwise, it returns the integer 0.
isdigit (int)	Returns a nonzero integer if the argument is any 0–9 digit. Otherwise, it returns the integer 0.
isgraph (int)	Returns a nonzero integer if the argument is any printing character except space (' '). Otherwise, it returns the integer 0.
islower (int)	Returns a nonzero integer if the argument is a lowercase letter. Otherwise, it returns the integer 0.
isprint (int)	Returns a nonzero integer if the argument is any printing character including space (' '). Otherwise, it returns the integer 0.
ispunct (int)	Returns a nonzero integer if the argument is any printing character other than space (' '), lower or uppercase letter, or 0–9 digit. Otherwise, it returns the integer 0.
isspace (int)	Returns a nonzero integer if the argument is space (' '), new line ('\n'), horizontal tab ('\t'), carriage return ('\r'), form feed ('\f'), or vertical tab ('\v'). Otherwise, it returns the integer 0.
isupper (int)	Returns a nonzero integer if the argument is an uppercase letter. Otherwise, it returns the integer 0.
isxdigit (int)	Returns a nonzero integer if the argument is any 0–9 digit, lowercase letters a–f, or uppercase letters A–F; that is, any hexadecimal digit. Otherwise, it returns the integer 0.
tolower (int)	If the argument is an uppercase letter, this function returns the corresponding lowercase letter. Otherwise, the argument is returned unchanged.
toupper (int)	If the argument is a lowercase letter, this function returns the corresponding uppercase letter. Otherwise, the argument is returned unchanged.

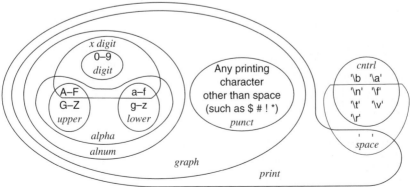

2. Find errors in this program:

```
#include <stdio.h>
void main(void)
{
char aa[10], bb[50];
strcpy(aa,'Dragon');
strcpy(bb, "Apple, pear, peach, plum");
strcpy(aa,bb);
}
```

3. Write a program to display the following output on the screen. The program should define

```
        char date[30], sender[50], status[40], page[10];
```

and show this output on the screen:

```
12345678901234567890123456789012345678901234567890901234
Fax Transaction Report              P.01
```

```
Date            Sender          Status
12/23/1997      Dell Kevin      OK
12/24/1997      Halton Bors     Out of paper
```

Solutions

1. a. **b = sizeof (double);**
    b. No error.
    c. **f=strlen("q");**
    d. **h=strlen(g);**
    e. **bb=sizeof(aa);**

LESSON 7.3 2-D CHARACTER ARRAYS

Topics

- Initializing strings in 2-D arrays
- Printing strings in 2-D arrays

We previously saw the usefulness of two-dimensional numeric arrays. In this lesson we examine two-dimensional character arrays.

A very common way to store strings is in two-dimensional character arrays. This is because each row of a 2-D character array can be considered a separate string if it is terminated with the null character, \0. This program has two sections. The first section illustrates initializing 2-D character arrays, and the second section illustrates printing them.

What you should observe about this lesson's program.

1. The two-dimensional character array aa[ ][ ] is declared to have 2 rows and 90 columns.
2. The two-dimensional character array bb[ ][ ] is declared to have 3 rows and 15 columns using constant macros.

3. In section 1, all of the strings in the aa[ ][ ] and bb[ ][ ] arrays are initialized with the strcpy( ) function.

In section 2,

4. The first for loop uses printf to print the strings of aa[ ][ ]. The output shows that all the strings are printed on the *same* line.
5. The second for loop uses puts to print the strings of bb[ ][ ]. The output shows that all the strings are printed on *different* lines.
6. The third for loop uses fputs to print the strings of bb[ ][ ] to an output file. The output file shows that all of the strings are printed on the *same* line.
7. The operator sizeof and the function strlen can be used with two-dimensional arrays and one set of brackets.

A question you should attempt to answer before reading the explanation. What is a fundamental difference between the way that puts and fputs print strings?

Source code

Insert cutfig. 7-6 here

```
#include <stdio.h>
#include <string.h>
#define NUM_ROWS 3
#define NUM_COLS 15

void main(void)
{
        char aa[2][90], bb[NUM_ROWS][NUM_COLS];
        int i, occupied, reserved;
        FILE *outfile;
        outfile=fopen("L7_3.OUT","w");




        printf("********** Section 1 - Initializing **********\n");
        strcpy(aa[0],"The aa[ ][ ] array ");
        strcpy(aa[1],"has 2 strings.\n");
        strcpy(bb[0],"The bb array ");
        strcpy(bb[1],"has ");
        strcpy(bb[2],"3 strings.");




        printf("********** Section 2 - Printing **********\n");
        for(i=0; i<2; i++) printf("%s",aa[i]);
        for(i=0; i<NUM_ROWS; i++) puts(bb[i]);
        for(i=0; i<NUM_ROWS; i++) fputs(bb[i],outfile);
```

Two-dimensional character arrays can be declared with constant macros. Constant macros allow a programmer to modify array sizes easily.

The first argument in each of the strcpy statements is an address indicated by the two-dimensional array name with just one set of brackets.

A different string is put into each row of the aa[ ][ ] and bb[ ][ ] arrays.

Because addresses must be used for the function arguments, array names with just one set of brackets are used.

Looping over the number of rows in an array is used to print all of the strings in a two-dimensional array.

```
        reserved = sizeof(bb[0]);
        occupied = strlen(bb[0]);
        printf("Reserved bytes for bb[0]=%d \nOccupied bytes for bb[0]=%d\n",
                          reserved, occupied);
}
```

sizeof operating on the beginning
of the row of a two-dimensional
array returns the number of
columns declared for the array.

strlen returns the number of characters
actually stored in the row.

Screen output

```
********* Section 1 - Initializing **********
********* Section 2 - Printing ***********
The aa[ ][ ] array has 2 strings.
The bb array
has
3 strings.
Reserved bytes for bb[0]=15
Occupied bytes for bb[0]=13
```

Output file L7_3.OUT

The bb array has 3 strings.

Explanation

1. Why do we often use multidimensional character arrays for storing strings?
One reason for using two-dimensional arrays, for instance, is that we are accustomed
to looking at 2-D representations of text, and our image of a 2-D array being rows
and columns fits what we commonly see. For instance, if this page of text had 50
lines down and 80 characters across, we could store the entire page in an array
aa[50][80].

We can extend this to three dimensions by thinking of a book. For instance, a
400-page book could be represented by a 3-D array aa[400][50][80]. When text is
stored in this manner it is easy for us to visualize what we are working with.

We must realize, though, that sometimes our row-column image of arrays is not
sufficient for us to work with them. In such cases, we must use a linear model of
the storage of multidimensional arrays.

2. How do we declare a 2-D character array? We declare 2-D character
arrays with the keyword `char` followed by an identifier with two sets of brackets.
The form is

        char   *array_name* **[**number_of_rows**] [**number_of_columns**];**

where *array_name* can be any valid identifier and *number_of_rows* and *number_
of_columns* must be positive integer constants. For example,

                 char aa[2][90];

declares aa[ ][ ] to be a character array with 180 character-size memory cells. These cells may be thought of as being 2 rows and 90 columns.

3. How do we initialize a 2-D character array? We describe this using examples from this lesson's program. For instance,

```
#define NUM_ROWS 3
#define NUM_COLS 15
char bb[NUM_ROWS][NUM_COLS];
strcpy(bb[0],"The bb array ");
strcpy(bb[1],"has ");
strcpy(bb[2],"3 strings.");
```

declares the array bb[ ][ ] to be 3 rows with 15 columns each. The rows are initialized to store the following:

T	h	e		b	b		a	r	r	a	y		\0	
h	a	s		\0										
3		s	t	r	i	n	g	s	.	\0				

Each double-quote enclosed string is stored in a row. Note that the length of each string in the list must be smaller than the declared size of the second dimension. In this case, the size of the second dimension, 15, is greater than the size of the longest string, "The bb array \0", which contains 14 characters (including the terminating null character).

When you are initializing your strings in this manner, it is important that you not forget about the terminating null character that must be put at the end of each string. The size of your second dimension must include this character. For instance, the minimum acceptable size for the array bb[ ][ ] above is bb[3][14], because 14 characters are in the first string.

4. How can we use printf to print a 2-D character array? We can use printf with a %s conversion specification. We illustrated that %s in a printf statement causes the printing to begin at the address specified and to continue until the null character is encountered. By specifying an address of the beginning of a row of a two-dimensional character array, the entire row is printed. For example, with the 2-D array aa[ ][ ], the loop

```
for (i=0; i<2; i++) printf("%s",aa[i]);
```

causes each of the two rows to be printed. Recall that this works because specifying the array aa[ ][ ] with only a single set of brackets indicates the address of the beginning of a row. For instance, aa[0] represents the address of the beginning of the first row and aa[1] represents the address of the beginning of the second row. Since each of the rows ends with \0, by looping through all the rows, all the strings are printed.

5. How do we use the sizeof operator with 2-D character arrays? Depending on what we use as the operand, we can determine the declared size of the entire array, a single row or a single element. For instance, for the 2-D array bb[ ][ ] in this lesson's program, the statement

<code>reserved = sizeof(bb[0]);</code>

evaluates the declared size (or number of columns) of the first row of bb[ ][ ], because the operand bb[0] has a single set of brackets, meaning that it indicates the address of the beginning of the first row. From the output we can see that this is 15, which is the declared number of columns.

If we use bb with no brackets as the operand, we can evaluate the size of the entire array. For instance, if we had used

<code>reserved = sizeof(bb);</code>

the value of `reserved` would be 45, which is the total number of bytes reserved for the entire bb[ ][ ] array. If we wanted to calculate the number of rows reserved for bb[ ][ ], we could have written

<code>reserved = sizeof(bb)/sizeof(bb[0]);</code>

This statement evaluates to be 45/15, which is 3, the reserved number of rows.

6. How do we use puts and fputs to print a two-dimensional character array? Both puts and fputs take an address as their argument and print until they come upon the null character. So, as an argument, we can use an array name with one set of brackets to print a single row of the array. For instance,

```
puts(bb[i]);
fputs(bb[i],outfile)
```

cause the *i*th row of array bb[ ][ ] to be printed. The function puts prints the output to the screen while fputs prints to the file pointed to by outfile. To print all the rows of the b[ ][ ] array, we can put these into loops, such as

```
for (i=0; i<NUM_ROWS; i++) puts(bb[i]);
for (i=0; i<NUM_ROWS; i++) fputs(bb[i],outfile);
```

7. In these loops, do puts and fputs print the same thing? No, the screen output produced by puts is

```
The bb array
has
3 strings.
```

while fputs has printed in the output file:

```
The bb array has 3 strings.
```

From this, we see clearly that puts has printed a new line character at the end of each string while fputs has not. You should be aware of these differences when you use these functions.

8. How can we copy the contents of one 2-D character array into the memory cells reserved for another 2-D character array? We have not shown it in this lesson's program, but we can use the strcpy function. For instance, if we had two arrays declared as

char cc[4][60], dd[4][40];

we can use a loop such as

for (i=0; i<4; i++) strcpy(cc[i], dd[i]);

to do it. This loop copies the dd[ ][ ] array contents, one row at a time, into cc[ ][ ]. To successfully copy the string, it is imperative for enough memory to have been reserved for the cc[ ][ ] array to accommodate each string stored in dd[ ][ ].

9. How can we use the strlen function with a 2-D character array? To do this, we need to pass to strlen the address of the beginning of each string, which is the address of the beginning of each row. Therefore, the statement

occupied = strlen(bb[0]);

in this lesson's program passes to strlen the address of the beginning of the first row of bb[ ][ ] and strlen returns the number of characters in that row (excluding the terminating null character). The integer return value from strlen is assigned to the integer variable occupied.

10. How do we represent 2-D character arrays in our table of variables? For illustration purposes, we show 2-D arrays in a row-column fashion with the 3-D illustration shown in Fig. 7.3. Each memory cell of a row is projected backward. In this figure, we have not shown all of the memory cells; however, the concept of many memory cells for each row is illustrated. Each cell is filled with a character. The null character is at the end of each string in a row. The values shown are the values of the first character in each row.

EXERCISES

1. Find the errors, if any, in the following statements:
   a. **char aa[2][10]**
      strcpy(aa[0],"aaa);
      strcpy(aa[1],"bbb");
      strcpy(aa[2],"ccc");
   b. **char bb[2][3]**
      strcpy(bb[0],"aaa");
      strcpy(bb[1],"bbb");
   c. **char cc[ ][25]**
      strcpy(cc[0],"Good");
      strcpy(cc[1] "morning");

FIG. 7.3
Variable table for the 2-D arrays used in this lesson's program. Only the first character for each row is shown in the Value column.

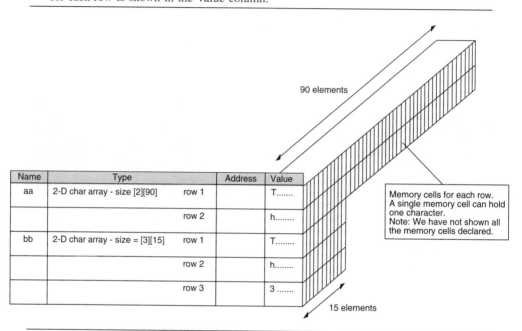

Name	Type		Address	Value
aa	2-D char array - size [2][90]	row 1		T.......
		row 2		h........
bb	2-D char array - size = [3][15]	row 1		T.......
		row 2		h........
		row 3		3 ....

Memory cells for each row. A single memory cell can hold one character. Note: We have not shown all the memory cells declared.

2. Find errors in the following program:

```
void main void()
{
        char a[][12] ={'aaa', 'bbb', 'ccc'};
        char b[2][2];
        strcpy(a[0],a['aaa']);
        strcpy(a[1],'bbb');
        strcpy(b[0],a[0]);
        a[0][0]=strlen(a);
        strcpy (b[0][1], a[0][1]);
        b[1][0]=a[1][0];
}
```

3. Write a program that contains a 2-D array named student[5][100]. The array should be used to store this information:

Name	Age	Math grade
John Kelly	21	3.3
Brian Jason	23	1.8
Mary Fox	19	4.0

The first column should have the student names, the second column should have their ages, and the last column should contain their grades. Display the table on the screen.

Solutions

1. a. ```
        char aa[2][10];
        strcpy(aa[0],"aaa");
        strcpy(aa[1],"bbb");
        ```
        *Maximum of two strings in aa[ ].*
    b.  ```
        char bb[2][4];
        strcpy(bb[0],"aaa");
        strcpy(bb[1],"bbb");
        ```
    c. ```
        char cc[2][25];
        strcpy(cc[0],"Good");
        strcpy(cc[1], "morning");
        ```

## LESSON 7.4 READING STRINGS FROM THE KEYBOARD AND FILES

**Topics**
• Reading strings from the keyboard
• Reading strings from files

To this point we have covered how to print strings but not how to read them from the keyboard or a file. In this lesson we describe reading both 1-D and 2-D character arrays.

*What you should observe about this lesson's program.*
1. Five character arrays have been declared a[ ], b[ ][ ], c[ ][ ], d[ ], and e[ ].
2. In section 1 for a[ ], a string has been read from the keyboard using the gets( ) function. The argument for the function is the address of the first element of a[ ].
3. In section 2 for b[ ][ ], a for loop is used to read strings from the keyboard. Again, the argument for gets( ) is an address (the address of the beginning of a row of the b[ ][ ] array).
4. In section 3 for c[ ][ ], the fgets( ) function is used for reading the strings for c[ ][ ] from a file.
5. In section 3, the puts and fputs functions have been used to print the strings in c[ ][ ].
6. The screen output for section 3 shows that a portion of it is double-spaced, whereas the file output for the same strings is not double-spaced.
7. In section 4 for d[ ] and e[ ], a string for d[ ] has been read using scanf whereas the string for e[ ] has been read using gets( ).
8. In section 4, the output shows only a portion of the string read using scanf is printed out.

*Questions you should attempt to answer before reading the explanation.*
1. What function has caused some of the output in section 3 to be double-spaced?
2. Why has scanf read only a portion of the string entered in section 4?
        ```

Source code

Insert cutfig. 7-8 here.

```
#include <stdio.h>
void main(void)
{
        char a[40],b[3][60],c[4][100],d[50],e[30];
        int i;
        FILE *infile;
        FILE *outfile;
        infile=fopen("L7_4.TXT","r");
        outfile=fopen("L7_4.OUT","w");

        printf("*********** Section 1 for Array a[ ] ************\n");
        printf("Enter a line of text for a[ ]\n");
        gets(a);
        puts(a);


        printf("\n\n*********** Section 2 for Array b[ ][ ] ************\n");
        printf("Type 3 lines for b[ ] and press return at the end of each one\n");
        for (i=0; i<3; i++)
                {
                gets(b[i]);
                puts(b[i]);
                }

        printf("\n\n*********** Section 3 for Array c[ ][ ] ************\n");
        printf("We will read c[ ] from a file\n");
        for (i=0; i<4; i++)
                {
                fgets(c[i],100 ,infile);
                puts(c[i]);
                fputs(c[i] ,outfile);
                }

        printf("*********** Section 4 for Arrays d[ ] and e[ ] ************\n");
        printf("Enter a line of text for d[ ] and press return\n");
        scanf ("%s",d);
        puts(d);
        gets(e);
        printf("This is the rest of the text entered\n");
        puts(e);
}
```

The arrays a[ ], d[ ], and e[ ] are 1-D arrays.
The arrays b[ ][ ] and c[ ][ ] are 2-D arrays.

The function gets reads the characters typed in at the keyboard until the Enter key is pressed. It stores the characters at the address indicated by its argument; in this case, the beginning of the a[ ] array.

A loop can be used to read many strings from the keyboard using gets and a 2-D array.

Address of the location in which the string is to be stored.

The function fgets reads a string from a file.

Pointer to file where string is stored.

Maximum number of characters to be read.

The function scanf with %s reads only until white space is encountered, so it does not read the entire string entered.

The function gets reads until the newline character is encountered. It reads the rest of the string entered.

Output

```
*********** Section 1 for Array a[ ] *******
Enter a line of text for a[ ]
This is a short string.
This is a short string.

*********** Section 2 for Array b[ ][ ] ****
Type 3 lines for b[ ] and press return at the end
of each one.
This is the first string.
This is the first string.
This is the second string.
This is the second string.
This is the third string.
This is the third string.

*********** Section 3 for Array c[ ][ ] ****
We will read c[ ] from a file
We read

four lines of

text from our

input file.

*********** Section 4 for Arrays d[ ] and e[ ] ******
Enter a line of text for d[ ] and press return
At first, not all of this string is printed.
At
This is the rest of the text entered
first, not all of this string is printed.
```

Keyboard input — (Section 1)
Keyboard input — (Section 2, line 1)
Keyboard input — (Section 2, line 2)
Keyboard input — (Section 2, line 3)
Keyboard input — (Section 4)

Input file L7_4.TXT

```
We read
four lines of
text from our
input file.
```

Output file L7_4.OUT

```
We read
four lines of
text from our
input file.
```

Explanation

1. How do we use the function gets to read a string typed in at the keyboard?
The form for the gets function is

$$\textbf{gets}(address);$$

where *address* is the address of the first memory cell into which the string is to be copied. In many cases, *address* will be the first element of an array or the first element of a row of an array. For instance, from this lesson's code, the declaration and gets function call

```
char a[40];
gets(a);
```

cause gets to take the string input from the keyboard and copy it into the memory location reserved for the array a[ ] beginning with a[ ]'s first memory cell. The function gets reads the string typed into the keyboard until the newline character (created by typing "Enter") is read and, in this case, has read all of the characters typed at the keyboard. The newline character, though, is not copied into memory by gets. Instead, gets discards the newline character and inserts a null character into its location. The null character terminates the string in memory. Thus, the following line

```
puts(a);
```

prints a[ ] to the screen. The function puts automatically appends a newline character to the string it reads. Thus, the advancement to a new line in the printing is caused by puts and not a newline character at the end of the a[ ] string stored in memory.

2. How can we use gets to at least partially fill each row of a 2-D array with input from the keyboard? We can use gets to read one line after another by putting the function in a loop. As we illustrated, gets reads characters until it encounters the newline character. For each time the Enter key is pressed on the keyboard, we need to execute a call to gets to read the characters typed. To partially fill all the rows of a two-dimensional array, we need a call to the function gets within a loop that loops over all of the rows. The argument for gets should be the address of the first element of a row of an array. In this lesson's program, the declaration and loop

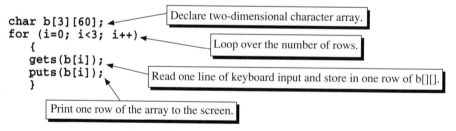

cause gets to read three lines of input, one for each time through the loop and store each line read into one row of the array b[ ][ ]. Note that the argument for gets is the address of the first element of a row of the b[ ][ ] array. Again, gets discards the newline character and inserts a null character immediately after the last ordinary character in each row.

From the output, it can be seen that, even though gets has discarded the newline character, the function puts has printed back exactly what was typed in. This is because, while gets *discards* the newline character, puts *adds* a newline character, making the output an echo of the input.

3. How do we read input from a file into a 2-D character array? C provides the functions fgets for reading from files. The function fgets has the following form:

fgets (*address, number_of_char, file_pointer*)**;**

where *address* is the address of the first element into which the characters are to be stored, *number_of_char* is one greater than the maximum number of characters to be read, and *file_pointer* represents the file from which input is to be read. For instance, the statement in this lesson's program

fgets(c[i],100,infile);

causes the first 99 characters (or until the new line character is read) to be read from the file pointed to by infile into the *i*th row of the array c[ ][ ]. If a newline character is read, fgets does not discard it. A null character is placed after the last character read, whether it is a newline or not. Therefore, if 99 characters are read, fgets puts a null character into the 100th memory cell.

The following declaration and loop cause each row of the c[ ][ ] array to be at least partially filled with character input:

```
char c[4][100];
for (i=0; i<4; i++)
        {
        fgets(c[i],100,infile);
        }
```

4. The loop for c[ ][ ] in this lesson's program printed the input exactly to the output file but added a blank line after each printed line to the screen. Why? This occurred because the function puts adds a newline character each time it prints a line to the screen, while fputs does not add a newline character for each line it prints to a file. The declaration and loop with comments is

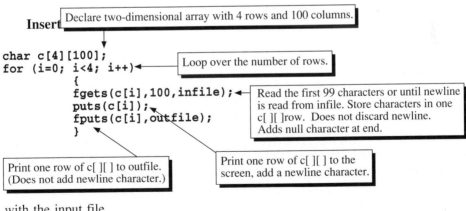

with the input file

```
We read\n
four lines of\n
text from our\n
input file.\n
```

On reading this input file, fgets adds a null character but does not discard (as gets does) any newline characters. Therefore, fgets put the following in the c[ ][ ] array:

W	e		r	e	a	d	\n	\0						
f	o	u	r		l	i	n	e	s		o	f	\n	\0
t	e	x	t		f	r	o	m		o	u	r	\n	\0
i	n	p	u	t		f	i	l	e	.	\n	\0		

The function puts adds a newline character at the end of each line and does not print the null character. So, puts printed the following:

W	e		r	e	a	d	\n	\n						
f	o	u	r		l	i	n	e	s		o	f	\n	\n
t	e	x	t		f	r	o	m		o	u	r	\n	\n
i	n	p	u	t		f	i	l	e	.	\n	\n		

Because there are two newline characters at the end of each line, we double-spaced our output. The lesson here is that C's string functions work differently with the newline and null characters. Be aware of this when you read and print strings.

5. Can we use scanf to read a single line of input from the keyboard? It can be done but it is not straightforward. The reason for this is that the %s specification causes scanf to read until white space (not a null or newline character) is encountered. So, scanf with one %s specification reads just one word, *not* an entire line. For example, the declaration and statement

```
char d[50];
scanf("%s",d);
```

with the keyboard input

At first, not all of this string is printed.

causes scanf to read only the word At and store it at the beginning of the d[ ] array. The rest of the sentence remains unread in the input buffer with the position indicator between the letter t in At and the following space:

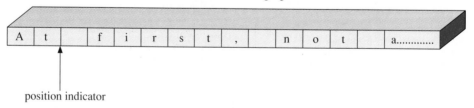

position indicator

Because scanf reads only one word, in most cases you will find it more convenient to use gets to read a single line of input from the keyboard.

Although not shown in this lesson's program, the function fscanf with the %s conversion specification also only reads until white space is encountered and, therefore, only one word.

6. In this lesson's program, how has the rest of the contents of the input buffer been read? The gets function has read the rest of the contents (from the position indicator to the newline character) of the input buffer and stored it in the array e[ ] with the statement:

```
gets(e);
```

7. What should we remember about the %s conversion specification? Remember that, with printf, %s indicates to print until a *null* character is encountered, whereas with scanf, %s means to read until a *white-space* character is encountered. Because strings end with a null character, it is very convenient to use printf with %s to print strings. However, it is not particularly convenient to use scanf with %s to read strings. These comments also hold for the companion functions of fprintf and fscanf.

Make sure that you do not confuse the way that printf and scanf treat %s. With %s, printf prints an *entire string,* whereas scanf reads only *one word!*

8. We have now covered input and output functions for characters and strings. Can we summarize them? Figure 7.4 illustrates character input and output functions.

FIG. 7.4
Character and string input and output functions

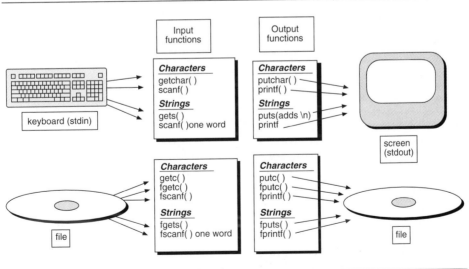

9. We saw previously that scanf and fscanf return EOF if an error occurs in the reading process and printf and fprintf return a negative integer if an error occurs in the printing process. Do the character and string I/O functions return special values when errors occur? Yes, Table 7.2 summarizes what is returned from these functions for both an error occurring or not occurring.

10. In Table 7.2, what is a null pointer and how can we use it? A pointer, of course, is an address. However, we can convert an address to int using the cast operator, (int). A null pointer is an address that evaluates to 0 when converted to int. A null pointer is guaranteed to compare unequally with any pointer that is not a null pointer.

One way to use these characteristics of a null pointer to evaluate whether or not an input error has occurred follows. For instance, we could have declared an int variable, j, in this lesson's program and written the lines:

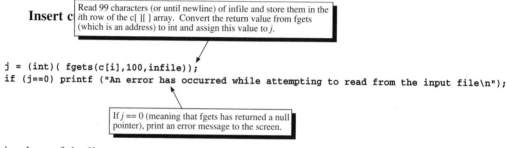

Insert c | Read 99 characters (or until newline) of infile and store them in the ith row of the c[ ][ ] array. Convert the return value from fgets (which is an address) to int and assign this value to j.

```
j = (int)( fgets(c[i],100,infile));
if (j==0) printf ("An error has occurred while attempting to read from the input file\n");
```

If $j == 0$ (meaning that fgets has returned a null pointer), print an error message to the screen.

in place of the line

```
                    fgets(c[i],100,infile);
```

Another way to use the fact that fgets and gets return a null pointer on error is to use the constant macro NULL, defined in stdlib.h, which is equivalent to a null pointer. We will illustrate the use of NULL in Application Program 7.4.

Procedures of this sort, that check for input errors, are an improved method of reading input. We did not use error checking methods in this lesson's program to

TABLE 7.2
Error alerts

Function	Returned if no error occurs	Returned on occurrence of error
getchar	Next character in input stream	EOF
gets	Address of first element of string	Null pointer
getc	Next character in input stream	EOF
fgetc	Next character in input stream	EOF
fgets	Address of first element of string	Null pointer
putchar	Character written	EOF
puts	Nonnegative integer	EOF
putc	Character written	EOF
fputc	Character written	EOF
fputs	Nonnegative integer	EOF

reduce the complexity of the program. However, in the application programs we use error checking methods. In this text, we have not used as many error checking methods as we could have, not because we do not endorse error checking (in fact we encourage it), but because we are focusing on other programming features.

EXERCISES

1. Find errors, if any, in the following statements:
   a. ```
      char a[10];
      gets( );
      ```
   b. ```
      char a[1],b[2];
      gets(a, b);
      ```
   c. ```
      puts("AAA BBB");
      ```
   d. ```
      char a[3];
      scanf("%f", a);
      ```
   e. ```
      char a[30];
      gets(a);
      ```

2. Find errors in this program:

```
void main( void)
{ char name[2][30], number[2][10];
printf("Please type your first name, a blank, and last name);
gets(name[0]);
scanf("    %s    ",name[1]);
printf("name=%s, %s\n", name[0], name[1]);

printf("Please type a number, press the return key, and
another number);
scanf("    %s    ", number[0]);
gets(    number[1]);
printf("number =%s, %s\n",number[0], number[1]);
}
```

3. Use various types of character input/output functions you have learned to read your name and mailing address from the keyboard and then
   a. Display them on the screen and to a file.
   b. Read the file just obtained, modify it, and make a hard copy of your business card.

4. Given the source code of Lesson 7.4, file L7_4.C, write a program to do the following:
   a. Use loop statements and the fscanf() function to read the file, one character at a time, with the %c format, and then use the printf() and fprintf() functions to display the file on the screen and save it in file L7_4.CA. Check whether file L7_4.C matches file L7_4.CA.
   b. Use loop statements and the fgetc() function to read the file, one character at a time, and then use the putc() and fputc() functions to display the file on the screen and save it in file L7_4.CB. Check whether file L7_4.C matches file L7_4.CB.
```

5. The program that follows shows some of the character I/O functions defined in ANSI C. Please copy and run the program to understand how to use them properly.

```c
#include <stdio.h>
#include <ctype.h>

void main(void)
 {char aa,bb,cc;
 FILE *inptr, *outptr;
aa='A';
bb='B';
cc='C';

 printf("\n\nWrite output file--------------------\n\n");
 outptr= fopen("7_4.OUT","w");

 fputc(aa,outptr);
 putc(bb,outptr);
 fprintf(outptr,"\nThis is output file 7_4.OUT, aa=%c,  bb=%c\n",aa,bb);
 fclose(outptr);

 inptr= fopen("7_4.OUT","r");

 aa=getc(inptr);
 printf("\n   d. Use getc() to read character aa=%c from a file\n",aa);

 bb=fgetc(inptr);
 printf("\n   d. Use fgetc() to read character bb=%c from a file\n",bb);

 printf("\n   f. Use fscanf() to read the rest of file 7_4.OUT\n");
 while( (fscanf(inptr,"%c",&cc) !=EOF) ) printf("%c",cc);
 fclose(inptr);
 }
```

Solutions
1. a. **gets(a);**
   b. **char a[2],b[2];**
       gets(a);
       gets(b);
   c. No error.
   d. **char a[3];**
       scanf("%s", a);
   e. No error.

▨ LESSON 7.5 POINTER VARIABLES VS. ARRAY VARIABLES

Topics

• Comparing and contrasting arrays and pointers
• Initializing and printing arrays and pointers

We have seen that the address of the first element of a string frequently is used when working with strings. If we have a large number of strings, it would be convenient to simply store the addresses of the first elements in an array.

For instance, suppose we had 10,000 sentences, each an individual string. If we stored the address of the first element of each in an array, a[10000], then it would be simple to print out each sentence by looping from $i = 0$ to $i = 9999$ and doing puts(a[i]) in the loop. While this is not exactly how it commonly is done, we often find it convenient to create an array of pointers to the beginning of strings.

Recall that, to declare a pointer variable, we use * in the declaration. For instance,

```
char *aa;
```

reserves space in memory for a single pointer variable named *aa*. Because the keyword *char* is used, aa is designated to hold the address of a character. A declaration such as

```
char *cc[5]
```

reserves five cells in memory. Each cell is designated to hold the address of a character. This form may look confusing at first. Simply memorize it and accept that C interprets this declaration to be an array of pointers.

Previously, we used 1-D and 2-D character arrays to work with strings. In this lesson, we illustrate how to use pointers and arrays of pointers to manipulate strings and characters in strings. We will see both similarities and differences between the forms used for 1-D and 2-D character arrays and pointers and arrays of pointers. There are three sections to this program: initializing strings, printing strings using puts, and printing strings using putchar.

What you should observe about this lesson's program.
1. The first four declarations are for a pointer variable (aa), a one-dimensional array (bb[ ]), an array of pointers (cc[ ]), and a two-dimensional array (dd[ ][ ]), respectively.
2. The fifth and sixth declarations (ee[ ] and ff[ ]) are 1-D character arrays in which we store strings with the first two strcpy statements. We manipulate these strings in this program.
3. The left side of the first assignment statement has the pointer variable aa and the right side has ee.
4. The first call to strcpy in section 1 uses bb as the first argument and ee as the second argument.
5. The second assignment statement has cc[0] (an element of the array of pointers) on the left side and ff on the right side.
6. The second call to strcpy in section 1 uses dd[0] (indicating the address of the first row of the two-dimensional dd[ ][ ] array) as the first argument and ff as the second argument.
7. In section 1, the pointer variables and pointer arrays are initialized with assignment statements and the 1-D and 2-D character arrays are initialized with strcpy.
8. In section 2, aa and bb are used with the puts function in a similar manner even though aa is a pointer variable and bb[ ] is an array.
9. In section 2, cc[ ] and dd[ ] are used with the puts function in a similar manner even though cc[ ] is an array of pointers and dd[ ][ ] is a 2-D array.

10. In section 3, aa[ ] and bb[ ] are used with the putchar function in a similar manner even though aa is a pointer variable and bb is an array. Note that array notation is used with the aa pointer variable.

11. In section 3, cc[ ][ ] and dd[ ][ ] are used with the puts function in a similar manner even though cc[ ] is an array of pointers and dd[ ][ ] is a 2-D array. Note that two-dimensional array notation is used with the cc array of pointers.

Questions you should attempt to answer before reading the explanation.

1. Why can we not use bb on the left side of an assignment statement?
2. Why should we not use aa as the first argument in a call to strcpy( )?
3. Why can we not use dd[0] on the left side of an assignment statement?
4. Why should we not use cc[0] as the first argument in a call to strcpy?

Source code

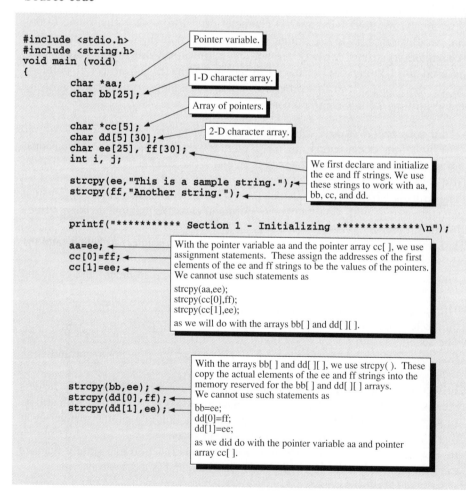

```
printf("*********** Section 2 - Printing using puts **************\n");

puts(aa);  ◄────── However, when we print using aa, we do it similarly to printing bb
puts(bb);  ◄────── even though aa is a pointer variable and bb[ ] is an array.
puts(cc[0]);
puts(dd[0]); ◄──── Also, when we print using cc, we do it similar to printing dd even
                   though cc[ ] is a pointer array and dd[ ][ ] is a 2-D array.


printf("*********** Section 3 - Printing using putchar **************\n");
```

We can access each element of the ee string using aa with 1-D array notation even though aa is a pointer variable. Note that aa was previously assigned the address of ee.

```
for(i=0; i<10; i++) putchar(aa[i]) ;
putchar('\n');
for(i=0; i<10; i++) putchar(bb[i]);
putchar('\n')   ;
```

We use bb[ ] and aa with putchar in a similar manner even though aa is a pointer variable and bb[ ] is a 1-D array.

Nested loops for using putchar for printing 2-D arrays.

We use dd[ ][ ] and cc[ ] with putchar in a similar manner even though aa is an array of pointers and dd[ ][ ] is a 2-D array.

```
for(i=0; i<2; i++)
        {
        for(j=0; j<10; j++) putchar(cc[i][j]);
        putchar('\n');
        }

for(i=0; i<2; i++)
        {
        for(j=0; j<10; j++) putchar(dd[i][j]) ;
        putchar('\n');
        }
}
```

We can access each element of the ee and ff strings using cc with 2-D array notation even though cc is an array of pointers. Note that cc[0] previously was assigned the address of ff and cc[1] was previously assigned the address of ee.

Output

```
*********** Section 1 - Initializing **************
*********** Section 2 - Printing using puts **************
This is a sample string.
This is a sample string.
Another string.
Another string.
*********** Section 3 - Printing using putchar **************
This is a
This is a
Another st
This is a
Another st
This is a
```

Explanation

1. Explain the difference in the declarations

```
char *aa;
char bb[25];
```

The first declaration reserves a *single* memory location for a pointer variable, aa. The second declaration reserves *25* locations for character values. However, all of these memory cells remain empty unless, in the program body, we fill them.

2. Explain the difference in the declarations

```
char *cc[5];
char dd[5][30];
```

The first declaration reserves five memory locations for *single* pointer variables. The second declaration reserves five memory regions with each containing cells for 30 characters. All of these memory cells remain empty unless, in the program body, we fill them.

3. Why are the statements aa = ee and cc[0] = ff, and cc[1] = ee, appropriate? These statements, of course, are assignment statements that say to take what is indicated by the expression on the right side of the assignment statement and put it in the memory location indicated by the left side. Because aa is a pointer variable and cc[0] and cc[1] are elements of an array of pointer variables, we can store addresses in any of these locations. Since ee and ff indicate addresses, the assignment statements can be executed easily. Thus we are storing addresses in memory cells that are prepared to take addresses.

4. Why are the statements bb = ee, dd[0] = ff, and dd[1] = ee not appropriate? Because bb is not a pointer variable, we cannot store an address in any memory location indicated by bb. This statement clearly makes no sense, because bb without brackets indicates the address of the first element of the array bb[ ]. This address is set during compilation, we cannot change it or try to change it in the program body. Also, because dd[0] and dd[1] are not elements of a pointer array, we cannot store addresses in these cells.

5. Why are the statements strcpy(bb,ee), strcpy(dd[0],ff), and strcpy(dd[1],ee) appropriate? These statements cause the actual characters of the strings indicated by ee and ff to be copied into the memory cells beginning with the addresses indicated by bb, dd[0], and dd[1]. This is appropriate because bb[ ] has 25 memory cells reserved for it and dd[0] and dd[1] have 30 memory cells reserved for each of them. This means that all of the characters of ee[ ] and ff[ ] can be stored in bb[ ] and dd[ ][ ].

6. Why are the statements strcpy(aa,ee), strcpy(cc[0],ff), and strcpy(cc[1],ee) not appropriate? We consider strcpy(aa,ee) first. Because we have not previously assigned an address to aa that is guaranteed to have at least 25 memory cells available behind it, we cannot use this statement. Had we previously assigned to aa an address with reserved memory, we could have used the statement strcpy(aa,ee).

Be aware that, if aa were previously assigned an address with fewer reserved memory cells than are required (25 in this case), other memory cells would be overwritten with the statement strcpy(aa,ee). This type of error could cause major problems in your programs. Note that the C compiler will not necessarily indicate an error if you use this. Later in this chapter, we illustrate how we can reserve memory space behind aa during execution and be able to safely utilize this sort of statement.

A similar situation exists for strcpy(cc[0],ff) and strcpy(cc[1],ee). Because cc[ ] is simply an array of five pointers, we have room to store only five addresses. This declaration reserves no space to store actual character elements. To safely use strcpy statements with pointer variables and pointer array elements as first arguments, we must make sure that the first arguments represent addresses with reserved memory.

7. What is our image of memory after executing section 1 of this program?
After executing section 1 of this lesson's program we have the image in Fig. 7.5. Note that aa contains the *address* of ee[ ] while bb contains a *duplicate* of ee[ ]. Also, cc[0] contains the *address* of ff[ ] while dd[0] contains a *duplicate* of ff[ ]. Further, cc[1] contains the *address* of ee and dd[1] contains a *duplicate* of ee. This illustrates a primary difference between pointer variables and arrays.

FIG. 7.5
Memory after section 1 executes

Name	Type	Address	Value
aa	pointer to char	FFF4	FF22
bb	array of char - size=[25]	FFD2	T
cc	array of pointers - size=[5]	FFEC	FF04
dd	array of char - size=[5][30]	FF3C	A
		FF5A	T
		FF78	
		FF96	
		FFB4	
ee	array of char- size=[25]	FF22	T
ff	array of char- size=[30]	FF04	A

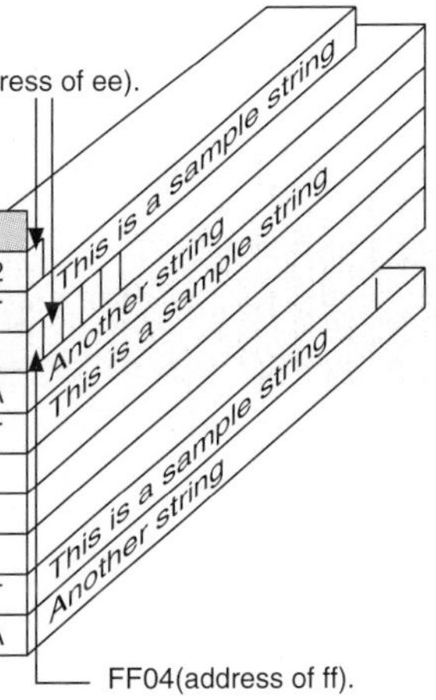

FF22(address of ee).

FF04(address of ff).

8. How can we use the aa pointer variable and puts( ) to print the string stored in ee? Because we have stored the address of the first character of ee[ ] in aa with the statement

```
aa=ee;
```

we can use puts (which requires an address as an argument) with aa as in the statement

```
puts(aa);
```

to print the string stored in ee.

9. How can we use the cc[ ] array of pointers and puts( ) to print the strings stored in ee and ff? Because we have stored the addresses of the first character of ff[ ] and ee[ ] in cc[0] and cc[1], respectively, with the statements

```
cc[0]=ff;
cc[1]=ee;
```

we can use puts with cc[0] and cc[1] as arguments

```
puts (cc[0]);
puts (cc[1]);
```

to print the ff and ee strings.

10. How can we use the aa pointer variable and putchar( ) to print the string stored in ee? Because putchar() requires an individual character element as an argument, we must access each individual element of the ee[ ] array using aa (which was assigned the address of ee[ ] with the statement aa = ee;). It may seem strange, but C allows the use of array notation with a pointer variable to access individual elements of an array. This means that aa[0] accesses ee[0], aa[1] accesses ee[1], and others can be accessed similarly. Therefore, the statement and loop

```
aa=ee;
for(i=0; i<10; i++) putchar(aa[i]) ;
```

print the first ten characters of the string stored in the ee[ ] array (which are "This is a").

11. How can we use the cc[ ] pointer array and putchar( ) to print the strings stored in ee and ff? Again, we must be able to access individual elements of the ee[ ] and ff[ ] arrays using the addresses stored in the pointer array cc[ ]. C allows array notation to do this. In other words, when we store the address of ff[ ] in cc[0] with the statement

```
cc[0]=ff;
```

cc[0][0] refers to the first character of the ff[ ] array, cc[0][1] refers to the second character in the ff[ ] array, and so forth through the entire array. Therefore, this loop

```
for(j=0; j<10; j++) putchar(cc[0][j]);
```

prints the first ten characters of the ff[ ] array. To print the first ten characters of both the ee[ ] and ff[ ] strings, we use the assignment statements and nested loop

```
cc[0]=ff;
cc[1]=ee;
for(i=0; i<2; i++)
          {
          for(j=0; j<10; j++) putchar(cc[i][j]);
          putchar('\n');
          }
```

12. Why does C allow array notation to be used with pointer variables? C allows array notation because, when a program is compiled, it converts array notation to pointer notation. You should recall that the unary * operator is used with pointer notation and brackets [ ] are used with array notation. With either of these notations, we can access an element at a particular address or a number of elements past a given address. For instance, in this lesson's program aa[5] accesses an address that is five elements past the address stored in the memory reserved for aa. We will not go into the details here but reserve it for a later lesson.

13. Can we use assignment statements with individual characters? Yes, although we have not shown it in this lesson's program, such assignment statements as

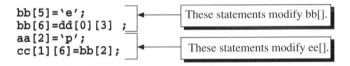

```
bb[5]='e';
bb[6]=dd[0][3] ;
aa[2]='p';
cc[1][6]=bb[2];
```

These statements modify bb[].

These statements modify ee[].

are perfectly acceptable. Only with strings do we need to use functions such as strcpy( ) for initialization or modification. If we had performed the previous assignments in this lesson's program after initialization, the strings would be

Array	String
bb[ ]	This et a sample string
ee[ ]	Thps ii a sample string

Remember, with a single pointer variable (aa for instance) C allows 1-D array type notation to access a single character. With a pointer array (cc for instance) C allows 2-D array type notation to access a single character.

14. What are the main points of this lesson? You should understand that
1. To initialize character arrays, we use strcpy. To initialize pointer variables and pointer arrays, we use assignment statements.
2. Even though we use *different* methods to *initialize* character arrays and pointers (as described in point 1), we use *similar* methods to *access* strings and individual characters. In other words, we can use array notation (brackets) with both pointers and arrays to access strings and individual characters.

EXERCISES

1. Based on these declarations and initializations:

```
char        aa[10], *bb, *cc[2], dd[2][10], ee[10], ff[10];

strcpy(aa,"Apple");
strcpy(ee,"Cat");
strcpy(ff,"Cow");
strcpy(dd[0],"Dog");
strcpy(dd[1],"Doll");
```

find the errors, if any, in the following statements:
a. `bb[0]=dd[0];`
b. `cc[2]=dd[0];`
c. `strcpy(aa,dd[1]);`
d. `aa[1]=dd[1][1];`
e. `strcpy(aa[1], cc[1]);`
f. `strcpy(dd[1][1],aa);`

2. Find errors in the following program:

```
main()
{
        char a[10], *b,  *c[2], d[2][10];

        strcpy(aa,"Apple");
        strcpy(d[1],"Dog");
        d[0][0]=strlen(b);
        d[0][1]='\0';
        b=D[0];
        c[0]=a;
        strcpy(c[1],b);
}
```

3. A file directory contains the following information:

File name	File size	Date
AAA.C	1234	08-12-94
XXX.TXT	5678	12-11-96
DDD.C	9876	01-12-97
BBB.TXT	4455	06-08-93

Write a program that uses a 2-D char array to read the contents of this directory, then sort the directory based on
a. File size
b. File date
c. File type and name (i.e., the files should be in the order of AAA.C, DDD.C, BBB.TXT, and XXX.TXT)

Display the output on the screen. You are not allowed to use strcpy() function.

Solutions
1. a. `bb=dd[0];`
   b. `cc[1]=dd[0];`
   c. No error.
   d. No error.
   e. `cc[1]=aa;`
   f. `strcpy(dd[1],aa);`

▧ LESSON 7.6 INITIALIZING WITHIN A DECLARATION

Topics
- Initializing strings in a declaration
- Differences between initializing in a declaration and initializing in the program body

In this program, we use four different types of declarations to store strings and initialize those strings within the declaration. We have deliberately postponed the discussion of doing this seemingly simple operation because its form leads to considerable confusion. Before we explain it, we wanted you to have an understanding of how to properly work with arrays, pointers, and arrays of pointers, which means that this is a later lesson than in most texts. With the background you have developed, we believe you will find the concepts in this lesson straightforward.

What you will learn with this lesson is that, within a declaration, characters enclosed in quotes sometimes do *not* represent addresses. This is the case when we initialize ordinary character arrays. When we initialize pointers and pointer arrays quote-enclosed characters *do* represent addresses. The form used to both declare and initialize ordinary character arrays uses =, which gives the appearance of being an assignment statement. This is confusing to students learning C programming because assignment statements are not allowed for initializing strings in ordinary arrays in program bodies.

In this lesson, we illustrate where in memory string constants are stored and contrast this with the locations used for arrays. We show that, to access a string constant, we must use its address because no name is associated with it. Read the program and look at the output to understand initializing strings in declarations.

What you should observe about this lesson's program.
1. The four different declarations are

   aa is a pointer variable
   bb is a one-dimensional character array
   cc is an array of pointers
   dd is a two-dimensional character array

2. Strings have been initialized in all of the declarations.
3. Single strings are initialized for aa and bb. The array size for bb[ ] is not explicitly given.

4. More than one string is initialized for cc and dd. Complete array sizes are not explicitly given.

5. The printf statement prints the addresses stored in the pointer variable aa and pointer array cc[ ], along with the addresses of the ordinary character arrays bb[ ] and dd[ ][ ].

6. The output shows that the addresses of the ordinary arrays begin with F and the addresses stored in the pointer variable and pointer array begin with 0. This means that different regions of memory are used for the two cases.

7. The puts( ) function is used to print all of the strings used in the program.

Questions you should attempt to answer before reading the explanation.

1. Even though aa and bb are treated similarly in the program body, is memory being used in the same manner for both of them?

2. Even though cc and dd are treated similarly in the program body, is memory being used in the same manner for both of them?

Source code

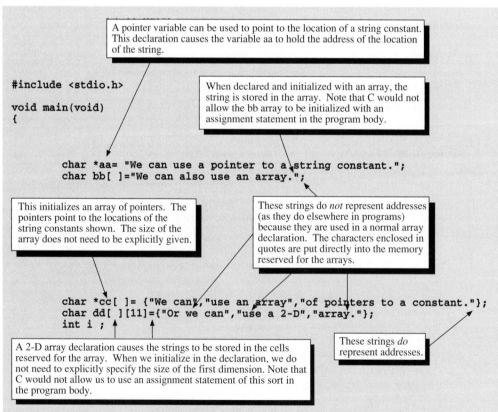

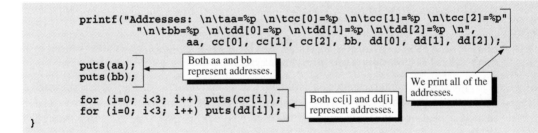

```
printf("Addresses: \n\taa=%p \n\tcc[0]=%p \n\tcc[1]=%p \n\tcc[2]=%p"
       "\n\tbb=%p \n\tdd[0]=%p \n\tdd[1]=%p \n\tdd[2]=%p \n",
          aa, cc[0], cc[1], cc[2], bb, dd[0], dd[1], dd[2]);

puts(aa);
puts(bb);
```
Both aa and bb represent addresses.

We print all of the addresses.

```
for (i=0; i<3; i++) puts(cc[i]);
for (i=0; i<3; i++) puts(dd[i]);
```
Both cc[i] and dd[i] represent addresses.

```
}
```

Output

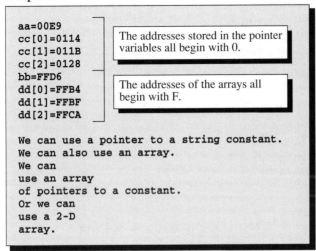

```
aa=00E9
cc[0]=0114
cc[1]=011B
cc[2]=0128
bb=FFD6
dd[0]=FFB4
dd[1]=FFBF
dd[2]=FFCA
```
The addresses stored in the pointer variables all begin with 0.

The addresses of the arrays all begin with F.

```
We can use a pointer to a string constant.
We can also use an array.
We can
use an array
of pointers to a constant.
Or we can
use a 2-D
array.
```

Explanation

1. How can we both reserve memory for a pointer variable and specify the string to which it points (that is, initialize it) in a declaration? We can declare and initialize a pointer variable to point to a string constant. For instance, the declaration

```
char *aa={"We can use a pointer to a string constant."};
```

declares the char pointer variable aa to hold the address of the first character of the string, "We can use a pointer to a string constant".

2. Where in memory is the string pointed to by aa located? We have printed out its address: 00E9. Note that, unlike other addresses we have looked at, this address does not begin with F. This means that this string constant is stored in a region of memory different from the variables.

3. In terms of how they cause strings to be stored, how are the following two declarations different?

```
char *aa={"We can use a pointer to a string constant."};
char bb[ ]={"We can also use an array."};
```

The first one, using the pointer variable aa, causes the *address* of the string constant, "We can use a pointer to a string constant.", to be stored in aa's memory cell. The second declaration causes the *string,* "We can also use an array." actually to be stored in the memory cells reserved for the bb[ ] array. The image of memory is shown in Fig. 7.6. We can see that, although both declarations cause strings to be stored in memory, the method of storage is very different. Note that no name is associated with the string constant "we can use a pointer to a string constant." The only way we can access this string is with its address.

4. Given these storage methods, how can we print the strings stored? We can use the puts function. Recall that the puts function uses the address of the first character of the string to be printed as its argument. In this program, the two statements

```
puts(aa);
puts(bb);
```

FIG. 7.6
Image of memory for aa and bb for this lesson's program

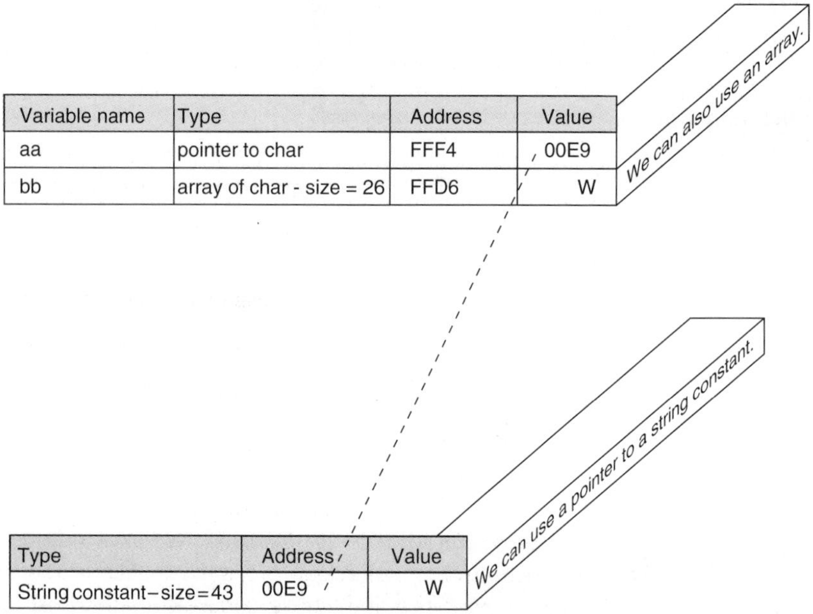

Variable name	Type	Address	Value
aa	pointer to char	FFF4	, 00E9
bb	array of char - size = 26	FFD6	W

Type	Address	Value
String constant – size = 43	00E9	W

print the strings to the screen. Note the similarities between the two statements. They work in both cases because both aa and bb represent addresses. Because aa is a pointer variable, the expression aa represents the *value* of the variable aa, which is an address, the address of the beginning of the string constant. Because bb is an array name without brackets, it also represents an address, the address of the beginning of its first memory cell. Therefore, the statements and the results from them look similar.

5. How can we access the individual characters of the string pointed to by aa? We can use array notation as we did in the previous lesson. For instance, aa[3] represents the fourth character of the string, 'c'.

6. How can we declare and initialize an array of pointer variables to point to many strings? We can declare and initialize an array of pointers with each element of the array pointing to a different string constant. For instance, the declaration

```
char *cc[ ]={"We can","use an array","of pointers to a constant."};
```

creates an array that contains three pointer values. Each value represents the address of the first element of one of the three strings given. The first pointer value is the address of the first character of "We can", the second pointer value is the address of the first character of "use an array", and the third pointer value is the address of the first character of "of pointers to a constant.". Again, even though these strings are not in the program body, they represent addresses.

Learn or memorize the notation given here. The form is

char *array_name**[ ]={*"string_1"*,   *"string_2"*,   *"string_3"*,
                as many strings as desired   **};**

You can aid your memorization by realizing that the single set of brackets indicates that we are declaring a one-dimensional array. The * indicates that we are storing addresses in the array.

In this lesson's program, we chose not to indicate the size of the array. C can determine the size automatically by counting the number of strings listed. Because we have listed three strings, C sizes the array to be *cc[3].

7. Where in memory are the strings pointed to by cc located? Again, these string constants are not located in memory with the variables, but with the constants.

8. In terms of how they cause strings to be stored, how are the following two declarations different?

```
char *cc[ ]={"We can","use an array","of pointers to a constant."};
char dd[ ][11]={"Or we can","use a 2-D","array."};
```

The first one, using the pointer array cc, causes the *addresses* of the first characters of the string constants, rather than the characters of the constants themselves, to be stored in its memory cells. Thus, the actual characters of the constants, "We can", "use an array", and "of pointers to a constant." are stored in memory where constants are stored. In contrast, the second declaration causes the actual *characters* of the strings, "Or we can", "use a 2-D", and "array." to be stored in the memory cells reserved for the array. The image of memory is shown in Fig. 7.7.

Observe from this that the addresses of the first elements of the string constants are stored in cc[]. The dashed lines indicate their connection. In other words, cc[0]=0114, cc[1]=011B, and cc[3]=0128.

Also observe that different amounts of memory are reserved for the two methods of declaration. The method for declaring dd[ ][ ] as a 2-D array causes a "rectangular" shaped memory region to be reserved, in this case 3 × 11. Because of this, wasted memory space lies at the end of the last row because the last string takes only seven memory cells. There is no way to use this type of declaration and avoid wasting space for the last line. However, the method used for declaring cc[ ] produces three separate 1-D arrays, allowing C to size each one correctly without wasting memory. Although the amount of memory wasted for dd[ ][ ] is of no practical importance, we point out this difference between the two methods because there may

FIG. 7.7

Image of memory for cc and add for this lesson's program

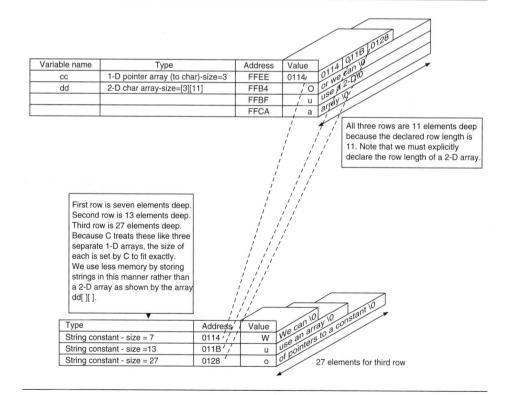

be times when the memory wasted by using a 2-D array is substantial. In terms of memory usage, the pointer array method of storage is superior.

9. Given these storage methods, how can we print the strings stored? We can use the puts function in a loop over the number of strings to be printed; in this case, three. Recall that the puts function uses the address of the first character of the string to be printed as its argument. Therefore, the arguments for puts of cc[0], cc[1], and cc[2] cause each of the three strings pointed to by cc[ ] to be printed. Also, recall that, for a 2-D array, the address of each individual row of the array is given by the array name with only one set of brackets. Thus, dd[0], dd[1], and dd[2] represent the addresses of each row of dd[ ][ ] and should be used as arguments to puts to print each row.

The following two loops cause these arguments to be used for puts and all three strings in both of these arrays to be printed:

```
for(i=0; i<3; i++) puts(cc[i]);
for(i=0; i<3; i++) puts(dd[i]);
```

Note how similar these two loops appear, even though the declarations for cc and dd are substantially different.

10. How can we access the individual characters of the strings pointed to by cc? We can use array notation as we did in the previous lesson. For instance, cc[1][4] represents the fifth character of the second string, 'a'. Notice that C allows 2-D array notation with cc[ ] even though cc[ ] is declared as a 1-D pointer array.

11. Could we have used the assignment statements

```
aa = "We can use a pointer to a string constant.";
bb = "We can also use an array";
```

in this lesson's program body? The answer is yes for the first of these but no for the second. The first is allowed because the strings enclosed in double quotes represent addresses, and we can assign an address to the pointer variable aa. The second is not allowed because bb was given an address during compilation (since it is an array). A new address cannot be given to bb during execution. C is confusing because it allows the declaration: `char bb[] = "We can also use an array";` but does not allow the assignment statement: `bb = "We can also use an array";`

EXERCISES

1. Find errors, if any, in the following statements:
   a. `char *paa="aa" "bb" "cc";`
   b. `char *pbb="abc[3]";`
   c. `char *pcc[3]={"a","b","c[3]"};`
   d. `char *pdd[2]={"aa" "bb" "cc"};`

2. Find errors in the following program:

```
#include <string.h>
void main(void)
 {char dd[3][8]={"Dog", "Donkey", "Dragon"}, *x[3]={"aa",'bb'};
 x[2]=dd[3];

 for (i=0;i<3;i++)
 printf("x[%d]=%s\n",i,x[i]);

 }
```

3. In a declaration, use a 1-D char array, `*name[4]`, to save the following names: peTer dodge, kEith hill, erIc randy, and lisa freDo. Write a program to change the names to
   a.   Peter Dodge, Keith Hill, Eric Randy, and Lisa Fredo
   b.   PETER DODGE, KEITH HILL, ERIC RANDY, and LISA FREDO

Solutions
1.   a.   No error, paa is initialized with the address of the string constant "aabbcc".
   b.   No error, pbb is initialized with the address of the string constant "abc[3]".
   c.   No error, pcc[2] is initialized with the address of the string constant "c[3]".
   d.   No error. However, only pdd[0] is initialized with string "aabbcc".

LESSON 7.7 PASSING STRINGS TO USER-DEFINED FUNCTIONS

Topics
• Calculating number of rows and number of elements in an array
• Passing arrays and pointers to functions

In this lesson, we illustrate how to write functions and function calls that accept and pass strings.

To this point we have used four different types of declarations for strings:

1.   One-dimensional character arrays
2.   Two-dimensional character arrays
3.   Pointers to strings
4.   Arrays of pointers to strings

While these are not the only options available to you, they represent a variety from which you can learn. In this lesson's source code, each of these is declared. All are passed to a function and printed out.

   What you should observe about this lesson's program.
1.   The following have been declared:

   aa[ ], a 1-D array
   bb[ ][ ], a 2-D array
   cc, a single pointer variable
   dd[ ], an array of pointers

2. The first four arguments as shown in the prototype for function1 are of the same type as listed in the previous observation.

3. The number of rows of bb[ ][ ] and the number of elements of dd[ ] have been calculated with the first two assignment statements. These also are arguments passed to function1.

4. Despite the different declarations, the first four arguments in the call to function1 are simply the array or variable names.

5. Within function1, the arrays and pointer variable are manipulated in a manner consistent with their declarations in the function header.

A question you should attempt to answer before reading the explanation. In the assignment statement for num_elems_dd, the denominator on the right side has an unusual operand, char *. Can you guess what this is supposed to represent and what the result of this operation should be?

Source code

```
#define LENGTH 20
#include <stdio.h>
#include <string.h>
void function1 (char ee[ ], char ff[ ][LENGTH], char *gg, char *hh[ ], int num_rows_ff, int
        num_elems_hh);
```

1-D array.

2-D array.

All of these are passed to a function.

```
void main(void)
{
        char aa[ ]= "One-dimensional array.";
        char bb[ ][LENGTH]={"Two-","dimensional ","array."};
        char *cc= "Pointer to string constant.";
        char *dd[ ]={"Array ","of pointers ","to string ","constants."};
        int num_rows_bb, num_elems_dd;
```

Pointer to string.

Array of pointers to strings.

```
        num_rows_bb = sizeof(bb)/LENGTH;
        num_elems_dd = sizeof(dd)/sizeof(char *);
```

The sizeof operator can be used to calculate the number of rows of bb[ ][ ] and the number of elements of dd[ ]. We pass these to a function to work with the arrays in the function.

All of these represent addresses even though they are declared very differently.

```
        function1(aa, bb, cc, dd, num_rows_bb, num_elems_dd);
}


void function1 (char ee[ ], char ff[  ][LENGTH], char *gg, char *hh[ ],  int num_rows_ff, int
        num_elems_hh)
{
        int i;
```

These declarations match the corresponding declarations in main.

```
        puts(ee);
        for (i=0; i<num_rows_ff ; i++) puts(ff[i]);
        puts(gg);
        for (i=0; i<num_elems_hh; i++) puts(hh[i]);
}
```

Within a function, we work with the variables and arrays according to the way they are indicated in the function header.

Output

```
One-dimensional array.
Two-
dimensional
array.
Pointer to string constant.
Array
of pointers
to string
constants.
```

Explanation

1. How have we written the four different types of string declarations? We previously examined all four types. In the declarations that follow, aa[ ] is a 1-D character array, bb[ ][ ] is a 2-D character array, cc is a pointer to a string constant, and dd[ ] is an array of pointers to string constants.

```
char aa[ ]= "One-dimensional array.";
char bb[ ][LENGTH]={"Two-","dimensional ","array."};
char *cc= "Pointer to string constant.";
char *dd[ ]={"Array ","of pointers ","to string ","constants."};
```

In this case, we have initialized each one in its declaration. The 1-D array is given one set of brackets. The brackets are empty because we have initialized the array. The 2-D array is given with two sets of brackets with the first set empty because we have initialized the array. The second set is given a size using the constant macro LENGTH. Even though the array is initialized, we must specify all but the leftmost size. The pointer to the string constant is indicated with * in the declaration. The array of pointers is indicated with both * and brackets. The brackets are empty because we have initialized the array.

2. How can we pass each of these four types to a function? With the function call

```
function1(aa, bb, cc, dd, num_rows_bb, num_elems_dd);
```

we have passed

    The address of the first element of aa[ ]
    The address of the first element of bb[ ][ ]
    The value of cc, which is an address since it is a pointer variable
    The address of the first element of dd[ ]

Note that, in passing these to function1, we simply have given the name of each one. So, even though each has been declared completely differently, we can pass all of them to a function in a very similar manner.

3. How have we written the header for function1? Writing the header correctly is a very important part of using functions in your programs. In this case, it is straightforward because it parallels the declarations for the corresponding arguments in main. In the following table, we show the declaration in main and the header for function1:

Header in main	Header in function1
char aa[ ]	char ee[ ]
char bb[ ][LENGTH]	char ff[ ][LENGTH]
char *cc	char *gg
char *dd[ ]	char *hh[ ]

Note the similarities between the left and right columns of this table.

4. In the call to function1, we indicated that we are passing four addresses. Each argument is very similar to the others in that it is simply the identifier or name of the item. Why then do the corresponding arguments in the function header need to be so different from each other? By making each argument in the function header different, we specify different *pointer arithmetic* for each of the items. We describe more about pointer arithmetic later in this lesson and in Lesson 7.9. At this point we just say that the function must have enough information to properly perform the pointer arithmetic to work with the correct memory cell.

In many cases, the program will not compile if the way the function call arguments are declared in main and the way that the arguments are given in the function header do not match. Therefore, if the compiler indicates a type mismatch between the function call and header or prototype, look at your header and prototype and make sure that there is a match between the function header and the declarator in main.

5. Why did we not have to worry about this before? Previously, we did not have to pay particularly close attention to this because we were working with only one-dimensional or multidimensional arrays of numeric values. It is reasonably straightforward with them to indicate that an address is being received by a function by using brackets in the function declarator. Now, we must work with arrays of pointers and this makes things more complicated.

6. What should we remember to help write our function calls, headers, and prototypes correctly? Remember, we can pass only an address or a value with each argument. We cannot pass an entire array to a function with a single argument. If we pass an address, we use an address in the function call. However, in receiving the address, in the function header, we must indicate more than simply that we are receiving an address. For the function to be able to work with the address, we must indicate enough information for C to properly perform what is called pointer arithmetic. For instance, we must indicate that the address is to the beginning of a 2-D array with the row length being a certain value.

The way we indicate that the function is receiving an address is by using brackets or * in the function header and prototype. The arrangement of the brackets and * indicates how the pointer arithmetic should be performed by C in the function. For instance, two sets of brackets indicate that the address is to a 2-D array and the value in the second set of brackets indicates the row length. This indicates the rules that the pointer arithmetic is to follow. The notation we use in the function body then causes this type of pointer arithmetic to be performed.

For example, in the header for function1 we have as shown in Fig. 7.8. In the diagram, we passed four addresses and two values to function1. These addresses and values have been copied into the memory region reserved for function1.

7. Given that we are passing addresses to function1, what do the memory regions for main and function1 look like? In Fig. 7.9, we show the way memory is arranged. Note from this drawing that the memory region for function1 contains (as values) four addresses. It has no arrays in its memory region. However, function1 is able to work with all of the arrays because the type indicated in the function header has given it enough information to properly work with the arrays.

Also note the connections between the memory region for main and function1. The dashed lines indicate that the addresses for aa, bb, and dd in main are stored in the memory cells for ee, ff, and hh in function1, respectively. And, the value of cc, which is an address, is stored in the memory cell for gg in function1. Note that because function1 has all these addresses, it can modify all these arrays. In your own functions, after passing addresses, you can work with the addresses using such functions as puts, gets, and others as we have described previously in this chapter. We previously described the connection between string constants and pointer arrays and will not discuss them further here.

FIG. 7.8

The declarator for function1 with notations of the meanings of each argument

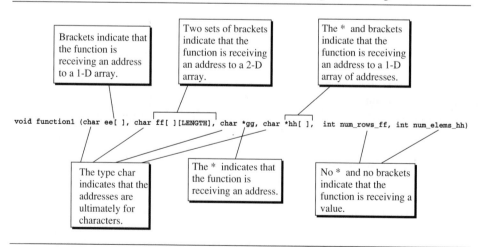

FIG. 7.9

Memory layout for this lesson's program showing main, function1, and string constants. Note that the memory region for function1 contains no arrays, only addresses and integers. However, it accesses arrays using the addresses.

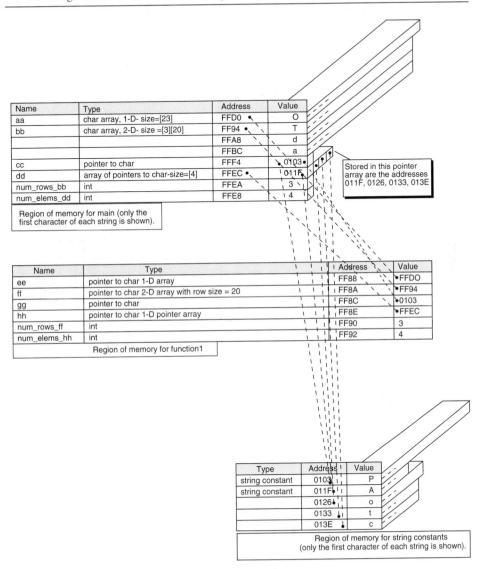

Name	Type	Address	Value
aa	char array, 1-D- size=[23]	FFD0	O
bb	char array, 2-D- size =[3][20]	FF94	T
		FFA8	d
		FFBC	a
cc	pointer to char	FFF4	0103
dd	array of pointers to char-size=[4]	FFEC	011F
num_rows_bb	int	FFEA	3
num_elems_dd	int	FFE8	4

Region of memory for main (only the first character of each string is shown).

Stored in this pointer array are the addresses 011F, 0126, 0133, 013E

Name	Type	Address	Value
ee	pointer to char 1-D array	FF88	FFD0
ff	pointer to char 2-D array with row size = 20	FF8A	FF94
gg	pointer to char	FF8C	0103
hh	pointer to char 1-D pointer array	FF8E	FFEC
num_rows_ff	int	FF90	3
num_elems_hh	int	FF92	4

Region of memory for function1

Type	Address	Value
string constant	0103	P
string constant	011F	A
	0126	o
	0133	t
	013E	c

Region of memory for string constants (only the first character of each string is shown).

8. How do we determine the number of rows of bb[ ][ ] and the number of elements of dd[ ]? We use the sizeof operator to get the declared size of each of these. For instance, sizeof(bb) and sizeof(dd) compute the total number of bytes reserved for bb[ ][ ] and dd[ ], respectively. For bb[ ], knowing that a character occupies 1 byte, we can get the number of rows by dividing the total size by the length of one row. This statement calculates the number of rows of bb[ ][ ]:

```
num_rows_bb = sizeof(bb)/LENGTH;
```

Because dd[ ] is a 1-D array, we get the number of elements by dividing the total size by the size of a single element. We realize that one element is an address to a char, not a numeric type such as int or double. As a result, we divide by sizeof(char *). Although odd looking, sizeof(char *) gives us the number of bytes reserved for storing an address of a char. The next statement calculates the number of elements of dd[ ]:

```
num_elems_dd = sizeof(dd)/sizeof(char *);
```

A typical size for char * is 2 bytes. This is a common size for int *, double *, and other addresses.

9. Why was it necessary to calculate the number of rows of bb[ ][ ] and the number of elements of dd[ ]? As mentioned when we discussed numeric arrays, it often is important to pass the size of an array through the parameter list. By doing this, a function can work efficiently with the array. Unlike numeric arrays, though, we need not pass the size of 1-D character arrays because these often are treated as strings and therefore not used element by element. However, for 2-D character arrays we should pass the number of rows, and for 1-D pointer arrays we should pass the number of elements.

10. What is pointer arithmetic and how is it done for the pointer array hh[ ]? Pointer arithmetic is arithmetic (such as addition and subtraction) performed with addresses. We did not describe this previously, so we describe it here. One can imagine, for instance, if you lived on a street where each house had an address number that differed by 1 from the house next to it, it would be simple to calculate the address of a house that is three away from your own. You would simply add 3 to your own address. C works similarly. To determine the address of a memory cell from one that is known, it adds to (or subtracts from) the address of the known memory cell. To illustrate pointer arithmetic, consider the following two, which C treats as equivalent:

```
hh[1]
*(hh+1)
```

The function1 header (Fig. 7.8) indicates that hh is an address to an array of addresses. Each element of the array occupies 2 bytes (as described). Therefore, C treats hh + 1 as equivalent to hh + 2 bytes. The unary * operator then takes us to the value at the address given by hh + 2 bytes. If we look at our figure for the use of memory for this lesson's program, we see that the value of hh is FFEC, which means that hh + 1 is FFEE.

Although not shown in the figure, FFEE is the address for the second element of dd[ ]. Thus, *(hh + 1) is the same as *(FFEE). This gives us the value of the second element of dd[ ], which is the address 0126.

Using puts as

```
puts (*(hh+1));
```
or
```
puts(hh[1]);
```

is equivalent to puts(0126), which prints the string "of pointers".

We describe more about pointer arithmetic and its use in Lesson 7.9.

EXERCISES

1. Find the value of *x* in the following program manually and then run the program to check your result:

```
#include <string.h>
#include <stdio.h>

void f1 (char a, char b[], char *c);
{
        a='a';
        strcpy(b,"bcde");
        b[0]=a;
        c[2]=*b+5;
}
void main(void)
{
        char x[25]="9876", *px;

        f1('1',x,&x[0]);
        printf("1. x=%s\n",x);

        f1('2',x,&x[1]);
        printf("2. x=%s\n",x);

        px=x;
        f1(*(px+1), x+2, px+2);

        printf("3. x=%s\n",x);
}
```

▓ LESSON 7.8 STANDARD CHARACTER STRING FUNCTIONS

Topics

- String manipulation functions
- String conversion functions

Most of the book to this point has been devoted to manipulating numeric data. Assignment statements are ideally suited for performing many of these manipulations. However, because assignment statements are awkward to use with strings, C provides library functions for manipulating them.

On numeric data, the operations of addition, subtraction, multiplication, and division commonly are performed in programs. However, you may wonder what types of operations commonly are performed on strings, since it makes no sense to multiply or divide them. The C library functions perform such tasks as

1. Taking one string and attaching it to the tail end of another.
2. Searching a string to find a particular character.
3. Searching a string to find a particular smaller string.
4. Converting a numeric portion of a string to int or double.
5. Copying the contents or a portion of the contents of one string to the memory cells of another (we have seen this with strcpy).
6. Comparing two strings to see which goes first in alphabetical order.
7. Breaking up a string into a series of substrings.

This lesson's program uses most of C's string manipulation functions. The ones in this program are the ones you will use most commonly. At this time, you need not study or carefully read this program. It is provided primarily as a reference so that when you want to use any of these functions, you have an example from which to work. However, read this lesson's explanation. Doing this will make you aware of the operations you can perform with these functions.

In looking at the table for this lesson (Table 7.3), be aware that the examples in the table follow the code in the program. To understand the details, look at both the table and the source code. We use the character array, hello[ ], to hold the contents of the string that we manipulate. This string is initialized to be "Good". With operations of the various functions it progresses to be "Good morning!", "Good morning! John.", and "Good morning! Linda.".

Many of the string functions require two strings as input arguments. In these examples, we frequently use the string hello[ ], whose address (represented by `hello` with no brackets) is an argument in the functions' parameter lists. A second argument frequently is a string constant indicated by a string enclosed in double quotes. Remember that both of these arguments are addresses and that any string in double quotes is an address! Should you want to use these functions in your programs you must use addresses as arguments. This means that, where we have shown strings in double quotes or the string indicated by `hello` with no brackets, you may use any other form of indicating an address as an argument, provided the address is an address of a character array terminating in a null character (that is, a string).

The form that we use for describing the functions does not fit the ANSI C standard way of describing them. However, we chose this form because we feel it is easier to understand than the formal language of the standard. Should you require more details or more formal language, please refer to the ANSI C standard.

Again, at this point you need not read the source code. Skip now to the explanation of this lesson's program. In reading the explanation, refer back to the source code. Also, when you want to use any of these functions in programs, use this source code as a guide.

Source code

```c
#include <string.h>
#include <stdio.h>
#include <stdlib.h>
#include <errno.h>

void main(void)
{
  int pos, len, ia, ib;
  char hello[50]="Good", token_separator[]="!,\n \t...";
  char *pa, *pb, *pc;
  double da;
  long la;
  unsigned long ula;
  FILE *infile;

  printf("/****** A - function atoi *******************************/\n");
  ia=atoi("-123.45xyz");
  printf("A--- atoi() converts -123.45xyz to ia=%5d\n\n\n",ia);

  printf("/****** B - function atof *******************************/\n");
  da=atof("-987.65E+01pqr");
  printf("B--- atof() converts -987.65E+01pqr to da=%8.2lf\n\n\n",da);

  printf("/****** C - function atol *******************************/\n");
  la=atol("-456.89abc");
  printf("C--- atol converts -456.89abc to la=%5ld\n\n\n",la);

  printf("/****** D - function strcat *******************************/\n");
  pa=strcat(hello," morning!");
  printf("D--- hello=%s $$$ String at pa=%s\n\n\n",hello, pa);

  printf("/****** E - function strchr *******************************/\n");
  pb=strchr(hello,'m');
  pos=pb-hello+1;
  printf("E--- Character 'm' is the %1dth character of string %s, string at "
            "pb=%s\n\n\n",pos,hello,pb);
  printf("/****** F - function strcmp *******************************/\n");
  ia=strcmp(hello,"Good xyz");
  if (ia<0)  printf("F--- ia=%2d, %s is less than    Good xyz!\n\n",ia, hello);
  if (ia==0) printf("F--- ia=%2d, %s is identical to Good xyz!\n\n",ia,hello);
  if (ia>0)  printf("F--- ia=%2d, %s is greater than Good xyz!\n\n\n",ia,hello);

  printf("/****** H - function strcpn *******************************/\n");
  pos = strcspn(hello, "dog");
  printf("H--- The first occurrence of any character in substring, dog,  \n"
      "     in string %s is the %1dnd character, o\n\n\n",hello,pos+1);

  printf("/****** I - function strerror *******************************/\n");
  errno=0;
  infile=fopen("abcdefgh.ijk","r");
  if (errno)
  {
  pa=strerror(errno);
  printf("I--- Does file  abcdefgh.ijk exist? strerror() says %s\n",pa);
  }

  printf("/****** J - function strlen *******************************/\n");
  len=strlen(hello);
  printf("J--- Not including the null character, %s has %2d characters\n\n\n",
      hello,len);
```

This source code is meant to be used as a reference. You do not need to study this source code until you want to use one of the string functions shown here.

```
printf("/****** K - function strncat ********************************/\n");
pb=strncat(hello," John. How are you!",5);
printf("K--- hello=%s $$$ String at pb=%s\n\n\n",hello, pb);


printf("/****** L - function strncmp ********************************/\n");
ib=strncmp(hello,"Good car",15);
if (ib>0) printf("L--- ib=%1d, %s is greater than Good car\n\n\n",ib, hello);

printf("/****** M - function strncpy ********************************/\n");
pa=strncpy(hello+14,"Linda. How are you!", 6);
printf("M--- hello=%s $$$ String at pa=%s\n\n\n",hello,pa);

printf("/****** N - function strpbrk ********************************/\n");
pb=strpbrk(hello,"dear");
printf("N--- hello=%s $$$ String at pb=%s\n\n\n",hello,pb);

printf("/****** O - function strrchr ********************************/\n");
pa=strrchr(hello,'m');
printf("O--- hello=%s, String at pa=%s\n\n\n\n",hello,pa);


printf("/****** P - function strspn ********************************/\n");
ia=strspn("Good year",hello);
printf("P--- The %dth character 'y' is the first character in oGdo year\n"
     "      that is not present in %s\n\n\n",ia+1, hello);

printf("/****** Q - function strstr ********************************/\n");
pa=strstr(hello,"Linda");
ia=pa-hello+1;
printf("Q--- Linda was found at position %d of %s @@@ String at pa=%s\n\n\n",
     ia,hello,pa);

printf("/****** R - function strtod ********************************/\n");
da=strtod("123.45abc",&pb);
printf("R--- Find double number %6.21f in 123.45abc $$$ String at "
        "pb=%s\n\n\n",da,pb);

printf("/****** S - function strtol ********************************/\n");
la=strtol("98765xyz",&pa, 10);
printf("S--- Find long number %61d in 98765xyz $$$ String at"
        "pa=%s\n\n\n",la,pa);

printf("/****** T - function strtoul ********************************/\n");
ula=strtoul("45678pqr",&pc,10);
printf("T--- Find unsigned long %61d in 45678pqr $$$ String at "
        "pc=%s\n\n\n",ula,pc);

printf("/****** U - function strtok ********************************/\n");
printf("hello=%s,     token_separator=%s\n",hello,token_separator);
pa=strtok(hello,token_separator);
while ( pa!=NULL)
     {
     printf("U--- String at pa=%10s pa=%5u\n",pa,pa);
     pa=strtok(NULL,token_separator);
     }
}
```

Output

```
****** A - function atoi *******************************
A--- atoi() converts -123.45xyz to ia= -123


****** B - function atof *******************************
B--- atof() converts -987.65E+01pqr to da=-9876.50


****** C - function atol *******************************
C--- atol converts -456.89abc to la= -456


****** D - function strcat ******************************
D--- hello=Good morning! $$$ String at pa=Good morning!


****** E - function strchr ******************************
E--- Character 'm' is the 6th character of string Good morning!,
     String at pb=morning!


****** F - function strcmp ******************************
F--- ia=-11, Good morning! is less than    Good xyz!


****** H - function strcspn *****************************
H--- The first occurrence of any character in substring, dog,
     in string Good morning! is the 2nd character, o


****** I - function strerror ****************************
I--- Does file abcdefgh.ijk exist? strerror() says No such file or
     directory


****** J - function strlen ******************************
J--- Not including the null character, Good morning! has 13 characters


****** K - function strncat *****************************
K--- hello=Good morning! John $$$ String at pb=Good morning! John


****** L - function strncmp *****************************
L--- ib=10, Good morning! John is greater than Good car


****** M - function strncpy *****************************
M--- hello=Good morning! Linda. $$$ String at pa=Linda.


****** N - function strpbrk *****************************
N--- hello=Good morning! Linda. $$$ String at pb=d morning! Linda.


****** O - function strrchr *****************************
O--- hello=Good morning! Linda., String at pa=morning! Linda.


****** P - function strspn ******************************
P--- The 6th character 'y' is the first character in oGdo year
     that is not present in Good morning! Linda.


****** Q - function strstr ******************************
Q--- Linda was found at position 15 of Good morning! Linda. @@@
     String at pa=Linda.
```

```
****** R - function strtod *********************************
R--- Find double number 123.45 in 123.45abc $$$ String at pb=abc

****** S - function strtol *********************************
S--- Find long number  98765 in 98765xyz $$$ String at pa=xyz

****** T - function strtoul ********************************
T--- Find unsigned long  45678 in 45678pqr $$$ String at pc=pqr

****** U - function strtok *********************************
hello=Good morning! Linda.,      token_separator=!,
    ...
U--- String at pa=     Good pa=65416
U--- String at pa=   morning pa=65421
U--- String at pa=     Linda pa=65430
```

Explanation

1. How are the functions in this lesson's program used? Table 7.3 lists all the functions compatible with the ANSI C standard used in this lesson's program. A brief explanation of how they are used in the program also is provided. Even if your compiler is not ANSI C compatible, it should support all of these functions. For other details about them please see your manual.

The primary string operated on in this program is hello[ ]. In the table, note the progression of changes in this string as the functions operate on it. Notice that the functions have been used in the program and are presented in the table in alphabetical order, except for the last function, strtok.

You need not memorize the table. However, read the entire table and get a general understanding of what each function does. Pay particular attention to the example given for each description. These descriptions give you an idea of what is available for your programming needs.

When you feel you want to use one of these functions, refer to this table and the implementation of the function in this lesson's program. With this information, you should be able to use any of these functions in your programs.

We use some of these functions in the Application Examples at the end of this chapter to give you further insight.

2. Can we group these functions in terms of use? Yes, Table 7.4 classifies the functions based on what they do.

3. How do we interpret function prototypes given for standard C functions? We now know enough to look intelligently at many of the function prototypes given for standard C functions and correctly interpret them. We show you these because, as you go further with C, you will see them quite frequently. When you first encounter them, they look somewhat cryptic; therefore, we describe a few of them in Table 7.5. Having understood these, you should be able to interpret many more. Read the function prototype in the left column of the table, then read the explanation in the right column, referring back to the left column as needed.

■   **TABLE 7.3**
Standard character string functions

Function name, example, and required header file	Explanation

A   atoi
 	
`ia=atoi("-123.45xyz");`
 	
`#include<stdlib.h>`
	Converts a character string in the form of *"whitespace sign digits"* to an int value. The function returns 0 if the input cannot be converted. In case of overflow, the return value is undefined. The example converts only the characters −123 to int. Therefore, `ia = -123`. Note: If any character other than *whitespace* is before *sign digits,* no conversion occurs.

B   atof
 	
`da=atof("-987.65E+01pqr");`
 	
`#include<stdlib.h>`
	Converts a character string in the form of *"whitespace sign digits .digits d\D\e\E sign digits"* to a double. The function returns 0.0 if the input cannot be converted. In case of overflow, the return value is undefined. The example converts all characters except pqr to double. Therefore, `da = -987.65E + 01`. Note: If any character other than *whitespace* is before *sign digits*, no conversion occurs.

C   atol
 	
`la=atol("-456.89abc");`
 	
`#include<stdlib.h>`
	Converts a character string in the form of *"whitespace sign digits"* to a long int value. The function returns 0 if the input cannot be converted. In case of overflow, the return value is undefined. The example converts only the characters −456 to long. Therefore, `la = -456`. Note: If any character other than *whitespace* is before *sign digits*, no conversion occurs.

D   strcat
 	
`pa=strcat(hello,`
`" morning!");`
 	
`#include<string.h>`
	Appends a copy of the second string, "morning!", to the first string indicated by the address, hello, and returns a pointer to the first string to which the second string has been concatenated (that is, connected or appended). The first character of "morning!" overwrites the null character of hello. The array, hello[ ], must have declared enough space to accommodate "*morning*". Note: Prior to executing this code, the string hello[ ] is, "Good". After executing the code, the string, hello[ ], is, "Good morning!"

E   strchr
 	
`pb=strchr(hello,'m');`
 	
The string hello[ ] is, "Good morning!"
 	
`#include<string.h>`
	Finds the specified character, m, in the string pointed to by hello. Returns a pointer, assigned to pb, to the first occurrence of 'm' in the string hello. A null pointer is returned if 'm' is not found in hello. The position of 'm' within the string hello can be calculated using the statement `pos=pb-hello+1;` where pos, converted to int, indicates the position of 'm' in the hello string.

■   **TABLE 7.3 (CONTINUED)**
Standard character string functions

Function name, example, and required header file	Explanation
F strcmp **ia=strcmp(hello, "Good xyz");** The string hello[] is, "Good morning!" **#include<string.h>**	Compares the first string, hello, and the second string, "Good xyz", lexicographically. Returns an int value, assigned to ia, as follows: ia < 0, first string is less than the second string ia = 0, first string is identical to the second string ia > 0, first string is greater than the second string In this example, the first five characters in both strings are identical, but the sixth characters are different. The characters are 'm' and 'x', which have ASCII values of 109 and 120, respectively. The return value ia is calculated as follows: **ia='m' - 'x' = 109-120 = -11.** which indicates that the string hello [ ] (which is "Good morning!") is less than the string "Good xyz!" meaning that "Good morning" comes before "Good xyz" in alphabetical order.
G strcpy **pa=strcpy(hello, "Good Morning!");** After execution, the string hello[ ] is, "Good morning!" **#include<string.h>**	Copies the second string into the memory cells reserved for the first string, hello. Returns a pointer, assigned to pa, that points to the first string. Note: This function was not used in the lesson's program. Please see Lesson 7.3 for its use.
H strcspn **pos = strcspn(hello, "dog");** The string hello[ ] is, "Good morning!" **#include<string.h>**	Finds the first occurrence of any character in the second string in the first string, hello [ ] (which is "Good morning"). Returns the position (meaning the location relative to the first character) of the first character in the first string that also appears in the second string. For example, two characters, d and o, in the second string can be found in the first string. Since the position of character 'o' is in front of the position of character 'd', the return value, pos, is the position of character 'o' in the first string, which is 2. Note that because white space is a character, two strings with white space may match white space and no other character.

I   strerror
**errno=0;
infile=fopen
("abcdefgh.ijk","r");
if (errno)
  {
pa=strerror(errno);
printf("%s=n",pa);**
	This function often is used with errno (implemented as a global variable in many C compilers, meaning that it is available to functions without being passed through the parameter list), which has information about it in errno.h. The global variable errno is one of the few used in C's library functions. Do not name any of your own variables errno because of this! C's functions will assign an integer value other than 0 to errno when they detect an error

▨   **TABLE 7.3 (CONTINUED)**
Standard character string functions

Function name, example, and required header file	Explanation
} *#include<string.h> for strerror* *#include<errno.h> for full use of errno*	(usually caused by improper arguments being passed). This integer value, when passed to the function strerror, determines the address of the string describing the error found by the function that set errno. So, strerror and errno commonly are used hand in hand, although they are not required to be. To use errno and strerror, first set errno to 0 (which indicates no error currently), then call a C library function. If the C library function detects an error, then it will assign a nonzero integer to errno. If the value of errno is nonzero, then strerror can be called with the value of errno. The function strerror returns the address of the string that gives a written description of the error. By printing this string, a user can read the error detected by the function. In the example, errno is set to 0, then the fopen function is called to open a file that does not exist for reading purposes. Because of this error, the function fopen sets the value of errno to be nonzero, and since errno is a global variable, it can be accessed by our program. The if statement determines that errno is nonzero and calls strerror with the errno value. The function strerror returns the address of the string that has a written description of the error. We then print out this written description. Note that each compiler may implement this differently, meaning the written error message corresponding to errno = 2 on one compiler may not be the same as it is on another. Also, errno may not be implemented as a global variable; it may be a macro. However, if you use strerror and errno as described here, your program will be portable as each compiler should be internally consistent.
J strlen **len=strlen(hello);** The string hello[ ] is, "Good morning!" *#include<string.h>*	Returns the length in bytes of string, hello [ ] (which is "Good morning!"). The length does not include the string terminating null character. In the example, len = 13.
K strncat **pb=strncat(hello,** **" John. How are you!",5);** *#include<string.h>*	Appends the first five characters of the second string to the first string, hello [ ]. The first character in the second string overwrites the terminating null character in the first string. A null character from the second string is not copied; however, a null character is added to the end of the concatenated string. This means that the concatenated string will not have two null characters at the end. The first string must have enough declared size to accommodate the second string being added to the end. Returns a pointer, pb, to the new concatenated string hello[ ]. Note: Before

■ **TABLE 7.3 (CONTINUED)**
Standard character string functions

Function name, example, and required header file	Explanation
	executing this code the string hello[ ] is, "Good morning!" After executing the code, the string, hello[ ], is, "Good morning! John."

L   strncmp
 	
`ib=strncmp(hello,`
`"Good car",15);`
 	
The string hello[ ] is, "Good morning! John."
 	
`#include<string.h>`
	Compares the first 15 characters of string hello; and the second string, "Good car", lexicographically. Returns an int value, assigned to ib, as follows:
	
	ib < 0, first string is less than the second string
	ib = 0, first string is identical to the second string
	ib > 0, first string is greater than the second string
	
	See Fig. 7.10. In this example, the first 5 (which is less than 15) characters in both strings are identical, but the sixth characters are different. The characters are 'm' and 'c', which have ASCII values of 109 and 99, respectively. The return value ib is calculated as follows:
	$ib = 'm' - 'c' = 109 - 99 = 10.$
	which indicates that the string hello[ ] (which is "Good morning! John.") is greater than the string "Good car".

M   strncpy
 	
`pa=strncpy(hello+14,`
`"Linda. How are you!", 6);`
 	
`#include<string.h>`
	Copies the first six characters, "Linda.", from the second string to the first string beginning at the address indicated by hello + 14. The expression *hello* + 14 indicates 14 characters past the first character. This sort of addition (with pointers) will be explained in more detail in the next lesson. This function returns a pointer, assigned to pa, to the first string hello. In the example, the first string at and after the hello+14 position is replaced by the string "Linda". After the replacement, a null character is added at the end of the new string hello[ ]. Note: Before executing this code, the string hello[ ] is, "Good morning! John." After executing the code, the string hello[ ] is, "Good morning! Linda."

N   strpbrk
 	
`pb=strpbrk(hello,"dear");`
 	
The string hello[ ] is, "Good morning! Linda."
 	
`#include<string.h>`
	Scans the first string, hello[ ], and determines whether it contains any character of the second string. If it finds a match, the function returns a pointer (which is assigned to pb) to the location of the first occurrence of the matching character in the first string. For example, the second string, "dear," contains the character *d,* which also is a component of the first string, "Good morning! Linda."; the pointer, pb, then is assigned the address of the character 'd' in "Good" of the first string. If no match is found, a null pointer is returned.

▨ **TABLE 7.3 (CONTINUED)**
Standard character string functions

Function name, example, and required header file	Explanation
O strrchr **pa=strrchr(hello,'o');** The string hello[ ] is, "Good morning! Linda." **#include<string.h>**	Scans the first string, hello[ ] (which is "Good morning! Linda."), for the last occurrence of the specified character, 'o'. If it is found, the function returns a pointer (assigned to pa) to the location of the last occurrence of the matching character in the first string. For example, the first string contains three 'o' characters. The pointer pa then is the address of the last character 'o' in the first string (that is, 'o' in "morning". If no match is found, a null pointer is returned.
P strspn **ia=strspn("oGdo year",hello);** The string hello[ ] is, "Good morning! Linda." **#include<string.h>**	Returns the length of the initial portion of the first string that contains only characters found in the second string. For example, the first five characters in the first string appear in the second string hello[ ] (which is "Good morning! Linda."); however, the character 'y' in the first string does not appear in the second string, hello[ ]. Therefore, ia = 5.
Q strstr **pa=strstr(hello,"Linda");** The string hello[ ] is, "Good morning! Linda." **#include<string.h>**	Finds the second string (excluding the null character) in the first string, hello[ ] (which is "Good morning! Linda."). Returns a pointer, assigned to pa, to the first occurrence of the second string in the first string. In the example, both strings contain "Linda.", therefore, pa points to the character "L" of the first string. If the string is not found, a null pointer is returned.
R strtod **da=strtod("123.45abc",&pb);** **#include<stdlib.h>**	Converts a character string (which must begin with the first non-white-space character being a sign, digits, or decimal point) to a double value. One difference between functions strtod() and atof() is that the strtod() argument list contains the address of a pointer variable, &pb. The value of pb is set by the function strtod to be the address of the character in the string where the function stops scanning. The function returns 0 if no conversion can be performed. For example, the function stops scanning at 'abc' because this is after the end of the numeric part of the string. Therefore, pb is equal to the address of 'a' in 'abc'. The value of da is 123.45.
S strtol **la=strtol("98765xyz", &pa, 10);** **#include<stdlib.h>**	Converts a character string (which must begin with the first non-white-space character being a sign, digits, or decimal point) to a long value. One difference between functions strtol() and atol() is that the strtol() function provides a number base for converting the string (10 for this example). In addition, the function has as one of its arguments the address of a pointer variable, &pa. The value of pa is set by the function strtol to be the address of the character

■   **TABLE 7.3 (CONTINUED)**
Standard character string functions

Function name, example, and required header file	Explanation
	in the string where the function stops scanning. The function returns 0 if no conversion can be performed. For example, the function stops scanning at 'xyz'. Therefore, pa is equal to the address of 'x' in 'xyz'. The value of la is 98765.

T   strtoul
 	
`ula=strtoul("45678pqr", &pc, 10);`
 	
`#include<stdlib.h>`
	Converts a character string (which must begin with the first non-white-space character being a sign, digits, or decimal point) to an unsigned long value. The function provides a number base for converting the string (10 for this example). In addition, the function has as one of its arguments the address of a pointer variable, &pc, where pc is set by the function to point to a character in the string where the function stops scanning. The function returns 0 if no conversion can be performed. For example, The function stops scanning at 'pqr'; therefore, *pc is equal to the address of the 'p' in 'pqr'. Note that the declaration and usage of pc is slightly different from the usage of pa and pb for the strtod and strtol examples, but it fulfills a similar function.

U   strtok
 	
`pa=strtok(hello,`
`        token_separator);`
`while (pa!=NULL)`
`    {`
`    printf("%10s\n",pa);`
`    pa=strtok(NULL,`
`        token_separator);`
`    }`
The string hello[ ] is, "Good morning! Linda."
Note: The declaration for token_separator is
`token_separator[]="!,\n \t...";`
 	
Therefore, strtok examines the string hello for any character in the preceding string.
`#include<stdlib.h> for NULL`
`#include<string.h> for strtok`
	Breaks a string into many strings delimited by the null character. The method is this: It defines a token in the first string, hello[ ] (which is "Good morning! Linda."), by finding the first character in hello[ ] that *is not* in the token_separator[ ] string. This becomes the beginning of the first token. It then searches for the first character that *is* in the token_separator string (called a *delimiter*). It replaces the delimiter in the string with the null character. It returns a pointer (assigned to pa) to the beginning of first token (which is the beginning of the string preceding the first token delimiter found). The function strtok internally saves a pointer to the character after a delimiter is found. To continue replacing each subsequent delimiter with a null character in the string hello[ ], we must make subsequent calls to strtok, not with our string of interest (hello[ ], in this case) but with NULL (the null pointer) as an argument. Because the call to strtok is with NULL, strtok uses its internally saved pointer as the beginning of the next search for a delimiter. After executing the statements shown, hello[ ] becomes:
	`Good\0morning\0Linda\0`
	If strtok reaches the end of the string hello without defining a new token, it returns the NULL pointer. This is the reason for the (pa!=NULL) test expression.

FIG. 7.10
Comparing strings using strncmp

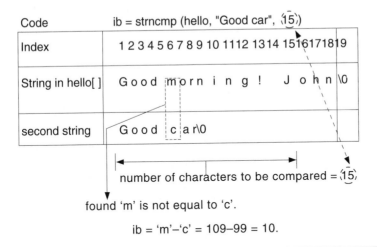

found 'm' is not equal to 'c'.

ib = 'm'–'c' = 109–99 = 10.

Realize that the function prototypes alone do not give enough information to use the functions effectively. We described all these functions previously in the book, so we do not go into them further here. To use other functions in the ANSI C standard you should read both the prototype and the accompanying description.

EXERCISES

1. Write a program to read the source code of this lesson (file L7_8.C) and then do the following:

   a. Find the total number of tokens used in the code. The token separator used contains the following characters:

      ```
      ""[](){}*,:=;...#\n\t\ and space
      ```

   b. Find the total number of string constants "Good" in the source code.

   c. Convert all strings that contain digit characters in the source code to doubles.

2. For each function shown in Table 7.4, develop a function that contains the same arguments and returns the same type of data. For example, the function

   ```
   pb=strncat(hello," John. How are you!",5);
   ```

   has three arguments; the function appends the first five characters of the second string to the first string, hello[ ]. The function returns a pointer, assigned to pb, to the new concatenated string hello[ ]. For this function, develop a function

TABLE 7.4
Operations and their functions

Type of operations	Function name	Comment
Conversion from string to numeric value	atoi	Returns int
	atof	Returns float
	atol	Returns long
	strtod	Returns double, passes address of end of numeric portion
	strtol	Returns long, passes address of end of numeric portion
	strtoul	Returns unsigned long, passes address of end of numeric portion
Copying one string or a portion of a string into memory cells reserved for another string	strcat	Copies second string onto the tail end of first string, returns address of first string
	strcpy	Copies second string onto the beginning of the first string, returns address of first string
	strncpy	Copies specified number of characters from second string to beginning of first string, returns address of first string
	strtok	Copies null character into first string at location of matching character in second string, returns address of location at which null character is placed
Finding the address or position of a particular character, string, or portion of a string in a given string	strchr (address)	Finds specified single character in string, returns address of first occurrence of character
	strcspn (position)	Finds first occurrence in first string of any character in second string, returns position of character in first string
	strpbrk (address)	Finds first occurrence in first string of any character in second string, returns address of character in first string
	strrchr (address)	Finds last occurrence in first string of any character in second string, returns address of character in first string
	strspn (position)	Finds the first character in first string that does not occur in second string, returns position of character in first string
	strstr (address)	Finds the first occurrence of the second string in the first string, returns the address of the beginning of the second string in the first string

TABLE 7.4 (CONTINUED)

Type of operations	Function name	Comment
Comparing two strings lexicographically	strcmp	Compares first string to second string, returns int indicating which string is greater
	strncmp	Compares specified number of characters from first string to second string, returns int indicating which is greater
Finding the length of a string	strlen	Returns number of characters in string excluding null character
Determining the address of a string describing an error	strerror	Returns address of string that describes error

named M_strncat(), where *M* represents mine. Test both functions using the same actual arguments:

```
pc=M_strncat(hello," John. How are you!",5);
```

The pointers pb and pc should point to the same object.

3. Given the source code of Lesson 7.8 (file L7_8.C) write a program to display any character used in the program and the number of occurrences of that character. The output should follow the format

```
Input file ----- L7_8.C
------------------------------
Character                number of occurrences
a                             ??
b                             ??
...
```

4. Given the source code of Lesson 7.8 (file L7_8.C) use all the ANSI C character type functions to write a program to determine the number of occurrences of various types of characters. The output should follow the format

```
Input file ----- L7_8.C
----------------------------------
Character type      number of occurrences
Alphanumeric                  ??
Alphabetic                    ??
...
```

5. Write a program to convert a character string in the form of

whitespace sign digits

to an int value. Use the atoi() function to check if the conversion is done correctly.

6. Write your own version of strchr() function. However, the function should return the location of the character found. If no character is found, the program should return −999. Use strchr() to check your output.

▓ **TABLE 7.5**
   Function prototypes

Function prototype (as given by the ANSI C standard)	Explanation
int puts(const char *s)	One argument is given in parentheses, *s*, which represents an address because it has * next to it. This means that we must pass an address when we call puts. The char indicates that the address must be an address to a char. The const is the qualifier, which means that the function puts cannot change the characters accessed by the address given by s. The int before puts indicates that puts returns an int value (recall that puts returns a nonnegative integer unless a write error occurs, in which case it returns EOF). The s in the prototype has no meaning in itself. It simply is the typical identifier ANSI C has used to represent a string. Therefore, this prototype tells us that, to use puts, we can have in our program *integer_variable* = **puts**(*address_of_char*);
double sin (double x)	Because the argument x has neither * nor brackets next to it, we must pass a value to the sin function. The double inside the parentheses indicates that we should pass a double value to sin. The double before sin indicates that the sin function returns a double value. The x has no particular meaning except it is the identifier ANSI C uses to represent a double value. From this prototype, we can use the sin function in the following way: *double_variable* = **sin**(*double_value*);
char *strcpy(char *s1, const char *s2)	This prototype indicates two arguments, s1 and s2. Both are addresses because they are preceded by *. Both also are addresses to char because they both are preceded by char. The second one, s2, has the const qualifier, meaning that the function strcpy cannot modify the characters accessed through the address s2. The * before strcpy indicates that strcpy returns an address. The char that precedes all this indicates that strcpy returns an address to a char. The s1 and s2 have no particular meaning. They are just identifiers ANSI C has chosen to use. This function prototype indicates that we can use strcpy in the following way: *char_pointer_variable* = **strcpy**(*address_of_char*, *address_of_char*);
double atof(const char *nptr)	The argument is nptr, which is an address to a char as indicated by char *. The const means that the function atof cannot modify the characters accessed through nptr. The nptr has no particular meaning except it is the identifier ANSI C sometimes uses to represent a pointer variable. The double indicates that atof returns a double value. We can use atof in the following way: *double_variable* = **atof**(*address_of_char*);

▨ LESSON 7.9 POINTER NOTATION VS. ARRAY NOTATION

Topic

• Using pointer notation to access array elements and strings

We have alluded to but not used the fact that C allows array data to be accessed using pointer notation as well as array notation. This lesson illustrates that.

What you should observe about this lesson's program.

1. Three arrays (1-D, 2-D, and 3-D) have been declared. All have been initialized at least partially.
2. A single pointer variable has been declared.
3. The first two putchar statements in section 1 print the first character of the aa[ ] array.
4. The fourth and fifth putchar statements in section 1 print the 17th character in the aa[ ] array.
5. These two observations illustrate that a single unary * operator can be used to access an element of a 1-D array.
6. The first two putchar statements in section 2 print the sixth character of the fourth row of the bb[ ] array.
7. The first puts statement in section 2 prints the third string in the bb[ ] array.
8. The previous two observations indicate that using two unary * operators accesses a single character in a 2-D array, whereas using a single unary * operator accesses a string in a 2-D array.
9. In section 2, dd (which is a single pointer variable) is assigned the address of the single first element of the bb[ ] [ ] array.
10. The fourth putchar statement in section 2 prints a single character of the 2-D bb[ ][ ] array even though only one unary * operator is used.
11. The first two putchar statements in section 3 print the same character of the cc[ ][ ][ ] array.
12. The first puts statement in section 3 prints one string of the cc[ ][ ][ ] array.
13. In section 3, dd (which is a single pointer variable) is assigned the address of the single first element of the cc[ ] [ ] array.
14. In section 3, using dd with a single unary * operator accesses a single character of the cc[ ][ ][ ] array.
15. In section 3, using dd with no unary * operator accesses a portion of a single string of the cc[ ][ ][ ] array.
16. The last printf statement in the program and the output from it illustrate the differences in meaning between cc+1, *cc+1, and **cc+1.

Questions you should attempt to answer before reading the explanation.

1. With a 1-D array, how many unary * operators are needed to access a single character?
2. With a 2-D array, how many unary * operators are needed to access a single character?
3. With a 3-D array, how many unary * operators are needed to access a single character?

Source code

```
#include <stdio.h>
void main(void)
{
        char aa[35]={"This is a one-dimensional array"};
        char bb[5][40]={"We can","use both","array and pointer",
                        "notation to access","one and two-dimensional arrays"};
        char cc[2][3][20]={"A three ","dimensional ","array ","is ","shown ","also"};
        char *dd;


        printf("**************** Section 1   1-D array ***************\n");
        putchar(aa[0]);
        putchar(*aa);
        putchar('\n');


        putchar(aa[16]);
        putchar(*(aa+16));
        putchar ('\n');



        printf("**************** Section 2   2-D array ***************\n");

        putchar(bb[3][5]);
        putchar(*(*(bb+3)+5));
        putchar('\n');

        puts(*(bb+2));

        dd=&bb[0][0];
        putchar(*(dd+125));
        putchar('\n');




        printf("**************** Section 3   3-D array ***************\n");

        putchar(cc[1][2][3]);
        putchar(*(*(*(cc+1)+2)+3));
        putchar('\n');


        puts(*(*(cc+1)+2));


        dd=&cc[0][0][0];
        putchar(*(dd+80));
        putchar('\n');
        puts(dd+80);
```

1-D array.

2-D array.

3-D array.

Array notation for accessing single character in 1-D array.

Pointer notation for accessing single character in 1-D array.

Array notation for accessing single character in 1-D array.

Pointer notation for accessing single character in 1-D array.

Array notation for accessing single character in 2-D array.

Pointer notation for accessing single character in 2-D array.

Pointer notation for accessing string in 2-D array.

If we assign the address of the beginning of an array to a separate pointer variable, we can use pointer notation with just one * to access a single character in a 2-D array. We can do this because pointer arithmetic is done differently with a single pointer variable. Note: dd = bb; will not work.

Array notation for accessing a single character in a 3-D array.

Pointer notation for accessing a single character in a 3-D array.

With a 3-D array, pointer notation with two *s indicates the address of the beginning of a string.

If we assign the address of the beginning of an array to a separate pointer variable, we can use pointer notation with just one * to access a single character in a 3-D array. Note: dd = cc; will not work.

Pointer arithmetic with a single pointer variable, dd, allows us to access a string in a 3-D array without the use of *.

```
printf("Address of cc[0][0][0]=%p, cc+1=%p, *cc+1=%p, **cc+1=%p\n",
       &cc[0][0][0],cc+1,*cc+1,**cc+1);
}
```

> &cc[0][0][0], cc, *cc, and **cc all indicate the
> address of the first element of cc[ ][ ][ ]. However,
> adding 1 in pointer arithmetic to cc, *cc, and **cc
> gives different results.

Output

```
*************** Section 1  1-D array ***************
TT
mm
*************** Section 2  2-D array ***************
ii
array and pointer
i
*************** Section 3  3-D array ***************
oo
also
s
shown
Address of cc[0][0][0]=FE92, cc+1=FECE, *cc+1=FEA6, **cc+1=FE93
```

Explanation

1. How can we use pointer notation to access the elements of a 1-D character array? As we have seen with the one-dimensional character array aa[ ] we can access its elements using either array or pointer notation. To access the very first element of the array (aa[0]), the pointer notation *aa can be used. To access the element indicated by aa[16], the pointer notation *(aa+16) can be used. Therefore, the two statements in this lesson's program

```
putchar(aa[16]);
putchar(*(aa+16));
```

are equivalent.

2. What logic underlies this notation? This notation involves pointer arithmetic in the form of adding an integer to an address. C performs this type of operation by first noting the type of address. For instance, in the expression

aa+16

aa is the address of the beginning of a 1-D array of characters. Because characters occupy 1 byte of memory and we are working with a 1-D array, C adds 16 bytes to the address indicated by aa. Thus, this expression gives us the address of the character indicated by a[16], the 17th character. Using the unary * operator, the expression *(aa+16) gives us the value of the 17th character. This expression can be used as an argument to the function putchar.

3. How can we use pointer notation to access the elements of a 2-D array? In this lesson's program, we use the 2-D character array bb[5][40]. We can access its elements using array notation or pointer notation. For instance,

```
putchar(bb[3][5]);
putchar(*(*(bb+3)+5));
```

both print the same character in the bb[ ][ ] array.

The second expression again works through pointer arithmetic. We describe it beginning with the innermost pair of parentheses. To perform this addition, C determines the type of the address. This becomes a little tricky with multidimensional arrays because C regards them as arrays of arrays. In other words, the bb[ ][ ] array in this lesson's program is considered to be five 1-D arrays of size 40 (because bb was declared to be bb[5][40]). The address expressed as bb is regarded an address of the beginning of one of the rows. Thus, an individual element of that address is considered to have a size of 40 not 1! Therefore, when we add bb and 3, we add $3 \times 40 = 120$ bytes to the address indicated by bb.

With a 2-D array, the unary * operator also works somewhat differently. The expression

$$*(bb+3)$$

represents the address (not the value, as occurs when we work with 1-D arrays or single pointer variables) of the beginning of the fourth row. So, the expression

$$*(bb+3)+5$$

is the addition of an address, *(bb+3), and integer, 5. Again, C has to consider the type of the address (which must be done with the declaration in mind) before the addition can be performed. Because it is a 2-D char array and the address is to the beginning of the row using one unary * operator, the integer causes the addition of 5 bytes to the address. This takes us to the address of the individual character, i, in the bb[ ][ ] array. Using the unary * operator on this expression gives

$$*(*(bb+3)+5)$$

which is the value of the character. Therefore,

```
putchar(*(*(bb+3)+5));
```

causes the character i to be printed. Note that, to access a single value using pointer notation with a 2-D array, we need to use two unary * operators. To access a single value using pointer notation with a 3-D array we need to use three unary * operators.

4. With the 2-D array bb[ ][ ], what does *(bb+2) represent? It is equivalent to bb[2], which represents the address of the beginning of the third row in the bb[ ][ ] array. This serves as an acceptable argument for the puts function, and the statement

```
puts(*(bb+2));
```

causes the string "array and pointer" to be printed.

5. With the declaration char *dd, what does

dd+125

indicate? The variable dd indicates an address, and because dd is declared as a simple char pointer variable, the 125 causes 125 bytes to be added to the address indicated by dd. Also, because dd is a simple pointer variable, only one unary * is needed to access a value. So,

```
dd=&bb[0][0];
putchar(*(dd+125));
```

prints the character i, which is the 126th character in the bb[ ][ ] array.

Remember, to determine the 126th character, we must use the declared size of the array, bb[5][40], not the characters that are initialized. In other words, the 126th character is the sixth character of the fourth row, which is i.

6. Could we have used the statement dd = bb; instead of dd = &bb[0][0] in this lesson's program? No, because C considers it to be a type mismatch. Even though bb represents an address and dd is a pointer variable, bb is the address of a 2-D array. Because dd was declared as a single char pointer variable, it cannot take on the characteristics of an array address. When we use &, the "address of" operator, with the single bb[0][0], we get an acceptable type that can be stored in dd.

7. Would putchar(*(bb+125)); print the 126th character of the bb[ ][ ] array? No; first, the statement would not compile because using bb directly means we must use two unary * operators to point to a single character. Also, pointer arithmetic would not advance us 125 bytes but $125 \times 40 = 5000$ bytes! So, even if it were to compile, we would point to the wrong location.

8. How can we use pointer notation to access the elements of a 3-D array? In this lesson's program, we used the 3-D character array, cc[2][3][20]. We can access its elements using array notation or pointer notation. For instance,

```
putchar(cc[1][2][3]);
putchar(*(*(*(cc+1)+2)+3));
```

both print the same character in the cc[ ][ ][ ] array.

The second expression works through pointer arithmetic, which we describe beginning with the innermost pair of parentheses. Again, to perform this addition, C determines the type of the address. Because C regards 3-D arrays as arrays of arrays, cc[ ][ ][ ] (which was declared to be cc[2][3][20]) is considered two 2-D arrays of size [3][20]. The address expressed as cc is regarded an address of the beginning of one of the 2-D arrays. So, an individual element of that address is considered to have a size of $3 \times 20 = 60$, not 1. When we add cc and the integer 1, we add $1 \times 3 \times 20 = 60$ bytes to the address indicated by cc.

Again, because we are working with a multidimensional array, a single unary *
operator does not indicate that a value is represented, it indicates that an address is
represented. Therefore, the expression

```
*(cc+1)
```

represents the address of the beginning of the second 2-D array. The expression

```
*(cc+1)+2
```

is the addition of an address, *(cc+1), and an integer, 2. Again, C has to consider the
type of the address (which must be done with the declaration in mind) before
the addition can be performed. Because it is a 3-D char array and the address is to
the beginning of the next 2-D array using one unary * operator, the integer 2 causes
the addition of 2 × 20 = 40 bytes. The expression

```
*(*(cc+1)+2)
```

represents the address of the beginning of the third row (because we have +2) of
the second (because we have +1) 2-D array. When we have the expression

```
*(*(cc+1)+2)+3
```

we are adding the integer 3 to an address. Because this is a 3-D array and we have
used two unary * operators, the 3 means to add 3 bytes to the address represented
by *(*(cc+1)+2). Applying another unary * operator, we get

```
*(*(*(cc+1)+2)+3)
```

Because we have three unary * operators and this is a 3-D array, the expression represents a single character value and we can use it in a putchar statement

```
putchar(*(*(*(cc+1)+2)+3));
```

and print out a single character.

9. With the 3-D array cc[ ][ ][ ], what does *(*(cc+1)+2) represent? As
already indicated, it represents the address of the beginning of the third row (because
we have +2) of the second (because we have +1) 2-D array. We can use this with
puts to print the row this address indicates. The statement

```
puts(*(*(cc+1)+2));
```

prints this row.

**10. How can we use putchar with a single unary * operator to print a charac-
ter from the cc[ ][ ][ ] array?** By assigning the address of the first character of the
cc[ ][ ][ ] array to the single pointer variable, dd, we can use a single unary * operator to access individual values of the cc[ ][ ][ ] array. For instance, the statements

```
dd=&cc[0][0][0];
putchar(*(dd+80));
```

cause the 81st character to be printed. Using the declared size of the cc[2][3][20] array, the 81st character can be considered the 21st character of the second 3 × 20 two-dimensional array. Because 20 characters are in each row, the 21st character is the first character of the second row, which is s.

11. How can we use dd and puts to print a single row of the cc[ ][ ][ ] array? Because dd represents the address of a single character, to print a single row we must add to dd the correct number of bytes to get us to the row in which we are interested. For instance, with each of the rows having a length of 20, the follow statement

puts(dd+80);

goes past the first four rows and causes the string of the fifth row to be printed, "shown".

12. What do cc, *cc, and **cc represent? All three represent the address of the first element of the cc[ ][ ][ ] array. However, remember that, when we add integers to each of these, we get different results. For instance,

$$cc+1 = \text{address of first element} + 60 \text{ bytes}$$
$$*cc+1 = \text{address of first element} + 20 \text{ bytes}$$
$$**cc+1 = \text{address of first element} +1 \text{ byte}$$

For this lesson's program, this leads to:

Address of c[0][0][0]=FE92, cc+1=FECE, *cc+1=FEA6, **cc+1=FE93

If you run this program yourself, you will likely get different addresses, but the relationships between the addresses should be the same.

13. What visual image can we use for the pointer notation used on the bb[ ] [ ] and cc[ ][ ][ ] arrays in this lesson's program? Figure 7.11 illustrates the actions indicated by the pointer notation. In this figure, we show only the first character of each row of each array. As with other figures of this sort, the other characters are behind the first one of each row.

14. How can we visualize using the pointer variable dd accessing the cc[ ][ ][ ] array? Because dd is a single char pointer variable, adding 1 causes 1 byte to be added. This is illustrated most easily by showing the cc[ ][ ][ ] array in the manner that we have used for 1-D arrays. In Fig. 7.12, we show cc[2][3][20] as a 1-D array with six 20-byte segments. The action of *(dd+80) follows the path shown, which indicates moving 80 bytes along the cc[ ][ ][ ] array.

15. Can we use other forms of this notation to access array elements? Yes, we can use a combination of array and pointer notation. For instance, *(*(bb+3)+5) can be written as *(bb[3]+5). This works because of the similarity of the meanings of *(bb+3) and b[3]. We will not use this particular notation in this text; however, you may see it in other programs.

FIG. 7.11
Pointer notation for this lesson's program

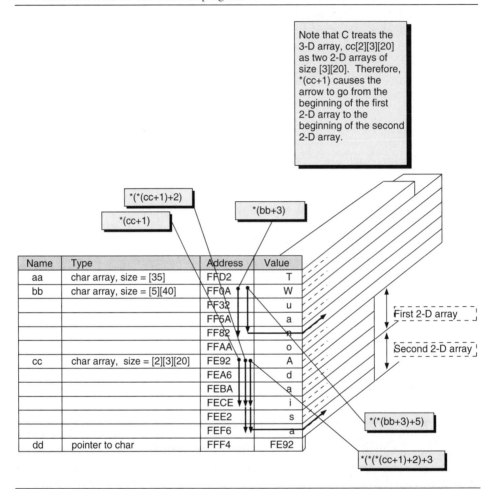

16. Why can we use so many different types of notations? Many C compilers actually convert array notation to pointer notation during compilation. So, even if you have written array notation, it is converted to pointer notation in your object code. In most cases, using array or pointer notation makes little difference to your program's speed of execution.

17. How does pointer arithmetic work if we are using types other than char? We must be aware that an individual element of other types occupies more than 1 byte. Recall that we previously listed common amounts of storage for other data types. While ANSI C defines the size of char to be 1 byte, the size of the other types is not

FIG. 7.12
Visual representation of *(dd+80)

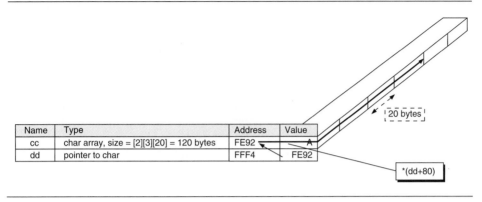

Name	Type	Address	Value
cc	char array, size = [2][3][20] = 120 bytes	FE92	A
dd	pointer to char	FFF4	FE92

20 bytes

*(dd+80)

defined by the standard. However, type int often is 2 bytes, float often is 4 bytes, and double often is 8 bytes. If we have the following declarations for 1-D arrays

```
int aa[50];
float bb[100];
double cc[70];
```

the pointer arithmetic is, for example:

Declaration	Number of bytes per element	Operation	Action
int aa[50]	2	aa+20	adds (2 bytes $\times$ 20) = 40 bytes
float bb[100]	4	bb+12	adds (4 bytes $\times$ 12) = 48 bytes
double cc[70]	8	cc+9	adds (8 bytes $\times$ 9) = 72 bytes

Because C uses the array type in performing pointer arithmetic, there is a direct correspondence between pointer notation and array notation, even for int, float, double, and other data types. For instance, the following are equivalent:

Pointer notation	Array notation
*(aa+20)	aa[20]
*(bb+12)	bb[12]
*(cc+9)	cc[9]

If we have multidimensional arrays with int, float, and double, the rules for using pointer notation just described also apply, meaning that C treats multidimensional arrays as arrays of arrays.

Remember, in determining what pointer notation means, you must look at the declaration closely and understand how C uses the declaration in interpreting the pointer notation. If you do not look at the declaration, you will not understand what the notation is indicating.

This works conversely, too. For C to be able to perform pointer arithmetic properly, it must have the correct declaration. This becomes especially important when dealing with user-defined functions, because the function header must be correct for C to properly perform pointer arithmetic in the function body. Luckily, C often will indicate an error on compilation if the header is incorrect. However, by understanding pointer arithmetic, we understand how C uses our commands in functions. The next lesson addresses the topic of user-defined functions.

18. If *(dd + 2) had been used at the end of this lesson's program, it would represent the character 't' (the third character in the cc[ ][ ][ ] array). What is the meaning of *dd + 2? Without parentheses the first action is evaluating *dd, which is the character 'A'. Because C treats characters as integers, adding 2 to 'A' gives the character 'C' in the ASCII character set. The lesson from this is that parentheses are important in using pointer notation.

EXERCISES

1. Based on these declarations and statements,

    ```
    char y[4]={"321"}, z[2][4]={"CAT","DOG"}, *py, *pz;
    py=&y[0];
    pz=&z[1][0];
    ```

    determine whether each of the following statements is true or false:
    a. Both &y[0] and &y represent the address of y[0].
    b. Both &y[0] and y represent the address of y[0].
    c. Both &y[2] and y+2 are equal to the address of y[2].
    d. y[0] is equal to y.
    e. y[0] is equal to py.
    f. y[0] is equal to *py.
    g. *pz is equal to 'D'.
    h. *(pz+1) is equal to 'O'.
    i. *pz+1 is equal to 'E'.
    j. *y+2 is equal to '5'.
    k. *(y+2) is equal to '1'.
    l. *py+2 is equal to '5'.
    m. *(py+2) is equal to '1'.

2. Based on these declarations

    ```
    char y[4]={"321"}, z[2][4]={"CAT","DOG"}, *py, *pz;
    ```

    find the error(s), if any, in the following statements:
    a. `py= y[1];`
    b. `py = &y[4];`
    c. `pz = &z[1][2];`
    d. `*(*z+2) = *y+2;`

3. Find the values of a, b, and c manually and then run the program to check your answers:

```c
#include <stdio.h>
void main(void)

{
        char x[5]="ABCD", y[2][6]={"EFGH", "JKL"}, *px, *py;
        char a, b, c;

        px=x;
        py=&y[1][0];
        a = *px+2;
        b = *x + 10 ;
        c = 10+ *(*(y+1));
        printf("a=%c, b=%c, c=%c\n",  a,b,c);
}
```

4. Find the value of *x* in the following program manually and then run the program to check your result:

```c
#include <string.h>
#include <stdio.h>

void f1 (char a[][10], char *b[], char *pa)
{
        int I;

        pa=&b[1][2];
        strcpy(a[0],b[0]);
        *a[1]=*pa;
        *(*(a+1)+1)= *(pa+1);
        a[1][2]='\0';
}

void main(void)
{
        char x[10][10],*y[10]={"abcde","wxyz"}, *px;

        px=y[1];
        f1(x,y, px);
        printf("x[0]=%s\nx[1]=%s\n",x[0],x[1]);
}
```

Solutions

1. a (true), b (true), c (true), d (false), e (false), f (true), g (true), h (true), i (true, because *pz is 'D' and C treats characters as integers; adding 1 to 'D' gives 'E'), j (true, because *y is '3'; adding 2 to *y gives the character '5'), k (true), l (true), m (true)

2. a. Statement is in error because py is a pointer, y[1] is a character '2'.
   b. The subscript of y must be <=3.
   c. No error.
   d. No error, statement replaces z[0][2] with the character '5'.

■ LESSON 7.10 DYNAMIC MEMORY ALLOCATION

Topics
- Reserving memory during execution
- Using calloc, malloc, and realloc

As mentioned at the beginning of this chapter, there may be a problem if you try to use fixed-size arrays for the maximum possible problem that your program may solve. The reason for this is that your program may work on some computer systems that have more than enough memory to handle the fixed-size arrays but not on others that lack enough memory. Your program would not work at all on some systems, even if you want to use the program for handling only a small amount of data.

To make your programs work on systems that have both large and small amounts of memory, rather than use fixed-size arrays, you can reserve memory based on the size of problem you are running. In other words, if you are using character arrays, for instance, to store both large and small engineering reports, you can reserve a large amount of memory when the reports are long and small amounts of memory when they are short. By doing so, the program written can work on systems with both large and small amounts of memory. However, a particular execution of the program may fail if a user attempts to store long reports on a computer with only a small amount of memory.

In this lesson, we illustrate the use of the functions (calloc, malloc, realloc, and free) that can be used to perform this dynamic memory allocation. In addition, Application Program 7.4 shows how to make use of the function calloc in a practical problem.

For their effective use in a program, it is necessary to declare data structures that allow us to utilize dynamic memory features. In this lesson's program, we declare two arrays for dynamic storage: a 1-D character array, aa[19], and a 1-D pointer array, bb[22]. These two arrays contrast with the one fixed-size array, cc[22][19], that would be used if only standard storage methods were in the program. In this program, we use arrays aa[ ] and bb[ ] to store all the information contained in cc[ ][ ]. In the Explanation, we illustrate that, with aa[ ] and bb[ ], only slightly more than the amount of memory actually needed is reserved, whereas with cc[ ][ ], a large amount is reserved but not occupied.

We have deliberately sized cc[ ][ ] to be larger than actually needed in this specific example, because we typically size fixed arrays to handle the largest possible problem we envision, not what is needed for a particular case. We will see that, with dynamic storage, it is necessary to reserve only slightly more memory than is actually needed, making a more efficient use of memory than standard storage methods.

No modern computer would have any trouble handling an array of size [22][19]. Therefore, this example is somewhat artificial, in that dynamic storage would not be needed to store the fixed-size array. We simply ask that, for the purpose of this lesson, you visualize the character array cc[22][19] to be very large, one that would push the limits of the memory of your computer, and so one that might need to be replaced by the character array aa[19] and the pointer array bb[22] to make more efficient use of memory. Note that the amount of memory needed to accommodate the two 1-D arrays of aa[19] and bb[22] is far less than that needed for the 2-D array cc[22][19].

Dynamic memory allocation in C uses the functions of calloc( ) and malloc( ). The arguments included in the calls to these functions specify how much memory is to be reserved. This enables us, in programs, to specify only the amount of memory needed when we call calloc or malloc. These functions then return the address of the beginning of the memory reserved. With this address, we can access the memory and store in it the desired information. Make the observations listed next and read the explanation to understand how to use dynamic storage methods.

What you should observe about this lesson's program.

1. Three arrays are declared, an ordinary character array aa[ ], a pointer array bb[ ], and a 2-D character array cc[ ][ ].

2. The cc[ ][ ] array is initialized with three strings. These three strings occupy only a very small amount of the memory reserved for cc[ ][ ].

3. In section 1, the five statements perform the following five steps:

   Copy the first row of cc[ ][ ] to aa[ ].

   Determine the length (meaning number of characters) of the string stored in aa[ ].

   Call calloc with the arguments being the length of the string stored in aa[ ] and the number of bytes to store a single character of the string (indicated by sizeof(char)). The value returned by calloc (an address) is made to be an address to a string (with the cast operator (char *)), which is stored in the first element of the bb[ ] array.

   Copy the string stored in aa[ ] into the allocated space indicated by the address stored in bb[0].

   Print the string stored at the address indicated by bb[0].

4. In section 2, the six statements perform the following six steps:

   Copy the second row of cc[ ][ ] to aa[ ].

   Determine the length (meaning number of characters) of the string stored in aa[ ].

   Determine the number of bytes of the string stored in aa[ ].

   Call malloc with the argument being the number of bytes of the string. The value returned by malloc (an address) is made to be an address to a string (with the cast operator (char *)), which is stored in the second element of the bb[ ] array.

   Copy the string stored in aa[ ] into the allocated space indicated by the address stored in bb[1].

   Print the string stored at the address indicated by bb[1].

5. In section 3, the seven statements perform the following seven steps:

   Copy the third row of cc[ ][ ] to aa[ ].

   Determine the length (meaning number of characters) of the string stored in aa[ ].

   Determine the number of bytes of the string stored in aa[ ].

Call realloc with the arguments being the number of bytes of the string and the address stored in bb[1]. The value returned by realloc (an address) is made to be an address to a string (with the cast operator (char *)), which is stored in the second element of the bb[ ] array.

Copy the string stored in aa[ ] into the allocated space indicated by the address stored in bb[1].

Print the string stored at the address indicated by bb[1].

Free the memory (deallocate or unreserve it).

Source code

```
#include <stdio.h>
#include <stdlib.h>
#include <string.h>
void main(void)
{
        char aa[19];
        char *bb[22];
        char cc[22][19]={"Example", " String 1 ", "Words"};
        int xx, yy;



        printf("*********** Section 1 - Using calloc ************\n");

        strcpy(aa,cc[0]);
        xx=strlen(aa);
        bb[0]=(char *)calloc(xx,sizeof(char));


        strcpy(bb[0],aa);
        puts(bb[0]);




        printf("*********** Section 2 - Using malloc ************\n");

        strcpy(aa,cc[1]) ;
        xx=strlen(aa);
        yy=xx*sizeof(char);
        bb[1]=(char *)malloc(yy) ;
```

Initializing 3 of the declared 22 strings.

Temporarily storing the first string of cc[ ][ ] in aa[ ].

Determining the length of the first string of cc[ ][ ].

Reserving memory of the size of the first string in cc[ ][ ] and storing the address of the memory in bb[0]. At this point the memory is empty but reserved.

Copying the first string in cc[ ][ ] into the memory indicated by bb[0].

Temporarily storing the second string of cc[ ][ ] in aa[ ].

Determining the length of the second string of cc[ ][ ].

Calculating the number of bytes needed to store the second string of cc[ ][ ].

Reserving memory of the size of the second string in cc[ ][ ] and storing the address of the memory in bb[1]. At this point the memory is empty but reserved.

```
strcpy(bb[1],aa);
puts(bb[1]) ;
```

Copying the second string in cc[ ][ ] into the memory indicated by bb[1].

```
printf("********** Section 3 - Using realloc ************\n");
strcpy(aa,cc[2]);
xx=strlen(aa);
yy=xx*sizeof(char);
```

Same three first steps listed for Section 2, but using the third string of cc[ ][ ].

```
bb[1]=(char *)realloc(bb[1],yy);
strcpy(bb[1],aa);
puts(bb[1]) ;
free(bb[1]);
```

Reserving memory of the size of the third string in cc[ ][ ] at the location previously used to store the second string (if possible). If it is not possible, memory is reserved elsewhere automatically by realloc. The address of the reserved memory is stored in bb[1]. Previous contents are copied to the new memory location if a new location is created.

Copying the third string in cc[ ][ ] into the memory indicated by bb[1].

Unreserving the memory indicated by bb[1].

```
}
```

Output

```
********** Section 1 - Using calloc *************
Example
********** Section 2 - Using malloc *************
String 1
********** Section 3 - Using realloc *************
Words
```

Explanation

1. What does the function calloc do? The function calloc reserves memory during program execution. The amount of memory reserved is determined by the arguments passed to calloc. The form for calling calloc is

calloc (*number_of_elements, bytes_per_element*)

The amount of memory reserved is equal to the product of the integers *number_of_elements* and *bytes_per_element*. For example, in this lesson's program,

calloc(xx,sizeof(char));

causes memory for xx elements (where xx in the execution of this program is 8) of size sizeof(char) (which is 1 byte) to be reserved (it also initializes all bits in this memory region to be 0). In other words, this call to calloc reserves 8 bytes of memory.

The function calloc returns the address of the beginning of the first element reserved. For instance,

bb[0]=(char *)calloc(xx,sizeof(char));

causes the address of the beginning of the memory reserved to be stored in the first element of the pointer array, bb[ ]. The cast operator (char *) causes the pointer returned by calloc to be a pointer to a character, which is the same type as the elements of bb[ ]. However, should we be dealing with numeric arrays, the cast operator (int *) or (double *) can be used with calloc.

2. What does the function malloc do? The function malloc also reserves memory during program execution. The amount of memory reserved is determined by the single argument passed to malloc. The form for calling malloc is

malloc (*number_of_ bytes*)

The amount of memory reserved is equal to the *number_of_bytes*. To reserve the proper amount of memory, it is necessary for the programmer to code the calculation for the *number_of_bytes*. For example, in this lesson's program, the statement

yy=xx*sizeof(char);

multiplies xx (which has a value equal to the length of the string to be stored) by the number of bytes required to store a single character (which is 1 byte). Therefore, yy represents the number of bytes needed to store the string. Then,

malloc(yy);

causes memory for yy bytes to be reserved. The function malloc returns the address of the beginning of the first element reserved. For instance,

bb[1]=(char *)malloc(yy);

causes the address of the beginning of the memory reserved to be stored in the second element of the pointer array bb[ ]. The cast operator (char *) causes the pointer returned by malloc to be a pointer to a character, which is the same type as the elements of bb[ ]. However, should we be dealing with numeric arrays, the cast operator (int *) or (double *) can be used with malloc.

3. What does the function realloc do? The function realloc modifies the amount of memory previously reserved by a call to either malloc or calloc. The form is

realloc (*pointer, number_of_bytes*);

The amount of memory reserved is equal to the *number_of_bytes*, which must be calculated prior to calling realloc. The location of the memory reserved is indicated by *pointer*. The argument *pointer* must have been returned by a previous call to either calloc or malloc. For example,

realloc(bb[1],yy);

causes memory for yy bytes to be reserved at the location indicated by bb[1]. The address stored in bb[1] previously was returned by malloc and so is an appropriate argument for the call to realloc.

If the *number_of_bytes* is less than the *number_of_bytes* previously reserved by calloc or malloc for that pointer, then realloc can successfully reserve the space in memory indicated by *pointer*. In this case, the contents of the memory remain unchanged after the call to realloc. However, if the *number_of_bytes* is greater than the *number_of_bytes* previously reserved by calloc or malloc for that *pointer*, then realloc may not be able to reserve the same block of memory. If this is the case, then realloc reserves memory at a new location. The function returns the pointer to that new location. Therefore,

```
bb[1]=(char *) realloc(bb[1],yy);
```

causes memory to be reallocated and causes the address of the beginning of the block of memory to be stored in bb[1]. If the address stored in bb[1] after the execution of this statement is different from the address stored in bb[1] before execution of this statement, realloc copies the contents of memory from the first address to the second.

As with calloc and malloc, the cast operator (char *) causes the pointer returned by realloc to be a pointer to a character which is the same type as the elements of bb[ ]. However, should we be dealing with numeric arrays, the cast operator (int *) or (double *) can be used with realloc.

4. What does the function free do? The free function cancels the reservation for memory previously made by calloc, malloc, or realloc. The form is

free (*pointer*)

where *pointer* is the address of the region of memory that is to be unreserved. For instance,

```
free(bb[1]);
```

causes the block of memory previously reserved by realloc at the address indicated by bb[1] to be available for use by other calls to calloc or malloc. The address stored in b[1] remains unchanged after the call to free; however, we cannot use this address to store information, because memory at this address is no longer reserved. If we wanted to use bb[1] in further operations, we would need to call calloc or malloc (but not realloc) and store the returned value in bb[1].

It is good practice to free memory once it is not required. Do not rely on the operating system to free memory after termination of execution.

5. What happens if calloc, malloc, or realloc cannot reserve the memory as requested? They return a null pointer. It is worthwhile checking the return pointer from these functions to make sure that memory has been successfully allocated. We have not done so in this lesson's program only to keep it simple.

6. Conceptually, where in memory is the space reserved? ANSI C does not specify how and in what arrangement memory should be allocated using the dynamic memory allocation functions. So, various compilers handle this differently. Here, we describe a common image of memory that, although not precise, allows you to visualize some memory issues and their implications for programming.

The memory management functions can be thought of as reserving space in a region of memory that is sometimes called the *heap*. Conceptually, C can be considered to break up memory into four regions:

A region called the *stack*.
A region called the *heap*.
A region for global variables.
A region for program instructions.

The last three of these are in memory cells that have low addresses and the first one has high addresses. Regions 3 and 4 do not grow or shrink during execution, but regions 1 and 2 do. An image that you can use to visualize this is shown in Fig 7.13.

In this figure, the program instructions and global variables are on the left side, where low addresses in memory are located. These have been enclosed in solid lines, representing a fixed size during program execution. The heap sits immediately adjacent to these and the stack is located at the far right, at the end of the high addresses. As memory is reserved in the heap by the memory management functions, the heap grows to the right. As memory is freed, it shrinks.

The stack grows to the left as functions are called and memory is reserved to accommodate the variables and data structures associated with the functions. After execution of the functions, unless otherwise specified, this memory is deallocated and the stack shrinks.

FIG. 7.13
Conceptual image of regions of memory

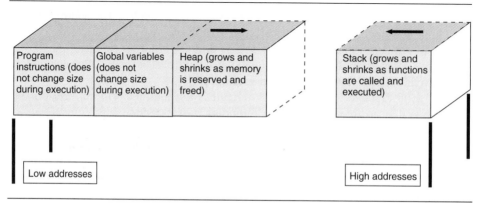

If too much memory is required by either the heap or the stack, the two of them can meet, which may cause an abnormality to occur and execution to terminate. This particular type of system, though, allows both regions of memory to grow independently and allows for the potential of all useful memory to be occupied.

7. How does the storage created with the memory management functions differ from the storage created with a fixed size array? In Fig. 7.14, we illustrate our image of memory with a fixed-size 2-D array. A characteristic of a fixed-size array is that, although a considerable amount of memory may be reserved, not all of it may be occupied for a particular problem. The reason for this is that the size of the block of memory is established at the time of compilation and this size should be large enough to accommodate the largest array of practical use. An array of size [22][19], similar to cc[ ][ ] used in this lesson's program, is illustrated in the figure. We outline a region of this array that may be occupied for a particular execution of a program, with the rest of the array being vacant. This leads to having a considerable

FIG. 7.14
Two-dimensional array storage using ordinary storage methods. The outlined region represents the occupied portion of memory. Note that a considerable amount of memory is unused.

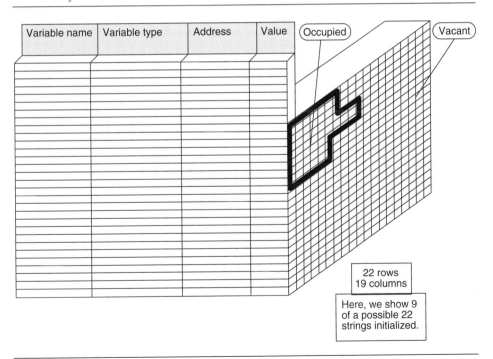

portion of memory unused. This may be acceptable if all systems on which the program is to run are large enough to accommodate all fixed-size arrays.

However, if this is not the case, then dynamic storage such as is illustrated in Fig. 7.15 is better, because it uses nearly all of the memory it reserves. In this figure, we show how only the memory needed is reserved in the heap, while the addresses to the memory in the heap are stored in a pointer array in the stack. Also, a standard array (type char, for instance) in the stack is shown. This array is used to temporarily store a row of information that later will be moved to the heap.

If we compare Figs. 7.14 and 7.15, we see that both storage schemes hold the same amount of information. However, we also see that, because the fixed-size array of Fig. 7.14 has a considerable portion of reserved space that is vacant, the dynamic allocation system of Fig. 7.15 uses substantially less memory.

8. What steps are involved in using dynamic memory allocation in this lesson's program? We illustrate a portion of the process in Fig. 7.16. Read steps 1–6 and follow the diagram in this figure. First we store a string from cc[ ][ ] temporarily in the character array aa[ ]. Then, we determine the length of this string and send the length to calloc. The function calloc reserves memory of sufficient

FIG. 7.15

Two-dimensional array storage using dynamic storage methods. Note that all of the heap memory is occupied, meaning that little memory is wasted. The amount of memory reserved is much less than that for the standard method shown in Fig. 7.14.

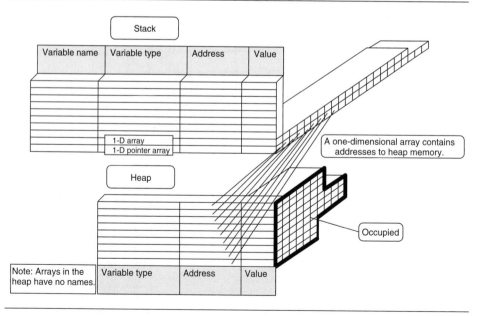

FIG. 7.16

Using strlen and calloc to create dynamic storage for character strings. After having reserved memory using this method, it is necessary to load the information into the reserved cells. This step is not shown in the diagram.

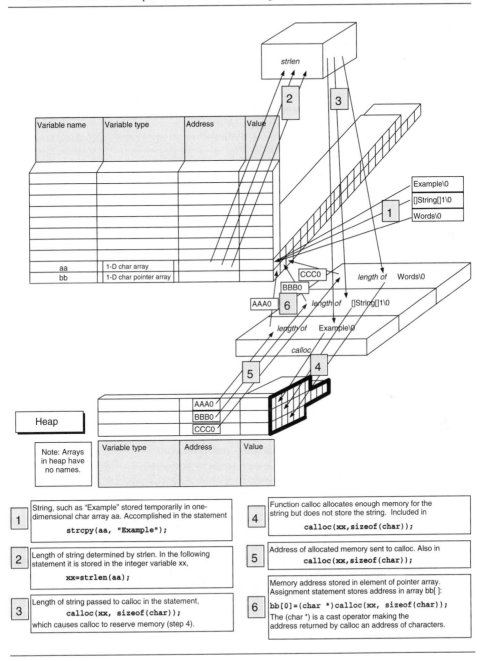

		Variable name	Variable type	Address	Value
		aa	1-D char array		
		bb	1-D char pointer array		

	Variable type	Address	Value
Note: Arrays in heap have no names.			

Heap

1. String, such as "Example" stored temporarily in one-dimensional char array aa. Accomplished in the statement
   ```
   strcpy(aa, "Example");
   ```

2. Length of string determined by strlen. In the following statement it is stored in the integer variable xx,
   ```
   xx=strlen(aa);
   ```

3. Length of string passed to calloc in the statement,
   ```
   calloc(xx, sizeof(char));
   ```
   which causes calloc to reserve memory (step 4).

4. Function calloc allocates enough memory for the string but does not store the string. Included in
   ```
   calloc(xx, sizeof(char));
   ```

5. Address of allocated memory sent to calloc. Also in
   ```
   calloc(xx, sizeof(char));
   ```

6. Memory address stored in element of pointer array. Assignment statement stores address in array bb[ ]:
   ```
   bb[0]=(char *)calloc(xx, sizeof(char));
   ```
   The (char *) is a cast operator making the address returned by calloc an address of characters.

length in the heap and sets the address of the beginning of the reserved portion. This address is then stored in the pointer array bb[ ]. The following statements perform these operations:

```
strcpy(aa,cc[0]);
xx=strlen(aa);
bb[0]=(char *)calloc(xx,sizeof(char));
```

After executing these three statements, the memory in the heap is reserved but nothing is stored in it. We store the string aa[ ] in the heap at address b[0] using the statement

```
                        strcpy(bb[0],aa);
```

Thus, we have been able to utilize C's dynamic memory management capabilities with only four statements.

EXERCISES

1. True or false:
   a. C's memory management functions reserve memory during program execution.
   b. Memory is arranged for fixed-size arrays during compilation.
   c. C's memory management functions cause the stack to grow and shrink during execution.
   d. The function realloc can be the first memory management function called by a program.
   e. The function malloc initiates the bits of the reserved memory to be 0.
   f. The function calloc initiates the bits of the reserved memory to be 0.

2. Given these declarations

   ```
   char aa[10], *bb, cc[5][50], *dd[8];
   int xx, yy, *zz;
   ```

   find the errors, if any, in the following statements:
   a. `bb = calloc (xx, sizeof(char));`
   b. `bb = (char *)malloc(xx, sizeof(char));`
   c. `aa[0] = (char *) calloc(xx, sizeof(char));`
   d. `zz = (int *)calloc(xx, sizeof(int));`
   e. `cc[0] = (char *) malloc(xx);`

3. Write a program that uses calloc to reserve memory to store an equivalent of a [400][80] character array. Fill the array with only [10][30] characters. Print these characters to the screen.

4. Write a program that uses malloc to reserve memory to store an equivalent of a [400][80] integer array. Fill the array with only [10][30] integers. Print these integers to a file.

Solutions

1. a (true), b (true), c (false), d (false), e (false), f (true)
2. a. `bb=(char *)calloc (xx, sizeof(char));` need (char *) to cast pointer type.
   b. `yy=xx*sizeof(char);`
      `bb = (char *)malloc(yy);`
   c. `dd[0] = (char *) calloc(xx, sizeof(char));` must use pointer variable on left side.
   d. No error.
   e. `dd[0] = (char *) malloc(xx);`

▧ PROGRAM DEVELOPMENT METHODOLOGY

To this point, we have used a four-step method of program development. However, because we have introduced more complex data structures such as arrays and arrays of pointers, we will add a step for deciding on a method of data storage. We will find that, as programs become more complex, we will have several possible data structures to use and choosing one over the others will simplify the programming process and reduce the likelihood of bugs. In addition, we may be able to reduce memory requirements and increase the speed of execution. Our steps in program development are

1. Assemble relevant equations and background information.
2. Work a specific example.
3. Decide on the major data structures to be used.
4. Develop an algorithm, structure charts, and data flow diagrams.
5. Write source code.

▧ APPLICATION PROGRAM 7.1 CREATION OF A SPREADSHEET TYPE PROGRAM; INDIVIDUAL CHARACTER OPERATIONS

Problem statement

Write a simple spreadsheet type program capable of reading a table of data from a file and performing addition, subtraction, multiplication, or division on the rows. This input should consist of a file with a table of eight rows and five columns of numbers. The first column is column A, the second is column B, continuing with the last column being column E. Prompt the user to enter an equation with five operands and four operators from the keyboard involving these columns. For instance, A + B − C + E × D, which indicates the operations to be performed on each column of numbers. The program should be capable of creating a results column that represents the result of the user input equation. Print this result to the screen.

Solution

RELEVANT EQUATIONS AND BACKGROUND INFORMATION

A spreadsheet program works with tables of data and performs operations column by column (among many other things). An example of such a program is Microsoft Excel™. For instance, we may have a table of student grades:

Name	Score, Exam 1	Score, Exam 2	Score, Exam 3
John Doe	68	43	81
Mary Green	87	76	98
Mark White	76	73	83
Sam Brown	92	65	97
Susan Jones	83	56	92

In a spreadsheet program, we can create input data in a tabular form that looks very much like the table just shown. We then write simple equations as input to create new columns containing our desired information. For instance, we could use a spreadsheet program on the table of data to get a fifth column that contains the average of the three exam scores. To get this, we would simply add the scores of exams 1, 2, and 3 (which are in columns 2, 3, and 4) and divide the sum by 3.

In this example, we write a very simple version of a program that can do operations such as these. With the data in a file, we read it into memory. Then, at the keyboard, we write an equation representing what to do with each column and have our program perform the operations written in the equation.

Because we are doing only addition, subtraction, multiplication, and division, the equations are rather trivial. So, we will not dwell on them in this section. Instead we will illustrate the operations with a sample calculation.

SPECIFIC EXAMPLE

For a data file consisting of (the A, B, C, D, and E are not in the file, only the numbers)

A	B	C	D	E
1	3	2	3	4
2	5	4	6	8
3	7	6	9	12
4	9	8	12	16
5	11	10	15	20
6	13	12	18	24
7	15	14	21	28
8	17	16	24	32

we can perform the action of $A - C + B \times E - D$, where the variables A, B, C, D, and E represent their respective columns of numbers. The associativity of the action will be from left to right with no order of precedence of the operators. In other words, the multiplication in this expression will not be carried out prior to the addition or subtraction.

This result of this for the first row would be $1 - 2 + 3 \times 4 - 3 = 5$. Continuing for the rest of the rows gives the following result:

Result of $A - C + B \times E - D$
5
18
39
68
105
150
203
264

DATA STRUCTURES

Now that you have begun to develop experience in programming, you need to draw on the experience to help you make early decisions that will affect the rest of the programming process. Specifically, you need to decide how to store your data (for instance, what dimension of arrays should be used) because this will affect how you work with the data in your programs.

Before you get going, you must make these decisions, then work with the structures on which you decided. If the structures become too cumbersome to implement, you can change them. However, changing your data structures may mean rewriting a considerable amount of code. Therefore, it is worthwhile to spend time in choosing your data structures as well as you can before going further in the programming process.

In this program, we have two categories of data to consider:

1. The columns of input data.
2. The result column.

We have several data structure options from which to choose:

1. Since we have five columns of input data we could create five 1-D arrays, a[ ], b[ ], c[ ], d[ ], and e[ ]; that is, one for each column. We could create a sixth array with the results, called result[ ]. All of these arrays would have size of 8 because we have eight rows in our table. We would declare these as

    ```
    double a[8], b[8], c[8], d[8], e[8], result[8];
    ```

2. Since we have eight rows, we could create eight 1-D arrays, row1[ ], row2[ ], row3[ ], . . . , row8[ ]. If we choose this option, we would have a size of 6 for each of these because we have five data columns and a result column. These would be declared

    ```
    double row1[6], row2[6], row3[6], row4[6], row5[6], row6[6],
    row7[6], row8[6];
    ```

3. Since we have what can be called a table or a matrix, we could choose a 2-D array to represent the columns and rows. In addition, we could have a 1-D array

represent the result. For the 2-D array, we have eight rows and five columns. Together with the eight rows, for the result column, we have the following declaration:

```
double a[8][5], result[8];
```

4. Last, we could put the result column in as the last column in the 2-D array rather than having it separate. This gives us six columns in our table, leading to the following declaration:

```
double a[8][6];
```

For the particular program specified for this example, *any* of these options can be used. Given the time and effort, we could write a successful program with any of these data structures. In a sense, then, there is no wrong answer in choosing one of them. However, in most cases, given a choice between a series of 1-D arrays and a 2-D array, the 2-D array will be easier to implement and more flexible for making future changes. Therefore, we can narrow our choices down to numbers 3 and 4. Of these two, we will choose number 3. Number 4 offers flexibility advantages, but we will choose number 3 because the result is clearly separate from the input information. However, number 4 is perfectly acceptable and easily could be used.

We recommend that, with your programs, you make a similar list of options of data structures. Having them listed in front of you allows you to consider all the possibilities. With more experience, you will feel more comfortable choosing one of the options. However, after choosing one, you may realize that you have made a bad decision. This may become obvious if you find that the programming becomes too difficult. If this occurs, consider another structure on the list, because it may solve some of your programming problems. Remember, even after you have written a considerable amount of code, if you see a better way of writing your program, you should start over and try again.

ALGORITHM
The general algorithm can be stated as follows:

1. Read the input data from the data file, which consists of the numbers in the table with eight rows and five columns.
2. Prompt the user to enter an equation in the form of "operand operator operand operator operand operator operand operator operand". For example, "A + B * C − D * E."
3. Take the first operand from the entered equation and set the initial values in the result [ ] array to be the values in the column indicated by the operand.
4. Repeat for the remaining four operators and four operands:

    Read an operator and an operand.
    Use the operator to determine the operation to be performed.
    Perform the operation using the result array and the column indicated by the operand.
    Put the result of the operation in the result [ ] array.

5. Print the result [ ] array to the screen.

We describe the development of the code for each of the steps except step 2, which is trivial.

1. Read the input data, which consists of the numbers in the table with eight rows and five columns. The first impact of our choice of data structure being a two-dimensional array is in the way that we read in the data. Since we have previously covered using a two-deep nested loop to read a 2-D array of data, we will not dwell on it here but simply show the loops and note that the outer loop loops over the number of rows and the inner loop loops over the number of columns, which we define in preprocessor directives.

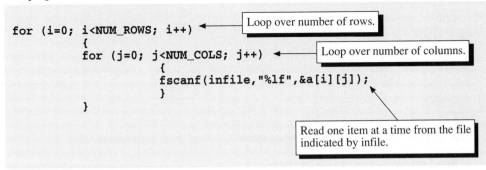

3. Read the first operand from the keyboard and set the initial value of our result [ ] array to be the values of the column indicated by the operand. Since we know that the first character input from the keyboard is an operand (a letter, from A to E), we begin by reading an operand (and in this case we have chosen the variable name to be operand) using the getchar function. The column chosen by the letter input becomes our initial result [ ] array. Because eight elements are in the result [ ] array we must loop over this number. Given that we previously initialized the result [ ] array to be 0, we add the correct column of the a[ ][ ] array to the result [ ] array. We determine the correct column by converting the character read to its integer value. Recall that for the ASCII or EBCDIC codes, an integer corresponds to each character. For instance the ASCII codes for A–E are

Character	ASCII code	Second subscript of a[ ][ ] array
A	65	0
B	66	1
C	67	2
D	68	3
E	69	4

Also shown in this table is the second subscript (because the subscripts are in [row] [column] order) corresponding to the column indicated by the character. For instance, column A in our spreadsheet is the first column, which corresponds to the second subscript 0. When we enter the letter C, for example, we need to be able to compute the number 2 to get the correct second subscript for our a[ ][ ] array. We note that $67 - 65 = 2$, which leads to $(int)('C') - (int)('A') = 2$. So, to get the correct subscript we do $(int)operand - (int)('A')$. The following code performs these operations:

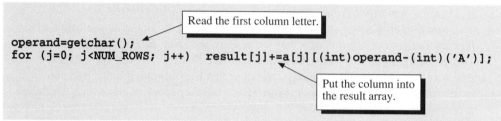

```
                          Read the first column letter.
operand=getchar();
for (j=0; j<NUM_ROWS; j++)   result[j]+=a[j][(int)operand-(int)('A')];

                                     Put the column into
                                     the result array.
```

Remember, the user has typed in an entire equation and pressed the return key. This action sends a sequence of characters to the input buffer. These characters remain in the buffer until the program directs that they be extracted. Each getchar call causes a character to be read. At the end of execution of this code, we have read one character. Eight characters of interest to us remain in the buffer. In step 4, we extract those characters.

4. Repeat for the remaining four operators and four operands:

Read an operator and an operand.
Use the operator to switch to the operation to be performed.
Perform the operation using as operands the result [ ] array and the column indicated.
Put the result of the operation in the result [ ] array.

We use a loop because we are repeating operations. The first two steps take the following form:

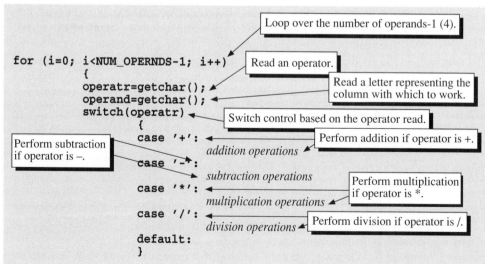

```
                                    Loop over the number of operands-1 (4).

for (i=0; i<NUM_OPERNDS-1; i++)
        {                         Read an operator.
        operatr=getchar();                  Read a letter representing the
        operand=getchar();                  column with which to work.
        switch(operatr)           Switch control based on the operator read.
            {
            case '+':                        Perform addition if operator is +.
                addition operations
Perform subtraction
if operator is –.
            case '-':
                subtraction operations
                                             Perform multiplication
            case '*':                        if operator is *.
                multiplication operations
            case '/':                        Perform division if operator is /.
                division operations
            default:
            }
```

Recall that the symbols $+$, $-$, $*$, and $/$ are considered to be characters. A switch structure is capable of distinguishing those characters. The *operations* in the switch structure consist of taking the entire column specified and adding, subtracting, multiplying, or dividing it with the result array. Note that we need to use a break statement to bypass the other cases of the switch structure. Therefore, the *operations* in the preceding code are represented by a loop of the sort that follows for multiplication (as an example):

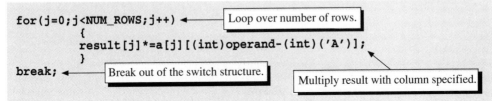

```
for(j=0;j<NUM_ROWS;j++)          Loop over number of rows.
         {
             result[j]*=a[j][(int)operand-(int)('A')];
         }
break;              Break out of the switch structure.

                                 Multiply result with column specified.
```

Note that in this code, the value of the second subscript for the a[ ][ ] array is determined using the method described under step 3. Similar loops can be written for the other operators.

5. Print the result array to the screen. The result can be printed with a heading first. Then, in a loop over the number of rows, a printf statement with \n in the string literal causes the results to be printed in column form.

```
                          Print heading.


printf ("Result \n");
for (i=0; i<NUM_ROWS; i++) printf("%lf \n",result[i]);


                          Print result on each line, making a column.
```

SOURCE CODE

Each of the preceding steps has been put into the complete source code that follows. Read the code line by line to make sure you understand how it operates.

```
#include <stdio.h>
#define NUM_COLS 5
#define NUM_ROWS 8
#define NUM_OPERNDS 5
void main(void)
{
        char operatr,operand;
        int i,j;
        double a[NUM_ROWS][NUM_COLS], result[NUM_ROWS]={0.};
        FILE *infile;
```

```
infile = fopen ("SPREAD.DAT", "r");
for (i=0; i<NUM_ROWS; i++)
        {
        for (j=0; j<NUM_COLS; j++)
                {
                fscanf(infile,"%lf",&a[i][j]);
                }
        }

printf ("Enter a command of the sort A+B*C-D*E:\n");
operand=getchar();
for (j=0; j<NUM_ROWS; j++)  result[j]+=a[j][(int)operand-(int)('A')];

for (i=0; i<NUM_OPERNDS-1; i++)
        {
        operatr=getchar();
        operand=getchar();

        switch(operatr)
                {
                case '+':
                        for(j=0;j<NUM_ROWS;j++)
                                {
                                result[j]+=a[j][(int)operand
                                -(int)('A')];
                                }
                        break;
                case '-':

                        for(j=0;j<NUM_ROWS;j++)
                        {
                        result[j]-=a[j][(int)operand-(int)('A')];
                        }
                        break;
                case '*':
                        for(j=0;j<NUM_ROWS;j++)
                                {
                                result[j]*=a[j][(int)operand
                                -(int)('A')];
                                }
                        break;
                case '/':
                        for(j=0;j<NUM_ROWS;j++)
                                {
                                result[j]/=a[j][
                                (int)operand-(int)('A')];
                                }
                        break;
                default:
                        break;
                }

        }

printf ("Result \n");
for (i=0; i<8; i++) printf("%lf \n",result[i]);
}
```

INPUT FILE SPREAD.DAT

1	3	2	3	4
2	5	4	6	8
3	7	6	9	12
4	9	8	12	16
5	11	10	15	20
6	13	12	18	24
7	15	14	21	28
8	17	16	24	32

OUTPUT

Keyboard input	```Enter a command of the sort A+B*C-D*E:```
	```A-C+B*E-D```
	```Result```
	```5.000000```
	```18.000000```
	```39.000000```
	```68.000000```
	```105.000000```
	```150.000000```
	```203.000000```
	```264.000000```

COMMENTS

To keep this program relatively simple, we deliberately neglected doing any data checking. If the user accidentally types in something incorrect, the program will fail. This is unacceptable for a professionally written program. In a later example, we will illustrate how to do data checking to avoid this problem.

Our program has been written such that it works for both ASCII and EBCDIC character sets. This is because, in determining the column subscript from the character input, we subtracted the value (int)('A'). Both ASCII and EBCDIC have character values that increase by 1 in alphabetical order. In other words, *B* is always one greater than *A*, and *C* is always 1 greater than *B*. We can use this characteristic not only in the way that we have in this program but in alphabetizing a list of characters. Because of the correspondence between the letters and number order, we can use the sorting techniques we illustrated in Chapter 6 to alphabetize. The conversion is straightforward. One of the exercises assigns the task of developing an alphabetizing program. In alphabetizing, one point is a source of difficulty, though: when lowercase and uppercase letters are mixed in a list. When this occurs the relationships just described no longer hold. In other words, the letter *b* is not 1 greater than the letter *A*. To get the alphabetizing correct, we need at least temporarily to make all the letters the same case. With them all the same case, standard sorting techniques can be applied.

Modification exercises

1. Modify the program so that it can handle an equation with ten operands and nine operators. In other words, an acceptable input equation would be C / D − E + A / C − B / E − B + C + D.

2. Modify the program so that it can handle an input table with ten columns (A–J). Change the input file so that it has ten columns. Have the program accept ten operands and nine operators.

3. Modify the program so that it can use a variable number of operands. Have the program automatically determine the number of operands entered.

4. Modify the program so that it prints an error message if an unacceptable operand or operator has been typed. Make sure that the program does not stop executing. Allow the user to type in a new equation. Repeat the process until the user types in an acceptable equation.

5. Modify the program so that it can handle one set of parentheses. In other words, an acceptable equation would be A * (B + C + D − E). Although this sounds simple, it does involve a fair amount of rewriting of the program.

APPLICATION PROGRAM 7.2 UNITS CONVERSION PROGRAM; STRING OPERATIONS

Problem statement

Write a program that converts measurements of length or weight in one set of units to another set of units. Input should be typed in from the keyboard in the form of (for example)

`5.2 cm to inches`

Output should be to the screen giving the original measurement and the measurement in the new units. For example,

`5.2 cm is 2.047 inches`

Solution

RELEVANT EQUATIONS

We will use the following form to convert units:

$$x_{new} = x_{old}c_1 / c_2 \tag{7.1}$$

where

x_{new} = value in new units
x_{old} = value in old units
c_1 = number of old units per base unit
c_2 = number of new units per base unit

To obtain values of c_1 and c_2, we need to select a base unit for both length and weight. For length, we choose meters and for weight, we choose newtons. Note,

▨ **TABLE 7.6**
  Length

Unit	c_1 or c_2 (number of meters in one unit)
mm (millimeter)	0.001
cm (centimeter)	0.01
m (meter)	1.0
km (kilometer)	1000
inch (inch)	0.0254
ft (feet)	0.3048
yd (yard)	0.9144
miles (English mile)	1609

▨ **TABLE 7.7**
  Weight

Unit	c_1 or c_2 (number of newtons in one unit)
N (newton)	1.0
kN (kilonewton)	1000
MN (meganewton)	1,000,000
ounces (16 ounces = 1 lb)	0.2782
lb (pound)	4.452
ton (English ton = 2000 lb)	8905

though, that we could have chosen other base units. This leads to the following Tables 7.6 and 7.7 of values of c_1 and c_2.

SPECIFIC EXAMPLE

To convert 5.2 cm to inches, using equation (7.1), we get

$$x_{new} = (5.2)(0.01)/(0.0254) = 2.047 \text{ in.}$$

where the old units are cm (which gives c_1 from the table) and the new units are inches (which gives c_2 from the table).

Similarly, to convert 5000 lb to kN, we have

$$x_{new} = (5000)(4.452)/(1000) = 22.26 \text{ kN}$$

DATA STRUCTURES

In this program, we work with strings instead of single characters (which we used in the previous application example). With strings, we must use arrays.

In the program, our strings are single words (the units involved) instead of entire lines or sentences. We have eight different length units to store and six different weight units to store. We choose to put them all in one 2-D array. (Note: we could have put each in a separate 1-D array, but this is unwieldy.) This leads to

```
char units[14][7]={"mm", "cm", "m", "km", "inches", "ft", "yd",
        "miles", "N", "kN", "MN", "ounces", "lb", "ton"};
```

The size of the second subscript is set to accommodate the largest word in the list.

Since we put all the units in one array, we should put all of the conversion factors into one array and in an order corresponding to the order for the units[ ][ ] array. This leads to the following:

```
double conv_fc[14] = {0.001,0.01,1.0,1000.,
             0.0254,0.3048,0.9144,1609.,
        1.0,1000.,1.0e+06,0.2782,4.452,8905.};
```

Remember that, because this is a numeric array, it is only one dimension. Because of the correspondence between the arrays units[ ][ ] and conv_fc[ ] they are said to be *parallel arrays.*

Finally, we choose to put the units entered from the keyboard into the 1-D arrays:

```
char from_units[7], to_units[7];
```

because they are unrelated. Both of them have a size of 7 to accommodate the longest word in the units list.

ALGORITHM
The following steps are used in this program:

1. Prompt the user for a request.
2. Read the one line input from the user.
3. Find a match between the units that the user entered and the units in the program list.
4. From the match, use the corresponding conversion factors to compute the units conversion using equation (7.1).
5. Print the results to the screen.

These steps are described next, except step 1, which is trivial.

2. Read the one line input from the user. An example of one line of input is

```
5.2 cm to inches
```

which can be broken down into the following variable names:

```
from_value   from_units   word   to_units
```

The one line input has both double and char values in it. Because a space separates each of the input values, the scanf function with the %s conversion specifier works well for reading this information.

This leads to the following scanf statement:

```
scanf ("%lf%s%*s%s",     &from_value, from_units, to_units);
```

Note that the double variable from_value has & whereas from_units and to_units have no & (because they are arrays and array names without subscripts are addresses). Also, recall a feature of the conversion specifier in the format string that we have not discussed earlier. The * symbol placed between % and type specifier causes input to be read but discarded. Here, %*s causes the word to to be read but discarded. We do not need it so we choose to discard it.

3. Find a match between the units the user has entered and the units in the program list. Because we want to compare strings, we use the strcmp function. First, we want to compare from_units to the list of units in the units[ ][ ] array. To do this, we need to create a loop. Each cycle through the loop compares from_units to one of the units in the list. The goal is to find the matching units and save the array index that gives the match. The following loop accomplishes this:

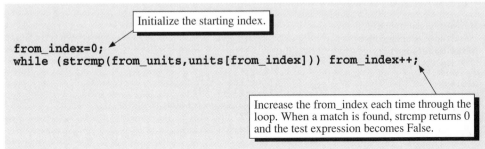

```
from_index=0;
while (strcmp(from_units,units[from_index])) from_index++;
```

Initialize the starting index.

Increase the from_index each time through the loop. When a match is found, strcmp returns 0 and the test expression becomes False.

In this loop, when a match occurs between from_units and units[from_index], the function strcmp returns 0 and causes the while loop to terminate. On termination of the loop, the value of from_index is equal to the index that produces the match. We will use this index in further calculations.

A similar loop can be used for the to_units:

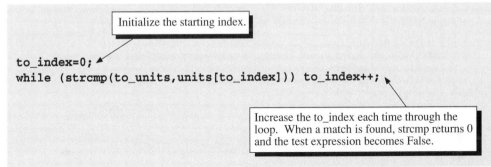

```
to_index=0;
while (strcmp(to_units,units[to_index])) to_index++;
```

Initialize the starting index.

Increase the to_index each time through the loop. When a match is found, strcmp returns 0 and the test expression becomes False.

4. Compute the units conversion using equation (7.1). Since the preceding loops give us the correct indices for both the from_units and to_units (being from_index and to_index), we can access the correct conversion factors. Recall that the conversion factor array is an array parallel to the units array. Therefore, the indices that apply to the first subscript in units[ ][ ] are the indices of interest for conv_fc[ ]. The following assignment statement uses equation (7.1) to get the new units value:

```
to_value = from_value/conv_fc[to_index]*conv_fc[from_index];
```

5. Print the results to the screen. Because we are printing single words and mixing double and char types in the output, the printf function with a %s conversion specifier for char and %lf for double is a good choice for the printout:

```
printf("%lf %s is equal to %lf %s\n",from_value, from_units,
                to_value, to_units);
```

Note that we could have used a mixture of printf( ) and puts( ) to do the printing. However, puts( ) is difficult to use in this situation because it adds a newline character at the end of the string it prints. Using puts( ) makes the printout extend over two lines rather than just one. Therefore, printf( ) is the preferred choice.

SOURCE CODE
This source code uses the code blocks just described.

```c
#include <stdio.h>
#include <string.h>
void main(void)
        {
        double conv_fc[14] = {0.001,0.01,1.0,1000.,
                       0.0254,0.3048,0.9144,1609.,
                       1.0,1000.,1.0e+06,0.2782,4.452,8905.};
        char units[14][7] ={"mm","cm","m","km",
                       "inches","ft","yd","miles",
                       "N","kN","MN",
                       "ounces","lb","ton"};
        char from_units[7], to_units[7];
        double from_value, to_value;
        int from_index, to_index;

        printf ("Enter request in the form \"5.2 cm to inches\":\n\n"\
               "\tAllowable units are mm, cm, m, km, inches, ft, yd, miles,\n"
               "\tN, kN, MN, ounces, lb, and ton.\n");
        scanf ("%lf%s%*s%s",&from_value,from_units,to_units);

        from_index=0;
        to_index=0;
        while (strcmp(from_units,units[from_index])) from_index++;
        while (strcmp(to_units,units[to_index]))     to_index++;

        to_value = from_value/conv_fc[to_index]*conv_fc[from_index];

        printf("%lf %s is equal to %lf %s\n",
               from_value, from_units, to_value, to_units);

        }
```

OUTPUT

	Enter request in the form "5.2 cm to inches": Allowable units are mm, cm, m, km, inches, ft, yd, miles, N, kN, MN, ounces, lb, and ton.
Keyboard input	5000 lb to kN 5000 lb is equal to 22.260000 kN

COMMENTS
Note that we have included string.h because we used strcmp( ) in the program. Again, in the interest of simplicity, we have done no data checking in this program. If the units entered are different from those on the list, the program fails. After you have read our example on data checking, you can modify this program so that it does not take any unacceptable data.

Modification exercises

1. Modify the program so that it can calculate length unit conversions using decimeters and micrometers.

2. Modify the program to handle weight unit conversions using kg-force and kilo-pounds.

3. Modify the program to calculate pressure unit conversions. Use psi (pounds per square inch), kPa (kilo Pascals), MPa (mega Pascals), psf (pounds per square foot), and ksf (kilopounds per square foot).

4. Modify the program so that it checks the units typed in by the user. If the units are not on the list of possible units, print an error message to the screen.

5. Create a modular design for the program. Make three functions—one for reading, one for printing, and one for calculating the converted units.

APPLICATION PROGRAM 7.3 PIPE FLUID VELOCITY, CHECKING INPUT DATA, MODULAR DESIGN

Problem statement

Write a program that thoroughly checks keyboard input data. After the data has been found to be acceptable, have the program calculate the velocity, V, of a fluid at a section in a pipe. The keyboard input data is the velocity, v, of the fluid at another section, and the diameters of the pipe at the two sections, d and D. The program is to allow the input in the following manner; for example,

```
d=12.3
v=23.4
D=34.5
```

The program should allow any order of entry. In other words, v could be entered before d. If a letter other than these three is typed in, the program should display an error message and prompt the user to enter another. If a negative number or nondigit value after the = sign is entered, the program also should display an error message and prompt the user to enter another value. The program should allow the user to input no more than five bad entries, and no spaces are to be allowed. Use a modular program design.

Solution

RELEVANT EQUATIONS

Fluid flow through a pipe is governed by the equation that, at all sections of the pipe, the product of the velocity and cross-sectional area is a constant:

$$vA = \text{constant} \tag{7.2}$$

where

$$v = \text{fluid velocity}$$
$$A = \text{cross-sectional area}$$

Because the cross-sectional area of a circular pipe is $\pi d^2/4$, we find the following also true:

$$vd^2 = \text{constant} \qquad (7.3)$$

We can use this equation to calculate the velocity at various locations along a pipe. The calculations are straightforward. If we are given a fluid velocity, v, and pipe diameter, d, at one location and a pipe diameter, D, at another location, and we want to calculate the velocity, V, at the D location, then

$$vd^2 = VD^2$$

solving for V, we get

$$V = vd^2/D^2 \qquad (7.4)$$

SPECIFIC EXAMPLE
For a pipe with (see Fig. 7.17)

$$d = 20 \text{ cm}$$
$$D = 3 \text{ cm}$$

and the fluid flow

$$v = 50 \text{ cm/sec}$$

the velocity V at the 3 cm section of the pipe (using equation (7.4)) is

$$V = (50 \text{ cm/sec})(20 \text{ cm})^2/(3 \text{ cm})^2$$
$$V = 2222.22 \text{ cm/sec}$$

You can see that the velocity in the smaller section of pipe is considerably greater than in the larger section. This is the effect you see when you put a nozzle on the end of a hose.

FIG. 7.17
Fluid flow in pipe of varying diameter

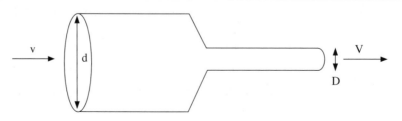

DATA STRUCTURES

For this type of problem, the data structure involved depends on the way we check the data for its validity. A technique commonly used is to read the entire line of data into a character array because a character array can accept any key typed. Then, we can evaluate each element of the character array to see if it meets our requirements. If any character is unacceptable, we can print an error message and ask the user to retype the input.

For this problem, then, we choose a 1-D character array (data_line[ ]) of size 30 to represent a string entered at the keyboard of the sort: $v = 23.4$. Because correctly entered data should have only one letter (*d, v,* or *D*) and one equal sign, an array length of 30 allows for 28 characters in representing the numeric portion of the input data. This is believed to be more than enough, so we will declare a size of 30 for our character array.

ALGORITHM

The calculations of fluid flow for this program are not particularly difficult. The difficult part of this program is checking the data and using a modular design to do so; therefore, the algorithm development we present focuses on these aspects of the program.

1. List general tasks. For this program we list the following tasks:

1. Prompt for, read, and store the input string.
2. Analyze the first part of the string (the letter portion). If not valid, then print an error message and prompt for another entry. If valid, then send to analysis unit for the second part (the numeric portion).
3. Analyze the second part of the string. If not valid, then print an error message and prompt for another entry. If valid, then store the numeric data.
4. After having received a complete set of valid entries, perform the calculation of the fluid velocity and print the result.

This leads to the structure chart of Fig. 7.18. The information going into and out of each function is shown in the data flow diagram of Fig. 7.19.

We describe the development of each function next.

FIG. 7.18
Structure chart for fluid flow program

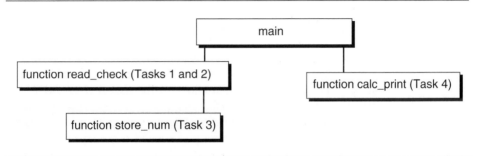

FIG. 7.19

Data flow diagram of pipe fluid flow program. Note that flags and counters are typically passed among functions. In this program we use the variables, flag, count, and bad_entry as flags and counters.

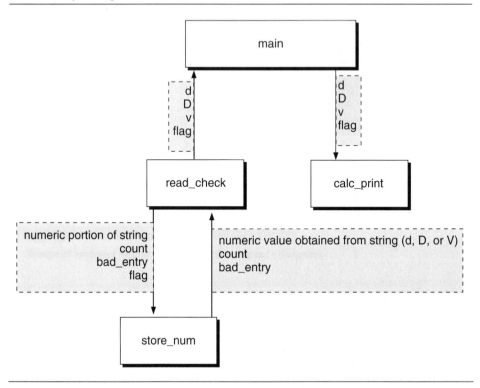

Function read_check. We choose to break the input string into two parts, the "letter=" part and the numeric part. The natural place to begin the development of the algorithm is in reading and evaluating the validity of the first portion of the string, which we decided will take place in the function read_check. You will find that, as you develop data checking routines, they involve a combination of looping and decision making. Some of the combinations may become quite complicated. To help envision the program flow, you may find it useful to draw 3-D type flow diagrams such as we showed in Chapter 4. We will show some of them here.

There are four possible categories of data entry for the first part of the string. They can be broken down into three acceptable entries and an unacceptable entry. These and the action that is to be taken after receiving such an entry is shown next:

Entry	Acceptability and action
d=	Acceptable, proceed to evaluate numeric portion
v=	Acceptable, proceed to evaluate numeric portion
D=	Acceptable, proceed to evaluate numeric portion
Anything that is not exactly as typed here	Unacceptable, print error message, prompt for new entry

Because we are selecting among one of four sets of operations, we choose to use an if-else-if control structure. Within this structure, if the input is acceptable, we call function store_num with only the numeric portion of the data_line[ ] string. If it is not acceptable, we set the flag to 0 (flag=1 means a valid entry), increase our count of the number of bad entries with bad_entry ++, and print an error message. The code for this is the following (using the array, data_line[ ] as the input character array and count as the counter for number of good entries):

```
if (data_line[0]=='d' && data_line[1]=='=')
        {
        store_num(data_line+2, flag, &count, d, &bad_entry);
        if (*flag==1) printf("d entered, d=%lf\n",*d);
        }
else if (data_line[0]=='v' && data_line[1]=='=')
        {
        store_num(data_line+2, flag, &count, v, &bad_entry);
        if (*flag==1) printf("v entered, v=%lf\n",*v);
        }
else if (data_line[0]=='D' && data_line[1]=='=')
        {
        store_num(data_line+2, flag, &count, D, &bad_entry);
        if (*flag==1) printf("D entered, D=%lf\n",*D);
        }
else
        {
        *flag=0;
        bad_entry++;
        printf("Error on input.  Try typing your data again\n");
        }
```

Annotations:
- If first two typed characters are *d* and =.
- Call store_num with only the numeric part.
- If store_num returns a valid entry indicated by flag, print it.
- Call store_num with only the numeric part.
- If first two typed characters are *v* and =.
- If store_num returns a valid entry indicated by flag, print it.
- If first two typed characters are *D* and =.
- Call store_num with only the numeric part.
- If the first two typed characters are not valid.
- Set the flag to 0.
- If store_num returns a valid entry indicated by flag, print it.
- Increment the bad entry counter.
- Print error message and prompt for another entry.

For the moment, do not concern yourself with the use of & and * in this code. Their use becomes clear when viewing the entire source code and studying the passing of information between functions.

Note that we called the function store_num using data_line+2 as an argument. As we know with pointer arithmetic, this is the address of the third character stored in the array data_line[ ]. Alternatively, we could have used &data_line[2] to pass the same information. By passing data_line+2 rather than data_line, we pass only the part of the array after the equal sign, that being the numeric part.

We must put the if-else-if control structure into a looping structure to read a new string after evaluating its validity and storing the numeric value of the string read. The resulting structure is shown in Fig. 7.20. From this figure, observe that we read

FIG. 7.20
Illustration of flow for function read_check. Note that ## means invalid entry.

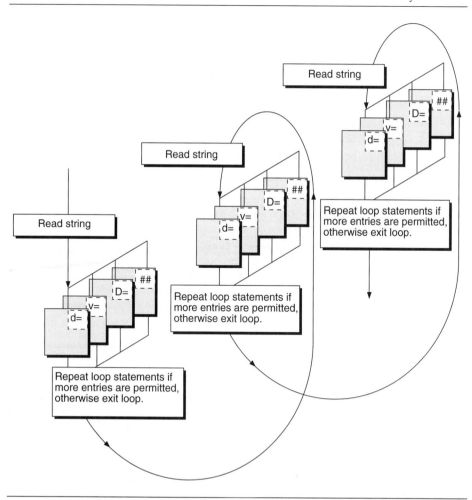

a new string each time through the loop. With each new string, we decide into which of the four categories it fits (## symbolizing invalid for this figure only). After having categorized the entry, we check the repetition condition to decide whether we should read another string or exit the loop. The repetition condition states that the loop is to repeat if fewer than three valid entries or fewer than five invalid entries have been entered. This figure shows reading only three strings; however, if the repetition condition is met, more strings can be read.

The code that creates this loop is

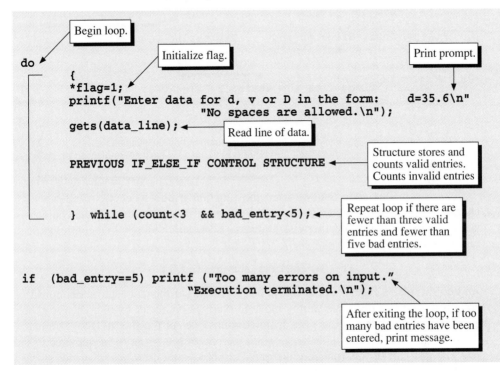

Function store_num. This function evaluates the validity of the numeric portion of the input in a manner similar to what we used for evaluating the first portion of the entry. By treating the data as a string rather than double, each element of the string can be evaluated. For instance, we accept a decimal point and any digit but reject any negative sign, letter, or symbol. We illustrate this method because it is very versatile. Once you have learned this technique, you can adapt it to a large number of situations. For instance, should we want to accept negative numbers for fluid velocity (meaning flow is in the opposite direction), we easily can do it with this method.

The structure of the primary part of this function is similar to that of the function read_check, in that it is basically an if control structure within a loop. We loop through all of the individual string elements and, using an if-else control structure, check to see

562 Chapter 7: Strings and Pointers

if each character is a valid digit or decimal point. Because we already described this structure, we will not cover it in detail here. The commented code follows:

```
do          Begin loop.                    If array element is a digit, decimal point, or
  {                                         string terminator, increment the array index.
    if(isdigit(num_string[i]) || num_string[i]=='.'
       || num_string[i]=='\0')  i++;
    else                                    else
        {
        (*bad_entry)++;                     Increment the bad entry counter.
        *flag=0;
        printf("Error on input.  Try typing your data"
               " again\n");
        }                 Set the flag to 0.
  } while (*flag==1 && num_string[i-1]!='\0');     Print an error message.
```

Repeat the loop only if the character read is valid (flag = 1) and we have not hit the end of the string (encountered the null character).

Note from this code that we used the function isdigit to check whether each character is or is not a digit. This function returns a nonzero integer if the character is a digit (meaning True). Otherwise, it returns 0 (meaning False). In a control structure with the beginning

```
if    ( isdigit(num_string[i] )
```

we get a result of True or False. In the previous loop, if the character under consideration indeed is a digit, we simply move to the next character by incrementing the array index. If it is not a digit (or decimal point), we set the flag to 0 to terminate the loop. We also terminate the loop when the end of the string is encountered.

The last part of this function interprets a double value from the character string if it has been determined to be valid. We do this using the function atof, which converts a string to double as described in Lesson 7.8. The following code performs this operation:

```
if (*flag==1)          If the input is valid.
    {
    *value=atof(num_string);      Convert the string to a double.
    (*count)++;                   Store the result in *value.
    }
```

Increment the valid input counter.

We use a pointer variable, value, to store the numeric value of what has been typed in because we return this value to the function read_check. Also, note that we have used (*count)++ rather than *count++. Given the precedence rules, we must use the parentheses, because without them we increment the address, count, and then use the unary * operator on this address. Instead, we want to take the value indicated by *count and increment it. Therefore, we must write (*count)++.

Function calc_print. This function calculates the fluid velocity using equation (7.2). We do this calculation only if all the input has been found to be valid. This occurs if the last string interpreted has a flag of 1. The code for this follows:

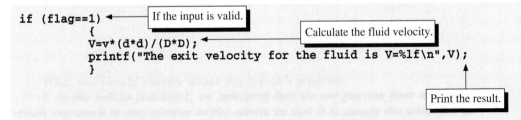

```
if (flag==1)  ◄──────┤ If the input is valid.
      {
      V=v*(d*d)/(D*D);  ◄──────────┤ Calculate the fluid velocity.
      printf("The exit velocity for the fluid is V=%lf\n",V);
      }
                                                          ┤ Print the result.
```

Function calls and declarators. An important part of a modular design is the transfer of information between modules. This is done using function calls and headers (which should match the function prototypes). For this program, we show the calls and declarators for each of the functions (Fig. 7.21) in an image similar to the data flow diagram. From this figure, note that each address of a single variable in a function call is matched with a pointer variable in the function prototype. Also, we used the const qualifier for num_string[ ] in function store_num, because this function does not modify the string but simply stores it. Remember, we can pass only a single address or a single value with each argument in a function call. If an address is passed to a function, then within that function we must decide whether we want to work with the address or value for each line of code we write. On the other hand,

FIG. 7.21
Function calls and headers for fluid flow program

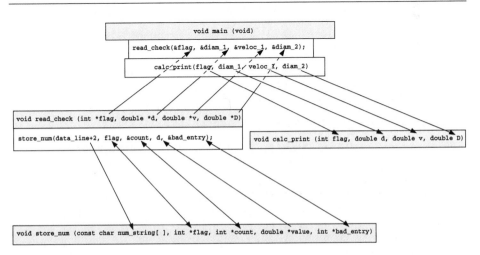

if a value is passed to a function, then within that function we only can work with the value. This rule leads to the choices we made in using & and * in each function.

SOURCE CODE

The code just described is used in the following program. Because we have thoroughly described it, we have not put comments in the code.

```c
#include <stdio.h>
#include <string.h>
#include <ctype.h>
#include <stdlib.h>

void store_num (const char num_string[ ], int *flag, int *count,
                        double *value, int *bad_entry);
void read_check (int *flag, double *d, double *v, double *D);
void calc_print (int flag, double d, double v, double D);

void main(void)
{
        int flag;
        double diam_1, veloc_1, diam_2, veloc_2;

        read_check(&flag, &diam_1, &veloc_1, &diam_2);
        calc_print(flag, diam_1, veloc_1, diam_2);
}

void read_check(int *flag, double *d, double *v, double *D)
{

        int count=0, bad_entry=0;
        char data_line[30];

do
{
    *flag=1;
    printf("Enter data for d, v or D in the form:     d=35.6\n"
                        "No spaces are allowed.\n");
    gets(data_line);

    if(data_line[0]=='d'  &&  data_line[1]=='=')
        {
        store_num(data_line+2, flag, &count, d, &bad_entry);
        if (*flag==1) printf("d entered,"
        "d=%lf\n",*d);
        }
    else if(data_line[0]=='v'  &&  data_line[1]=='=')
        {
        store_num(data_line+2, flag, &count, v, &bad_entry);
        if (*flag==1) printf("v entered,"
        "v=%lf\n",*v);
        }
    else if(data_line[0]=='D'  &&  data_line[1]=='=')
        {
```

```
                store_num(data_line+2, flag, &count, D, &bad_entry);
                if (*flag==1) printf("D entered,"
                "D=%lf\n",*D);
                }
        else
                        {
                        *flag=0;
                        bad_entry++;
                        printf("Error on input.  Try typing your data
                        again\n");
                        }

        } while (count<3  && bad_entry<5);

    if (bad_entry==5) printf ("Too many errors on input. Execution
    terminated.\n");
}
void calc_print(int flag, double d, double v, double D)
{
        double V;
        if (flag==1)
                    {
                    V=v*(d*d)/(D*D);
                    printf("The exit velocity for the fluid is
                    V=%lf\n",V);
                    }
}
void store_num(const char num_string[ ], int *flag, int *count,
                                        double *value, int
                                            *bad_entry)
{
        int i;

        i=0;
        do
        {
                if(isdigit(num_string[i]) || num_string[i]=='.'
                        || num_string[i]=='\0')  i++;

                else
                        {
                        (*bad_entry)++;
                        *flag=0;
                        printf("Error on input.  Try typing"
                        "your data again\n");
                        }

                } while (*flag==1 && num_string[i-1]!='\0');

                if (*flag==1)
                {
                *value=atof(num_string);
                (*count)++;
                }
}
```

OUTPUT

```
                    Enter data for d, v or D in the form: d=35.6
                    No spaces are allowed
Keyboard input      d=20
                    d entered, d=20.000000
Keyboard input      f=50
                    Error on input. Try typing your data again.
Keyboard input      v=50
                    v entered, v=50.000000
Keyboard input      D=3
                    D entered, D=3.000000
                    The exit velocity for the fluid is V=2222.222222
```

COMMENTS

In the interest of keeping this program reasonably simple, we have not made it capable of operating flawlessly. For instance, if one were to enter the information for *d* three times without error, the program would conclude that three valid entries have been made and attempt to perform the calculations and crash. For this example, we will accept this defect; however, for commercial software this would be unacceptable.

Also, in writing a program of this sort, it is important to clearly define what is valid and invalid input. For instance, in this program, we decided not to accept spaces, negative signs, or scientific notation, even though including these could be regarded as being perfectly acceptable. We simply wanted to show the programming principles in this example, so we chose to not include these possibilities. In your own software, though, you may need to consider a large range of possible entries, making the program more complex.

Modification exercises

Modify the program to do the following:

1. Accept negative fluid velocities (meaning the fluid flows in the opposite direction).

2. Be capable of recognizing error if the same variable (*d, v,* or *D*) is entered more than once.

3. Accept spaces as input.

4. Accept scientific notation as input (using *e* or *E*).

5. Accept *V, v,* and *d* to be the entered data and have the program calculate *D*.

▓ APPLICATION PROGRAM 7.4 EARTHQUAKE ANECDOTAL REPORT ANALYSIS, STRING OPERATIONS AND USING DYNAMIC STORAGE

Problem statement

Write a program that can help develop a modified Mercalli earthquake intensity map from a moderate earthquake using anecdotal data from people who felt the earthquake.

Available to you are several sentence descriptions from a large number of individuals of what each of them felt during the earthquake shaking. All descriptions are in one file. Each description begins with the individual's location city. This is followed by a description of the shaking they underwent and terminated with a # sign.

The output from your program should consist of a table of cities and intensities. The number of those experiencing each intensity in each city should be listed in the table.

Solution

RELEVANT EQUATIONS AND BACKGROUND INFORMATION

The Mercalli intensity in an earthquake can be compared with the depth of water in a flood. For a given flood, the water is deeper in some locations than others. After a major flood, a map can be drawn indicating the various water depths in the affected areas. This map can be used to help predict locations of deep water in future floods. Thus, it can be used by engineers and city planners to help reduce the impact of future floods.

Similarly, after a major earthquake, information is collected regarding the severity of shaking at different locations. For a given earthquake, the shaking is greater in some locations than others. A map can be drawn indicating the amount of shaking at various locations in the affected region. This map can be used to help predict locations of strong shaking in future earthquakes. Thus, it can be used by engineers and city planners to help reduce the impact of future earthquakes.

Because the number of sites with earthquake measuring equipment in an area is small, engineers rely on anecdotal descriptions by individuals experiencing the motion to help them estimate the level of shaking that took place at sites without instruments. A large volume of information may result because a large number of people experience the shaking in a moderate earthquake in an urban area. A computer evaluation of the anecdotal comments may improve efficiency in evaluating the information.

For this example, we give a simplified version of the actual modified Mercalli intensity scale. It works in the following manner. If someone uses the word *strong* in describing the shaking, then we say that person experienced a modified Mercalli intensity (MMI) of 8. If the person uses *weak*, we give it an MMI = 4. A full table of what we use for descriptive words and MMI follows:

Modified Mercalli intensity	Descriptive words
4	mild, weak, slow
6	moderate, medium, tempered
8	strong, powerful, sharp
10	violent, destructive, extreme

We solicit data from five cities in the region of a recent moderate earthquake: San Francisco, Berkeley, Palo Alto, Santa Cruz, and San Jose. Realizing that each city may not have just one MMI, we simply tally all the MMIs sent to us from each city.

SPECIFIC EXAMPLE

Suppose we have the following anecdotal descriptions of a particular earthquake. Note that we have the city given first, then one or two comments, terminated with a # sign.

```
San Francisco.
I felt a strong shock followed by rolling waves.
It lasted a long time.#
Berkeley.
It was mild shaking, rattling windows.
It frightened my dog.#
Palo Alto.
The shaking was very violent.#
Santa Cruz.
The earthquake was very destructive.
It knocked down the chimney on my house.#
Palo Alto.
The extreme shaking made me feel like I was on a
boat in rough sea.#
San Francisco.
I slept right through it. It was much weaker than our
last earthquake.#
```

Using our table we can assign an MMI to each of these descriptions. We get

City	Descriptive word used	MMI
San Francisco	strong	8
Berkeley	mild	4
Palo Alto	violent	10
Santa Cruz	destructive	10
Palo Alto	extreme	10
San Francisco	weak	4

We can tally these responses in a grid of MMI vs. City:

	Number of responses of each MMI				
MMI	San Francisco	Berkeley	Palo Alto	Santa Cruz	San Jose
4	1	1	0	0	0
6	0	0	0	0	0
8	1	0	0	0	0
10	0	0	2	1	0

This table can help us assess the intensity of shaking in each city. It is an example of the final product that we expect to have from this program.

DATA STRUCTURES

Dynamic data structure. The fundamental storage issue for this program is in handling the reports. Because we need to size our arrays, we need to decide how many reports and how many characters per report, maximum, we will have. By putting these sizes in constant macros, we easily can change them if we want to later. Despite this feature, we need to make a decision at this point.

We will decide on 1000 reports and 800 characters per report, maximum. With each character being 1 byte, we have $1000 \times 800 = 800,000$ bytes $= 800$kB. This should be handled easily by today's memory capabilities. However, this is the maximum we will accommodate and not what actually is needed in a specific case. This is because, after some earthquakes, fewer than 1000 people may report and each description may be fewer than 800 characters. Rather than requiring the maximum memory be reserved each time the program is run, we choose to reserve only what is needed by using dynamic memory allocation, meaning our program will ask for only the amount of memory needed to store each report that is submitted. To do this, though, we need to reserve space for 1000 pointers and for at least one report of 800 characters. In summary, even if you are using dynamic data structures as we are describing them, there are still limits on the amount of memory that your program can use. In writing your programs, you set these limits with a consideration of the hardware on which the program will run.

We illustrate the use of dynamic structures by contrasting them with the ordinary array structures we have used to this point. For instance, if we were going to store all of the reports in an ordinary array structure, then we would declare a single 2-D array to handle the maximum of 1000 reports 800 characters long in a declaration of the sort that follows:

```
char all_reports[1000][800];
```

However, because we are using dynamic data structures, we define two different 1-D arrays, one capable of handling a single report of 800 characters,

```
char indiv_rept[800];
```

and one array of pointers that point to a maximum of 1000 reports,

`char *report[1000];`

We use constant macros to define the sizes of arrays, so in this program we have the following for preprocessor directives and declarations for these arrays:

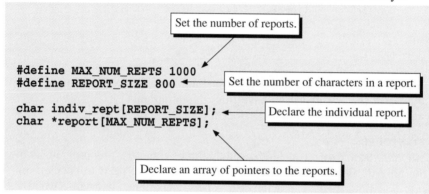

The way we use this data structure in the program is as follows:

1. We read the first report from the report file (that is, the characters in the input file up to the first # sign) and store it temporarily in the indiv_rept[ ] array. Note that this array is capable of storing the maximum size of an individual report, 800 characters.
2. We count the actual number of characters in the array using the strlen( ) function. We now know the actual amount of memory needed for storing this report.
3. We request to reserve the amount of memory needed for this report using the calloc or malloc function. In doing so, we receive back from calloc or malloc the address of the block of memory reserved.
4. We store this address as the first element of the pointer array report[ ]. We have later access to this block of memory using the identifier report[0].
5. Using the strcpy( ) function, we copy the array indiv_rept[ ] into the memory block indicated by the address returned from calloc, which is stored in report[0]. This first report is now in memory and we can access it later using its address.
6. We read the second report from the report file and store it temporarily in the indiv_rept[ ] array. This has the effect of overwriting the first report stored in this array; however, this is acceptable because we previously copied the first report into the memory block indicated by report[0]. We now

    Count the characters in the report using strlen( ).
    Use calloc or malloc to reserve memory for this length of array.
    Store the address of this memory block in the pointer array (using report[1])
    Copy the report from indiv_rept[ ] to the address indicated by report[1].

7. We repeat what is listed in step 6 for all of the other reports in the file (we terminate reading the file when EOF is encountered). Observe that, because we

work only with the reports that are in the file, we do not reserve memory for 1000 reports, as we would if we had used an array declaration. We reserve memory only for the reports that actually exist and only for the number of characters they actually have.

In many cases, to create a dynamic data structure in your programs, you should

1. Determine the size of the ordinary 2-D array you would use. This gives you an understanding of your maximum memory requirements.
2. Instead of creating this 2-D array, create two 1-D arrays:

   An ordinary 1-D array equal to the size of the second subscript of your step 1 array.
   A 1-D pointer array equal to the size of the first subscript of your step 1 array.

3. Temporarily store a portion of what you want stored in the ordinary 1-D array of step 2.
4. Find the actual amount that is filled in this array (for instance, using strlen()).
5. Use calloc or malloc to allocate only the memory needed for storing this array.
6. Store the address returned by calloc or malloc in one element of the 1-D pointer array in step 2.
7. Copy the ordinary 1-D array into the memory pointed to by the array element in step 6.
8. Repeat steps 3–7 a number of times, until all the information you want is stored.
9. Access any of this information later in your program using an array element in step 6.

We have described steps 3–6 in Lesson 7.10 and illustrated them in Fig. 7.14.

Other data structures for this program.
We choose

```
char city[NUM_CITIES][15]={"San Francisco", "Berkeley",
                    "Palo Alto", "Santa Cruz", "San Jose"};
char descriptor[NUM_DESCRIPTORS][15]={"mild","weak","slow",
        "moderate", "medium", "tempered", "strong", "powerful",
        "sharp", "violent", "destructive", "extreme"};
```

to represent our cities and descriptors.

We also decided to arrange the descriptor array in such a way as to show a correlation between the order of the descriptors listed and the MMI. For instance, the first three words, "mild", "weak", and "slow", correspond with MMI = 4. The next three words correspond with MMI = 6, the next three for MMI = 8, and the last three for MMI = 10. So, the first subscript for this array is directly related to the MMI that the descriptor represents. We use this correspondence in our program. We could have decided on making this a 3-D array (with one dimension having a size of 4 to account for the four MMIs); however, we find that a 2-D array is adequate.

Our result is a table with four rows (one for each MMI) and five columns (one for each city). For reasons described for other examples, we choose to use a 2-D integer array that we call tally[ ][ ]. The declaration is

```
#define NUM_CITIES 5
int tally [4] [NUM_CITIES];
```

ALGORITHM

Because we chose a different program for this chapter's modular design example and because we want to focus on dynamic data storage, we will not use a modular design for this program.

The general algorithm can be deduced from the specific example as

1. Read a report from the input file.
2. Store the report information using the dynamic memory allocation method described under data structures.
3. Repeat steps 1 and 2 until all of the reports have been successfully read (until EOF is returned).
4. Search for the city name in a report.
5. Search for the descriptive word that gives the MMI in the report.
6. Increase the tally by one for the given MMI and city.
7. Repeat steps 4–7 for each report.
8. Print the results in the form of a table.

We describe the development of the source code for each of these steps except 8, which is considered to have been illustrated sufficiently in previous lessons.

1. Read a report from the input file. Each report is expected to have numerous words and cover several lines. In the file white space will separate the words and newline characters break a report into several lines. Also, the # sign is used to separate reports. We use the # sign as a sentinel value (which means we search for it to indicate a separate report).

We could use fscanf (with the %s conversion specification) in a loop to read one word at a time into a 2-D array. However, we are searching for the # sign to end a report. Therefore, we choose to use getc to input the text one character at a time into a 1-D array (indiv_rept[ ]) and check whether the character read is # or if EOF is returned by getc. If either value is returned from getc, we break out of the loop.

Because each report is allowed only 800 characters, we stop reading at 799 or less so that we can terminate the string by adding \0 at the end. The following loop performs these operations:

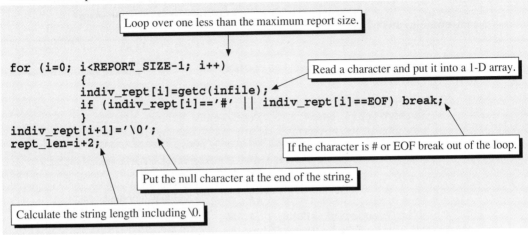

In the Comments section of this example, we illustrate another way to write this for loop without using a break statement.

2. Store the report information using the dynamic memory allocation method described under data structures. In step 1 we calculated the report length (rept_len). We use this to reserve sufficient memory with the statement

```
report[j]=(char *)calloc(rept_len, sizeof(char));
```

After executing the statement, there is no information stored in the memory location begun with the address stored in report[j]. We can remedy this by copying into that memory location an individual report (read in step 1) stored at the address indiv_rept. The following statement does this:

```
strcpy (report[j], indiv_rept);
```

Should we wish to access this individual report later in the program, we use report[j].

3. Repeat steps 1 and 2 until all the reports have been successfully read (until EOF is returned). The two statements in step 2 (along with the data structures) are the heart of the dynamic memory allocation scheme used in this program. However, we need to do two more things:

Loop over all the reports to be stored.
Check to make sure that calloc has successfully allocated memory.

The following code does these additional operations:

```
while(indiv_rept[i] != EOF  && j<MAX_NUM_REPTS)
       {

       STEP 1 CODE. READ AN INDIVIDUAL REPORT
       INTO indiv_rept[ ] . rept_len = REPORT LENGTH.

       report[j]=(char *)calloc(rept_len, sizeof(char));


       if (report[j]==NULL)
                {
                printf("Memory not allocated for report number %d", j+1);
                }
       else
                {
                strcpy (report[j], indiv_rept);
                }
       puts(report[j]);
       j++;
       }
```

Loop until we read EOF or reach the max number of reports.

Loop previously shown for reading an individual report (step 1).

Reserve memory. Store address of memory in report[j]. However, calloc returns NULL if it cannot allocate memory.

Print an error message if memory cannot be allocated.

Print report to screen.

Otherwise copy report into location pointed to by report[j].

Increment loop counter.

Note that each time through the loop, we increment j, the index for the pointer array. In other words, each time through the loop, we create a new memory address to store a new report. Using strcpy we store a new report at that address.

Although not required, we have chosen to print each report to the screen to verify that each report indeed has been stored in the location pointed to by report[j]. Because the function puts prints a string until it encounters \0, it is well suited for our purpose. Recall that we added \0 to the end of each report in step 1. Here, puts works much better than using printf or putchar in a loop.

4. Search for the city name in a report. Now that we read all the reports into memory and established all the pointers to the beginnings of the reports, we easily can access each report individually. For instance, we can set an array index, i, and search in the string pointed to by report[i] for information of interest to us. We start by determining which city is referenced in the report. The list of cities we are studying is given in the array that follows:

```
char city[NUM_CITIES][15]={"San Francisco", "Berkeley",
        "Palo Alto","Santa Cruz", "San Jose"};
```

We can check if one of these cities appears in a report using the strstr function, as illustrated in the following loop:

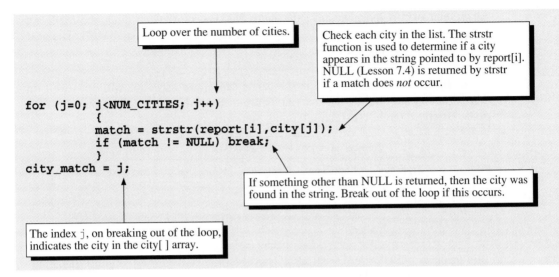

Loop over the number of cities.

Check each city in the list. The strstr function is used to determine if a city appears in the string pointed to by report[i]. NULL (Lesson 7.4) is returned by strstr if a match does *not* occur.

```
for (j=0; j<NUM_CITIES; j++)
        {
        match = strstr(report[i],city[j]);
        if (match != NULL) break;
        }
city_match = j;
```

If something other than NULL is returned, then the city was found in the string. Break out of the loop if this occurs.

The index j, on breaking out of the loop, indicates the city in the city[ ] array.

We note that `match` is declared as a pointer variable in the program because strstr returns a pointer. We cannot have an ordinary variable for `match`.

5. Search for the descriptive word that gives the MMI in the report. We do a similar search in determining which descriptive word is used in the report. The descriptive word in the report indicates the MMI felt by the observer. Here is an array of descriptors:

```
char descriptor[NUM_DESCRIPTORS][15]={"mild","weak","slow",
   "moderate", "medium", "tempered", "strong", "powerful",
        "sharp","violent", "destructive", "extreme"};
```

The next loop searches for a match between one of these words and the words in the report string:

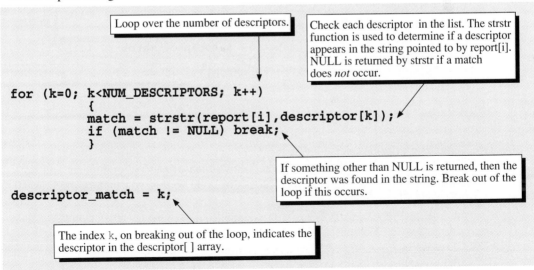

Loop over the number of descriptors.

Check each descriptor in the list. The strstr function is used to determine if a descriptor appears in the string pointed to by report[i]. NULL is returned by strstr if a match does *not* occur.

```
for (k=0; k<NUM_DESCRIPTORS; k++)
        {
        match = strstr(report[i],descriptor[k]);
        if (match != NULL) break;
        }



descriptor_match = k;
```

If something other than NULL is returned, then the descriptor was found in the string. Break out of the loop if this occurs.

The index k, on breaking out of the loop, indicates the descriptor in the descriptor[ ] array.

At the end of this loop, we save the index k. This index determines the MMI we have, which enables us to increase the tally for the matching city and MMI. The next step illustrates finding the MMI.

6. Increase the tally by 1 for the given MMI and city. We will not dwell on this particular step because it does not introduce anything new about string operations. Briefly though, our final table has the form

	Number of responses of each MMI				
MMI	San Francisco	Berkeley	Palo Alto	Santa Cruz	San Jose
4	1	1	0	0	0
6	0	0	0	0	0
8	1	0	0	0	0
10	0	0	2	1	0

This leads to a 2-D array for tally[ ][ ], in the order tally [row][column], meaning, for example, tally[0][0] indicates MMI = 4 and city = San Francisco and tally[3][2] indicates MMI = 10 and city = Palo Alto. Trace the flow of the next code to see how we increase the tally for the matching city and descriptor.

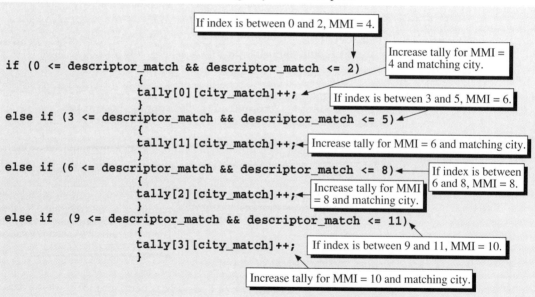

The complete source code is given below. Please read the comments and note the locations of the above described blocks of code so that you understand how all of the code fits together.

```
#include <stdio.h>
#include <string.h>
#include <stdlib.h>
#define MAX_NUM_REPTS 1000
#define NUM_CITIES 5
#define NUM_DESCRIPTORS 12
#define REPORT_SIZE 800
```

We use macros to size our arrays.

match is a pointer variable.

```
void main (void)
        {
        int i, j, k, rept_len, tally[4][NUM_CITIES]={0}, city_match,
                     descriptor_match, mm, undefined, num_repts;
        char *report[MAX_NUM_REPTS], *match;
        char indiv_rept[REPORT_SIZE]={0};
        char city[NUM_CITIES][15]={"San Francisco", "Berkeley",
              "Palo Alto", "Santa Cruz", "San Jose"};
        char descriptor[NUM_DESCRIPTORS][15]={"mild","weak","slow",
                    "moderate", "medium", "tempered",
                    "strong", "powerful", "sharp",
                    "violent", "destructive", "extreme"};
        FILE *infile;

        infile = fopen ("EQREPT.TXT","r");
        rept_len=1;
        j=0;
        i=0;
        undefined=0;
```

Initializing flags and counters.

```
/******************************************************************
**This loop reads all of the reports in the file. It stores them in memory
**using the dynamic memory allocation system using calloc.
******************************************************************/
        while(indiv_rept[i] != EOF  && j<MAX_NUM_REPTS)
               {
        for (i=0; i<REPORT_SIZE-1; i++)
                   {
                   indiv_rept[i]=getc(infile);
                   if (indiv_rept[i]=='#' || indiv_rept[i]==EOF) break;
                   }
        indiv_rept[i+1]='\0';
        rept_len=i+2;
        report[j]=(char *)calloc(rept_len, sizeof(char));

        if (report[j]==NULL)
                   {
                   printf("Memory not allocated for report number %d",
                       j+1);
                   }
        else
                   {
                   strcpy (report[j], indiv_rept);
                   }
        puts(report[j]);
        j++;
        }

        num_repts = j;
```

Read the input file character by character using getc. Stop at # or EOF. After executing this loop, the indiv_rept[ ] array holds a single report.

Reserve memory to store a single report.

If memory cannot be reserved (indicated by calloc returning NULL) print an error message.

Otherwise, copy the report into the memory location reserved by calloc.

Print a report to the screen.

At this point, all of the reports can be accessed using the report[ ] array.

```
for(i=0; i<num_repts; i++)
    {
        for (j=0; j<NUM_CITIES; j++)
            {
            match = strstr(report[i],city[j]);
            if (match != NULL) break;
            }
        for (k=0; k<NUM_DESCRIPTORS; k++)
            {
            match = strstr(report[i],descriptor[k]);
            if (match != NULL) break;
            }
        city_match = j;
        descriptor_match = k;
        if (0 <= descriptor_match && descriptor_match <= 2)
                        {
                        tally[0][j]++;
                        }
            else if (3 <= descriptor_match &&
                        descriptor_match <= 5)
                        {
                        tally[1][j]++;
                        }
            else if (6 <= descriptor_match &&
                        descriptor_match <= 8)
                        {
                        tally[2][j]++;
                        }
            else if (9 <= descriptor_match &&
                        descriptor_match <= 11)
                        {
                        tally[3][j]++;
                        }
    }

printf ("Final tally of Modified Mercalli intensities:\n\n");
printf ("Modified\nMercalli\nIntensity ");
for (i=0; i<NUM_CITIES; i++) printf("%12s", city[i]);

mm = 2;
for (i=0; i<\#60>4; i++)
        {
        mm+=2;
        printf("\n\n%6d",mm);
        for (j=0; j<NUM_CITIES; j++)
                {
                printf("%12d",tally[i][j]);
                }
        }
}
```

Loop over number of reports.

Determining the city in the report.

Determining the descriptor in the report.

Incrementing the tally for the city and descriptor.

Printing the tally in table form using a nested loop.

INPUT FILE - EQREPT.TXT

San Francisco.
I felt a strong shock followed by rolling waves.
It lasted a long time.#
Berkeley.
It was mild shaking, rattling windows.
It frightened my dog.#
Palo Alto.
The shaking was very violent.#
Santa Cruz.

```
The earthquake was very destructive.
It knocked down the chimney on my house.#
Palo Alto.
The extreme shaking made me feel like I was on a
boat in rough sea.#
San Francisco.
I slept right through it.  It was much weaker than our
last earthquake.#
```

OUTPUT (NOT INCLUDING ECHO OF INPUT FILE)

Modified Mercalli Intensity	San Francisco	Berkeley	Palo Alto	Santa Cruz	San Jose
4	1	1	0	0	0
6	0	0	0	0	0
8	1	0	0	0	0
10	0	0	2	1	0

Final tally of Modified Mercalli intensities:

COMMENTS

This is not the perfect program to do this type of analysis. For instance, if a report were to have "the motions were not strong" we would focus only on the word *strong* and neglect the word *not*. Thus, we would misclassify the meaning of the report. Therefore, this program would not work in practice. However, we show it to illustrate dynamic data structures and some string handling operations. The program would be better with a modular design and an array structure for descriptor[][] that more closely parallels the MMI.

Once again, for simplicity, we have done little data checking in this program. As a minimum, we should print error messages or somehow flag reports that lack a correct city or proper descriptor words.

It is possible to write a for loop that reads each character in a report without using a break statement, as we have done here. A loop with no break statement is considered to be more structured. We show you the somewhat odd-looking for loop in Fig. 7.22. Decipher the meaning of this because it gives you added insight into the operation of a for loop.

This for loop works because of the roles that the initialization, condition, and "increment" statements play in the execution of the loop. The order of execution of the loop control statements and body are shown in Fig 7.23. For the loop in Fig. 7.22, we have the following steps:

1. The initialization in this loop causes one character to be read.
2. Then the condition is checked and, if True, the loop body is executed. The condition checks to see that the character read is not # or EOF and that the array index is less than the allowed number of report characters.
3. In this case, the loop body causes the character read to be stored in the array, indiv_rept[ ]. After it is stored in the array, the array index is incremented (remember the rule for the postincrement operator).

FIG. 7.22

A for loop that requires no break to read each character of a report

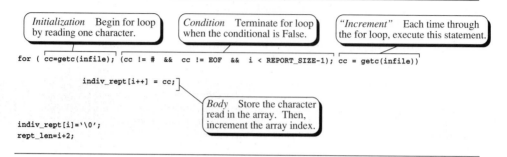

4. The so-called increment expression of the loop control statements then is executed. In this loop, the "increment" expression is not an increment at all. It causes another character to be read.
5. The condition, body, and increment are executed repeatedly until the condition becomes False, as shown in Fig 7.23.

Because the "increment" expression is executed repeatedly, the file is read character by character. The body causes the characters to be stored in the character array indiv_rept[ ]. Both of these continue to occur until the condition becomes False.

FIG. 7.23

Order of operation of for loop in Fig. 7.22. The "initialization" and "increment" statements are the same, both read a single character from the file. The body of the loop puts the character into the report array.

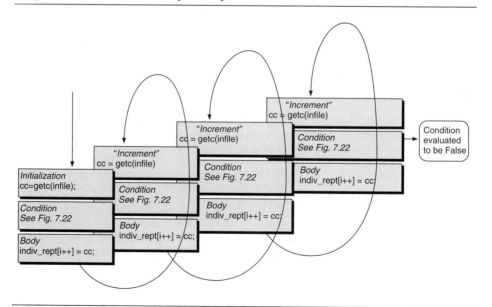

Thus, this loop performs the same task as the one we used in the program but without a break statement. Even though the loop shown here seems to be more unorthodox than what we have used in the application program, the loop here is considered to be more structured because it has no break statement.

Modification exercises

1. Change the array structure for descriptor[ ][ ] such that it is a 3-D array rather than two dimensional. Make the declaration

```
#define NUM_DESCRIPTORS 3
char descriptor[4][NUM_DESCRIPTORS][15]={"mild","weak","slow",
        "moderate", "medium", "tempered","strong", "powerful",
        "sharp", "violent", "destructive", "extreme"};
```

where in this case NUM_DESCRIPTORS is equal to the number of descriptors per modified Mercalli intensity level. In this case, there are three descriptors per level. Change the way that descriptor[ ][ ] is used in the body of the program to accommodate this declaration.

2. After making the preceding change, add the following descriptors:

```
MMI                Descriptor
4                  feeble
6                  firm
8                  jolting
10                 devastating
```

3. Modify the program to print error messages (but continue executing) if no proper cities or descriptors are used in a report.

4. Make the program check for multiple acceptable descriptors in a given report. Print an error message if the descriptors cause conflicting MMIs to be interpreted.

5. Without adding features to the program, modify it to have a modular design. Draw a structure chart and data flow diagram to fit your design. Three functions other than main probably are sufficient for this program.

APPLICATION EXERCISES

1. You have the following clients for your engineering consulting firm:

Client	Business type
Acme Construction	Machinery design
Johnson Electrical	Switch manufacturing
Brown Heating and Cooling	Boiler design
Smith Switches	Switch manufacturing
Jones Computers	Computer sales
Williams Cleaning Equipment	Machinery sales

To keep track of your clients in an orderly manner, you need a program that can arrange them alphabetically by name or by business type. Write such a program using a bubble sort. The input specifications are to

Read the client and business from a file.
Read the requirement of sorting according to name or business from the keyboard.

The output specification is to put the arranged list into an output file.

2. At times in engineering you may be required to show your work to a third party to illustrate your capabilities. However, to protect the privacy of the client for whom the work was done, you need to eliminate some of the details. Write a program that can take a text report and replace all of the cost figures (preceded by $) with **** and replace all instances of the client's name with "Client X." The input specifications are to

Read the text report from a file.
Read the client's name from the keyboard.

The output specification is to put the censored report into a file.

3. Write a program that can add equations with variables. For instance, the two equations

$$3a + 9b + 10c - 8d = 6$$
$$9a - 2b + 3c - 4d = 4$$

sum to

$$12a + 7b + 13c - 12d = 10$$

The input specification is to read two equations from the keyboard. The output specification is to print the result to the screen.

4. The equation relating current, voltage, and resistance in a circuit is $I = V/R$, where I = current, V = voltage, and R = resistance. Write a program that accepts two of the three from the keyboard, computes the third, and prints it to the screen. Enter the data in the form $I = 2.3$. Allow input in any order.

5. The kinetic energy, E, of a body of mass, m, moving at a constant velocity, v, is $E = 0.5mv^2$. Write a program that reads the values of m and v with their units from the keyboard in the form $m = 5$ kg. Acceptable units for m are kg, gm, and mg. Acceptable units for v are m/sec, km/hr, cm/sec, and mm/sec. Print the output energy in joules where 1 joule = 1 kg m^2/sec^2.

6. We want to determine the allowable force to be applied to a cable without the cable stretching an excessive amount. The change in length, Δ, of the cable is

found from the equation $\Delta = (PL)/(AE)$, where P = tensile force in the cable, A = cross-sectional area of the cable, L = initial cable length, and E = modulus of elasticity of the cable material. The value of E depends on the type of material from which the cable is made. For instance, this table gives some values of E (where 1 MN = 225,000 lb):

Material	E (MN/m^2)
Steel	200,000
Iron	100,000
Aluminum	190,000
Brass	100,000
Bronze	80,000
Copper	120,000

Write a program that computes the change in length of a cable given the input data of

Material type
Cable length
Cross-sectional area
Tension in cable

The input specifications are read in from the keyboard. The user should be prompted for each input value. The output specification is to print the result to the screen.

7. Rewrite problem 6, but have the user type in all of the information in the form "Material=Brass" in any order.

8. The period of a simple pendulum is

where l = pendulum length
                           g = acceleration due to gravity

Write a program that computes either the pendulum length or period, given that the other is entered as input data. The input specifications are to

Prompt the user to enter the data at the keyboard as being either $T = \ldots$ or $l = \ldots$.
Check the input data and do not accept negative or alphabetic values for T or l. If illegal data is entered, ask the user to enter the data again.

The output specification is to print the result to the screen.

9. A 2-D char array has the following string elements:

```
cc  c   ccc   cccc
ee  e   eee   eeee
aa  a   aaa   aaaa
dd  d   ddd   dddd
bb  b   bbb   bbbb
```

Write a program to do the following:
a. Sort the array using an exchange maximum sort so that it will look like

```
e    ee   eee   eeee
d    dd   ddd   dddd
c    cc   ccc   cccc
b    bb   bbb   bbbb
a    aa   aaa   aaaa
```

b. Sort the array using a bubble sort and rearrange the elements so that it will look like

```
aaaa   aaa   aa   a
bbbb   bbb   bb   b
cccc   ccc   cc   c
dddd   ddd   dd   d
eeee   eee   ee   e
```

10. Write a text-formatting program that reads a given text file and generates another text file for which the length of each line is specified by the user. The keyboard input specifications are

The name of the original text file
The maximum length, L, allowed in each line
The name of the output file

The program specification is that the program should contain at least one 1-D character array and one character pointer. The output specifications are to print

The original text file.
A line to separate the original text file and the new formatted text file. The line should contain a total of L digit characters (0 through 9) starting from 1. For example, if L is 25, then the line should be

```
1234567890123456789012345
```

The new formatted text file.
The program should not break any word and should not combine two paragraphs into one.

11. Modify problem 10 so that the output aligns the paragraph at the left indent.

12. Modify problem 10 so that the output aligns the paragraph at the right indent. You may need to add blank characters between words.

13. Modify problem 10 so that the output centers the paragraph between the left and right indents. You may need to add blank characters between words.

14. Modify problem 10 so that the output aligns the paragraph at both the right and the left indents. You may need to add blank characters between words.

15. Write a program to find if a file contains a specified word. The keyboard input specifications are

The name of the original text file.
The word to be found.

The screen output specification is to display the line that contains the specified word. If the file contains more than one specified word, all of them should be displayed on the screen.

16. Write a program to replace a misspelled word with a correct one. The keyboard input specifications are

The name of the original text file.
The misspelled word to be replaced.
The correct word that replaces the wrong one.
The name of the new text file.

The output specifications are to

Display the line that contains the misspelled word.
Display this line after the correction is made.
Save the file that has been corrected.

17. Write a program that can perform a simple cut and paste operation for a given file. The keyboard input specifications are

The name of the original text file.
The boundary of the cut section, which is represented by the starting word(s) and the ending word(s). Note that the word to be used for the cut and paste boundary must be unique. If the word is not unique, you may need to use more than one word to define the boundary.
The location to start the paste section, which is represented by a specified word(s).
The new output file name.

The program specifications are to

Call a function to display the line that contains the starting word(s) for the cut section. The function must be able to receive the contents of the line that was saved in a 1-D array in the main() function.
Call a function to display the line that contains the ending word(s) of the cut section. The function must be able to receive the contents of the line that was saved in a char pointer in the main() function.

Display the line that contains the word for placing in the pasted section.
Save the file that has been cut and pasted.

18. Write a program to split a text file into several equal length files. The keyboard input specifications are

The name of the original text file.
The total number of split files.
The names of each split file.

The output specifications are to

Display the name of the original text file.
Display the total number of split files.
Display the names of each split file.
Save each split file.

19. Write a program that can correctly sort a file of words with both upper- and lowercase first letters. For instance, suppose the words in the file consist of the words on this page. Use an exchange maximum sort to rearrange the words with no influence from the first letter's case. For instance, the word *For* should come before the word *influence*. To do this you will have to temporarily convert the case of the first letter to lower case, alphabetize it with the other words, then return the case to its original type.

8

Data Structures and Large Program Design

Data structures are used in programming because they enable data to be accessed and manipulated in a relatively easy manner that fits with particular problems to be solved. Recall that the first data structure we discussed was an array, which is a set of like-type data stored in contiguous and increasing memory locations. Arrays are referenced by using either subscripts (array notation) or pointer notation. Arrays are valuable for use in programming because they group information that is related and allow for easy access and manipulation.

We will find that there are other ways in which we would like to group or access data. For instance, a record is a grouping of information that may include more than one type of data. A stack is a data structure for which it is easy to access the last piece of information entered. This stack should not be confused with the region of memory we called the *stack* (although both work somewhat similarly).

In this chapter, we examine some of the data structures either built into the C language or created by a programmer. The following table lists different types of structures and the form in which they exist in C:

Data structure	Form
Array	Built in
Record	Built in
Linked list	Must be created
Stack	Must be created
Queue	Must be created
Graph	Must be created
Tree	Must be created

We also describe all of these data structures except graphs (because they are beyond the scope of this text) and arrays (since they have already been covered). Our goal in generating the structures that must be created is to develop a level of data abstraction. *Data abstraction* is a term used to describe a concept of working with data. If the concept is implemented correctly, the details of the abstraction are invisible to the user and the abstraction can be used by simply knowing the concept. In this book, of course, we show you the details. By knowing the details, you will be able to create a level of data abstraction for your programs.

Also in this chapter, we look at some issues involved with developing large programs. We do this because it is likely that, if you go into commercial software development, you will be working with programs much larger than what we have developed to this point. Large program development requires certain skills that you may need, such as how to use multiple source code files and how to develop header files.

We begin by looking at what C calls *structures*. A structure is C's version of the data structure that generically is called a *record*. From this point forward we use the term *structure* only.

▓ STRUCTURES IN C

Recall that the data types we used to this point have been int, float, and double, among others. A *structure* in C is a data type that is a grouping of data types. The programmer defines a structure to fit the program's needs. So, a structure is called a *derived data type*. Within a program you may define many structures. Once a structure is defined, we treat it much the same as we treat C's data types int and double.

Structures are useful and convenient for storing and manipulating information, because in engineering we find we often need to group unlike information. For instance, suppose, in an effort to develop an understanding of the yearly use of water in the world, we find it necessary to evaluate the following information, city by city:

City name
Year
Usage

We collect these data for as many cities as we can. While we can handle working with this information with it not connected, it is much easier for us to work with it if it is connected.

For example, in the same way that we create arrays of int or double, we can create arrays of a structure. We could create an array where each element of the array consists of all three of the components: city name, year, and usage. Each component is known as a *field*. This is illustrated in Fig. 8.1a.

If we wanted to sort this information in alphabetical order by city names, we could do so using one of the sorting techniques we developed previously. If we store the information in a structure, we find that, after rearrangement into alphabetical order, the year and usage still are connected to the appropriate city automatically as shown in Fig. 8.1b. Therefore, we need not program further manipulations to make the year and usage match the city name. In doing this manipulation, we have used the city name as a *key*, a term used to represent the field being used as a basis for manipulation.

Also, if instead we wanted to arrange the information in order of increasing usage (meaning that usage is the key), the city and year still are correctly connected

FIG. 8.1

Five elements of an array of a structure type, where the structure consists of city name, year, and water usage

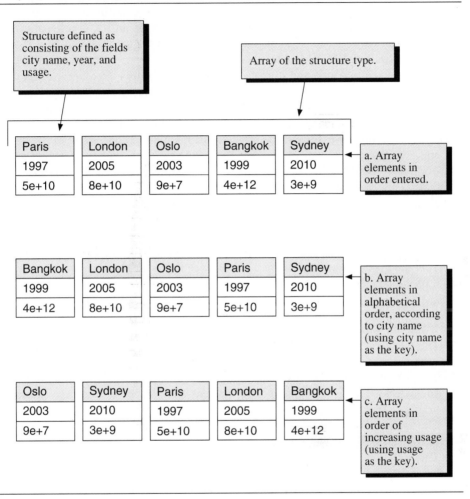

Structure defined as consisting of the fields city name, year, and usage.

Array of the structure type.

Paris	London	Oslo	Bangkok	Sydney
1997	2005	2003	1999	2010
5e+10	8e+10	9e+7	4e+12	3e+9

a. Array elements in order entered.

Bangkok	London	Oslo	Paris	Sydney
1999	2005	2003	1997	2010
4e+12	8e+10	9e+7	5e+10	3e+9

b. Array elements in alphabetical order, according to city name (using city name as the key).

Oslo	Sydney	Paris	London	Bangkok
2003	2010	1997	2005	1999
9e+7	3e+9	5e+10	8e+10	4e+12

c. Array elements in order of increasing usage (using usage as the key).

even after the rearrangement, as illustrated in Fig. 8.1c. The connection between the data (caused by using a structure to store the data) makes sure that, even after rearrangement, the three data types within the structure are connected.

We begin our discussion in this chapter not with arrays of structures but just structures. After understanding structures and how to access individual members (or fields) of structures, we move to a description of arrays of structures and other important topics related to using structures in subsequent lessons.

LESSON 8.1 STRUCTURES—DECLARING, INITIALIZING, READING, AND PRINTING

C treats a structure much like it treats a data type such as char, int, float, or double. For instance, if a memory location is designated to contain an int, C internally recognizes that 2 bytes (commonly) of memory are connected together and treated as a single unit. Similarly, for a double, C recognizes that 8 bytes (commonly) of memory are connected together, and it treats them as a unit. With a structure, we can create a new data type that may have 50 or 100 bytes, all connected. C allows us to define this data type, and we set roughly how many bytes are to be used for it.

We do this by specifying that a single structure consists of a number of standard data types or arrays. For instance, a structure can contain a char array size [20], an int, and a double (for a total of 20 + 2 + 8 = 30 bytes occupied), or a char array size [20], three doubles, and another char array size [30] (for a total of 20 + 3 × 8 + 30 = 74 bytes occupied).

We created structures of both of these in the program for this lesson. In this program, we define two structures, declare a number of variables that are of the structure types, and print the values of the structure members of the variables. The dot operator (.) is important in the use of structures in programs. It allows access to the structure member memory cells. We use it between the name of the variable and the name of the member. Look for its use in the program.

What you should observe about this lesson's program.

1. The name of the first structure defined is Consumption. The keyword *struct* is used before the name.
2. This structure has three members: city[ ], year, and usage. The members are enclosed in braces and a semicolon appears after each of them and after the braces.
3. The second structure defined is Resource.
4. Resource has five members: material[ ], longitude, latitude, quantity, and units[ ].
5. The structure definitions have been written outside the body of main or any other function.
6. Within main, three lines of declarations are given. These lines declare the variables metal, fuel, and wood to be of type Resource, and the variables water and power to be of type Consumption.
7. The dot operator (.) is used with the Resource structure variable, metal, in the first three assignment statements in the program. These statements initialize some members of metal.
8. The function strcpy is used to copy strings into metal.material[ ] and metal.units [ ]. These names without brackets represent addresses.
9. In the first printf statement, all of the members of metal are printed.
10. The first scanf statement reads values for the members of water and power. The "address of" operator (&) is used in a manner similar to the way it is used for ordinary variables.

A question you should attempt to answer before reading the explanation.
What do you think the location of the structure definition indicates about the scope
of the structure definition?

Source code

```
#include <stdio.h>
#include <string.h>
                          Structure tag

struct Consumption
        {
        char city[20];
        int year ;                Structure members        Structure definition
        double usage;
        };


struct Resource
        {
        char material[30] ;
        double longitude;
        double latitude;          Structure members        Structure definition
        double quantity;
        char units[20];
        };
                                  Initialization of all of the structure
                                  members for the variable wood.
void main (void )
{
        struct Resource metal, fuel;
        struct Resource wood = {"Oak", 32.5, 13.2, 5e+8, "hectares"};
        struct Consumption water, power;
                                         Declarations of variables
                                         of structure types.
        metal.longitude = 57.3 ;
        metal.latitude = 32.1 ;
        metal.quantity = 3e+10;
        strcpy (metal.material, "Iron");      Initialization of structure
        strcpy (metal.units, "cubic meters");  variable's members.

        printf ("The metal information is:\n%s\n%4.1lf degrees longitude\n"
                        "%4.1lf degrees latitude\n%4.0e %s\n\n",
                        metal.material, metal.longitude, metal.latitude,
                        metal.quantity, metal.units);


        printf("Enter water and power:\ncity, year and usage\n");
        scanf ("%s%d%lf%s%d%lf", water.city, &water.year, &water.usage,
                        power.city, &power.year, &power.usage);


        printf ("\n\nThe water and power are:\n%s\n%d\n%4.0lf million liters\n\n"
                        "%s\n%d\n%4.0lf mega watts\n",
                        water.city, water.year, water.usage,
                        power.city, power.year, power.usage);

}
                                         Reading and printing
                                         structure members.
```

Output

```
                        The metal information is:
                        Iron
                        57.3 degrees longitude
                        32.1 degrees latitude
                        3.00000 e + 10 cubic meters


                        Enter water and power:
                        city, year and usage
   Keyboard input       Paris 2003 120 Chicago 2010 50000


                        The water and power are:
                        Paris
                        2003
                        120 million liters


                        Chicago
                        2010
                        50000 mega watts
```

Explanation

1. How do we define a structure in C? We define a structure with a name (called a *structure tag* or *tag*) and a list of the names and types of the members. For instance,

```
struct Consumption
        {
        char city[20];
        int year;
        double usage;
        };
```

defines a structure with the tag Consumption to consist of a character array (city[20]), an integer (year), and a double (usage). Also,

```
struct Resource
        {
        char material[30];
        double longitude;
        double latitude;
        double quantity;
        char units[20];
        };
```

defines a structure with the tag Resource to consist of two character arrays (material[30] and units[20]), and three doubles (longitude, latitude, and quantity).

In general, the form is

```
struct Tag
        {
        type member_1;
        type member_2;
          .
          .
          .
        };
```

where the *Tag* can be any valid identifier. The *type* can be any valid C data type, including int, float, double, and others. Also, *member_1* and *member_2* are valid identifiers and may include arrays, pointers, or even structures. Note that a semicolon is required at the end of the line for each member type and after the closing brace of the structure.

2. Is it necessary to make the first letter of a structure tag a capital letter? No; however, we follow the convention of using a capital letter for the first letter of the tag. This is not required by C and is not followed by all programmers, but it is done in some circles. Check with your employer or instructor for the convention used at your university or company.

3. How do we declare variables to be of a particular structure type? We do it in a manner similar to what we do for declaring variables to be of int or double type. At the beginning of the function (for variables with function scope), we write the type and follow it with the variable names. For instance,

```
struct Resource metal, fuel;
struct Consumption water, power;
```

declare the variables metal and fuel to be of type struct Resource and the variables water and power to be of type struct Consumption. The general form is

```
struct Tag variable_1, variable_2, variable_3;
```

where *Tag* is the tag for a structure previously defined, and *variable_1, variable_2,* and *variable_3* are identifiers for variables created by the programmer.

4. How do we initialize a structure in its declaration? We can initialize each of a structure variable's members in a declaration using = and braces after the variable name. For instance,

```
struct Resource wood = {"Oak", 32.5, 13.2, 5e+8, "hectares"};
```

initializes for the variable wood, the following:

Member for structure variable wood	Value
material[30]	Oak
longitude	32.5
latitude	13.2
quantity	5e + 8
units[20]	hectares

Note that there is a correspondence between the order in the structure definition and the initialization list.

5. How do we access the members for a particular structure variable? We use the dot operator (.), which also is called the *structure member operator.* The dot operator is placed between the variable name and member name. For instance,

`metal.longitude`

refers to the member longitude of the structure variable metal. The variable metal already has been declared to be of type struct Resource, which has been defined to have a member longitude. Similarly, metal.latitude and metal.quantity access other structure members of the variable, metal.

6. How do we access the address of the beginning of character arrays contained within structures? Using the dot operator, we can access the beginning of strings. For instance, with the member material declared as

`char material[30];`

the following

`metal.material`

refers to the address of the beginning of the member array material[ ] for the variable metal. Thus, we can use:

`strcpy (metal.material, "Iron");`

to copy the string, "Iron", into the memory reserved for the member material of the variable metal.

7. How can we access an individual character of a character array that is a structure member? We did not show it in this lesson's program, but if we wanted to print the first character of the character array member material[ ], we could use the following C statement for the structure variable metal:

`putchar (metal.material[0]);`

8. How will we represent our table of variables for structures and how is the dot operator illustrated? We will show the name of each structure variable in the left column with the structure members in the type column. The values are given for each individual member for each structure variable. This is shown in Fig. 8.2 for this lesson's program.

9. How do we refer to the address of numeric data types contained as members of structures? We use both the & and dot operators. For instance, &water.year refers to the address FF0E as shown in Fig. 8.2. It has been used in the scanf statement:

```
scanf ("%s%d%lf%s%d%lf", water.city, &water.year, &water.usage,
                    power.city, &power.year, &power.usage);
```

FIG. 8.2

Table of variables for structures and illustration of dot operator

> Illustration of dot operator.
> Operation shown is metal.latitude

Name	Type		Address	Value
metal	struct Resource		FFAC	
	material	char[30]	FFAC	I
	longitude	double	FFD2	57.3
	latitude	double	FFCA	32.1
	quantity	double	FFDA	3e+10
	units	char[20]	FFE2	c
fuel	struct Resource		FF62	
	material	char[30]	FF62	not initialized
	longitude	double	FF88	not initialized
	latitude	double	FF80	not initialized
	quantity	double	FF90	not initialized
	units	char[20]	FF98	not initialized
wood	struct Resource		FF18	
	material	char[30]	FF18	O
	longitude	double	FF3E	32.5
	latitude	double	FF36	13.2
	quantity	double	FF46	5e+8
	units	char[20]	FF4E	h
water	struct Consumption		FEFA	
	city	char[20]	FEFA	P
	year	int	FF0E	2003
	usage	double	FF10	120
power	struct Consumption		FEDC	
	city	char[20]	FEDC	C
	year	int	FEF0	2010
	usage	double	FEF2	50000

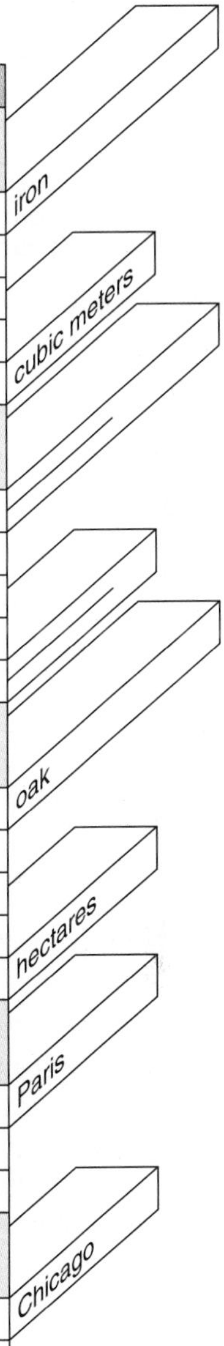

iron

cubic meters

oak

hectares

Paris

Chicago

to have the function scanf read data from the keyboard and store it in the memory cell FF0E.

10. Is there any other way to define and declare structures in C? Yes; however, what we have described currently is the preferred method. C also allows variable names to be included with the structure definition. For instance, in this lesson's program, we could have used

```
struct Consumption
        {
        char city[20];
        int year;
        double usage;
        }water, power;
```

to both define the structure Consumption and declare the variables water and power. If this form is used, C allows the structure tag to be omitted; therefore,

```
struct
        {
        char city[20];
        int year;
        double usage;
        }water, power;
```

also is a valid definition and declaration for a structure. Not using a tag, though, prevents further declarations of variables of this structure type in the program.

When using definitions or declarations of this form, care must be taken in paying attention to the scope rules. If these are written outside any function, then we have declared water and power to be global variables, which reduces the modularity of the program. On the other hand, if this form of structure definition is written within a function, then the structure applies only to that function. Because of these shortcomings, we will not use these types of structure definitions.

11. Does C create structures with no unreserved bits of memory between members? ANSI C does not specify that all structure members be contiguous in memory and therefore allows unreserved bits to exist between members. So, we cannot strictly take the sum of the memory needed by the individual members to determine the memory required for a particular structure variable. In the introduction to this lesson's program, when we described structures with 30 and 74 bytes occupied, a structure with the members listed may require more than 30 or 74 bytes. The actual amount of memory required depends on the particular C compiler.

EXERCISES
1. True or false:
    a. *Structure* is the C word for "record."
    b. We can use the dot (.) operator to access structure members.
    c. Structure members are required by ANSI C to be stored in contiguous memory locations.

   d.   Structure definitions are required to have tags.

   e.   The first letter of a structure tag commonly is a capital letter.

2.   Given the structure definition and declarations

```
struct Force
        {
        char name[40]
        int point_number;
        double xforce;
        double yforce;
        }

struct Force wing, fuselage;
```

find errors, if any, in the following statements:

   a.   **wing.xforce=15.3;**

   b.   **wing.xforce.yforce=28.5;**

   c.   **fuselage.name = 'v';**

   d.   **strcpy (wing.name, "DC 10");**

   e.   **fuselage.point_number = 85;**

3.   Write a program that reads a structure from a data file and prints it to the screen.

Solutions

1.   a (true), b (true), c (false), d (false), e (true)

2.   a.   No error.

     b.   **wing.yforce = 28.5;**

     c.   **strcpy (fuselage.name, "v");**

     d.   No error.

     e.   No error.

■   LESSON 8.2 STRUCTURES AND POINTERS AS STRUCTURE MEMBERS

In this lesson's program, we look more closely at structure members. In the source code, we define two structures, declare a number of variables to be structure types, initialize the variables, and print them.

What you should observe about this lesson's program.

1.   The first structure, Xxx, has an integer array as a member.

2.   The second structure, Yyy, has a member that is a struct Xxx type.

3.   Yyy also has a pointer variable and an array of pointers as members.

4.   In the declarations, qq and nn have been initialized. Braces within braces have been used to initialize the members.

5.   In the program body, pp is initialized member by member.

6. The dot operator is used twice on the left side of the first three assignment statements.

7. The last assignment statement copies all of the members of pp into rr. The last printf statement prints all of them out for verification.

Questions you should attempt to answer before reading the explanation.
1. What is the value of qq.bb[1]?
2. What is the value of nn.mm.bb[1]?

Source code

```
#include <stdio.h>
#include <string.h>

struct Xxx
        {
        char aa[30];
        int bb[2];
        double cc;                    Structure as a member of a structure.
        };
struct Yyy                                              Structure definitions.
        {
        char dd[50];
        struct Xxx mm;
        char *ee;                 Pointer variable as a structure member.
        char *ff[3];
        };                     Pointer array as a structure member.
                                                 Braces used to separate initializations of structure members.

void main (void)
{
        struct Xxx qq={{"Sample"},{0,1},{5.4}};
        struct Yyy nn={{"String constant"},{{"Text"},{7,8},{12.3}},{"Address"},{"a","b","c"}};
        struct Yyy pp, rr ;

        strcpy (pp.dd,"Structure ");
        strcpy (pp.mm.aa,"in structure");
        pp.mm.bb[0]=10;                            Accessing members of structure within a structure.
        pp.mm.bb[1]=12;
        pp.mm.cc=57.8;
        pp.ee="Pointer and ";
        pp.ff[0]="array ";                     Accessing pointer members of a structure.
        pp.ff[1]="of ";
        pp.ff[2]="pointers. ";

        rr=pp;                         Copying the values of all of the members of one structure variable
                                       into the memory for members of another structure variable.

        printf("%s%s %d %d %lf\n%s%s%s%s\n",rr.dd,rr.mm.aa,rr.mm.bb[0],
                    rr.mm.bb[1],rr.mm.cc,rr.ee,rr.ff[0],rr.ff[1],rr.ff[2]);
}
```

Output

```
Structure in structure 10 12 57.800000
Pointer and array of pointers.
```

Explanation

1. Can the member of a structure be another structure? Yes; in this lesson's program, the structure Yyy has members as shown in Fig. 8.3.

2. How do we access the members of structures that are themselves members of structures? We use the dot operator twice. For instance, for the struct Yyy variable pp, we access all of the members in the following way:

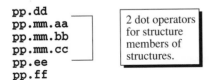

```
pp.dd
pp.mm.aa
pp.mm.bb
pp.mm.cc
pp.ee
pp.ff
```

2 dot operators for structure members of structures.

3. How can we use braces in the initialization of structure members? Braces are not required, although some compilers will give warnings if they are absent. If they are used, they can separate the individual members from each other. This is especially helpful when arrays or other structures are structure members. For instance, in this lesson's program, the braces in the initialization indicate the following divisions:

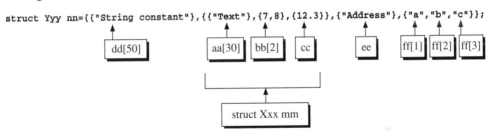

```
struct Yyy nn={{"String constant"},{{"Text"},{7,8},{12.3}},{"Address"},{"a","b","c"}};
```

dd[50]   aa[30]   bb[2]   cc   ee   ff[1] ff[2] ff[3]

struct Xxx mm

4. How do we access pointer members of structures? Single pointer variables can be accessed using a dot operator and the structure variable name, such as

```
pp.ee
```

Pointer arrays that are structure members are accessed using a dot operator and an array subscript; for instance,

```
pp.ff[0]
```

5. How can we copy all the members of one structure variable into the memory reserved for another structure variable? We can use a single assignment statement with no need to reference individual members of the functions or elements of arrays within the structures. For instance,

```
rr=pp;
```

FIG. 8.3

Members of struct Yyy. Observe that mm also has members

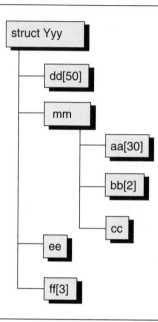

copies all of the values of the members of the structure variable pp into the memory reserved for the structure variable rr. It is very simple to copy structures. Note, though, that we cannot perform manipulations on the structure in this manner. In other words,

```
rr = 3*pp;
```

is not a legal C statement because we cannot apply the multiplication operator on a structure variable without referencing the individual members of the structure.

EXERCISES

1. True or false:
   a. C does not allow structure variables to be members of structures.
   b. To copy all the members of one structure variable to another same type structure variable, we must copy member by member.
   c. To access members of structures within structures, we use the dot operator twice.
   d. Braces are required in initializing structures in declarations.
   e. Structures cannot have pointers as members.

2. Given these structure definitions and declarations

```
struct Xxx
        {
        int aa;
        double *bb;
        char *cc[30]
        };
struct Yyy
        {
        int *dd;
        double ee;
        };

struct Xxx mm, nn;
struct Yyy pp, qq;
```

find errors, if any, in the following statements:

a. `mm=pp;`
b. `strcpy(nn.cc,  "Test");`
c. `Xxx.aa= 15;`
d. `Yyy.ee = 43.8;`
e. `mm=nn;`

Solutions
1. a (false), b (false), c (true), d (false), e (false)
2. a.   Cannot assign structures of different types.
    b.   `nn.cc[0]= "Test";`
    c.   `mm.aa = 15;`
    d.   `qq.ee =43.8;`
    e.   No errors.

▩   LESSON 8.3 POINTERS TO STRUCTURES

We also can access members of structure variables using pointers. To do so, we must create a pointer variable that can store an address to the structure. The program for this lesson defines a structure, declares an ordinary structure variable and a pointer to a structure, initializes the members of the ordinary variable, and then changes these member values using the pointer variable.

What you should observe about this lesson's program.
1. The Xxx structure has two members.
2. The variable mm is declared to be of type struct Xxx.
3. The variable pp is declared to be a pointer to type struct Xxx.
4. The members of mm are initialized in the first two assignment statements.
5. In the third assignment statement, pp is given the address of mm.
6. In the fourth assignment statement, pp is used to change the aa member of mm with the unary * operator.

7. In the fifth assignment statement, pp is used to change the bb member of mm with the arrow (->) operator.
8. The printf statement and the output verify that mm.aa and mm.bb have been changed.

A question you should attempt to answer before reading the explanation. Why do you think parentheses have been used around *pp in the fourth assignment statement?

Source code

```
#include <stdio.h>


struct Xxx
        {
        int aa;                         Structure definition.
        double bb;
        };


void main (void)
{
        struct Xxx mm;
        struct Xxx *pp;         Declaring a pointer to a structure (a variable that can store
                                the address of the beginning of an Xxx structure).

        mm.aa=8;
        mm.bb=23.2;


        pp=&mm;                 Assigning an address to the pointer.


        (*pp).aa = 12 ;         Using the pointer to access the structure variable's members.
        pp->bb   = 97.2;        Two methods are shown. Both accomplish the same task.


        printf("%d %lf\n", mm.aa, mm.bb);
}
                                Printing the mm members. The output
                                shows that the assignment statements using
                                the pointer operations indeed did modify
                                the members of mm.
```

Output

```
12 97.200000
```

Explanation

1. How do we declare a pointer variable to a structure? We use the keyword *struct,* the structure tag, *, and the variable name. For instance,

```
struct Xxx *pp;
```

declares the variable pp to be able to store the address of a struct Xxx type variable.

2. How can we assign an address to a structure pointer variable? We use the address of operator. For instance,

```
pp = &mm;
```

assigns the address of the beginning of the mm structure to the pointer variable pp.

3. What is the address of the beginning of the structure? It is the address of the first structure member for the variable. For example, &mm is &mm.aa for this lesson's program.

4. How do we access structure members using pointers? We have illustrated two methods. We can use the unary * operator. For instance, in this lesson's program,

```
pp = &mm;
(*pp).aa = 12 ;
```

causes the member aa of the structure variable mm to be set equal to 12. The parentheses are necessary because the dot operator has a higher precedence than the unary * operator. Therefore, *pp.aa=12; will not cause mm.aa to be set equal to 12.

Or, we can use the structure pointer operator ->, also called the *arrow operator.* For example,

```
pp = &mm;
pp->bb  = 97.2;
```

causes the member bb of the structure variable mm to be set equal to 97.2. The form for using this operator is

pointer_to_structure -> member_name

where no spaces are allowed between the - and the >. It is more common to use the arrow operator than the *. combination.

An image of these actions is shown in Fig. 8.4. Note that the value of &mm is stored in the pointer variable pp. It can be seen that &mm is the address of the first member of the structure.

FIG. 8.4

Accessing structure member using pointers

Name	Type		Address	Value
mm	struct Xxx		FFEC	
	int	aa	FFEC	12
	double	bb	FFEE	97.2
pp	pointer to struct Xxx		FFEA	FFEC

Image of action: (*pp).aa
This is the same as pp->aa

Image of action: pp->bb
This is the same as (*pp).bb

EXERCISES

1. True or false:
   a. The arrow operator is used to access structure members with pointer variables.
   b. We also can use the unary * and dot (.) operators with pointer variables to access structure members.
   c. The * and dot operators are used more commonly than the arrow operator to access structure members.
   d. We need not use parentheses with the * and dot operators to access structure members.

2. Given these structure definitions and declarations

```
struct Xxx
        {
        int aa;
        double bb;
        char cc[12];
        };
struct Xxx mm, nn, *pp, *qq;
```

   find the errors, if any, in the following statements:
   a. `pp = mm;`
   b. `qq = &nn;`
   c. `mm->bb = 54.2;`
   d. `*nn.aa =5;`
   e. `*qq.cc = "Sample";`

Solutions

1. a (true), b (true), c (false), d (false)
2. a. `pp = &mm;`
   b. No errors.
   c. `pp->bb=54.2;`
   d. `nn.aa=5;`
   e. `qq=&nn;`
      `strcpy(qq->cc, "Sample");`

■ LESSON 8.4 STRUCTURES AND FUNCTIONS

In the previous lesson, we learned how to work with pointers to structures. In this lesson, we use that knowledge to pass structures to functions and to work with the structures within a function to which a structure has been passed.

What you should observe about this lesson's program.
1. The structure Xxx has four different types of members: an integer, an array of doubles, a character pointer, and an array of pointers.
2. The call to function1 has two arguments. Both arguments are addresses of struct Xxx type variables.
3. The function1 header has qq and rr as pointers to struct Xxx.
4. Within function1, qq and rr have been used with both the unary * operator and the arrow operator to modify the values of the members of mm and nn.

Source code

```
#include <stdio.h>


struct Xxx
        {
        int aa;
        double bb[2];
        char *cc;
        char *dd[3];
        };


void function1 (struct Xxx *qq, struct Xxx *rr);


void main (void)
{

        struct Xxx mm, nn;


        function1 (&mm, &nn);


        printf ("%d %lf %lf %s%s%s%s\n", mm.aa, mm.bb[0], mm.bb[1],
               mm.cc, mm.dd[0], mm.dd[1], mm.dd[2]);
        printf ("%d %lf %lf %s%s%s%s\n", nn.aa, nn.bb[0], nn.bb[1],
               nn.cc, nn.dd[0], nn.dd[1], nn.dd[2]);


}

void function1 (struct Xxx *qq, struct Xxx *rr )
{
```

Passing addresses of first members of structure variables to function1.

Function header has struct Xxx pointer variables. On calling the function, these contain the addresses of the structures mm and nn.

```
        (*qq).aa = 12;
        (*qq).bb[0] = 23.4;
        (*qq).bb[1] = 34.5;
        (*qq).cc = "Structure ";
        (*qq).dd[0] = "passed ";
        (*qq).dd[1] = "to ";
        (*qq).dd[2] = "function.";


        rr->aa = 15 ;
        rr->bb[0] = 45.6  ;
        rr->bb[1] = 67.8  ;
        rr->cc = "Pointer ";
        rr->dd[0] = "operators ";
        rr->dd[1] = "can ";
        rr->dd[2] = "be used.";
}
```

Filling pointer array mm.dd[ ].

Because qq contains the address of mm, these assignment statements fill the mm structure using the * and dot operators.

Filling pointer array nn.dd[ ].

Because rr contains the address of nn, these assignment statements fill the nn structure using the -> operator.

Output

```
12 23.400000 34.500000 Structure passed to function.
15 45.600000 67.800000 Pointer operators can be used.
```

Explanation

1. How can we pass a structure to a function so that the function can modify the structure members? We can pass the structure's address (which is the address of the structure's first member) to the function so that the function has access to the memory cells reserved for the structure variable's members. For instance, the call to function1 in this lesson's program

function1 (&mm, &nn);

passes the addresses of the struct Xxx variables mm and nn.

2. Within the function, how do we access the structure members? We can use the indirection, *, and dot operators or we can use the structure pointer operator, ->. For instance, for function1 in this lesson's program,

(*qq).aa = 12;
rr->aa = 15 ;

both access the member aa. Because of the correspondence in the function call and header, (*qq).aa accesses mm.aa and rr->aa accesses nn.aa.

3. Which method is preferred? Most programmers use the structure pointer operator, ->, rather than the indirection and dot operators. You will see that its resemblance to an arrow makes it more intuitive as we use this operator to access members when we create data structures such as linked lists.

EXERCISES

1. True or false:
    a. Structures should not be used with functions.
    b. By passing the address of a structure variable, we can use a function to modify the structure's contents.
    c. We can pass structures to functions by either value (one member at a time) or reference (using the structure's address).

Solutions
1.   a (false), b (true), c (true)

■ LESSON 8.5 ARRAYS OF STRUCTURES

As we mentioned in the introduction to this chapter, one benefit of having a structure type in C is that we can create arrays of structures. With arrays of structures, we will see that we can solve a number of problems much more quickly and easily than if we were unable to create arrays of structures.

For instance, suppose we had a file that contained nine time/voltage measurements (treated as *x-y* coordinates) in random order, such as,

```
12 87
7 43
10 22
5 56
29 89
3 34
0 10
14 3
8 65
```

where the first number in each row is the time in seconds and the second number is the voltage measured at that time in millivolts. Also, suppose our goal is to rearrange and print these values in order of increasing time.

A very efficient way to do this is to create an array of a structure, where the structure could be defined as

```
struct coord
        {
        double x;
        double y;
        }
```

and then declare an array of this structure; for instance,

```
struct coord  volt_time[1000];
```

In doing so, we create an array of connected members that can be illustrated as

Time (sec)	Voltage (millivolts)
12	87
7	43
10	22
5	56
29	89
3	34
0	10
14	3
8	65

By rearranging this array in increasing order according to the time, the array becomes

Time (sec)		Voltage (millivolts)
0		10
3		34
5		56
7		43
8		65
10		22
12		87
14		3
29		89

Notice that, in rearranging the first column of the array, the second column of the array is rearranged as well, maintaining the correspondence between the time and voltage data. This would not have happened had we not used a structure.

For instance, if we had put the time into an ordinary array, x[9], and the voltage into another ordinary array, y[9], then on rearranging the x[ ] array, we would not have automatically rearranged the y[ ] array. This would eliminate the time-voltage correspondence. Of course, we could write code to rearrange the y[ ] array; however, it is clearly much more efficient to use an array of a structure and have it done automatically.

This is an illustration of one of the many valuable features of arrays of structures. In this lesson, we show how arrays of structures are implemented in programs. In the source code, we define the structure Xxx and declare an array of struct Xxx. We initialize the values of the first element of the array and copy the values of the first element into the memory reserved for the second element. We then print out the values of the second element.

What you should observe about this lesson's program.
1. The structure Xxx contains a character array, an integer, and a double.
2. An array of ten struct Xxx has been declared (mm[10]).
3. We initialize all the members of mm[0].
4. We copy all the members of mm[0] to mm[1] and print them out.
5. The output shows that mm[1] has the values assigned to mm[0].

Source code

```
#include <stdio.h>
#include <string.h>
struct Xxx
        {
        char aa[30];
        int bb;
        double cc;
        };
```

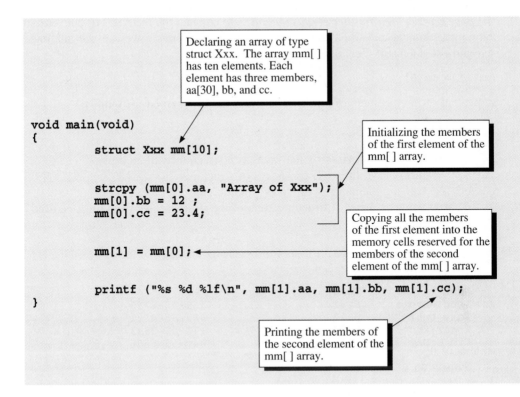

Declaring an array of type struct Xxx. The array mm[ ] has ten elements. Each element has three members, aa[30], bb, and cc.

Initializing the members of the first element of the mm[ ] array.

Copying all the members of the first element into the memory cells reserved for the members of the second element of the mm[ ] array.

Printing the members of the second element of the mm[ ] array.

```
void main(void)
{
        struct Xxx mm[10];

        strcpy (mm[0].aa, "Array of Xxx");
        mm[0].bb = 12 ;
        mm[0].cc = 23.4;


        mm[1] = mm[0];

        printf ("%s %d %lf\n", mm[1].aa, mm[1].bb, mm[1].cc);
}
```

Output

```
Array of Xxx 12 23.400000
```

Explanation

1. How do we declare an array of structures? A declaration of an array of structures looks like a declaration of an array of int or double. For instance,

`struct Xxx mm[10];`

declares mm to be an array of size [10] of struct Xxx.

2. How do we access a member of one of the elements of an array of a structure? We use both array notation and the dot operator. For example,

`mm[0].bb`

accesses the member bb of the first element of the mm structure array.

3. How do we copy all the members of one structure array element into the memory reserved for another array element? To copy all the members of one structure array element into the memory for another array element, we can use just one assignment statement. For instance,

```
mm[1] = mm[0];
```

copies all of the members of mm[0] into the memory reserved for mm[1].

EXERCISES

1. True or false:
    a. An array of a structure is the same as a structure with an array.
    b. Engineers have little use for arrays of structures.
    c. We can sort elements of an array of structures.
    d. In sorting an array of structures, one of the structure members must be used as a key.

2. Write a program to read a file as follows using an array of structures. Display this file neatly on the screen.

Name	Height(ft)	Age	SSN
Jean Garcia	5.61	21	123-45-6789
Tony Lutz	6.12	36	987-65-4321
Roger Ron	5.87	87	111-22-3333
Jim McKay	3.14	4	444-55-6666

    Solutions

1. a (false), b (false), c (true), d (true)

▨ LESSON 8.6 CREATING A LINKED LIST

Topics
*   Defining a structure to use for a linked list
*   Reserving memory for each node of a linked list
*   Initializing each node
*   Printing the contents of each node

A structure is ideal for creating a linked list because it is capable of storing more than one type of data in a block of memory. Thus, it is simple for us to store text, numeric information, and addresses (using pointer variables) in a structure. To create a linked list, it is necessary that we store an address in a pointer variable in the structure that we create. This address is the address of the next block of memory included in the linked list and therefore the link between blocks. In this lesson, we first describe the concepts behind a linked list, then we illustrate a program that creates a linked list.

We call a small block of memory a *node* in a linked list and represent it with a rectangle. Also, we call the address that is stored in the pointer variable in the block of memory for the structure a *link* and represent it with a dark block as shown in

FIG. 8.5

Representation of a linked list. Each rectangle represents a block of memory. The shaded portion represents the portion of the memory block that holds the address of the next block of memory indicated by the arrow.

5. For this lesson's program, how do we create and fill the linked list?

Fig. 8.5. An arrow points to the next block of memory indicated by the address stored in the pointer variable. Within the rectangle, we show text or numeric information that is stored in the block of memory along with the address stored in the pointer variable.

Figure 8.5 shows five blocks of memory. Each block of memory contains the name of a client company for an engineering consulting firm in addition to the address of the block of memory in which the next company is stored (indicated by the darkened rectangle).

Fundamentally, a linked list does not work like an array. We will see that to access the 500th node in a linked list, we must go node by node from the first to the 500th. (Note: It is not obvious to you right now why it is necessary to do this. When we look at the source code for this lesson's program, you will understand why a linked list works in this manner.) With arrays, we can go directly to the 500th element by using the correct subscript. This is an advantage that arrays have over linked lists.

However, because we must go node by node in a linked list, we find that it is easy to delete a node from a list. For instance, if we wanted to delete Hytech Co. from our linked list we could do so by simply changing the address that points to it (currently the one stored in the pointer variable with Natgren Co.'s structure). If we make this address point to Mibad Inc., we effectively have deleted Hytech Co. from the list, because as we go from node to node in the list, we never reach Hytech Co. This is illustrated in Fig. 8.6. Note that deleting an element in an array takes many more operations because we must shift all of the array elements to close the "empty hole" that may exist when an element is deleted.

FIG. 8.6

By changing the address associated with Natgren Co., we delete Hytech Co. from the linked list. This works because, to traverse a linked list, we must go node by node. Although the name Hytech Co. still may exist in memory, we never get to it because the address in the pointer variable with Natgren Co. structure points directly to Mibad Inc.

FIG. 8.7
We have inserted Leyedbak Inc. into the list by changing the address stored in the
pointer variable in the structure for Natgren Co. and having the address stored in the
pointer variable for Leyedbak Inc. be the address of Mibad Inc.

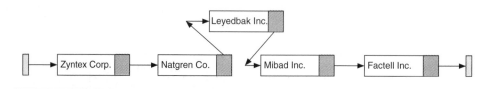

Also, we easily can insert a node into a linked list by changing the address
stored in the pointer variable for the node preceding the location of the insertion.
This is illustrated in Fig 8.7. Here, we inserted Leyedbak Inc. into the list by changing the address stored in the pointer variable in the structure for Natgren Co. and
having the address stored in the pointer variable for Leyedbak Inc. be the address of
Mibad Inc. These few operations have caused the new list to appear, as shown in
Fig. 8.8. Again, inserting an element in an array takes many more operations because
we must shift all of the array elements to make space for a new element.

We also find it simple to move a node from one location to another in a linked
list. For instance, if we wanted to move Natgren Co. to be the last element in the
list, we could change the pointing of the arrows (or the addresses stored, which indicates changing the locations to which the arrows point) to what is shown in Fig. 8.9.
The sketch given in Fig. 8.9 also can be shown as in Fig. 8.10, which clearly illustrates that Natgren Co. has moved to the end of the list. We note that to move Natgren Co. to the end of the list we changed three addresses, the ones in the pointer
variables for the structures for Zyntex Corp, Factell Inc., and Natgren Co. Even if
our original list had 1000 nodes and we wanted to move a node to the end of the
list, we could do it by changing just three addresses.

The few operations that it takes to do this is another advantage that linked lists
have over arrays. If we had an array with 1000 elements and wanted to move the
tenth element to the end, we would have had to remove the tenth element and move
990 elements forward one place and then put the removed element at the end. Thus,
to perform the same task with an array requires many more operations.

Because of some of the advantages that linked lists have over arrays, we find
that sometimes linked lists created with structures are preferable to arrays of struc-

FIG. 8.8
This is the list after inserting Leyedbak Inc. The number of operations to do this was
very few compared to the number it would take to do this with an array structure.

FIG. 8.9

By changing the addresses to which the pointer variables point, the order of the list effectively is changed. This list is the same as that shown in Fig. 8.10.

FIG. 8.10

This figure is the same as that shown in Fig. 8.9. We moved Natgren Co. to the end of the list by changing just three addresses.

tures. In this lesson, we show how to create a linked list in C. The program in this lesson's source code is different from programs that we used in other lessons in that it illustrates more than just syntax and form. It shows how to create a linked list (it does not show inserting and deleting); therefore, you need to follow the logic and understand the flow in detail to be able to create your own linked lists. In this introduction, we point out some of the highlights. The details are given in the explanation following the source code. This program is not a complete application of a linked list. We deliberately eliminated some details to focus on creating nodes and the connections between the nodes.

The program creates a linked list of client names *and* hours billed for an engineering consulting firm. Each node of the linked list holds a character array, client[ ], and an integer variable, hours_billed. Read the annotations in the source code and the explanation to understand how to create a linked list.

Source code

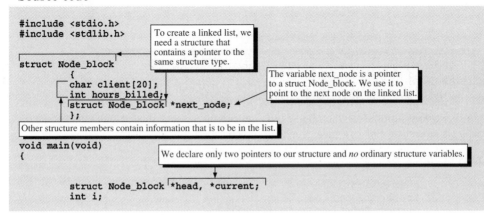

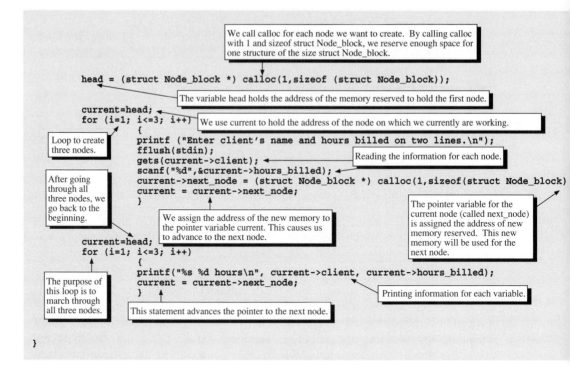

We call calloc for each node we want to create. By calling calloc with 1 and sizeof struct Node_block, we reserve enough space for one structure of the size struct Node_block.

```
head = (struct Node_block *) calloc(1,sizeof (struct Node_block));
```

The variable head holds the address of the memory reserved to hold the first node.

```
current=head;
for (i=1; i<=3; i++)
```

We use current to hold the address of the node on which we currently are working.

Loop to create three nodes.

```
{
printf ("Enter client's name and hours billed on two lines.\n");
fflush(stdin);
gets(current->client);
```

Reading the information for each node.

```
scanf("%d",&current->hours_billed);
current->next_node = (struct Node_block *) calloc(1,sizeof(struct Node_block));
current = current->next_node;
}
```

After going through all three nodes, we go back to the beginning.

We assign the address of the new memory to the pointer variable current. This causes us to advance to the next node.

The pointer variable for the current node (called next_node) is assigned the address of new memory reserved. This new memory will be used for the next node.

```
current=head;
for (i=1; i<=3; i++)
{
printf("%s %d hours\n", current->client, current->hours_billed);
current = current->next_node;
}
```

The purpose of this loop is to march through all three nodes.

This statement advances the pointer to the next node.

Printing information for each variable.

```
}
```

Output

Explanation

1. How do we define a structure for a linked list? A structure for a linked list can have any number of members, but one member must be a pointer to the same type of structure. For instance, the structure defined as

```
struct Node_block
    {
    char client[20];
    int hours_billed;
    struct Node_block *next_node;
    }
```

contains the member next_node that is a pointer to another struct Node_block. The other members of the structure, client[20] and hours_billed, are variables for holding information to be contained in the list. Here, we chose to hold in the list a character array and an integer, but we could have had many more members and types of members, including more character arrays, doubles, or numeric arrays.

In general, a structure definition for a linked list has the form

struct *Tag*
     {
     type member_1
     type member_2
      .
      .
     struct *Tag *pointer_name*
     }

where *Tag* is the tag chosen by the programmer to represent the linked list, *type* is any valid C data type, *member_1* and *member_2* are identifiers representing data for the structure to be stored in each node of the list, and *pointer_name* is an identifier used to represent the pointer to the next node in the list.

2. What visual image can we have of a linked list in memory? We can imagine a linked list as a number of blocks of memory (representing nodes in the linked list) connected to each other by the address stored in one of the cells of the block, as illustrated in Fig. 8.11. Each block contains the address of the beginning of another block. The program works with one block at a time (the current block) and the only way to get to the next block is to use the address stored in the current block.

Note that an array is very different from a linked list, because array elements are stored in contiguous memory locations whereas blocks of memory in linked lists are not at all contiguous. With arrays, we know the addresses of all of the elements if we know the address of just one. With a linked list, knowing the address of just one block of memory tells us nothing about the addresses of any other blocks used for the linked list. With an array, for instance, we can move from the first element to the 20th element by simply adding the number of bytes for 20 elements to the address of the first element. To move from the first block to the 20th block in a linked list, we must read the address of the second block, move to the second block and read the address of the third block, move to the third block and read the address of the fourth block. This must continue, block by block, until we reach the 20th block. Clearly, it takes more operations to move from node to node in a linked list than it does to go from element to element in an array. However, because of the ease of inserting and deleting nodes, linked lists have advantages over arrays for certain operations.

3. With an array, all of the memory for the elements of the array is reserved with a declaration. How is memory reserved for the nodes of a linked list? We make a separate call to calloc or malloc to reserve a block of memory for each node of a linked list. If a linked list has 1000 nodes, then we have 1000 calls to calloc or malloc. In a program that uses a linked list, the declarations do *not* reserve blocks of memory for the nodes of the list.

FIG. 8.11
An image of a linked list in memory

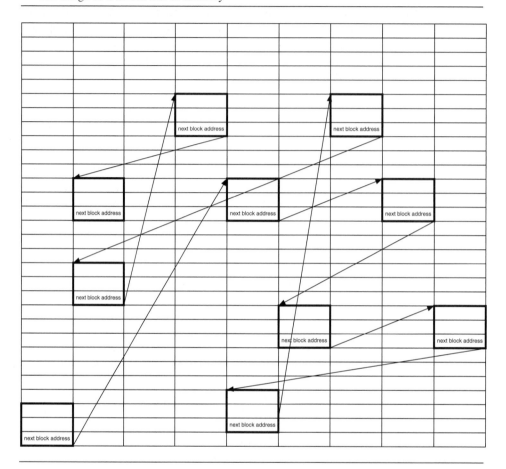

4. If the declarations do not reserve memory for the nodes of the list, what declarations do we need to create a linked list? We need to declare only pointer variables that will contain addresses to the blocks of memory for the list's nodes. For this lesson's program, we need to declare only two pointer variables, head and current, with the declaration

```
struct Node_block *head, *current;
```

For this program, we create a linked list with three nodes. However, with just these two pointer variables, we easily could have created a linked list with 1000 nodes. You will find that, when you write programs with linked lists, you need not declare many variables or reserve much memory with your declarations even for very long lists.

Note that we do not even declare an ordinary variable of the structure type. We declare only pointer variables because we need to know only the address of the block of memory that holds the node's information. With the address of the block, we access the structure members of the node's block.

We generally need only a few pointer variables because we work with only one or two nodes at a time; therefore, we need to hold only the addresses of the blocks of memory for these nodes. As we move from node to node to work with the information at each node, we simply replace the address stored in our pointer variable, current, with the address of the next node with which we will work.

5. For this lesson's program, how do we create and fill the linked list? We begin by creating the first node block and storing the address of this node block in the pointer variable head with the statement

```
head = (struct Node_block *) calloc (1, sizeof(struct Node_block));
```

Here, we use the cast operator (struct Node_block *) so that the address returned by calloc is of the correct type to point to a block of memory for a node of our linked list. The first argument of calloc is the integer 1 because we are creating memory for just one block. The sizeof operator is used to pass to calloc the number of bytes required for one block of memory for the structure.

After this block is reserved, we store the address in the pointer variable current with the statement

```
current = head;
```

We use the pointer variable current to hold the address of the block of memory on which we are working. Therefore, we have the address of the first block of memory stored in current and enter the loop. To understand the loop, go through it statement by statement. We number the statements here and show them in Fig. 8.12, which illustrates how the variables change each time through the loop.

```
   for (i=1; i<=3; i++)
1    {
       printf ("Enter client's name and hours billed on two lines.\n");
       fflush(stdin);
2      gets(current->client);
       scanf("%d",&current->hours_billed);
3      current->next_node = (struct Node_block *) calloc(1,sizeof(struct Node_block));
       current = current->next_node;
4    }
```

Statements 1 and 2 read the client name and number of hours billed, respectively, from the keyboard. These statements also store the information in the block of memory reserved for the node on which we are currently working. Statement 3 reserves memory of the size to hold all the information for a node. It also stores the address of this memory (which is the value returned from calloc) within the block of memory with which we are currently working. Statement 4 makes the address of the newly reserved block of memory (from statement 3) become the address with which we next work.

Note that we execute these statements three times (once for each time through the loop). Figure 8.12 shows the actual values of the variables each time through the loop and the statement numbers that create these values.

FIG. 8.12

Actions of the first for loop in this lesson's program

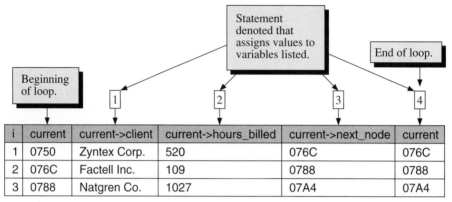

i	current	current->client	current->hours_billed	current->next_node	current
1	0750	Zyntex Corp.	520	076C	076C
2	076C	Factell Inc.	109	0788	0788
3	0788	Natgren Co.	1027	07A4	07A4

We did not print the addresses in the output for this lesson's program; however, we show them in Fig. 8.12. Note that one function of the loop is to march from one block of memory to the next, and we accomplish this, because the value of current has changed from 0750 to 076C to 0788, as we go through the loop.

Because we go through the loop three times, the last time through the loop we reserve a fourth memory block. This is one more block of memory (at address 07A4) than we need. We could have exited the loop early using a break statement, but for simplicity we chose to reserve the memory realizing that it does not occupy much space. Depending on the actions of your program, you may want to exit such a loop early to save memory.

Using a looping image, the first loop in this lesson's program is shown in Fig. 8.13. At the end of going through the loop, we have filled the memory as shown in Fig. 8.14.

Note that, in this table, nothing is in the Name column. To access the values in memory, we *must* use addresses. We have done this in the program, using the pointer variables head and current to store the addresses in which we are interested. Also, the only way we can get the address in which we are interested is from the pointer variable next_node in another memory block. In Fig. 8.14 arrows indicate the connections between nodes.

6. How can we delete a node in a linked list? Because we can access a node only through its address, which is stored in another node block, we can delete a node by removing its address from any node block. For instance, if we wanted to delete the second node in our linked list, we would eliminate the address 076C from the first node block. After doing this, we would find that no node block has the address 076C. Thus, the second node block effectively would be deleted from the list.

However, to maintain continuity of the linked list we must insert another address into the first node block. By inserting the address of the third node, we can complete

FIG. 8.13

The actions of the first for loop in this lesson's program

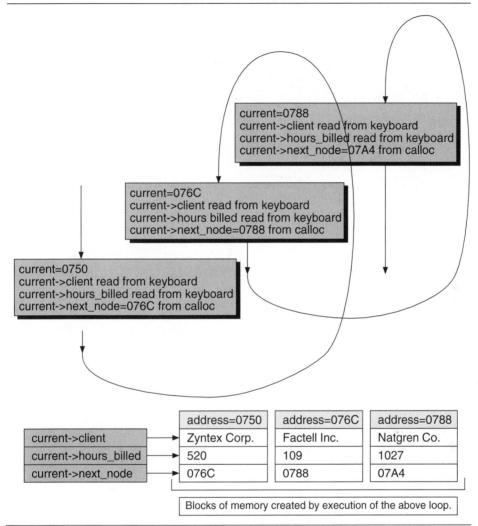

current=0788
current->client read from keyboard
current->hours_billed read from keyboard
current->next_node=07A4 from calloc

current=076C
current->client read from keyboard
current->hours billed read from keyboard
current->next_node=0788 from calloc

current=0750
current->client read from keyboard
current->hours_billed read from keyboard
current->next_node=076C from calloc

	address=0750	address=076C	address=0788
current->client	Zyntex Corp.	Factell Inc.	Natgren Co.
current->hours_billed	520	109	1027
current->next_node	076C	0788	07A4

Blocks of memory created by execution of the above loop.

the list. Figure 8.15 illustrates our memory blocks with the second node deleted from the linked list. In Fig. 8.15, note that nowhere is the address 076C stored. As we march from node to node, we will never enter the second node block and, therefore, have deleted the second node from our list even if all of the information remains stored in its memory block. It is good programming practice, though, to free the memory of a deleted node so it can be used to store other information.

7. Is the method shown in this lesson's program the only way that a linked list can be created? No; in fact it is possible to create another type of a linked list that actually uses arrays. The arrays are parallel, meaning that they correspond element

FIG. 8.14
Memory blocks after going through the loop

Name	Type		Address	Value
	struct Node_block		0750	
	char	client[20]		Zyntex Corp.
	int	hours_billed		520
	struct Node_block*	next_node		076C
	struct Node_block		076C	
	char	client[20]		Factell Inc.
	int	hours_billed		109
	struct Node_block*	next_node		0788
	struct Node_block		0788	
	char	client[20]		Natgren Co.
	int	hours_billed		1027
	struct Node_block*	next_node		07A4

by element. In this form, an array element's subscript is treated in the manner that we have treated an address of a node. One of the parallel arrays is used to hold subscript values. By changing the order of the subscript values stored in this array, we can change the order of the linked list. Space does not permit us to describe this method of creating a linked list.

EXERCISES

1. True or false:
   a. With a linked list, we can use only the memory needed.
   b. We need to use an array to get a linked list to work.
   c. The structure for a linked list must have a pointer variable to the same type of structure.
   d. The declarations for the nodes of a linked list must involve both an ordinary and a pointer variable of the structure.
   e. It is good practice to free the memory of a node deleted from a linked list.

2. Given these structure definitions and declarations

```
struct Xxx
        {
        int aa;
        double bb;
        struct Xxx *next_node;
        };

struct Xxx *current, *head;
```

FIG. 8.15
Memory blocks after the second node is deleted

Name	Type		Address	Value
	struct Node_block		0750	
	char	client[20]		Zyntex Corp.
	int	hours_billed		520
	struct Node_block*	next_node		0788
	struct Node_block		076C	
	char	client[20]		Factell Inc.
	int	hours_billed		109
	struct Node_block*	next_node		0788
	struct Node_block		0788	
	char	client[20]		Natgren Co.
	int	hours_billed		1027
	struct Node_block*	next_node		07A4

find the errors, if any, in the following statements:
a. `*current = *head;`
b. `head = calloc(1, sizeof(struct Xxx));`
c. `current = current.next_node;`
d. `scanf("%d", current->aa);`

Solutions
1. a (true), b (false), c (true), d (false), e (true)
2. a. `current=head;`
   b. `head = (struct Xxx *) calloc(1, sizeof(struct Xxx));`
   c. `current = current->next_node;`
   d. `scanf("%d", &current->aa);`

LESSON 8.7 STACKS

Topics
- Creating a stack
- Pushing onto a stack
- Popping from a stack

Like the previous lesson, this lesson is unusual in that we describe more than syntax and form. In this lesson, we describe a stack, which is a data structure that can be created using the linked list model illustrated in Lesson 8.6. In fact, the type of stack that we create in this lesson can be considered a special case of a linked list.

With a stack we perform only two fundamental operations:

1. Insert a node immediately after the head (this action is called a *push*)
2. Retrieve and delete the node immediately after the head (this action is called a *pop*)

Using the linked list from the previous lesson we show the actions of push and pop in Figs. 8.16 through 8.20. Read these figures and observe how the arrows change with each push or pop. In creating a stack, we use what can be considered to be dummy nodes for the head and the tail. These are useful because we know that the node at the top of the stack immediately follows the head and the node at the bottom of the stack points to the tail. Thus, we know where to find the top of the stack and when we have reached the bottom of the stack.

Because we work only with nodes at the top of the stack, a commonly used analogy for a stack data structure is a stack of plates in a cafeteria line, as shown in Fig. 8.21. Fresh plates are placed at the top of the stack in the cafeteria line. This is sim-

FIG. 8.16
Original stack

FIG. 8.17
Popping from the stack. Zyntex Corp. is removed from the stack by making head point to the second node.

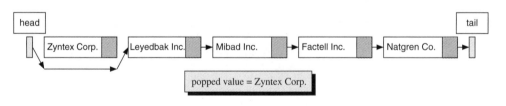

FIG. 8.18
Stack after one pop. Zyntex Corp. has been completely removed with the action shown in Fig. 8.17.

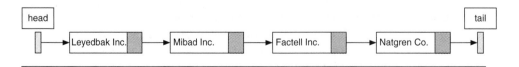

FIG. 8.19

Pushing a value onto the stack. Katman Corp. is added to the stack by making head point to Katman Corp. and Katman Corp. point to Leyedbak Inc. (the previous top node)

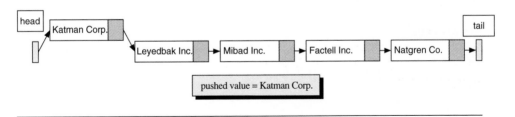

FIG. 8.20

Stack three successive pops after Fig. 8.19. The nodes at the head have been removed.

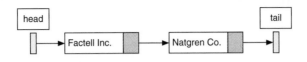

FIG. 8.21

Analogy of push and pop using a stack of plates

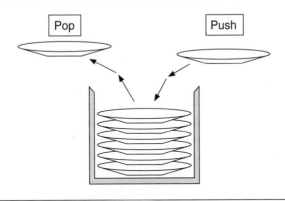

ilar to a push or many pushes. Each person in the line takes a plate. This is equivalent to a pop. This is similar to the stack data structure, because we are allowed to place plates only at the top of the stack and remove plates from the top of the stack. We have no access to plates in the middle of the stack without removing the top plates. Similarly, we have no access to nodes in the middle of a stack data structure without removing nodes at the head.

Why do we create such a data structure? In addition to simulating some real life actions, a stack is useful for helping us implement more complex data structures, such as trees. Read the annotations in the source code and the explanation to understand how to create a stack.

Source code

```
#include <stdio.h>
#include <stdlib.h>


struct Node_block
       {
       int key;
       struct Node_block *next_node;
       };


void push(int value,  struct Node_block *head);
int pop(struct Node_block *head);

void main(void)
{
       struct Node_block *head, *tail;
       int i,value;

       head=(struct Node_block *)calloc(1,sizeof(struct Node_block));
       tail=(struct Node_block *)calloc(1,sizeof(struct Node_block));
       head->next_node=tail;


       for(i=1; i<=3; i++)
              {
              printf("Enter an integer.\n");
              scanf ("%d",&value);
              push(value, head);
              }

       printf("\n\nThe stack from top to bottom is:\n");
       for(i=1; i<=3; i++)          printf("%d\n",pop(head));
}

void push(int value,  struct Node_block *head)
{
       struct Node_block *new_node;

       new_node=(struct Node_block *)calloc(1,sizeof(struct Node_block));
       new_node->key=value;
       new_node->next_node=head->next_node;
       head->next_node=new_node;
}
```

Annotations:

- Structure similar to what we used for the linked list in Lesson 8.6.
- Function push( ) for pushing onto the stack.
- Function pop( ) for popping off of the stack.
- Initializing the stack by creating a head and a tail and having the head point to the tail. The node after the head will be the top of the stack.
- Reserving memory only for the addresses of the head and tail nodes.
- Loop to push three values onto the stack.
- Reading the value to be put on the stack.
- Pushing the value onto the stack.
- Popping the top value from the stack.
- Reserving the memory for one node block.
- Putting the value in the key for the new node block.
- Pushing the value onto the stack by having new_node point to where head was pointing and making head point to new_node.

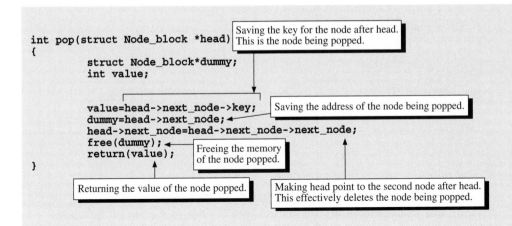

Output

```
              Enter an integer
Keyboard input  27
              Enter an integer
Keyboard input  83
              Enter an integer
Keyboard input  97
              The stack from top to bottom is
              97
              83
              27
```

Explanation

1. What form of structure do we use to create a stack? Although there are many ways to create a stack, we chose to use the same method used in creating a linked list in the previous lesson. Please look back at Lesson 8.6 to see the structure needed for a stack. The structure for this lesson's program is struct Node_block.

2. How did we begin creating a stack? We created a head node and a tail node. These nodes allow us to find the top (head) and bottom (tail) of the stack. The statements

```
head=(struct Node_block *)calloc(1,sizeof(struct Node_block));
tail=(struct Node_block *)calloc(1,sizeof(struct Node_block));
head->next_node = tail;
```

create the form shown in Fig. 8.22.

FIG. 8.22
The beginning of a stack

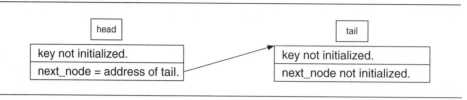

3. How did we use the function push to change the stack? We used the following four statements in function push to add a node to the top of the stack:

1. `new_node=(struct Node_block *)calloc(1,sizeof(struct Node_block));`
2. `new_node->key=value;`
3. `new_node->next_node=head->next_node;`
4. `head->next_node=new_node`

The actions of each of these statements is described in Fig. 8.23.

4. How did we use the function pop to change the stack? We used the following five statements in function pop( ):

1. `value=head->next_node->key;`
2. `dummy=head->next_node;`
3. `head->next_node=head->next_node->next_node;`
4. `free(dummy);`
5. `return(value) ;`

The actions of each of these statements is described in Fig. 8.24.

5. Should we have done more in this lesson's program? Yes, we should have checked to make sure that we did not pop from an empty stack. If we find that the head points to the tail, then we have an empty stack and should not attempt to pop from it. We did not include this check so that we could focus on the creation of the stack. However, in your programs, check for popping from an empty stack.

6. What is LIFO? LIFO stands for "last in, first out." This describes the stack data structure in which the last node pushed onto the stack is the first node popped off of the stack. LIFO contrasts with FIFO, which stands for "first in, first out." A queue (pronounced "cue") is a FIFO type structure in which the first node inserted into the queue is the first node removed from the queue. We discuss this data structure in the next lesson.

FIG. 8.23

The actions of the four statements in push( ). The dashed arrow represents what head pointed to before push( ) was called.

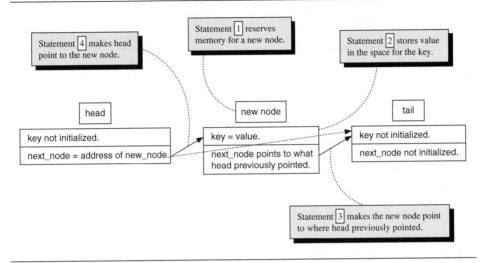

EXERCISES

1. True or false:

   a. A stack can be created only using the method shown in this lesson.

   b. A stack can be considered to be a special case of a linked list.

   c. A stack is a FIFO type structure.

   d. It is good practice to release the memory of a node popped from a stack.

   e. It is good practice to check whether the stack has nodes before attempting to pop from it.

2. Given these definitions and declarations for creating a stack, where head represents the head node and tail represents the tail node,

   ```
   struct Xxx
           {
           double aa;
           struct Xxx *next_node;
           };
   struct Xxx *head, *tail;
   double value;
   ```

   find the errors, if any, in the following statements:

   a. `push(value,tail);`

   b. `int push(int value, struct Xxx *tail)` (as a function header)

   c. `head->aa=23.4;`

FIG. 8.24
The actions of the four statements in pop( )

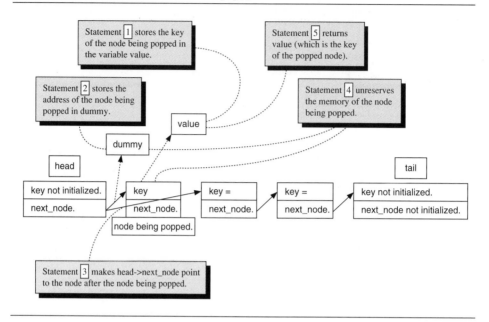

Solutions

1. a (false), b (true), c (false), d (true), e (true)
2. a. `push(value,head);`
   b. `void push(double value, struct Xxx *head)`
   c. The head node points only to the top node on the stack, so no value should be stored in it.

LESSON 8.8 QUEUES

Topics

- Creating a queue
- Inserting into a queue
- Removing from a queue

We all experience queues in everyday life. When we want to buy groceries, we need to wait in a queue. We enter the queue at the end. At the front, people are removed from the queue by having their transactions processed. The queue data structure works similarly. Nodes are inserted at the tail of the queue and removed from the head of the queue. This is a FIFO type structure.

Like we did with a stack, we can create a queue with the linked list form described in Lesson 8.6. We show a queue in Fig. 8.25. It must have both a head

FIG. 8.25
Original queue. Note that the tail node points to the node before it.

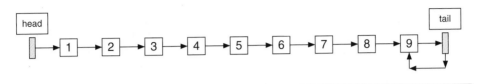

node and a tail node. The process of removing nodes at the head is similar to popping from a stack as shown in Fig. 8.26. Because we covered this in the previous lesson, we will not discuss it in detail here.

Unlike a stack, though, the tail node must point to the node immediately before it as shown in Fig. 8.25. Because the tail points to the last node, we know where to find the last node so that we can change its pointer. In Fig. 8.26, we show the process of adding a node at the tail. Adding a node involves making the previous last node point to the new node instead of the tail. We also make the new node point to the tail and change the tail pointer to point to the new last node.

This lesson's program illustrates how to perform these operations. Much of the code is very similar to the code for the stack in Lesson 8.7; therefore, the annotations focus on the differences between the two codes. Read them and the explanation.

Source code

```
#include <stdio.h>
#include <stdlib.h>


struct Node_block
        {
        int key;                              Linked list/Stack/Queue type structure.
        struct Node_block *next_node;
        };
                                              The function insert adds a node at the tail.

void insert(int value,  struct Node_block *tail);
int remov(struct Node_block *head);

                        The function remov is identical to the function pop in Lesson 8.6.
void main(void)
{
        struct Node_block *head, *tail;    Initializing a queue by creating both a head
        int i,value;                       and a tail and having them point to each other.


        head=(struct Node_block *)calloc(1,sizeof(struct Node_block));
        tail=(struct Node_block *)calloc(1,sizeof(struct Node_block));
        head->next_node=tail;
        tail->next_node=head;
```

```
        for(i=1; i<=3; i++)
                {
                printf("Enter an integer.\n");
                scanf ("%d",&value);
                insert(value, tail);
                }


        printf("\n\nThe queue is:\n");
        for(i=1; i<=3; i++)                printf("%d\n",remov(head));
}
```

Inserting the values read at the end of the queue.

The function insert creates a new node and puts value into its key.

```
void insert(int value,  struct Node_block *tail)
{
        struct Node_block *new_node;


        new_node=(struct Node_block *)calloc(1,sizeof(struct Node_block));
        tail->next_node->next_node=new_node;
        tail->next_node=new_node;
        new_node->key=value;
        new_node->next_node=tail ;
}
```

Causing the node before the tail (the one that the tail points to) to point to the new node.

Making the tail point to the new node.

Putting value into the new node.

Causing the new node to point to the tail.

```
int remov(struct Node_block *head)
{
        struct Node_block *dummy;
        int value;


        value=head->next_node->key;
        dummy=head->next_node;
        head->next_node=head->next_node->next_node;
        free(dummy);
        return(value);
}
```

The function remov is identical to the function pop in Lesson 8.7.

FIG. 8.26

Removing from a queue at the head and inserting into a queue at the tail

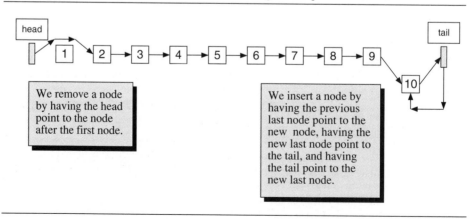

We remove a node by having the head point to the node after the first node.

We insert a node by having the previous last node point to the new node, having the new last node point to the tail, and having the tail point to the new last node.

Output

	Enter an integer
Keyboard input	27
	Enter an integer
Keyboard input	83
	Enter an integer
Keyboard input	97
	The queue is
	27
	83
	97

Explanation

1. How does the function insert add a value to the end of a queue? After reserving a block of memory and storing its address in the variable new_node with the statement

`new_node=(struct Node_block *)calloc(1,sizeof(struct Node_block));`

the function uses the following four statements:

1. `tail->next_node->next_node=new_node;`
2. `tail->next_node=new_node;`
3. `new_node->key=value;`
4. `new_node->next_node=tail ;`

The actions of these statements are described in Fig. 8.27. After executing these statements, a new node is inserted at the tail.

2. Are there any other types of queues? Yes, here are some:

- *Dequeue,* short for "double-ended queue" and pronounced "dekk." With a dequeue, nodes can be inserted at either end or removed from either end.
- *Output restricted dequeue* allows insertion at both ends of the queue but removal at only one end of the queue.
- *Input restricted dequeue* allows removal at both ends of the queue but insertion at only one end of the queue.
- *Priority queue* has a priority set for each node. Nodes with highest priority are processed first. Within the priorities, those that entered the queue first are processed first. It works similar to the way that some airlines allow boarding and unboarding of their flights. First-class passengers are given the highest priority and allowed to board and unboard first. Business class is given second priority and allowed to board and unboard second. All other passengers are allowed to board and unboard last. Also, the first first-class passenger in the queue boards first and the second first-class passenger in the queue boards second. Thus, within equal priorities, the first in the queue is processed first.

You can imagine that other variations of these can be created as well.

FIG. 8.27
The actions of the four statements in insert( ). The dashed arrows represent pointing before insert( ) is called.

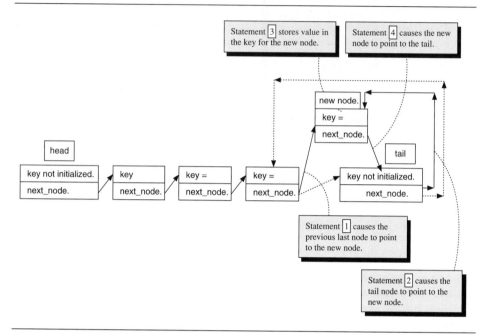

EXERCISES

1. True or false:
    a. An ordinary queue has a FIFO type structure.
    b. In an ordinary queue, we insert a new node at the head of the queue.
    c. In an ordinary queue, we remove a node from the tail.
    d. A priority queue is commonly used to determine the order that jobs should be printed on shared printers.
    e. The only way that a queue can be implemented is with the method we have shown for this lesson's program.

Solutions

1. a (true), b (false), c (false), d (true), e (false)

LESSON 8.9 BINARY TREES

Topics

• Creating a binary tree
• Using a stack with a tree

A tree is a particular type of what is called a *graph* and a binary tree is a particular type of a tree. In this text we develop programs only for binary trees and not general graphs or trees. However, to understand how binary trees fit into the bigger picture, we start with a discussion of graphs.

A graph, as we use it, is not related to what you previously may have viewed as a graph (like an x-y plot) in your math classes. In our context, a graph is a data structure that includes nodes similar to what we used in creating linked lists. We have seen how nodes can be connected using linked lists; however, a linked list has only one path to follow from beginning to end. This is because each node has only one pointer and it points to the next node in the list. If a node can have more than one pointer, then we can create a more general graph. Figure 8.28 shows a linked list and a graph.

We will not give formal definitions of graphs but simply general descriptions. Some terminology is

1. *Nodes* also can be called *vertices* or *points.*
2. Connections between the nodes are called *edges* or *arcs.*
3. Two nodes are considered *adjacent* or *neighbors* if an edge connects them.
4. A *path* between one node and another is indicated by a list of connected nodes between the two.
5. A *simple path* is a path that has no repeated nodes.

Two different graphs are shown in Fig. 8.29. Note that the graph on the left of the figure is the same as in Fig. 8.28. Although at first glance they look different, they are the same because the connections between the nodes are the same. For example, for both graphs, node 4 is connected to nodes 7 and 3. We can say this about every node in the graphs. Thus, the graphs are identical.

FIG. 8.28
Comparing a linked list and a graph

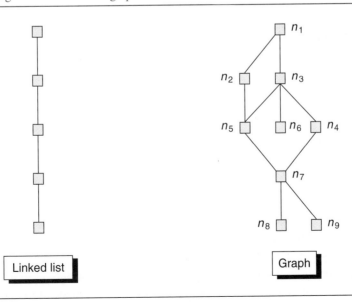

Linked list

Graph

FIG. 8.29

Two graphs. Note that the one on the left is the same as the one in Fig. 8.28 because the connections between the nodes are the same.

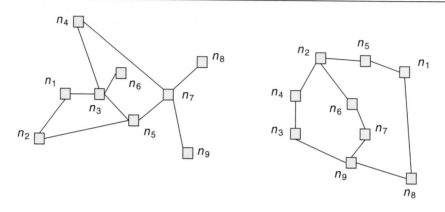

FIG. 8.30

Two trees. The one on the right is a binary tree because each node has no more than two children. There is only one simple path between any two nodes. Contrast this with the graphs of Fig. 8.29, in which more than one simple path can be constructed between two nodes.

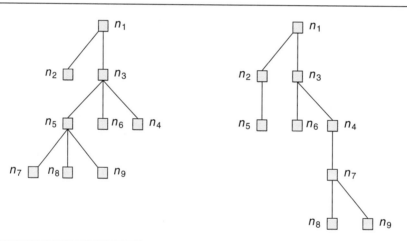

The most obvious real-life use of graphs is in simulating a map with the nodes being cities and the edges being roads between the cities. A *weighted graph* can be created to model the distance between the cities. With a weighted graph, we assign a *length* or *cost* to each edge. By summing the lengths of the edges in a path, we get the distance between two cities.

A *tree* is a graph in which there is only one path between any two nodes, and that path is a simple path. This condition leads to the images shown in Fig. 8.30. A *rooted tree* is a tree that has one node specified to be the root. The root of a tree traditionally is shown at the top of a diagram, as in Fig. 8.31. Any node can be specified as the root. Once a root is specified, the relationships between the nodes are classified in the following ways, using the tree at the left part of Fig. 8.31 as an example:

1. The root of the tree is n_1.
2. The children of n_3 are n_6 and n_4.
3. The left child of n_3 is n_6 and the right child is n_4.
4. The parent of n_6 and n_4 is n_3.

In a binary tree no node has more than two children. Although not required, to make sure that each regular node of a binary tree has two children, we can add terminating nodes as shown in Fig. 8.32.

This lesson's program illustrates how a binary tree can be created. Binary trees are useful to model the combination of two elements to create another element, which is used with yet another element to create a new element. For instance, if we were to construct an automobile, we might find that we could illustrate the process using a binary tree such as the one shown in Fig. 8.33. Because we are using two parts to create a third part, which then is used to create another part, we find that a tree can model this process. If we were interested in determining the time schedule of completion, with a tree we clearly see the dependence of the completion of the upper level elements on the completion of the lower level elements.

FIG. 8.31
Two binary trees that are identical (because the connections for each node are the same) but with different roots

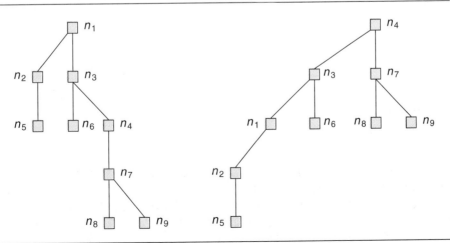

FIG. 8.32

A binary tree for which each regular node has two children. Terminating nodes have been added to assure this. The tree is the same as that shown at the left of Fig. 8.31 but with terminating nodes added.

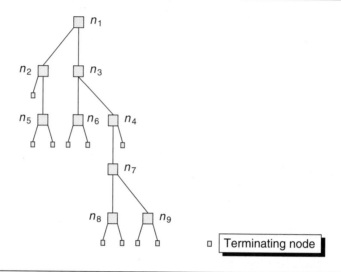

FIG. 8.33

A binary tree illustrating combining different parts to create a final product. This can be used to model the time schedule of completion or the costs of production.

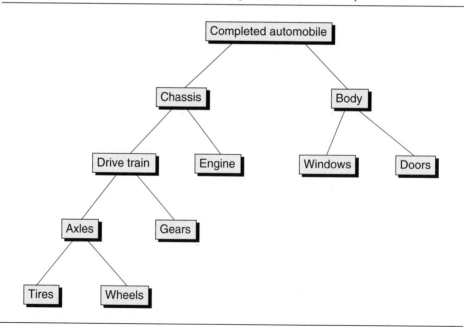

Similarly, we can model an arithmetic calculation using a binary tree. This is shown in Fig. 8.34. In this figure, we have modeled the equation

$$[(5 + 6) / (7 - 9)] \times 3 = -16.5$$

Binary trees are well suited to this type of representation because each operation has two operands. More complex equations may not be capable of being modeled with a binary tree.

In this lesson's program, we create the binary tree for this calculation. To create this equation without using parentheses, we use what is commonly called *reverse Polish notation*, named after the Polish mathematician who invented it. It also can be called *postfix notation*. If you have a calculator that operates with this method, then you know that the order of entering the input would be

$$5 \ 6 + 7 \ 9 - / \ 3 *$$

In this lesson's program, we use this input to create the binary tree of Fig. 8.34. We do not solve the equation here, but instead solve it in Application Program 8.4.

There are many different ways to create a binary tree. We use a procedure that involves four steps:

1. Reserve memory for the node, including space for the values (numbers or operators for this lesson's program) and the addresses of left and right children.
2. Fill the node's memory cells with the values to be stored in each node.
3. Fill the node's memory cells with the addresses of the left and right children.
4. Assign the address of the node itself to be the left or right child of another node (unless it is the root).

We also utilize a stack to help us perform steps 3 and 4 of the procedure. As a result, we have both a stack and a tree structure (for stack and tree nodes) in our program.

FIG. 8.34
A binary tree illustrating the calculation $[(5 + 6)/(7 - 9)] \times 3$

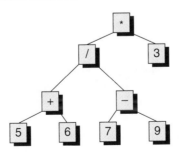

Also, from Fig. 8.34 note that all of the number nodes have no children, whereas each operator node has two children. To make sure that we have a complete binary tree, we put terminating nodes for both the left and right children onto all of the number nodes. This is shown in Fig. 8.35.

What you should observe about this lesson's program.

1. You can compare the stack and tree structures that are defined. The tree structure has two tree pointer variables as members, because each tree node has two tree-node children, a left child and a right child. The stack structure has only one stack pointer because each stack node points to only one other stack node. However, the stack structure has another pointer member, a tree pointer. This is because we keep the addresses of tree nodes on our stack. In the Explanation portion of this lesson, we describe exactly how this works.

2. Like our stack program of Lesson 8.7, we use the functions push and pop. However, because we pop an address rather than an integer or double, the function type of pop is a pointer. Also, one of the arguments for push is a tree address because we push a tree address onto the stack.

3. In this program, we create a tree with nine nodes, therefore the loop loops nine times. The user enters two pieces of information about each node, the node type (*n* for number or *o* for operator) and the node value (the value of the number or the operator character). As a result, the loop has two parts created by an if-else control structure. The true block of this structure creates the number nodes and the false block creates the operator nodes.

4. We use the pointer variable end to represent a terminating node. In the for loop, we also use this node as both the left and right children of a number node. Note that, for convenience, we use the same terminating node for all the numbers.

FIG. 8.35
Terminating nodes added to the binary tree of Fig. 8.34

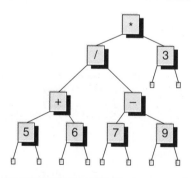

Source code

If a node contains a number, we store its value in the member operand. If the node contains an operator, it is stored in the member operat. Whether we have an operator or an operand stored in a node is reflected in the member type.

```c
#include <stdio.h>
#include <stdlib.h>

struct Tree_block
       {
       char type, operat;
       double operand;
       struct Tree_block *left_child, *right_child;
       };
```

The tree node structure includes nodes for left and right children.

```c
struct Stack_block
       {
       struct Tree_block *tree_address;
       struct Stack_block *next_node;
       };
```

In Lesson 8.7, the stack held integers. In this lesson, the stack holds addresses to tree nodes. So, the stack structure has a struct Tree_block * member (along with the address of the next node in the stack).

The function push pushes an address of a tree node onto the stack.

```c
void                   push   (struct Tree_block *address,  struct Stack_block *head);
struct Tree_block *    pop    (struct Stack_block *head);

void main(void)
{
       int i;
       double number;
       char optr, tp;
       struct Tree_block *current, *end;
       struct Stack_block *head, *tail;

       head=(struct Stack_block *)calloc(1 ,sizeof(struct Stack_block));
       tail=(struct Stack_block *)calloc(1,sizeof(struct Stack_block));
       end=(struct Tree_block *) calloc (1, sizeof(struct Tree_block));

       head->next_node=tail ;
```

The function pop pops an address of a tree node from the stack.

Keeping track of the current node for the tree and setting the end (terminating) node to be the same for all terminating nodes.

For the stack, we need the head and tail nodes.

Reserving memory for the head and tail for the stack, and the end for the tree.

Looping over a combined total of nine operators and operands.

Reserving memory for a new tree node.

```c
       for (i=1; i<=9; i++)
               {
               current=(struct Tree_block *)calloc(1, sizeof(struct Tree_block));
               printf ("Enter 'n' for number or 'o' for operator then return\n");
               fflush(stdin);
               tp = getchar();
               fflush(stdin);

               if (tp == 'n')
                       {
```

Reading the type, either 'n' for number or 'o' for operator.

```
                                    printf("Enter a number\n");
                                    scanf ("%lf", &number);
        ┌──────────────┐            current->type='n';
        │ Creating a tree│           current->operand=number;
        │ node for a number.│        current->left_child = end;  ◄──
        └──────────────┘            current->right_child = end;  ◄──
                                    }
                             ┌──────────────────────────────────────────┐
                             │ Both children of numbers are terminating nodes.│
                             └──────────────────────────────────────────┘

              if (tp =='o')
                             {
                                    printf("Enter an operator\n");
                                    optr=getchar();
        ┌──────────────┐            current->type='o';
        │ Creating a tree│           current->operat=optr;
        │ node for an operator.│      current->right_child = pop(head);  ◄──
        └──────────────┘            current->left_child = pop(head);  ◄──
                                    }
                    push(current, head);   ┌──────────────────────────┐
                    }                      │ Both children of operators are│
}                                          │ popped from the stack.       │
   ┌──────────────────────────────────────┘└──────────────────────────┘
   │ Pushing the new tree node onto the stack.│
   └──────────────────────────────────────┘

void push(struct Tree_block *address,  struct Stack_block *head)
{                               ┌────────────────────────────────────────────┐
        struct Stack_block *new_node;  │ The push function is nearly identical to the push function│
                                │ used for Lesson 8.6.  The main difference is that we push  │
                                │ a tree address instead of an int onto the stack.          │
                                └────────────────────────────────────────────┘
        new_node=(struct Stack_block *)calloc(1,sizeof(struct Stack_block));
        new_node->tree_address=address;
        new_node->next_node=head->next_node;
        head->next_node=new_node;
}
   ┌──────────────────────────────────────────────────────────────────┐
   │ We pop an address of a tree node so the type of the function pop is the address of Tree_block.│
   └──────────────────────────────────────────────────────────────────┘

struct Tree_block *pop(struct Stack_block *head)
{
                                ┌────────────────────────────────────────┐
        struct Stack_block *dummy;  │ Accessing the tree_address member of the│
        struct Tree_block *address; │ node following the head of the stack.   │
                                └────────────────────────────────────────┘

        address=head->next_node->tree_address;
        dummy=head->next_node;
        head->next_node=head->next_node->next_node;
        free(dummy);               ┌────────────────────────────────────────┐
        return(address);  ◄────────│ Returning the address of the node popped.│
}                                  └────────────────────────────────────────┘
```

Output

```
                Enter 'n' for number or 'o' for operator then return
Keyboard input  n
                Enter a number
Keyboard input  5
                Enter 'n' for number or 'o' for operator then return
Keyboard input  n
                Enter a number
Keyboard input  6
                Enter 'n' for number or 'o' for operator then return
```

```
Keyboard input  o
                Enter an operator
Keyboard input  +
                Enter 'n' for number or 'o' for operator then return
Keyboard input  n
                Enter a number
Keyboard input  7
                Enter 'n' for number or 'o' for operator then return
Keyboard input  n
                Enter a number
Keyboard input  9
                Enter 'n' for number or 'o' for operator then return
Keyboard input  o
                Enter an operator
Keyboard input  -
                Enter 'n' for number or 'o' for operator then return
Keyboard input  o
                Enter an operator
Keyboard input  /
                Enter 'n' for number or 'o' for operator then return
Keyboard input  n
                Enter a number
Keyboard input  3
                Enter 'n' for number or 'o' for operator then return
Keyboard input  o
                Enter an operator
Keyboard input  *
```

Explanation

1. What structure definitions do we need to create a binary tree using the method illustrated in this lesson's program? We need to define both a tree structure and a stack structure. In this lesson's program, for the tree we used

```
struct Tree_block
    {
    char type, operat;
    double operand;
    struct Tree_block *left_child, *right_child;
    };
```

In this structure, the member type stores whether a node is a number or an operator. The member operat stores the operator ($+$, $-$, $*$, or $/$) for operator nodes. The member operand stores the number for number nodes. The members left_child and right_child store the addresses of the children nodes for a given node.

In general, the form of a tree structure is

```
struct Tree_tag
    {
    type key1;
    type key2;
    ...
    struct Tree_tag *left_child, *right_child;
    }
```

Where *Tree_tag, key1, key2, left_child,* and *right_child* are any valid identifiers. *Tree_tag* represents the tag for the tree structure, *key1* and *key2* are variable names for the values to be stored in the tree, and *left_child* and *right_child* represent addresses of the two children of a node.

For the stack we used

```
struct Stack_block
        {
        struct Tree_block *tree_address;
        struct Stack_block *next_node;
        };
```

In this structure, the member next_node is used to store the address of the next node on the stack. The member tree_address is used to store the address of a tree node. It is the addresses of the tree nodes that are pushed and popped from the stack.

In general, the form for a stack used to help create a tree is

```
struct Stack_tag
        {
        struct Tree_tag *tree_address;
        struct Stack_tag *next_node;
        };
```

Where *Stack_tag, tree_address,* and *next_node* are any valid identifiers. *Tree_tag* is the tag for the tree structure. The variable *tree_address* is used to hold the address of a tree node, and *next_node* is used to hold the address of the next node on the stack.

2. What structure variables do we need to create a binary tree? Like we found with our linked list structure, we need pointer variables of our tree structure. Because we use a stack to create our tree, we need pointer variables of both our stack and tree structures. In this lesson's program we used

```
struct Tree_block *current, *end;
struct Stack_block *head, *tail;
```

Notice that we did not need to create ordinary variables of either our stack or tree structures. For the tree, we need current to keep track of the node currently being considered and end to know the address of the end (terminator) nodes. For the stack we need both the head and tail addresses, as illustrated in the lesson on stacks.

3. How did we create a tree node for a number in this lesson's program? First, we reserved memory for two tree nodes, current and end, with

```
end     = (struct Tree_block *)  calloc (1, sizeof(struct Tree_block));
current = (struct Tree_block *) calloc(1, sizeof(struct Tree_block));
```

We used current to hold the address of the node with which we were working and end to hold the address of the terminating node.

To fill the memory cells for a number node, we used the following statements:

```
printf("Enter a number\n");
scanf ("%lf", &number);
current->type='n';
current->operand=number;
current->left_child = end;
current->right_child = end;
```

The printf statement prompts the user to enter a number and the scanf statement reads it. The assignment statements fill the block of memory reserved for current. The value of current->operand is assigned the number entered. Because the children of a number node both are terminating nodes, we assign the pointer variable end to both current->left_child and current->right_child. After executing the statements, a new number node is created with two children being terminating nodes.

4. How did we create a tree node for an operator in this lesson's program? To fill the memory cells for a node containing an operator, we used the following statements:

```
printf("Enter an operator\n");
optr=getchar();
current->type='o';
current->operat=optr;
current->right_child = pop(head);
current->left_child = pop(head);
```

The printf statement prompts the user to enter an operator and the getchar function reads it. The operator is assigned to current->operat. The children of a node containing an operator are not terminating nodes. We get them, as shown in the statements, by popping the stack. For this to work properly, we must make sure that the stack contains the proper nodes.

5. How do we make sure that the stack contains the correct nodes? Figure 8.36 shows the actions performed each time through the loop and how the stack changes. Note that, after completion of the loop, all the nodes have the correct children as shown in Fig. 8.35. Therefore, the tree has been correctly created.

6. Are there any special types of binary trees? Yes; one commonly used special binary tree is a binary search tree. We illustrate the image of a binary search tree using an example. Suppose each node of a tree has a key that is an integer. The tree shown in Fig. 8.37 is in the form of a binary search tree because it has the following property: The value of the key at a particular node is greater than the value of the key of each node in its left subtree but is less than or equal to the value of the key of each node in its right subtree.

7. How can we create a binary search tree? We will not go through the process in detail in this text. However, we can say that it involves creating a node and

FIG. 8.36

Loop creating nodes for the tree for this lesson's program. Observe how the stack changes each time through the loop. After completion of the loop, each node created has the left and right children corresponding to those shown in Fig. 8.35.

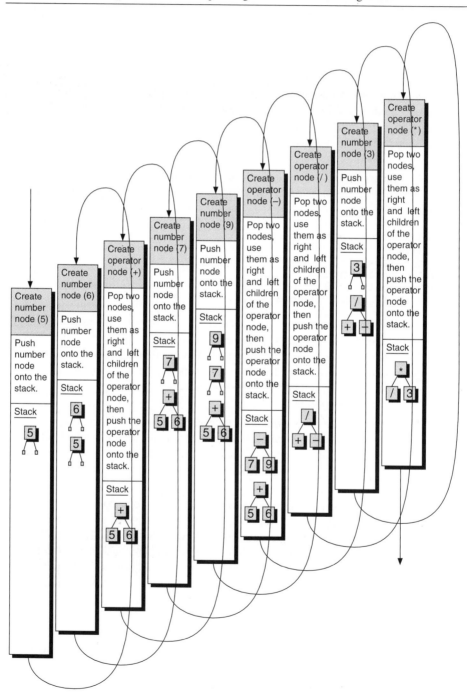

■ **FIG. 8.37**
A binary search tree with the key for each node being an integer

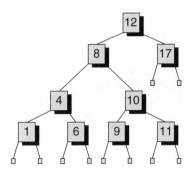

comparing the key for that node to the key for other nodes that already exist to find the correct "place" for the new node. Once the correct place is found, by changing pointers in a manner similar to what we describe for inserting into a linked list, we can insert the new node into the binary search tree.

8. Why would we want to create a binary search tree? Once a binary search tree is created, it is very simple to find a particular node of interest. For instance, if we wanted to delete the node with the value 11 in the tree of Fig. 8.37, we could use the following procedure to first find the node (beginning our search at the root):

1. The value at the root is 12; because 11 is less than 12, we know to continue searching in the left subtree.
2. The next node encountered, then, is 8. Because 11 is greater than 8, we know to proceed to the right subtree of 8.
3. The next node encountered is 10. Because 11 is greater than 10, we know to proceed to the right subtree of 10.
4. The next node encountered is 11 and we have found our node. We now can delete it from the tree.

Note that, with this procedure, we needed to visit only four nodes to find the node we wanted. Had we used another method of visiting nodes, more steps would have been needed. For large trees, a considerable savings in the number of steps can be obtained using a binary search tree.

EXERCISES
1. True or false:
    a. We can use a stack to create a binary tree.
    b. A tree node can have only two children.
    c. A binary tree node can have only two children.
    d. A graph is a special instance of a tree.
    e. The root of a tree typically is drawn at the bottom of the tree.

Solutions
1. a. (true), b (false), c (true), d (false), e (false)

■ LESSON 8.10 FUNCTIONS WITH ONE RECURSIVE CALL

Topics
- Concept of recursion
- Tracing the flow of a function with one recursive call
- Creating a function with one recursive call

In working with trees after they have been created, we find that the mechanism of recursion is very useful. In this lesson, we describe the simplest form of recursion: recursion with a single call. In the Application Programs, we illustrate recursion with two calls, which is needed for working with trees. To understand recursion, we must describe the actions of a recursive function. We do this using the log function as an illustration.

Suppose you were interested in finding the natural log of the log of the log of the log of 5239.7. You could write a single line of code that simply said "x = log (log (log (log (5239.7))));". Recall that the associativity of operations for evaluating an expression is from left to right. Therefore, in evaluating the right side of the assignment statement, the operation denoted by the leftmost log would be performed first. This operation is to call the math function log. However, when the math function log looks at the argument with which it is working, it sees the next call to the function log. So, it calls the function log again, at which point the third log is seen and log is called a third time, whereupon the fourth log is seen and log is called again. At this point, the function log has been called four times yet no logarithm has been calculated. As the function is executed on the fourth call, the function realizes that the argument no longer is another call to log but the real value of 5239.7.

Now, the process reverses. The log of 5239.7 is evaluated to be 8.56402. This value is returned to the third call to log and the log of this value is found to be 2.14757. This value is returned to the second call to log and the log of this value is found to be 0.76434, which is returned to the first log call to give -0.26875 as a final result. This value then is assigned to x in the assignment statement.

This is a long-winded explanation of a relatively simple calculation, but it introduces an important aspect of the concept of recursion: There essentially are three phases of operations in recursive calling. The first is the calling phase, whereby the function is called repeatedly. During this phase, there may or may not (depending on your design) be many calculations or operations performed. This phase ends when the function encounters the simplest case and actually performs one complete pass through the operations of the function, which is the reversal phase. Then the return phase begins. The values returned allow the earlier function calls to complete their operations and return values back to the very beginning. When the first function call has completed its operations, the process is finished.

The log function is not considered a recursive function in C because it does not automatically call itself. In our short example, we use the log function in a recursive manner by having it call itself repeatedly to illustrate what happens in recursive calls. In this lesson, though, we illustrate a function that automatically calls itself and thereby is called a *recursive function*. You can recognize a recursive function in C by looking in the function body for a line of code that calls a function by the function's own name. For instance, if the name of the function is average, it is prototyped as follows:

double average (double a, double b);

Then, within the body of function average, a call to average should occur. For instance,

median += average(c,d);

In the source code for this lesson, we have the function function1. Find the line in the function body that tells you that this is a recursive function. The purpose of the program simply is to add the values of i and j a number of times. We guide you to observe what statements are executed with each call to the function, at reversal, and on the returns from the function. This is a simple program, designed to demonstrate the flow of programs with a single recursive call (meaning the function has only one statement that calls itself).

When we called the log function recursively, we determined in advance how many times we were going to call it. We decided to call it four times. A problem with recursive functions is that, since they automatically call themselves, they might continue to call themselves forever. In other words, it is necessary for us to build into any recursive function we write a portion of code that no longer calls the function, the reversal portion. Commonly, this portion of the function is built within an if control structure.

What you should observe about this lesson's program.

1. The body of main consists primarily of a call to function1.
2. The body of function1 consists primarily of an if-else control structure. Either the True block or False block is executed on calling function1.
3. Within the True block is a call to function1. Because function1 has a statement that calls itself, it is a recursive function.
4. Within the False block, there is no call to function1. When the False block is entered, function1 no longer calls itself and the recursive process ends.
5. The value of k determines whether the True or False block is entered.
6. The value of k is modified *outside of* the if-else control structure. The value of k therefore is modified every time function1 is entered.

Questions you should attempt to answer before reading the explanation.

1. What happens to the value of k each time function1 is called? Based on this, how many times do you think function1 will be called?
2. Remember two phases to recursion are the calling phase and the returning phase. Which statements in function1 are executed in the calling phase? Look at the output for assistance in answering this question.
3. What does the output show the value of k becomes with each step of the calling phase? What happens to tot during the calling phase?
4. From the output, what is the value of k at the reversal point? With this value of k, which statement block is executed? Does this statement block have a call to function1?

5. Given the values of i and j at reversal, what is the value returned?
6. Which statements are executed in the returning phase? What happens to the values of k and tot on the returning phase?
7. Why does the value of k seem to change during the returning phase, even though execution does not pass through any statements that change it? This is tricky. To answer the question, think about the fundamental way that memory is reserved for each call to a function. Hint: Since there are five calls to function1, five regions of memory hold the values of the variables in function1. These values remain until the function has completed execution.
8. The final result is sum = 125. What, in essence, has this program done? Can you think of easier ways to arrive at the same result?

Source code

```
#include <stdio.h>

int function1 (int i, int j, int k);          Prototype for function1.
void main (void)
{
        int a=10, b=15, n=5, sum ;            Call to function1.
        sum = function1 (a,b,n);
        printf ("\n\n The end result is sum = %d \n", sum);
}
```
Statements before the call to function1. These statements are executed repeatedly during the calling phase.

```
int function1 (int i, int j, int k)                                    True
{                                                                      block.
       int tot;              The variable k changes with each function call
       k--;                  because it is the variable used in the conditional
       if (k != 0)           expression of the if control structure.
                   {
                   printf ("Values in phase 1 - calling phase\n"
                           " i = %d  j = %d  k = %d  tot = %d \n",
                           i,j,k,tot);
                   tot = (i+j) + function1 (i,j,k) ;
                   printf ("Values in phase 2 - returning phase\n"
                           " i = %d  j = %d  k = %d  tot = %d \n",
                           i,j,k,tot);              Call to function1 within function1.
                                                   This is the recursive call and therefore
                   return (tot);                   makes function1 a recursive function.
                   }
       else
                   {
                   tot = i+j;
                   printf ("Values at reversal\n"
                           " i = %d  j = %d  k = %d  tot = %d \n"
                           i,j,k,tot);
                                                   Both the True and
                                                   False blocks have a
                   return (tot);                  return statement.
                   }
}
```
True block of function1.

False block. These statements are executed only once, at the reversal.

False block. Note that there is no recursive call in the False block. If control goes into the False block, the recursive calling phase stops.

Statements after the call to function1. These statements are executed repeatedly during the returning phase.

Output

```
Values in phase 1 - calling phase
i = 10   j = 15   k = 4   tot = 0
Values in phase 1 - calling phase
i = 10   j = 15   k = 3   tot = 0
Values in phase 1 - calling phase
i = 10   j = 15   k = 2   tot = 0
Values in phase 1 - calling phase
i = 10   j = 15   k = 1   tot = 0
Values at reversal
i = 10   j = 15   k = 0   tot = 25
Values in phase 2 - returning phase
i = 10   j = 15   k = 1   tot = 50
Values in phase 2 - returning phase
i = 10   j = 15   k = 2   tot = 75
Values in phase 2 - returning phase
i = 10   j = 15   k = 3   tot = 100
Values in phase 2 - returning phase
i = 10   j = 15   k = 4   tot = 125

The end result is sum = 125
```

Explanation

1. How do we know that function1 is a recursive function? Within the body of function1 is the statement

```
tot = (i+j) + function1 (i,j,k);
```

This statement within function1 calls the function itself, making function1 a recursive function.

2. What block of code prevents function1 from calling itself indefinitely? The False block of the if control structure in function1, which follows, gives a path through which control can flow without calling function1 again. This block simply causes the value tot to be returned.

```
if (k != 0)
        {...
        ...}
else
        {
        tot = i+j;
        printf ("Values at reversal\n"
                        " i = %d  j = %d  k = %d  tot = %d \n",
                        i,j,k,tot);

        return (tot);
        }
```

False block.

This block of code is executed at the reversal. It represents the base or simplest case. The block of code is executed only once, whereas the True block is executed four times.

Beginning programmers often write recursive structures with a test condition that never changes from True to False (or False to True). When you design your recursive structures, pay special attention to this. Make sure that program control enters the block of code that has no recursive call.

3. What is the flow of this program? A conceptual illustration of the recursive calls to function1 is given in Fig. 8.38. From this you can see the first four calls to function1, the reversal, and the return from the function calls. Use this illustration and the source code to follow the flow of the program, which we discuss next.

Consider the process step by step. We start immediately after main calls function1. On the first call to function1, the statements prior to the if control structure are executed. The value of k is decremented to 4 (changing from 5). Since the value of k is not 0, control goes to the True block of the if control structure.

When the assignment statement containing the function call is encountered in the True block (as with all assignment statements), the expression on the right side

FIG. 8.38
Flow of source code with recursive function

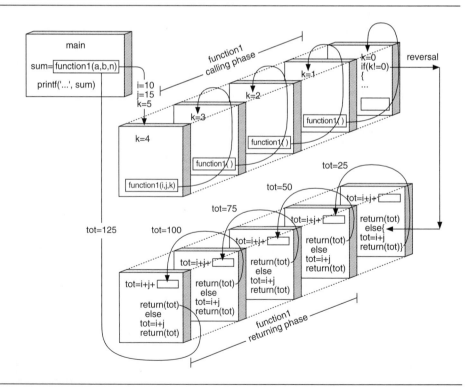

of the assignment statement is evaluated first. Since this expression contains a call to function1, the call to function1 takes place before the entire expression is evaluated. Thus, function1 is called at this point (that is, prior to the assignment to tot).

With this function call, control goes to the start of function1. The value of k is decremented to 3. The True block is entered and then the call to function1 is encountered. Control shifts again to the start of function1. The value of k is decremented to 2. The True block is entered and then the call to function1 is encountered. Control shifts again to the start of function1. The value of k is decremented to 1. The True block is entered and then the call to function1 is encountered. Control shifts again to the start of function1. The value of k is decremented to 0. All of this represents the calling phase.

Now, with this value of k, the False block is entered. This is the reversal. The assignment statement tot=i+j; is executed and tot becomes 25. The value of tot is returned to begin the returning phase. To where does it return?

The return location is to the statement in the True block of code:

```
tot = (i+j) + function1 (i,j,k);
```

Return location. At this point, this has a value of 25.
The statement is equivalent to tot = (i + j) + 25;

This return location is in the True block of code on the previous function call (the one where k = 1, the fourth function call). Now, the assignment statement tot = i + j + function1 (i, j, k); can be executed. The return gives a value of 25, i is 10, and j is 15, giving the value of tot to be 50. Now the other statements in the True block are executed (that is, the ones after the statement with the function call). The next statement is the return statement. To where does this value of tot return? Right back to its call (on the third function call):

```
tot = (i+j) + function1 (i,j,k);
```

Return location. At this point, tot has a value of 50.
The statement is equivalent to tot = (i + j) + 50;

This assignment statement can be completed with the values i = 10, j = 15, and a return value of 50, making tot equal to 75. The rest of the statements in the True block are finished returning a value of 75 to the same assignment statement (to the second function call). This gives a value of tot to be 100, which is returned again to the same assignment statement giving a value of tot of 125 (to the first function call). This is the last return and control goes to the calling statement in main, giving sum the value of 125, and the program execution halts.

In essence, the entire process is that, in the calling phase, the statements *prior* to the function call in function1 are executed repeatedly. The reversal is hit and the False block—the one without a function call—is executed. Then, in the returning phase the statements *after* the function call in function1 are executed repeatedly.

4. Why did the value of **k** *appear to be incremented with each step on the returning phase when no increment expression for* **k** *is encountered on the returning phase?* During the calling process, a region of memory is set aside for storing the values of all the variables in function1 for each function call as illustrated in Fig. 8.39. In other words, five different regions of memory are set aside for variables i, j, k, and tot because function1 was called five times.

This is also illustrated in Fig. 8.40. The variable values in the top five boxes represent the variable values on the calling phase. During the returning phase, these

FIG. 8.39
Reserving regions of memory during calls to function1

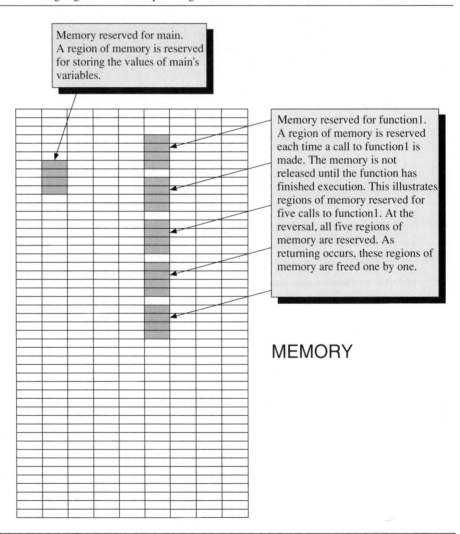

FIG. 8.40

Memory regions and variable values for both the calling and returning phases of this lesson's program. Note that because function1 is called five times, five memory regions are reserved.

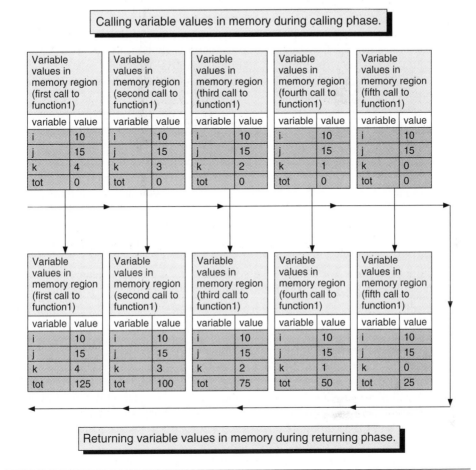

Calling variable values in memory during calling phase.

Variable values in memory region (first call to function1)		Variable values in memory region (second call to function1)		Variable values in memory region (third call to function1)		Variable values in memory region (fourth call to function1)		Variable values in memory region (fifth call to function1)	
variable	value	variable	value	variable	value	variable	value	variable	value
i	10	i	10	i	10	i	10	i	10
j	15	j	15	j	15	j	15	j	15
k	4	k	3	k	2	k	1	k	0
tot	0	tot	0	tot	0	tot	0	tot	0

Variable values in memory region (first call to function1)		Variable values in memory region (second call to function1)		Variable values in memory region (third call to function1)		Variable values in memory region (fourth call to function1)		Variable values in memory region (fifth call to function1)	
variable	value	variable	value	variable	value	variable	value	variable	value
i	10	i	10	i	10	i	10	i	10
j	15	j	15	j	15	j	15	j	15
k	4	k	3	k	2	k	1	k	0
tot	125	tot	100	tot	75	tot	50	tot	25

Returning variable values in memory during returning phase.

memory regions are returned to. Any values that are changed during the returning phase due to assignment statements being executed then are modified. However, values that do not change are remembered as what they were during the calling phase because that particular region of memory has not changed.

For example, consider the variable k. Figure 8.40 shows that, during the calling phase, k is changed with each call (due to the statement k--;). Therefore, each region of memory for each call has a different value of k. However, on the returning phase (during which the bottom portion of the True block of the if control structure in function1 is executed), no statements change k, but the value of k is remembered to be what it was on the calling phase. The values of k are the same in the corresponding top and bottom boxes of Figure 8.40. The variable `tot` is the only one that has different values in the top and bottom boxes, because only `tot` appeared on the left side of an assignment statement, which was executed on the returning phase. While it appears that we are changing `k` on the return, we actually are just remembering what it was when the function was called.

5. There must be an easier way to arrive at the same result. What is it? In many cases, you will find writing iterative structures (loops) will be much simpler than writing recursive structures. In addition, iterative structures often are more efficient than recursive structures, due to the so-called overhead with each function call, meaning that memory must be reserved and held until the function is closed and other sorts of peripheral operations done. Given the potential memory demands of function calls, it is important that the base case be reached before memory is exceeded. An improperly designed recursive structure easily can cause memory to be filled. Loops lack this overhead and therefore tend to be more efficient.

You can see that, with recursion, a number of function calls precede the base case or simplest condition. These function activations essentially are in limbo, waiting for the base case or reversal to occur. This is not the most efficient use of a computer system. In addition, it has been theoretically proven that recursive structures can be written as iterative structures. Recall that a for loop has three parts to it like a recursive function does.

All this being said, sometimes recursive functions are a very clean way to code a repeated process. They can be written without a large number of if control structures and therefore appear very elegant. In these cases, recursion is a valuable programming technique. Also, we will see in Application Program 8.4 that recursion works very well with trees.

EXERCISES

1. True or false:
   a. A call to a recursive function must appear at least once in its function body.
   b. A recursive function must have a control statement to prevent the program from running indefinitely.
   c. There is no void type recursive function.
   d. A recursive function must return a value to its calling function; otherwise, it cannot continue the recursive process.

2. Write a program that calls a function to read a one- to six-digit octal base number and convert it to a decimal base number.

3. Rewrite problem 2 using a recursive function.

4. Look in your mathematics textbooks for equations to calculate π. Use a recursive function to implement one of the formulas.

Solutions

1. a (true), b (true), c (false), d (false)

▨ LESSON 8.11 CREATING HEADER FILES

Topics
- Indicating that a new header file has been created
- What to put in a header file

As you begin to develop larger programs, you will find that properly managing the volume of code is essential to working efficiently on it. Different portions of the code commonly are put into different files. C allows us to create our own header files, giving us the option of having a separate file that contains code that is typically at the beginning of a source code file.

What you should observe about this lesson's program.
1. The very first line directs the preprocessor to include a header file that we created.
2. The header file name is surrounded by double quotes ("") rather than angle brackets (<>).
3. The constant macro MAX is used in main but not defined in File 1.
4. There is no prototype for function1 in File 1.
5. File 2 has directives to include header files, a macro definition, and a function prototype.

Source code

File 1

```
#include "header_1.h"
void main(void)
{
        int ii;
        double xx;


        ii=3;
        xx=44.7;


        function1(ii,xx,MAX);
}


void function1(int kk, double yy, int nn)
{
        double pp;


        pp=kk+log(yy)+nn;
        printf("pp=%lf\n",pp);
}
```

Header file we created. We use the extension .h and enclose the name in double quotes rather than < >.

Note that, in this file, there is no prototype for function1 and no definition of MAX.

File 2 header_1.h

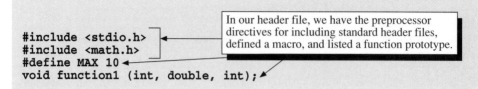

```
#include <stdio.h>
#include <math.h>
#define MAX 10
void function1 (int, double, int);
```

In our header file, we have the preprocessor directives for including standard header files, defined a macro, and listed a function prototype.

Output

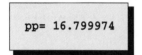

```
pp= 16.799974
```

Explanation

1. How do we create our own header files? We create a file with the extension .h and put it into a location where the compiler can find it when it looks for the header file. You should see your compiler documentation to determine where (for instance, which directories) the compiler searches for header files stored on disk and put your header file in that location.

2. What should we put into our header files? Such things as macros, function prototypes, comments, structure definitions, and inclusion of standard header files commonly are put into header files created by a programmer. At this point, you should understand the syntax and general meanings of much of the code in a standard header file, such as stdio.h. You can examine the contents of this file with any standard editor. We recommend that you look at this file to get an understanding of the types of statements sometimes put into header files.

3. How do we indicate that the header file being included is one we created? Enclosing the header file name in " " rather than <> does not explicitly tell the compiler that we have created a new header file, but it tells the compiler that the header file probably is not within the library of the standard header files. Therefore, the compiler searches for the header file elsewhere on disk storage to find it. If you put your header file in the header file library, you can use <> to enclose your header file name. However, we recommend that you use " " and put your header file with the other files of the source code for your program.

EXERCISES

1. True or false:
   a. We usually use " " instead of <> to enclose the name of a header file that we have created.
   b. Self-made header files can be no more than 20 lines in length.
   c. We often put constant macro definitions in header files.

Solutions

1. a (true), b (false), c (true)

LESSON 8.12 USE OF MULTIPLE SOURCE CODE FILES AND STORAGE CLASSES

Topics
- Using multiple source code files
- Global variables
- Extern, register, static, and auto storage classes
- Creating a personal library

In developing large programs, it is not practical to keep all of source code in a single file. And, with a modular design, it usually is unnecessary to work with more than just a few functions at a time. As a result, it is more manageable to create a large number of files, each combining related functions.

As you develop working versions of all the functions in a given file, the code in the file can be compiled, and the object code generated from this can be stored in your own personal library. As you write more functions in other files that use the working functions, they can be linked together using the linker of your compiler. One advantage of this procedure is that it does not require already working functions to be compiled repeatedly. This saves development time and reduces errors.

What you should observe about this lesson's program.
1. A global variable, aa, has been declared.
2. The keywords *register* and *auto* are used in the declarations for bb and cc.
3. The prototype for function1 is in File 1 but the definition of function1 is in File 2.
4. Within File 2, the keyword *extern* is used with the variable aa.
5. In the definition of function1, the keyword *static* is used in the yy declaration.

A question you should attempt to answer before reading the explanation. What do you think the keyword `register` means? Hint: Recall how registers are described in Chapter 1.

Source code

File 1

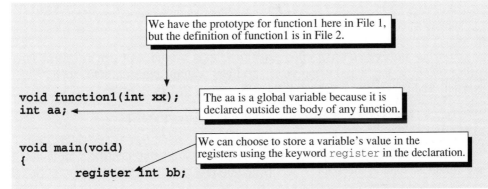

We have the prototype for function1 here in File 1, but the definition of function1 is in File 2.

```
void function1(int xx);
int aa;
```

The aa is a global variable because it is declared outside the body of any function.

```
void main(void)
{
        register int bb;
```

We can choose to store a variable's value in the registers using the keyword `register` in the declaration.

```
        auto int cc;
```

The default class for variables with function scope is auto. With this class, the memory is deallocated after control returns to the calling unit.

```
        aa=5;
        cc=10;
        bb=aa+cc;
```

```
        function1(bb);
```
We still can call function1, even though the definition of function1 is not in this file.

```
}
```

File 2

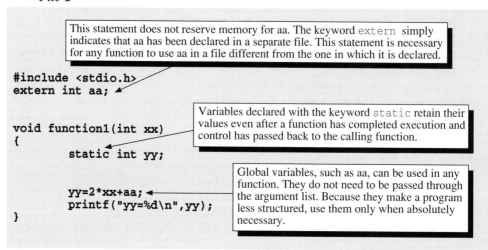

This statement does not reserve memory for aa. The keyword extern simply indicates that aa has been declared in a separate file. This statement is necessary for any function to use aa in a file different from the one in which it is declared.

```
#include <stdio.h>
extern int aa;
```

```
void function1(int xx)
{
        static int yy;
```
Variables declared with the keyword static retain their values even after a function has completed execution and control has passed back to the calling function.

```
        yy=2*xx+aa;
        printf("yy=%d\n",yy);
}
```
Global variables, such as aa, can be used in any function. They do not need to be passed through the argument list. Because they make a program less structured, use them only when absolutely necessary.

Output

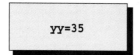

```
        yy=35
```

Explanation

1. What storage class specifiers are in C and what is the general form for using them? C has five storage class specifiers: extern, register, auto, static, and typedef. The general form for using them is

storage_class_specifier type identifier_1, identifier_2, identifier_n;

where *storage_class_specifier* is any of the listed specifiers, *type* is any valid C data type, and *identifier_1, identifier_2,* and *identifier_n;* are valid identifiers representing variables. One or more identifiers is permissible.

2. What are the meanings and usefulness of the storage class specifiers? We cover typedef in the supplemental material for this book (which is available to your instructor) because it is different from the other specifiers and will not cover it here. In order, the meanings of the other storage class specifers are as follows.

The specifier `extern` is for declaring global variables initially declared in other files. For instance, the variable aa is declared using extern in File 2 because it initially is declared in File 1. The declaration `extern int aa;` does not reserve memory for aa because the declaration `int aa;` in File 1 does. Using extern lets the compiler know that the global variable aa is declared in another file. If you do not use extern in declaring global variables already declared in other files, in most cases, your compiler will indicate an error for declaring multiple variables of the same name.

Using the specifier `register` in a declaration frequently will cause single variables to be stored in the register portion of the CPU although the ANSI C standard does not require it (it does suggest, though, that the register variables be given faster access than other variables). Large arrays may not be stored in the CPU due to lack of space. No matter which way a register variable or array actually is stored in an implementation, ANSI C does not allow the address to be referenced with the unary & operator. Because of the increased access speed of register variables, the register specifier is used most effectively for variables that are to be accessed frequently by a program, such as those referenced repeatedly in loops. The register specifier can be used only with local variables.

The `auto` specifier is the default specification for local variables and, so, is used less frequently than other storage class specifiers. Memory for local variables with the auto storage class is deallocated on completion of execution of the function. We described this process already in detail in this text.

A `static` local variable retains its value even after program control has returned to the calling function. On calling the function again, a static local variable has the same value as it had at the end of the previous call. Static local variables sometimes can be used in place of global variables. Therefore, before you use a global variable, consider the use of a static local variable. Using static local variables is preferable to global variables because they assure a more structured program.

A static global variable is a global variable accessible only to the functions within a given file. Declaring a global variable as static can reduce the likelihood, in a very large program, of having a function far removed from the one that uses the global variable accidentally changing its value. Before you declare an ordinary global variable, consider the possibility of creating a static global variable. If you can declare the global variable as static, you should do so.

3. What is the default storage class specification for a function? A function is given the default specification extern. Therefore, it is not necessary to give the prototype for function1 in this lesson's program as

```
extern void function1(int xx);
```

In each file that a function is called, a prototype for that function must be given.

EXERCISES

1. True or false:
    a. We often explicitly declare variables to be of storage class auto.
    b. The storage class specifier extern is used when a source code has multiple files.
    c. The default storage specification for a function is auto.
    d. The keyword `typedef` is classified by ANSI C as a storage class specifier.

Solutions

1. a (false), b (true), c (false), d (true)

■ LESSON 8.13 BITWISE MANIPULATIONS

Topics

- Bitwise operators
- Using hexadecimal notation in a C program
- Printing individual bits of integers
- Bit fields

C is a language capable of performing what can be considered to be very low level operations, such as manipulating individual bits in memory. C provides bitwise operators to perform these manipulations.

To work with the bitwise operators in a program, we need to consider the status of individual bits in a memory cell (being 1 or 0). In Chapter 1, we describe how hexadecimal and octal notation are convenient for representing bit patterns. In this lesson, we use hexadecimal notation in the source code itself and as the form of the output, so that we can view and understand the results of the bitwise operators. For convenience, we again show hexadecimal notation and the corresponding bit patterns in Table 8.1.

When a bit has a value of 1 we say that the bit is *set,* and when it is 0 we say that it is *clear.* These words also can be used as verbs so that sometimes we say that we want to *set* a bit (meaning making it 1) or *clear* a bit (meaning making it 0).

Six operators in C allow us to manipulate individual bits in a memory cell: & (bitwise AND), | (bitwise inclusive OR), ^ (bitwise exclusive OR, also called XOR), ~ (complement), >> (right shift), and << (left shift). The operator ~ is a unary operator (meaning it uses only one operand) while all of the others are binary operators (meaning that they need two operands).

The first two of these operators, & and |, work somewhat similarly to their counterparts && and ||, which we used in logical expressions. Recall that, in logical expressions, the value of True was represented as 1 or nonzero, and the value of False was represented as 0. In Chapter 4 we found that we had the following for && and ||:

1 && 1 = 1   (all other situations evaluate to 0; that is, 1 && 0 = 0, 0 && 0 = 0)

0 || 0   = 0   (all other situations evaluate to 1; that is, 1 || 0 = 1, 1 || 1 = 1)

■ **TABLE 8.1**
Hexadecimal notation and bit patterns

Decimal	Hexadecimal	Bit pattern
0	0	0000
1	1	0001
2	2	0010
3	3	0011
4	4	0100
5	5	0101
6	6	0110
7	7	0111
8	8	1000
9	9	1001
10	A	1010
11	B	1011
12	C	1100
13	D	1101
14	E	1110
15	F	1111

The bitwise AND and bitwise OR work with single bits but yield the same results:

1 & 1 = 1   (all other situations evaluate to 0; that is, 1 & 0 = 0, 0 & 0 = 0)
0 | 0 = 0   (all other situations evaluate to 1; that is, 1 | 0 = 1, 1 | 1 = 1)

With these operators, we go bit by bit to yield the following for a 4 bit example:

Bitwise AND

(hex A)	1	0	1	0
&				
(hex C)	1	1	0	0
(hex 8)	1	0	0	0

Bitwise OR

(hex A)	1	0	1	0
|				
(hex C)	1	1	0	0
(hex E)	1	1	1	0

Note that the & and | operations are commutative, meaning 1 & 0 = 0 & 1 and 1 | 0 = 0 | 1.

The complement operator (~) reverses all of the bits of its operand. Thus, ~(1010) = 0101. In the source code, we use these three operators.

The bitwise exclusive OR (^) gives the following results:

0 ^ 1 = 1   (all other situations evaluate to 0; that is, 0 ^ 0 = 0, 1 ^ 1 = 0)

Bitwise XOR

(hex A)	1	0	1	0
^				
(hex C)	1	1	0	0
(hex 6)	0	1	1	0

Again, notice that the operation is commutative, meaning that $1 \wedge 0 = 0 \wedge 1$.

The bitwise shift operators ($>>$ and $<<$) move all of the bits in a cell either to the right or left and add clear bits in the shift. For instance, if we shift the bit pattern 1011 to the right one place we get 0101. This is illustrated in Fig. 8.41. In this operation, the far right 1 is lost, and a 0 is placed in the leftmost bit location. All other bits are shifted one place to the right. (Note: In some implementations and under certain conditions, a 1 may be placed in the far left location on a right shift. However, we will not consider this possibility in this text.)

Similarly, if we shift the bit pattern 1011 to the left one place we get 0110. This is illustrated in Fig. 8.42. In this operation, the far left 1 is lost, and a 0 is placed into the rightmost bit location.

More than one place can be shifted in a shift operation. If we shift two places then two zeros are added, two bits are lost, and the other bits are shifted two places.

Also, you may want to print the bit representation of your integer type data to the screen. This can be done by creating what is called a *mask* and using it with the bitwise operators to isolate individual bits. After isolating a bit, it can be put into a full-size integer cell and printed using standard printf statements. In the Explanation for this lesson's program, we describe how the for loop in the program performs this operation.

In the program, we illustrate the use of all the bitwise operators. In addition, we show how to print all the bits of a given variable.

FIG. 8.41

Right shift one place causes the rightmost bit to be lost, the leftmost bit to be 0, and other bits to be shifted one place to the right

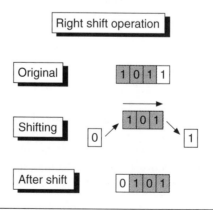

FIG. 8.42

Left shift one place causes the leftmost bit to be lost, the rightmost bit to be 0, and other bits to be shifted one place to the left

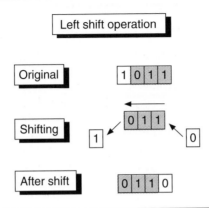

What you should observe about this lesson's program.

1. The assignment statements for the unsigned integers aa, bb, cc, dd, and ee begin with 0x on the right-hand side. Use a pencil and paper to write the bit patterns for all of them.

2. Recall that unsigned integers typically occupy 2 bytes or 16 bits.

3. The sixth through eleventh assignment statements use the bitwise operators. Use a pencil and paper to perform these operations by hand. The first printf statement prints the results of these operations. Compare the output to your hand calculations.

4. The assignment statement for mm gives it a single set bit with all of the other bits clear.

5. The for loop prints the individual bits of ee.

Questions you should attempt to answer before reading the explanation.

1. What are the bit representations for aa, bb, cc, dd, and ee? Use Table 8.2 to answer this.

2. What is the initial bit pattern for mm?

Source code

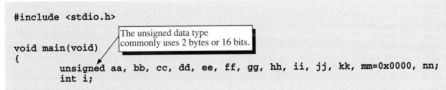

```
#include <stdio.h>

                    The unsigned data type
                    commonly uses 2 bytes or 16 bits.
void main(void)
{
        unsigned aa, bb, cc, dd, ee, ff, gg, hh, ii, jj, kk, mm=0x0000, nn;
        int i;
```

```
        aa=0xDFFF;
        bb=0x2840;         Hexadecimal notation begins with 0x. The four
        cc=0xFF7F;         hexadecimal symbols after 0x represent 16 bits.
        dd=0x0004;
        ee=0xA3C5;     Using bitwise operators AND, OR, exclusive OR, and complement.


        ff = aa & cc;
        gg = bb | dd;
        nn =  aa&(~dd);
        hh = aa ^ bb;
        ii = cc >> 1;     Shifting bits one place to the right.
        jj = dd << 3;     Shifting bits three places to the left.

        printf ("ff=%p,  gg=%p,  hh=%p,  ii=%p,  jj=%p,  nn=%p\n\n",ff,gg,hh,ii,jj,nn);


        printf ("The bits for ee (hex A3C5) are:\n");

                          Setting a bit in the far left location to make mm
        mm = 1 << 15;     a mask (for initially reading the far left bit).
        for (i=16; i>=1; i--)
                   {      Here, using & with a single set bit causes the bit for ee (that is
        Printing the   kk = ee & mm;   opposite the bit in the set bit position for mm) to be assigned to kk.
        individual bits  kk >>= (i-1);
        of ee.         printf ("%u ",kk);
                       mm >>= 1;      Shifting the bits of kk to the right (i – 1) places. This causes
                   }               the bit to be shifted to the far right (lowest) bit location.
        printf ("\n");
}
                     Shifting the bit in the mask mm one to the right.
```

Output

```
ff=DF7F,   gg=2844,   hh=F7BF,   ii=7BF,   jj=0020, nn=DFFB

The bits for ee (hex A3C5) are:
1 0 1 0 0 0 1 1 1 1 0 0 0 1 0 1
```

Explanation

1. How do we represent an integer in hexadecimal notation in a C program?
We represent a number using hexadecimal notation by preceding the notation with 0x or 0X. For example, 0xDFFF represents hexadecimal DFFF and 0xA3C5 represents hexadecimal A3C5.

2. How do we represent an integer in octal notation in a C program?
Although we did not use octal notation in this lesson's program, we can represent a number using octal notation by preceding the notation with 0. For example, 0364 represents octal 364 and 0751 represents octal 751.

3. What are the actions of all the bitwise operators? Tables 8.2 and 8.3 list the operators and their actions.

■ **TABLE 8.2**
Bitwise operator actions

Operator	Name	Type	Associativity	Example from program	Explanation
&	bitwise AND	binary	left to right	aa&cc	This operator sets a 1 in each bit position where its two operands have ones, and sets 0 for other cases
|	bitwise OR	binary	left to right	bb|dd	This operator sets a 0 in each bit position where its two operands have zeros, and sets 1 for other cases
^	bitwise eXclusive OR, or XOR	binary	left to right	aa^bb	This operator sets a 1 in each bit position where its two operands have different bits, and sets zeros where they have the same bits
~	bitwise complement	unary	left to right	~dd	This operator converts a 1 in each bit position to a 0 and a 0 in each bit position to a 1
<<	left shift	binary	left to right	dd<<3	This operator shifts the bits of the left operand to the left by the number of bits specified by the right operand
>>	right shift	binary	left to right	cc>>1	This operator shifts the bits of the left operand to the right by the number of bits specified by the right operand

■ **TABLE 8.3**
Bitwise operator results

Expression	Result	Expression	Result	Expression	Result	Expression	Result
1 & 1	1	1 | 1	1	1 ^ 1	0	~1	0
1 & 0	0	1 | 0	1	1 ^ 0	1	~0	1
0 & 1	0	0 | 1	1	0 ^ 1	1		
0 & 0	0	0 | 0	0	0 ^ 0	0		

4. Can we use bitwise operators on double or float data types? No, bitwise operators can be used only on integer data types (char, int, and modifications of these).

5. Is it guaranteed that a program that uses bitwise operators give the same results on every computer system? No; because all systems do not use the same bitwise representations, a given program may not give the same results on all systems.

This is an important point of which you should be aware when performing bitwise operations.

To effectively deal with individual bits in a program, it is necessary to be cognizant of the particular implementation you are using. In other words, you need to know how many bits are used for the data types (char, int, double, or other) and the type of character code being used (ASCII or EBCDIC, for example). Also, we have not covered it in detail in Chapter 1, but you may need to know more about the way that negative integers are represented. In our program, we deal only with unsigned integers and, so, do not need to be concerned with negative integers. In our implementation, positive integers are represented with the binary representation given in Chapter 1. We also use an implementation where unsigned integers occupy 2 bytes or 16 bits. What we describe here is based on this implementation.

6. What are the results of the bitwise operations in the first part of this lesson's program? The variables used in the first part of program and their bitwise representation are shown in Table 8.4. The operations give the results in bits as shown in Table 8.5. The hexadecimal representations of the results are given in Table 8.6. These are shown in the output.

7. What is a mask? As we use it here, a mask is a bit pattern used with the bitwise operators to modify another bit pattern. For instance, in the first two assignment statements in this lesson's program, we use cc and dd as masks on the variables aa and bb to create the new bit patterns ff and gg, respectively.

Frequently, we use a mask to either clear or set individual bits in a given pattern. Therefore, we must be aware of the methods used to clear and set bits.

8. How do we refer to different bits in a bit pattern? The rightmost bit is the first bit, the second from the right is the second bit, and so on from the right to the left. For instance, in the bit pattern

```
0001  0001  0010  1100
```

the third, fourth, sixth, ninth, and thirteenth bits are 1. The rest are 0.

TABLE 8.4

Bit representations of variables in this lesson's program

Variable name	Hexadecimal representation	Representation in bits
aa	DFFF	1101 1111 1111 1111
bb	2840	0010 1000 0100 0000
cc	FF7F	1111 1111 0111 1111
dd	0004	0000 0000 0000 0100
ee	A3C5	1010 0011 1100 0101

TABLE 8.5
Bitwise operations in this lesson's program

Operation	Result	Comment
ff = aa & cc;	1101 1111 1111 1111 = aa 1111 1111 0111 1111 = cc 1101 1111 0111 1111 = ff	We used cc as a mask to clear the eighth bit of aa. We stored the result in ff.
gg = bb | dd;	0010 1000 0100 0000 = bb 0000 0000 0000 0100 = dd 0010 1000 0100 0100 = gg	We used dd as a mask to set the third bit of bb. We stored the result in gg.
nn = aa&(~dd);	1101 1111 1111 1111 = aa 1111 1111 1111 1011 = ~dd 1101 1111 1111 1011 = nn	We used (~dd) as a mask to clear the third bit of aa. We stored the result in nn.
hh = aa ^ bb;	1101 1111 1111 1111 = aa 0010 1000 0100 0000 = bb 1111 0111 1011 1111 = hh	We used bb to reverse the seventh, twelfth, and fourteenth bits of aa. We stored the result in hh.
ii = cc >> 1;	1111 1111 0111 1111 = cc 0111 1111 1011 1111 = ii	We shifted cc to the right one bit location. We stored the result in ii.
jj = dd << 3;	0000 0000 0000 0100 = dd 0000 0000 0010 0000 = jj	We shifted dd to the left three bit locations. We stored the result in jj.

TABLE 8.6
Hexadecimal representation of the results shown in Table 8.5

Variable name	Representation in bits	Hexadecimal representation
ff	1101 1111 0111 1111	DF7F
gg	0010 1000 0100 0100	2844
nn	1101 1111 1111 1011	DFFB
hh	1111 0111 1011 1111	F7BF
ii	0111 1111 1011 1111	7FBF
jj	0000 0000 0010 0000	0020

9. How do we set a bit? We create a mask with ones in the locations that we want to set bits and zeros in the other locations. Then we use the bitwise OR operator (|) with the operands being the mask and the bit pattern we want to modify.

For instance, the bit pattern for dd (0000 0000 0000 0100), when used as a mask in this manner, causes the third bit to be set (see Table 8.5). When this mask was used on bb (0010 1000 0100 0000) with the bitwise OR operator (|), as in this lesson's program with the statement gg = bb | dd, we created a new bit pattern, gg (0010 1000 0100 0100). Note that the bit pattern for gg is the same as bb but with the third bit set. Thus, we have successfully used the mask to set a particular bit in a bit pattern.

10. How do we clear a bit? We can use one of two methods to clear a bit:

1.   We create a mask with zeros in the locations that we want to clear bits and ones
     in the other locations. Then, we use the bitwise AND operator (&) with the
     operands being the mask and the bit pattern we want to modify. For instance,
     the bit pattern for cc (1111 1111 0111 1111), when used as a mask in this man-
     ner, causes the eighth bit from the right to be cleared (see Table 8.6). When this
     mask was used on aa (1101 1111 1111 1111) with the bitwise AND operator
     (&), as in this lesson's program with the statement ff = aa & cc, we created a
     new bit pattern, ff (1101 1111 0111 1111). Note that the bit pattern for ff is the
     same as aa but with the eighth bit from the right cleared. Thus, we have suc-
     cessfully used the mask to clear a particular bit in a bit pattern.
2.   We create a mask with ones in the locations that we want to clear bits and zeros
     in the other locations. Then, we use the complement operator (~) to reverse all
     of the bits of the mask. After doing this, we use the bitwise AND operator (&)
     with the operands being the mask and the bit pattern we want to modify. For
     instance, the bit pattern for dd (0000 0000 0000 0100), when used as a mask in
     this manner, causes the third bit to be cleared (see Table 8.6). The first step is
     to reverse the bits of this mask with the complement (~) operator to give the bit
     pattern 1111 1111 1111 1011. When this mask was used on aa (1101 1111 1111
     1111) with the bitwise AND operator (&) as in this lesson's program with the
     statement nn = aa & (~dd), we created a new bit pattern, nn (1101 1111 1111
     1011). Note that the bit pattern for nn is the same as aa but with the third bit
     cleared. Thus, we have successfully used the mask to clear a particular bit in a
     bit pattern.

11. Is one method of clearing a bit more convenient than another? In many
cases, the second method of the two described is more convenient than the first. This
is because it is relatively simple to create a bit pattern with a 1 in a particular location.

***12. How, then, can we easily create a bit pattern with a 1 in a particular loca-
tion and zeros in the other locations?*** We can use the left shift operator. For
instance, the statement:

```
mm = 1 << 15;
```

causes the bitwise representation of the integer 1 (0000 0000 0000 0001) to be
shifted 15 places to the left giving the bit pattern 1000 0000 0000 0000. Note that
shifting 15 places causes a 1 to be in the 16th place from the right.
    Similarly, if we used

```
mm = 1 << 7;
```

we would create the bit pattern 0000 0000 1000 0000. Thus, we would have a 1 in
the eighth place.
    Because it is easy to create a bit pattern with a 1 in a particular location, we
often use the second of the two methods described for clearing a bit.

*13. Can we use compound assignment type operations with the bitwise opera-
tors?* Yes, all of them except the complement (~) operator; in other words, &=, |=,
^=, >>=, and <<= are legal operations, while ~= is not.

14. What do these operations mean? They act in a manner similar to the other
compound assignment operations we have used. For instance,

kk<<=7;

    is the same as

kk = kk<<7;

    and

kk &= aa;

    is the same as

kk = kk & aa;

15. Why can we not use ~= as an operator? We cannot use the complement
operator in a compound assignment is because it is a unary operator (meaning it has
only one operand). Only binary operators can be used in compound assignment.

*16. We know that we can complement all of the bits in a bit pattern using the
complement (~) operator. How can we complement only some of the bits in a bit
pattern?* We create a mask with ones in the locations of the bits that we want to complement and zeros in the other locations. Then, we use the bitwise exclusive OR operator (^) with the operands being the mask and the bit pattern we want to modify.

    For instance, the bit pattern for bb (0010 1000 0100 0000), when used as a mask
in this manner, causes the 7th, 12th, and 14th bits to be complemented. When this
mask was used on aa (1101 1111 1111 1111) with the bitwise XOR operator (^),
as in this lesson's program with the statement hh = aa ^ bb, we created a new bit
pattern, hh (1111 0111 1011 1111). Note that the bit pattern for hh is the same as
aa but with the 7th, 12th, and 14th bits complemented. Thus, we have successfully
used the mask to complement particular bits in a bit pattern.

*17. How can we check the status (that is, determine whether a bit is set or
clear) of the leftmost bit of an integer type variable?* We can perform the following operations:

1. Create a mask with a 1 in the leftmost position and all other bits 0.
2. Perform a bitwise & operation with the integer and the mask as operands and
   store the result in an integer variable.
3. Use the right shift operator to move the leftmost bit to the rightmost location in
   the integer variable.
4. Determine whether the integer has the value of 0 or 1.

The following code performs these operations with the mask, mm, being used to determine the status of the leftmost bit of the integer, ee:

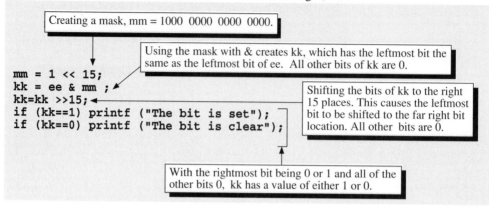

18. How can we determine and print the status of every bit in a bit pattern?
We can use the method and variables just described for the leftmost bit with the modification of repeating the operation for each bit with the mask changing on each repetition. If the mask has the 1 shifting to the right one place each time, as follows

```
1000 0000 0000 0000
0100 0000 0000 0000
0010 0000 0000 0000
...
...
0000 0000 0000 0010
0000 0000 0000 0001
```

then using & with the mask creates a variable, kk, that has zeros everywhere except at the location of the 1 in the mask. At this location, the bit matches the corresponding one in ee. We need only to shift the matching bit in kk to the rightmost location and print it using normal integer printing methods. The following code performs these operations:

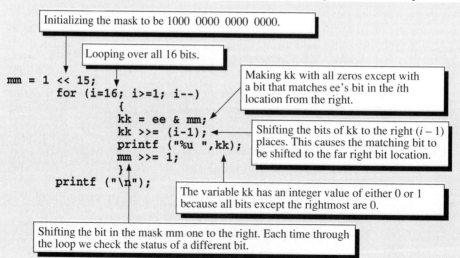

19. Can we do any arithmetic operations using bitwise operators? Yes; if we want to multiply or divide an integer by a power of 2, we can use the bitwise shift operators. For instance, the bit pattern 0000 0000 0010 1000 represents the integer 40. If we shift once to the left, the bit pattern becomes 0000 0000 0101 0000. This represents the integer 80, and we have multiplied by 2. If we shift the original bit pattern three times to the left, we get 0000 0001 0100 0000, which is 320, that is 2^3 or eight times our original. Shifting to the left n places is the same as multiplying by 2^n. We can do this provided that we do not shift ones off the left end.

If we shift our original bit pattern to the right one place, we get 0000 0000 0001 0100. This is the integer 20, and we have effectively divided by 2. If an integer is an odd number, the result is the same as dividing the integer minus 1 by 2. For instance, 0000 0000 0010 1001 is 41. Shifting once to the right causes the rightmost 1 to be lost and the result to be 0000 0000 0001 0100, which is 20. To divide by 8 (2^3), we shift to the right three places. Again, if we lose ones off the right end, we will not get the exact answer.

20. Why are bitwise operations used? There are a number of practical uses of the bitwise operators. They can be used in programs that control peripherals such as printers, monitors, disk drives, and modems, because in communicating with these devices it sometimes is necessary to clear or set individual bits.

Also, instead of using integers as True or False (1 or 0) flags, we can use individual bits as flags. If we do so, we can pack 16 different flags in one integer. This can save memory and allow faster communication. Also, such things as file encryption can be done using the bitwise operators.

Any array for which each array member has only two possible states can be handled instead with individual bits. For instance, if we are keeping track of the everyday presence or absence of students in a class of 32, we can specify that the first bit represents the first student in alphabetical order and the other bits represent the other students in alphabetical order. A 1 can indicate the presence of the student on a given day and 0 can represent absence. Thus, we need only 32 bits of memory for each day of class to have the entire attendance record. You can imagine that you can create similar representations for other situations in which two states are appropriate.

21. What are bitfields? We have not used them in this lesson's program, but C allows us to specify the number of bits to be used to store members of structures. For instance, the structure definition

```
struct Bitfield_str
              {
              unsigned aa : 3;
              unsigned bb: 4;
              unsigned cc:   2;
              };
```

sets the member, aa, to occupy 3 bits, bb to occupy 4 bits, and cc to occupy 2 bits. If, in a program, we have a declaration

```
struct Bitfield_str mm;
```

then we can access the individual members of the structure using mm.aa, mm.bb, and mm.cc. We will not go through it in detail, but using bitfields the size of just 1 bit is another way to control individual bits.

EXERCISES

1. For aa = 0xAD3F, bb = 0xCC43, cc = 0xAC23, dd = 0xFFFB, and ee = 0x23F2, find
    a.   aa & bb
    b.   cc | dd
    c.   dd ^ ee
    d.   ~cc
    e.   aa >> 5
    f.   dd << 4

    Express each of these in hexadecimal notation according to Table 8.1.

▨   LESSON 8.14 BINARY FILES

Topics

- Contrasting text and binary files
- Creating binary files
- Writing to binary files
- Reading from binary files
- Using the exit( ) function

All the files that we created prior to this lesson have been text files. In text files, the numbers (int or double) are not stored with the same bit pattern used in main memory. For instance, assuming the ASCII system is used (which we will do for this entire lesson), the representation of the integer 25 in main memory and in a text file follow. This table shows that the bit patterns for decimal 25 in main memory and in a text file are not the same:

Decimal representation	Main memory bit pattern	Text file bit pattern	
25	0000 0000 0001 1001	0011 0010	0011 0101
	Standard binary representation for decimal 25.	ASCII code for decimal 2.	ASCII code for decimal 5.

Because of the difference in bit patterns, it is necessary to perform operations on a number in order to take a value from main memory and copy it into a disk file in text format shown in Fig. 8.43 for the function fprintf. Similarly, when we read a number from a text file, we must convert it from the text format to the format used in main memory. This is illustrated in Fig. 8.44 for fscanf.

FIG. 8.43
Printing numbers to a text file using fprintf. Note that the binary representation of the number 25 in main memory is different from its representation in a text file.

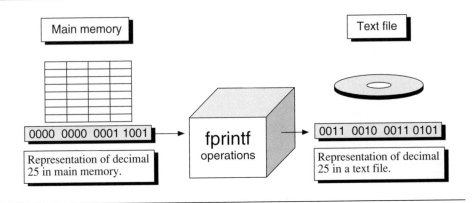

The operations required to convert between the main memory and text file formats can take considerable time when a significant amount of information is to be transferred. These operations can be eliminated if the same format is used in the disk file as is used in main memory, and this is the case for binary files. In Fig. 8.45 we illustrate storing and reading numbers in a binary file using the functions fwrite and fread. Because the writing and reading to binary files is much faster than writing and reading to text files, programs that require considerable movement in and out of files during an execution should use binary files instead of text files.

You may wonder why we would want to use text files. A primary reason is that editing software typically is designed to interpret standard text files. Therefore, an

FIG. 8.44
Reading numbers from a text file using fscanf. Note that the binary representation of the number 25 in main memory is different from its representation in a text file.

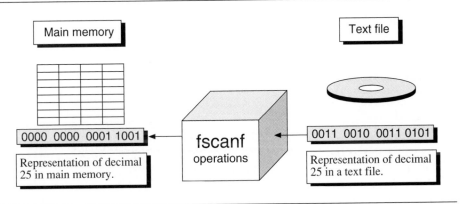

FIG. 8.45

Writing a number to a binary file using fwrite and reading from a binary file using fread. Note that the representations of the number 25 are the same in main memory and in the binary file. So, far fewer operations are needed for storing in a binary file than in storing in a text file.

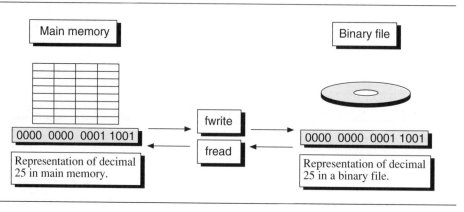

output file produced with fprintf can be read by editing software. If you try to read a binary file with editing software, it will show nonsensical symbols in most cases. As a result, a binary file cannot be printed using a printer with a text editor either. Also, a standard text editor can be used to create an input text file, but we cannot use a standard text editor to create a binary file for input. We need to write a program and use the function fwrite to create a binary file.

Text files are convenient for human-created input and human-interpreted output. Binary files are better for computer-created input and computer-interpreted output. In this lesson, we illustrate a program that creates a binary file, writes to it, and reads from it. We write an array and a structure to a binary file. We write much less code to copy arrays and structures to binary files than to text files. Read the annotated source code before you read the explanation.

Source code

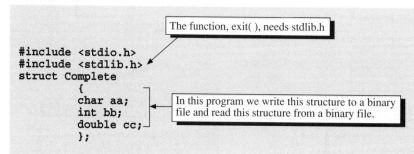

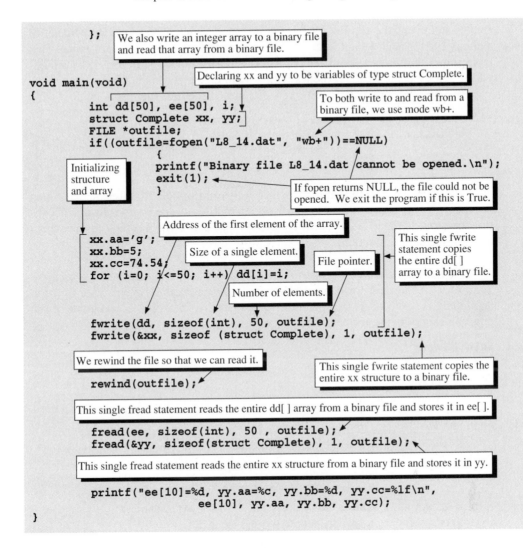

```
                };        We also write an integer array to a binary file
                          and read that array from a binary file.

                              Declaring xx and yy to be variables of type struct Complete.

void main(void)
{
                                                  To both write to and read from a
        int dd[50], ee[50], i;                    binary file, we use mode wb+.
        struct Complete xx, yy;
        FILE *outfile;
        if((outfile=fopen("L8_14.dat", "wb+"))==NULL)
                {
    Initializing    printf("Binary file L8_14.dat cannot be opened.\n");
    structure       exit(1);
    and array       }        If fopen returns NULL, the file could not be
                             opened.  We exit the program if this is True.

                    Address of the first element of the array.
                                                              This single fwrite
    xx.aa='g';                                                statement copies
    xx.bb=5;        Size of a single element.                 the entire dd[ ]
    xx.cc=74.54;                          File pointer.       array to a binary file.
    for (i=0; i<=50; i++) dd[i]=i;
                            Number of elements.

        fwrite(dd, sizeof(int), 50, outfile);
        fwrite(&xx, sizeof (struct Complete), 1, outfile);

    We rewind the file so that we can read it.          This single fwrite statement copies the
                                                        entire xx structure to a binary file.
        rewind(outfile);

  This single fread statement reads the entire dd[ ] array from a binary file and stores it in ee[ ].

        fread(ee, sizeof(int), 50 , outfile);
        fread(&yy, sizeof(struct Complete), 1, outfile);

  This single fread statement reads the entire xx structure from a binary file and stores it in yy.

        printf("ee[10]=%d, yy.aa=%c, yy.bb=%d, yy.cc=%lf\n",
                      ee[10], yy.aa, yy.bb, yy.cc);
}
```

Output

```
ee[10]=10, yy.aa=g, yy.bb=5, yy.cc=74.540000
```

Explanation

1. How do we open a binary file? We use the fopen function with the mode being a binary mode. For instance, the statement

```
outfile=fopen("L8_14.dat", "wb+");
```

creates and opens the file L8_14.dat with the mode wb+. The mode wb+ specifies a binary file for reading and writing. Table 8.7 summarizes the modes for fopen. This table shows that a mode with the letter *b* is a binary mode. Notice also that modes with *r* simply open files and modes with *w* both create and open files.

By using the fopen statement in an if control structure, such as

```
if((outfile=fopen("L8_14.dat", "wb+"))==NULL)
                {
                printf("Binary file L8_14.dat cannot be opened.\n");
                exit(1);
                }
```

we both open the file and check whether fopen returns NULL. Recall that fopen returns NULL if it encounters an error on attempting to open a file. In this if control structure, if fopen does return NULL, we print an error message to the screen and call the exit( ) function.

2. What does the exit( ) function do? The exit function causes the entire program to terminate and for control to return to the operating system. This function commonly is used when an operation fundamental to the proper execution of the program cannot be performed. The general form of the function is

exit (*code*)**;**

TABLE 8.7
Modes for fopen

Opening mode	Action
r	Opens text file for reading text
w	Creates and opens text file for writing text
r+	Opens text file for reading and writing text
w+	Creates and opens text file for reading and writing text
rb	Opens binary file for reading binary code
wb	Creates and opens binary file for writing binary code
rb+	Opens binary file for reading and writing binary code
wb+	Creates and opens binary file for reading and writing binary code
a	Opens or creates text file, causes further writes to take place at the current end of the file
a+	Opens or creates text file, causes further writes to take place at the current end of the file
ab	Opens or creates binary file, causes further writes to take place at the current end of the file
ab+	Opens or creates binary file, causes further writes to take place at the current end of the file

where *code* is an integer value passed to the operating system. For its meaning, see your operating system documentation, but in most cases the value 0 indicates normal termination and 1 indicates an error. In this lesson's program, we call exit( ) with the code 1 when it is not possible to open the file L8_14.dat, which is essential to operation of the program.

3. What is a fundamental difference between the functions fread and fwrite and the functions fscanf and fprintf? The functions fread and fwrite copy bit patterns without modification between main memory and disk files, whereas the functions fscanf and fprintf read bit patterns in one format and copy them in a different format. ANSI C categorizes fread and fwrite as direct input/output functions because of the way that they copy information.

4. How do we use fwrite to copy the contents of an array to a binary file? Unlike writing to a text file, it is not necessary to use a loop to write an entire array to a binary file. For instance, in this lesson's program, the single line

```
fwrite(dd, sizeof(int), 50, outfile);
```

writes the entire dd[ ] array to the file pointed to by outfile. The other information passed to fwrite using this statement is the number of bytes of a single element of the array with the argument sizeof(int) and the number of elements of dd with the number 50. The form of an fwrite statement for writing an entire array is

fwrite   (*address_of_array, bytes_per_element, number_of_elements, file_pointer*);

where *address_of_array* is the address of the first element of the array, *bytes_per_element* is the number of bytes needed to store one element, *number_of_elements* is the number of elements in the array, and *file_pointer* is the pointer to the binary file.

5. How do we use fwrite to write the values of the members of a structure to a binary file? Again, it is not necessary to specify each individual member of a structure to write all of the values of a structure to a binary file using fwrite. For instance, the line

```
fwrite(&xx, sizeof(struct Complete), 1, outfile);
```

copies the value of the structure members xx.aa, xx.bb, and xx.cc to the file pointed to by outfile. The form of an fwrite statement for writing a structure is

fwrite (*address_of_structure, bytes_for_structure, 1, file_pointer*);

where *address_of_structure* is the address of the first element of the structure, *bytes_for_structure* is the number of bytes needed for all the members of the structure, *1* is the number of structures being written, and *file_pointer* is the pointer to the binary file.

6. After writing an array and a structure to the binary file pointed to by out-file, why do we need to rewind the file before reading from it? At the end of writing to the file, the file position indicator is at the end of the file. By using rewind, we move the file position indicator to the beginning of the file. This puts it into the location we need to read from the file.

7. How do we use fread to read an array from a binary file? Unlike reading from a text file, it is not necessary to use a loop to read an entire array from a binary file. For instance, in this lesson's program, the single line

```
fread(ee, sizeof(int), 50, outfile);
```

reads 50 integers from the file pointed to by outfile and stores them in the memory reserved for the array ee[] (which is an array of size 50). The form of an fread statement for reading an entire array is

fread (*address_of_array, bytes_per_element, number_of_elements, file_pointer*)**;**

where *address_of_array* is the address of the first element of the array used for storing the information read, *bytes_per_element* is the number of bytes needed to store one element, *number_of_elements* is the number of elements to be read from the file, and *file_pointer* is the pointer to the binary file.

8. How do we use fread to read the values of the members of a structure in a binary file? Again, it is not necessary to specify each member of a structure to read all of the values of a structure from a binary file using fread. For instance, the line

```
fread(&yy, sizeof(struct Complete), 1, outfile);
```

copies values from the file pointed to by outfile and stores them in the memory reserved for the structure members yy.aa, yy.bb, and yy.cc. The form of an fread statement for reading values from a file and storing them in a structure is

fread (*address_of_structure, bytes_for_structure, 1, file_pointer*)**;**

where *address_of_structure* is the address of the first element of the structure, *bytes_for_structure* is the number of bytes needed to store all of the members of the structure, *1* is the number of structures being written, and *file_pointer* is the pointer to the binary file.

9. What has this lesson's program done? This lesson's program has copied the values of all the elements of an array and all the members of a structure to a binary file, rewound the file, and copied the array values from the binary file into memory reserved for another array and structure. It then printed out one of the values of the array and all the values of the structure to the screen to verify that we indeed successfully copied information using binary files.

10. Can we use a standard text editor to read the contents of the file L8_14.dat? No; if we if we attempt to do this we will see nonsensical symbols on the screen. To see what is in these files, we need a program that can properly interpret the bit patterns such as the one we have written in this lesson. To satisfy yourself that you cannot read L8_14.dat with a text editor, try it.

EXERCISES

1. a.   Input and output are faster using binary files than text files.
    b.   To write to a binary file, it is necessary to open the file in a binary mode.
    c.   The function, fscanf, is used to read from a binary file.
    d.   We can use exit( ) to terminate a loop.
    e.   Standard text editors can be used to read binary files.

2. Find the errors, if any, in the following statements (see the declarations in this lesson's program):
    a.   `if((outfile==fopen("L8_14.dat", "wb+"))=NULL)`
    b.   `fwrite(sizeof(int), dd, 50, outfile);`
    c.   `fwrite(&xx, sizeof (struct Complete), outfile, 1);`
    d.   `fread(ee, sizeof(double), 50 , outfile);`
    e.   `fread(yy, sizeof(struct Complete), 1, outfile);`

Solutions

1. a (true), b (true), c (false), d (false), e (false)
2. a.   `if((outfile=fopen("L8_14.dat", "wb+"))==NULL)`
    b.   `fwrite(dd, sizeof(int), 50, outfile);`
    c.   `fwrite(&xx, sizeof (struct Complete), 1, outfile);`
    d.   `fread(ee, sizeof(int), 50 , outfile);`
    e.   `fread(&yy, sizeof(struct Complete), 1, outfile);`

▦ LESSON 8.15 POINTERS TO FUNCTIONS AND FUNCTIONS RETURNING POINTERS

Space does not permit us to give all the details or cover all the topics related to pointers. However, after reading the last lessons in this chapter, you will gain an appreciation of the versatility and capabilities of pointers.

In the source code for this lesson, we create a pointer to a function and a function that returns a pointer. We call the function that returns a pointer and print out the value of the pointer. We also illustrate how to call a function using a pointer to a function. We include both of these in this lesson so that we can compare and contrast them.

What you should observe about this lesson's program.

1. The function that returns a pointer (function1) has a prototype outside the body of main.
2. The other function (function2) returns an integer.
3. In the declarations is a pointer to a function. This declaration looks very similar to the function1 prototype except that it has parentheses around the first portion of the declaration.

4. In the first assignment statement, the pointer variable gg is assigned the value of the pointer returned by function1.
5. In the second assignment statement, the pointer to a function (function_ptr) is on the left side while on the right side is function2 written without parentheses and arguments.
6. The third assignment statement has function_ptr on the right side with arguments.
7. The number and type of arguments for function2 and function_ptr are identical.

Source code

```
                              Function that returns a pointer.

#include <stdio.h>


int *function1(int, double);
int function2(int, double);


void main(void)                    Pointer to a function.
{
        int *gg, hh;
        int (*function_ptr)(int, double);

                                        Calling function1, which returns a pointer.
                                        The pointer is assigned to gg.
        gg=function1(12,97.5);
        printf("*gg=%d\n",*gg);
                                        Printing the value pointed to by gg.

        function_ptr=function2;
                                        Assigning the address of
                                        function2 to function_ptr.

        hh=function_ptr(23,58.3);
        printf("hh=%d\n",hh);           Calling function2 using function_ptr.
}                                       The arguments are 23 and 58.3.


int *function1(int aa, double bb)
{
        int *kk;
        printf("aa=%d, bb=%lf\n",aa,bb);
        kk=&aa ;
        return kk;                      Function1 returns the address of aa.
}


int function2(int cc, double dd)
{
        printf("cc=%d, dd=%f\n",cc,dd);
        return(5);                                      Function2 returns an integer.
}
```

Output

```
aa=12, bb=97.500000
*gg=12
cc=23, dd=58.300000
hh=5
```

Explanation

1. What is the difference between a prototype for a function returning a pointer and a declaration for a pointer to a function? The difference is in the parentheses that are used for the declaration for a pointer to a function. In this lesson's program we have

```
int     *function1        (int, double);
int     (*function_ptr)   (int, double);
```

These statements indicate that function1 is a function that returns a pointer to an int, and function_ptr is a pointer to a function that returns an int. The arguments for function1 are int and double, and the arguments for the function pointed to by function_ptr are int and double. You can see that, other than the identifiers function1 and function_ptr, the only difference between these two is the parentheses around *function_ptr.

Parentheses can be used in the prototype for function1 in the following way:

```
(int *)function1(int, double);
```

This prototype clearly illustrates that the type of the return value is (int *). However, the statement has the same meaning without the parentheses, and in practice, you most commonly will see this prototype without parentheses.

Further confusion between these two declarations can occur if the pointer to the function is declared as a global variable. Had this been done, the beginning of this lesson's program would have been

```
#include <stdio.h>

int *function1(int, double);
int function2(int, double);
int (*function_ptr)(int, double);

void main(void)
{
        int *gg, hh;
```

Beginning programmers looking at the function prototypes often confuse the meaning of the function_ptr declaration. The best way for you to avoid being confused by this is to memorize that, when a data type is followed by left parenthesis and *, a pointer is being declared not a function.

The general form of declaring a pointer to a function is

type   (function_ptr) (type,   type,   type,   type);*

where *function_ptr* is the identifier used for the function pointer. The *types* must agree with the types used for any function whose address is to be assigned to function_ptr. For instance, in this lesson's program, function2 returns an int and has arguments of int and double. Because the address of function2 is assigned to function_ptr, function_ptr must show that it is returning an int and must have int and double as argument types.

2. What is meant by the address of a function? In Chapter 1, we observed that, in addition to integers, floating point numbers, and characters, we store instructions in memory. These instructions are the source code of our program translated by the compiler into binary form or object code. Each function we have written is stored in memory. The address of the beginning of the region of memory that holds a function is indicated by the address of the function.

3. How can we reference the address of a function and how can we call a function with it? In C, we reference the address of a function by using the name of the function without following parentheses. We can assign this address to a variable that has been declared to be a pointer to a function. For instance,

```
function_ptr=function2;
```

assigns the address of function2 to function_ptr. After making this assignment, we can call function2 by using function_ptr. For example,

```
hh=function_ptr(23,58.3);
```

calls function2 with the arguments, 23 and 58.3. The return value from function2 is assigned to hh.

In executing this statement, we transfer program control to the location indicated by the function pointer. This effectively calls the function.

4. What is the usefulness of a pointer to a function? One use is that we can use pointers to functions as arguments to other functions. An example of this is found in using the library function, qsort.

5. What is qsort? This function is C's standard sorting function, the quicksort routine described in Application Program 8.3. It has the prototype:

```
void qsort (void *buf, int num, int size, int (*compare) (const void *, const void *));
```

where compare is a pointer to a function, buf is a pointer to the array being sorted, num is the number of elements in the array, and size is the number of bytes of a single element.

By using a pointer to the function compare, qsort is able to call a function that we define. The function pointed to by compare (that is, the function we define) must meet the following requirements:

1.   It must have two arguments, both of type const void *.
2.   It must return an integer.

3. The return integer must be dependent upon the relative magnitudes of the two arguments. To sort a list in increasing order, the return integer must be
    greater than 0 if the first argument is greater than the second
    less than 0 if the first argument is less than the second
    equal to 0 if the first and second arguments are equal

The short program that follows uses qsort to sort the array list[6]. We have defined the function sortinc, which conforms to the three requirements just given.

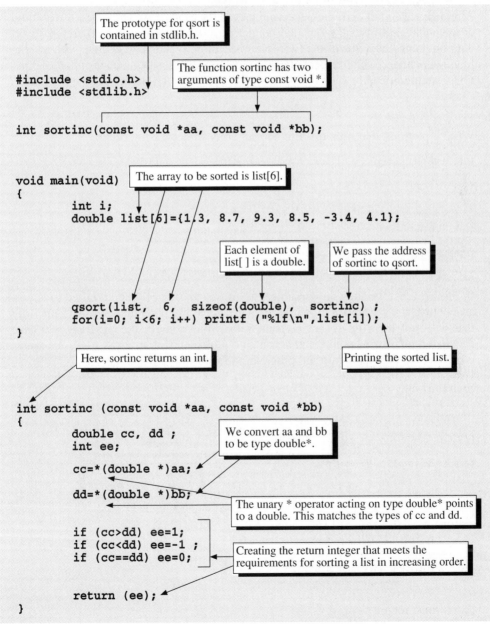

```c
                    The prototype for qsort is
                    contained in stdlib.h.

#include <stdio.h>         The function sortinc has two
#include <stdlib.h>        arguments of type const void *.


int sortinc(const void *aa, const void *bb);


void main(void)      The array to be sorted is list[6].
{
        int i;
        double list[6]={1.3, 8.7, 9.3, 8.5, -3.4, 4.1};

                              Each element of     We pass the address
                              list[ ] is a double. of sortinc to qsort.


        qsort(list, 6, sizeof(double), sortinc) ;
        for(i=0; i<6; i++) printf ("%lf\n",list[i]);
}
        Here, sortinc returns an int.                    Printing the sorted list.


int sortinc (const void *aa, const void *bb)
{
        double cc, dd ;       We convert aa and bb
        int ee;               to be type double*.

        cc=*(double *)aa;

        dd=*(double *)bb;
                        The unary * operator acting on type double* points
                        to a double. This matches the types of cc and dd.

        if (cc>dd) ee=1;
        if (cc<dd) ee=-1 ;
        if (cc==dd) ee=0;     Creating the return integer that meets the
                              requirements for sorting a list in increasing order.

        return (ee);
}
```

Output

```
-3.4
1.3
4.1
8.5
8.7
9.3
```

Although we do not here show the source code for qsort, if we could see it, we would observe that qsort uses the function pointer `compare` to call sortinc. Therefore, qsort does not know the name of the function it is calling. It is able to operate with just the function pointer. Using this technique of passing addresses of functions to other functions, we more readily can create function libraries for developing large programs.

EXERCISES

Write a program that uses qsort to sort an array in decreasing order.

LESSON 8.16 POINTERS TO ARRAYS

Topics
- Declaring pointers to arrays
- Accessing array elements using pointers to arrays

In this program we show how pointers to arrays can be declared and used to access array elements. To declare a pointer to an array, we use parentheses in the same way that we used them to declare a pointer to a function. In the program, we have declared a number of pointers and printed out elements of an array using the pointers. Read the annotations to the program and then the Explanation.

Source code

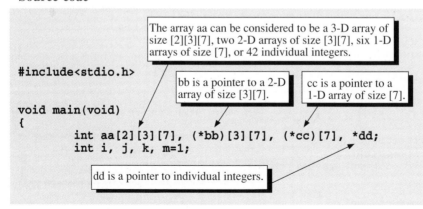

The array aa can be considered to be a 3-D array of size [2][3][7], two 2-D arrays of size [3][7], six 1-D arrays of size [7], or 42 individual integers.

bb is a pointer to a 2-D array of size [3][7].

cc is a pointer to a 1-D array of size [7].

```c
#include<stdio.h>

void main(void)
{
        int aa[2][3][7], (*bb)[3][7], (*cc)[7], *dd;
        int i, j, k, m=1;
```

dd is a pointer to individual integers.

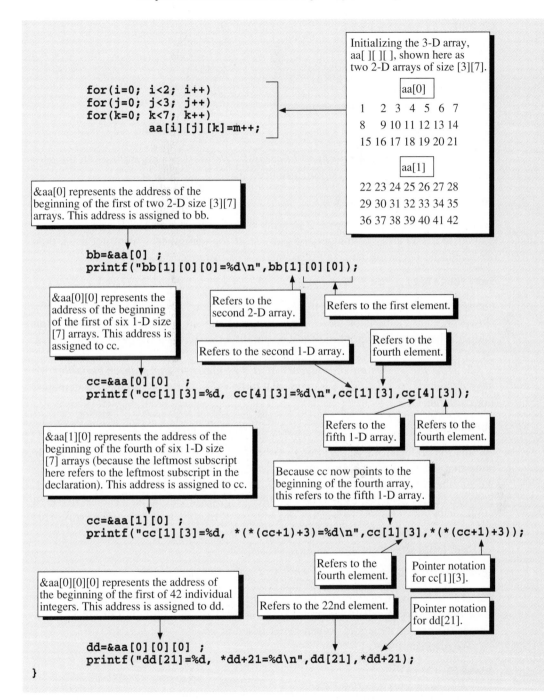

Initializing the 3-D array, aa[ ][ ][ ], shown here as two 2-D arrays of size [3][7].

aa[0]

```
 1   2  3  4  5  6  7
 8   9 10 11 12 13 14
15 16 17 18 19 20 21
```

aa[1]

```
22 23 24 25 26 27 28
29 30 31 32 33 34 35
36 37 38 39 40 41 42
```

```
for(i=0; i<2; i++)
  for(j=0; j<3; j++)
    for(k=0; k<7; k++)
      aa[i][j][k]=m++;
```

&aa[0] represents the address of the beginning of the first of two 2-D size [3][7] arrays. This address is assigned to bb.

```
bb=&aa[0] ;
printf("bb[1][0][0]=%d\n",bb[1][0][0]);
```

&aa[0][0] represents the address of the beginning of the first of six 1-D size [7] arrays. This address is assigned to cc.

Refers to the second 2-D array.

Refers to the first element.

Refers to the second 1-D array.

Refers to the fourth element.

```
cc=&aa[0][0]  ;
printf("cc[1][3]=%d, cc[4][3]=%d\n",cc[1][3],cc[4][3]);
```

&aa[1][0] represents the address of the beginning of the fourth of six 1-D size [7] arrays (because the leftmost subscript here refers to the leftmost subscript in the declaration). This address is assigned to cc.

Refers to the fifth 1-D array.

Refers to the fourth element.

Because cc now points to the beginning of the fourth array, this refers to the fifth 1-D array.

```
cc=&aa[1][0] ;
printf("cc[1][3]=%d, *(*(cc+1)+3)=%d\n",cc[1][3],*(*(cc+1)+3));
```

Refers to the fourth element.

Pointer notation for cc[1][3].

&aa[0][0][0] represents the address of the beginning of the first of 42 individual integers. This address is assigned to dd.

Refers to the 22nd element.

Pointer notation for dd[21].

```
dd=&aa[0][0][0] ;
printf("dd[21]=%d, *dd+21=%d\n",dd[21],*dd+21);
```

}

Output

```
bb[1][0][0]=22
cc[1][3]=11, cc[4][3]=32
cc[1][3]=32, *(*(cc+1)+3)=32
dd[21]=22, *dd+21=22
```

Explanation

1. How do we declare pointers to arrays? We must use parentheses in the declarations of pointers to arrays. For instance,

```
int (*bb)[3][7], (*cc)[7];
```

declares bb to be a pointer to a 2-D array of integers of size [3][7] and cc to be a pointer to a 1-D array of integers of size [7]. If the parentheses were not used, then we would be declaring an array of pointers. For instance,

```
int *bb[3][7], *cc[7];
```

declares bb to be a 2-D array of size [3][7] and cc to be a 1-D array of size [7]. However, each element of these arrays contains not an integer but a pointer to an integer. Therefore, it is important to use parentheses in declaring pointers to arrays.

2. How do we assign the address of a 3-D array to the pointer variables we have declared? It is critical for us to match the types in our assignment statements. For example,

```
bb=&aa[0];
```

assigns the address of the beginning of the aa array in this lesson's program to bb. Note that the number of subscripts used for aa is important. The factor determining the number of subscripts to be used lies in the way that bb and aa are declared. Because bb has been declared to be a pointer to a 2-D array and aa has been declared to be a 3-D array, &aa[0] represents the address of the beginning of a 2-D array. This is the correct type and so can be assigned to bb.

Similarly, we can assign

```
cc=&a[0][0];
```

Because cc has been declared to be a pointer to a 1-D array and aa has been declared to be a 3-D array, &aa[0][0] represents the address of the beginning of a 1-D array. This is the correct type and so can be assigned to cc.

We also can use

```
dd=&a[0][0][0];
```

Because dd has been declared to be a pointer to an integer and aa has been declared to be a 3-D array, &aa[0][0][0] represents the address of single integer. This is the correct type and so can be assigned to dd. An image of these assignments is shown in Fig. 8.46.

FIG. 8.46

The aa[ ][ ][ ] array and the way that bb, cc, and dd point to it. In this figure, we illustrate the 3-D array aa[ ][ ][ ] as a single line of memory cells.

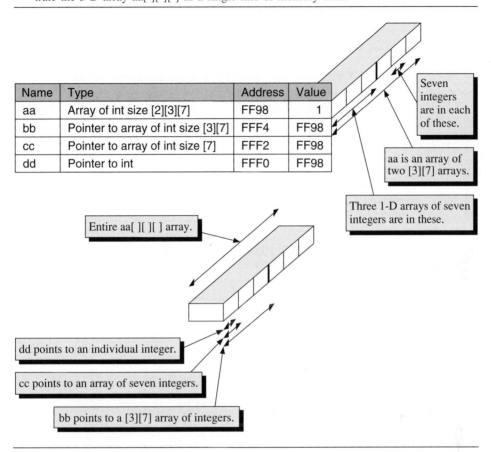

Name	Type	Address	Value
aa	Array of int size [2][3][7]	FF98	1
bb	Pointer to array of int size [3][7]	FFF4	FF98
cc	Pointer to array of int size [7]	FFF2	FF98
dd	Pointer to int	FFF0	FF98

Seven integers are in each of these.

aa is an array of two [3][7] arrays.

Three 1-D arrays of seven integers are in these.

Entire aa[ ][ ][ ] array.

dd points to an individual integer.

cc points to an array of seven integers.

bb points to a [3][7] array of integers.

3. How is pointer arithmetic performed on pointers to arrays? When we perform addition with a pointer to an array, we add the number of bytes represented by the pointer. For instance, assuming 2 bytes per integer, adding 1 to bb causes us to add $1 \times [3] \times [7] \times 2$ bytes = 42 bytes (because bb points to a [3][7] array of integers). Thus, bb + 1 takes us 42 bytes past bb. Similarly, because cc has been declared to be a pointer to an array of seven integers, cc + 1 takes us $1 \times [7] \times 2$ bytes = 14 bytes past cc.

4. How do we access the elements of the aa[ ][ ][ ] array using bb, cc, and dd? With bb we must use three subscripts. The first subscript indicates to which [3][7] integer array we are referring and the second and third subscripts access the element within that array. For cc we must use two subscripts. The first subscript indicates to which [7] integer array we are referring and the second subscript accesses the element within that array. For dd we must use one subscript, which indicates the element within the aa[ ][ ][ ] array. For example,

bb[1][0][0]   indicates the [0][0] element of the second 2-D array
cc[4][3]   indicates the [3] element of the fifth 1-D array
dd[21]   indicates the 22nd element of the array

5. Is there another way that we can reference the addresses of aa[ ][ ][ ] to make valid assignments to bb, cc, and dd? Yes, the following assignments can be used in place of the ones in the program:

```
bb = aa;
cc = aa[0];
cc = aa[1];
dd = aa[0][0];
```

These are permissible assignments because, as you may recall, the address of an array can be represented without using the & operator. However, the number of subscripts for aa in the assignments must match these to make sure that the types on the left and right sides of the assignment statements agree. Note that the number of subscripts for these assignment statements is one fewer than those used in the program with the & operator.

6. What are the implications of all of this for passing arrays to functions? This all means that we can pass arrays to functions in a number of different ways. For instance, we could have declared only aa in main and used the others in functions as shown below:

```
#include<stdio.h>
void function1  ( int (*bb)[3][7], int (*cc)[7], int *dd );


void main(void)
{
        int i, j, k, m=1;
        int aa[2][3][7];


        for(i=0; i<2; i++)
        for(j=0; j<3; j++)
        for(k=0; k<7; k++)
                aa[i][j][k]=m++;


        function1(aa, aa[0],  aa[0][0]);

}


void function1  ( int (*bb)[3][7], int (*cc)[7], int *dd )
{
```

The declarations in this function prototype match the declarations in main for this lesson's program.

This is the same array initialization as was used in main for this lesson's program.

We use a function call and prototype to make assignments similar to those in main of this lesson's program.

```
        printf("bb[1][0][0]=%d\n",bb[1][0][0]) ;
        printf("cc[1][3]=%d, cc[4][3]=%d\n",cc[1][3],cc[4][3]);
        printf("dd[21]=%d, *dd+21=%d\n",dd[21],*dd+21);
}
```

> These are the same printf statements used in main for this lesson's program.

Output

```
bb[1][0][0]=22
cc[1][3]=11, cc[4][3]=32
dd[21]=22,   *dd+21=22
```

Here, we see that the prototype for function1 uses the declarations we use for this lesson's program. Instead of using assignment statements, we use the function call and function prototype to effectively perform the same operations.

EXERCISES

Write a program that has a 4-D array. Use pointers to print the members of this array.

LESSON 8.17 POINTERS TO POINTERS

Topics

- Accessing array elements using pointers to pointers
- Method 1, using an array of pointers to help reference an array
- Method 2, using malloc to reserve memory for a copy of the array

We saw in the previous lesson that we can access arrays using pointers in a number of ways. In this lesson, we illustrate how to use pointers to pointers to access arrays.

In the program for this lesson, we declare and initialize the array aa[3][2]. We then illustrate how pointers to pointers can be used to access the aa[ ][ ] array and access a copy of the array stored in the heap using malloc. Read the annotations in the program and the Explanation to understand how to do this.

Source code

```
#include <stdio.h>
#include<alloc.h>
void main(void )
{
        int i, j, aa[3][2]={101,102,103,104,105,106};
        int *pa[3], **pb, **pc;
```

> aa is declared and initialized. It can be regarded as three rows and two columns.

> Array of [3] pointers.

> Both pb and pc are pointers to pointers.

> pa[i] is assigned the address of the beginning of each row of aa[ ].

```
        for (i=0; i<3; i++) pa[i]=aa[i];
```

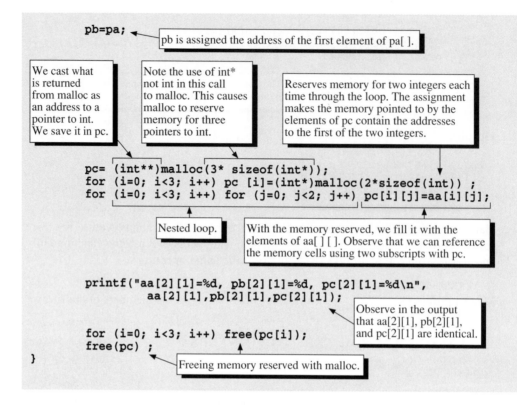

```
                pb=pa;  ◄──── pb is assigned the address of the first element of pa[ ].
```

We cast what is returned from malloc as an address to a pointer to int. We save it in pc.

Note the use of int* not int in this call to malloc. This causes malloc to reserve memory for three pointers to int.

Reserves memory for two integers each time through the loop. The assignment makes the memory pointed to by the elements of pc contain the addresses to the first of the two integers.

```
    pc= (int**)malloc(3* sizeof(int*));
    for (i=0; i<3; i++) pc [i]=(int*)malloc(2*sizeof(int)) ;
    for (i=0; i<3; i++) for (j=0; j<2; j++) pc[i][j]=aa[i][j];
```

Nested loop.

With the memory reserved, we fill it with the elements of aa[ ] [ ]. Observe that we can reference the memory cells using two subscripts with pc.

```
    printf("aa[2][1]=%d, pb[2][1]=%d, pc[2][1]=%d\n",
              aa[2][1],pb[2][1],pc[2][1]);
```

Observe in the output that aa[2][1], pb[2][1], and pc[2][1] are identical.

```
    for (i=0; i<3; i++) free(pc[i]);
    free(pc) ;
}
```

Freeing memory reserved with malloc.

Output

```
aa[2][1]=106, pb[2][1]=106, pc[2][1]=106
```

Explanation

1. How have we used a pointer to a pointer to access an array? We have used two methods. In the first method, we declare *pa[3], which is an array of pointers, and **pb, which is a pointer to a pointer. These are illustrated in the stack shown in Fig. 8.47. With the for loop

```
for (i=0; i<3; i++) pa[i]=aa[i];
```

the addresses of the beginning of each row of the 2-D array aa[ ][ ] are stored in the elements of pa[ ].

By storing the address of the first element of pa[ ] in pb with the statement

```
pb=pa;
```

we can use pb to access the elements of aa[ ][ ]. By following the arrows in Fig. 8.47, we see that the pointer notation *(*(pb + 2) + 1) accesses the last element of aa [ ][ ]. Because of the correspondence between array and pointer notation, we also

FIG. 8.47

Pointers to pointers being used for accessing array elements

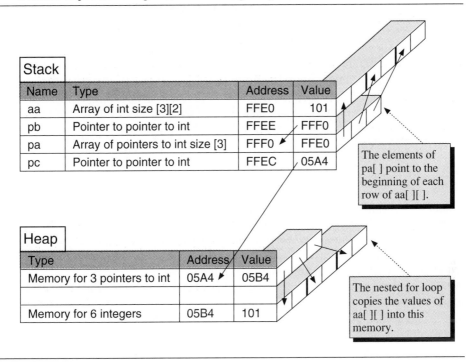

Stack

Name	Type	Address	Value
aa	Array of int size [3][2]	FFE0	101
pb	Pointer to pointer to int	FFEE	FFF0
pa	Array of pointers to int size [3]	FFF0	FFE0
pc	Pointer to pointer to int	FFEC	05A4

The elements of pa[ ] point to the beginning of each row of aa[ ][ ].

Heap

Type	Address	Value
Memory for 3 pointers to int	05A4	05B4
Memory for 6 integers	05B4	101

The nested for loop copies the values of aa[ ][ ] into this memory.

use pb[2][1] to perform the same task. The printf statement in this lesson's program uses the array notation.

The second method for using a pointer to a pointer to access array elements requires malloc to reserve memory in the heap that will be used to store a copy of the elements of aa[ ][ ]. First, we reserve memory for three pointers, using

```
pc= (int**)malloc(3* sizeof(int*));
```

Then, we fill that memory with pointers, using the loop

```
for (i=0; i<3; i++) pc [i]=(int*)malloc(2*sizeof(int)) ;
```

This above statement also reserves memory for integers. We fill this memory, using the loop

```
for (i=0; i<3; i++) for (j=0; j<2; j++) pc[i][j]=aa[i][j];
```

Executing these statements copies the elements of aa[ ][ ] into the heap. We can access these elements with pointer notation such as *(*(pc+2)+1) or array notation pc[2][1].

EXERCISES

Write a program that uses pointers to pointers to print the elements of an array of size [20][15].

▨ APPLICATION PROGRAM 8.1 UNDERGROUND POLLUTION PLUME BOUNDARIES—DETERMINING IF A POINT IS IN A POLYGON (NUMERICAL METHOD EXAMPLE)

Problem statement

Write a program that can determine whether a particular sampling point is located within an underground pollution plume. The plume of underground pollution is described by the *x-y* coordinates of 20 points on its boundary. Read the coordinates of the 20 points from a data file and the coordinates of the sample point from the keyboard. Write a message to the screen indicating whether the point is located in the plume.

Background

A region of underground pollution is called a *plume* because it is shaped like a plume of smoke. Figure 8.48 shows a map of a plume from a leaking storage tank. Engineers map plumes to understand their movement and behavior. Knowing exactly where a plume is allows engineers to make decisions regarding locations for sampling the soil or installing wells. In this example, we use an *x-y* coordinate system to establish points on the boundary of the plume.

In this process, it is necessary to determine whether a point is inside or outside the plume. This can be done by drawing a long horizontal line from the point considered and counting the number of times the line crosses the plume boundary. This is illustrated in Fig. 8.49. If the horizontal line crosses the boundary an even number of times, then the point is outside the plume. If the horizontal line crosses the boundary an odd number of times, then the point is inside the plume. The task with this program is to determine the number of times that the horizontal line crosses the plume boundary.

We represent the plume boundary with a series of straight line segments. Then we determine if the line segment represented by the horizontal dashed line in Fig.

▨ **FIG. 8.48**
Pollution plume from leaking storage tank. The tank is above ground. The plume is underground.

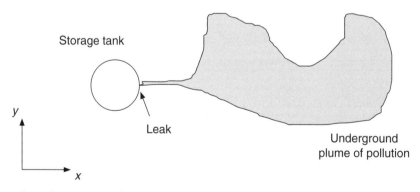

Coordinate system for mapping plume.

8.49 intersects each line segment of the plume boundary. To do this, we must consider the intersection of two line segments, a line segment representing a portion of the plume boundary and the horizontal dashed line segment of Fig. 8.49.

Solution

RELEVANT EQUATIONS

Determining the intersection point of two line segments is similar to determining the intersection point of two lines. The method that we show here consists of first finding the intersection point of the two lines. Then we determine if this point falls on the line segments. If it does, the line segments intersect. If it does not, the line segments do not intersect.

The slope and intercept of a line defined by a line segment are

$$m = (y_2 - y_1)/(x_2 - x_1) \tag{8.1}$$
$$b = y_2 - mx_2 \tag{8.2}$$

where the endpoints of the line segment are (x_1, y_1) and (x_2, y_2), m is the slope, and b is the intercept.

We covered the intersection of two lines in Application Program 4.1. The equations are

$$x = (b_2 - b_1)/(m_1 - m_2) \tag{8.3}$$
$$y = m_1 x + b_1 \tag{8.4}$$

FIG. 8.49
Pollution plume delineated by the points on the boundary

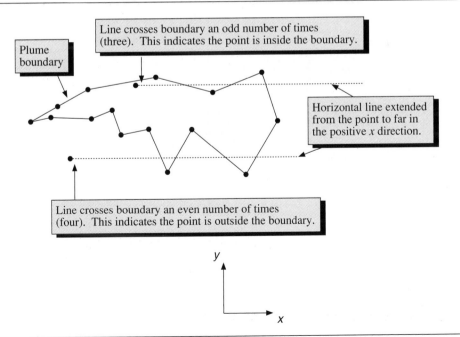

Line crosses boundary an odd number of times (three). This indicates the point is inside the boundary.

Plume boundary

Horizontal line extended from the point to far in the positive x direction.

Line crosses boundary an even number of times (four). This indicates the point is outside the boundary.

where x and y are the coordinates of the intersection point and m_1, m_2, b_1, and b_2 are the slopes and intercepts of the two lines.

To determine whether the intersection point (x, y) falls onto the line segments, we determine if it falls within the shaded region outlined by the line segments' endpoints. This is shown in Fig. 8.50. The shaded region is created by bracketing the x and y values of the endpoints.

FIG. 8.50
Criteria for determining whether the two line segments cross

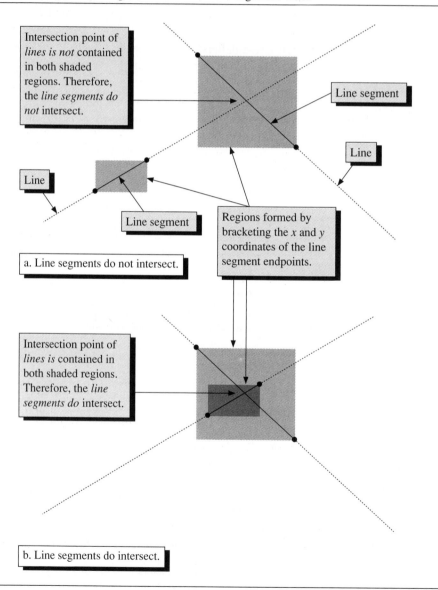

In this problem, we must be able to distinguish between the x and y coordinates of the two line segments. Therefore, we use $s1_{x1}$, $s1_{x2}$, $s1_{y1}$, and $s1_{y2}$ to indicate the x and y coordinates of the endpoints of the first line segment and $s2_{x1}$, $s2_{x2}$, $s2_{y1}$, and $s2_{y2}$ to indicate the x and y coordinates of the endpoints of the second line segment.

To use relational expressions to determine whether a point falls within the shaded region, we determine

$$
\begin{aligned}
x_{low1} &= \text{the lower of } s1_{x1} \text{ and } s1_{x2} = \text{smallest } x \text{ coordinate of line segment 1}\\
y_{low1} &= \text{the lower of } s1_{y1} \text{ and } s1_{y2} = \text{smallest } y \text{ coordinate of line segment 1}\\
x_{low2} &= \text{the lower of } s2_{x1} \text{ and } s2_{x2} = \text{smallest } x \text{ coordinate of line segment 2}\\
y_{low2} &= \text{the lower of } s2_{y1} \text{ and } s2_{y2} = \text{smallest } y \text{ coordinate of line segment 2} \quad (8.5)\\
x_{high1} &= \text{the greater of } s1_{x1} \text{ and } s1_{x2} = \text{greatest } x \text{ coordinate of line segment 1}\\
y_{high1} &= \text{the greater of } s1_{y1} \text{ and } s1_{y2} = \text{greatest } y \text{ coordinate of line segment 1}\\
x_{high2} &= \text{the greater of } s2_{x1} \text{ and } s2_{x2} = \text{greatest } x \text{ coordinate of line segment 2}\\
y_{high2} &= \text{the greater of } s2_{y1} \text{ and } s2_{y2} = \text{greatest } y \text{ coordinate of line segment 2}
\end{aligned}
$$

The line segments intersect (rather than just the lines intersect) when all of the following are true:

$$
\begin{aligned}
x_{low1} &<= x <= x_{high1}\\
x_{low2} &<= x <= x_{high2}\\
y_{low1} &<= y <= y_{high1} \quad\quad (8.6)\\
y_{low2} &<= y <= y_{high2}
\end{aligned}
$$

This is illustrated in Fig. 8.50 by the shaded regions. If the intersection point (x, y) falls within both shaded regions then these four relational expressions are true and the line segments intersect.

SPECIFIC EXAMPLE

We will determine whether the point (12, 27) falls within the plume defined by

x	y
0	0
3	5
4	16
7	35
10	22
19	44
20	55
25	31
24	4
10	3

Fig. 8.51 illustrates the plume created by these points.

We begin by creating a line segment using the first two points as endpoints:

$$
\begin{aligned}
s1_{x1} &= 0 \quad s1_{y1} = 0\\
s1_{x2} &= 3 \quad s1_{y2} = 5
\end{aligned}
$$

FIG. 8.51

Plume defined by endpoints of line segments. A long horizontal line is drawn from the point being considered.

x	y
0	0
3	5
4	16
7	35
10	22
19	44
20	55
25	31
24	4
10	3

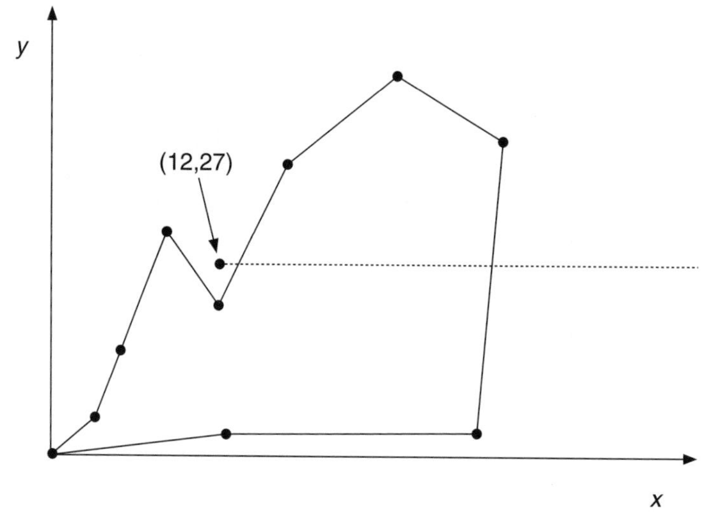

A second line segment is created with the points (12, 27) and (1.0E10, 27), which makes this a very long horizontal line segment, meant to intersect all possible plume boundaries. We call this line segment the *probe*. This is the dashed line in Fig. 8.51. With this line segment we have

$$s2_{x1} = 12 \qquad s2_{y1} = 27$$
$$s2_{x2} = 1.0E10 \quad s2_{y2} = 27$$

Using equations (8.1) to (8.4), we get

$$m_1 = (5 - 0)/(3 - 0) = 1.667$$
$$m_2 = (27 - 27)/(1.0E10 - 12) = 0.0$$
$$b_1 = 5 - 1.667(3) = 0.0$$
$$b_2 = 27 - 0.0(1.0E10) = 27$$
$$x = (27 - 0)/(1.667 - 0) = 16.2$$
$$y = 1.667(\ ) + 0.0 = 27$$

The intersection point of the two *lines* is (16.2, 27). To determine whether the two *line segments* intersect, we use the relational expressions of equations (8.5) and (8.6). By observation, we have

$$x_{low1} = 0$$
$$y_{low1} = 0$$
$$x_{low2} = 12$$

$$y_{low2} = 27$$
$$x_{high1} = 3$$
$$y_{high1} = 5$$
$$x_{high2} = 1.0E10$$
$$y_{high2} = 27$$

$$0 <= x <= 3$$
$$12 <= x <= 1.0E10$$
$$0 <= y <= 5$$
$$27 <= y <= 27$$

With $x = 16.2$ and $y = 27$, we see that the first and third of these relational expressions is false and therefore the intersection point is not contained on the line segments. We can also observe from Fig. 8.51 that the first line segment and the probe (dashed line) clearly do not intersect.

We continue by finding if each line segment on the plume boundary intersects with the horizontal dashed line. For each of the ten line segments, we have

Line segment	Intersection with horizontal dashed line?
1	No
2	No
3	No
4	No
5	Yes
6	No
7	No
8	Yes
9	No
10	No

The total number of crossings is two (line segments 5 and 8). Because this is an even number, the point is not located within the plume boundary.

DATA STRUCTURES

When working with line segments, it is common to create two structures, one for the points and one for the line segments. In this program, we use the following:

```
struct Point
      {
      double x;      The Point structure has both the x and y coordinates.
      double y;
      };


struct Line_seg
      {
      struct Point endpt1;      The Line_seg structure has two Point
      struct Point endpt2;      structures, one for each endpoint.
      double slope;
      double intcpt ;      Each line segment has a slope and
      };                   an intercept for the line it defines.
```

If we declare

struct Line_seg seg1;

then we reference the x and y coordinates of the first line segment, as shown in Fig. 8.52, using

seg1.endpt1.x
seg1.endpt1.y
seg1.endpt2.x
seg1.endpt2.y

Since we define the plume boundary with points, we create an array of type struct Point. For this program, we use 20 points to define the plume and so declare

struct Point bnd_pt[22];

This indicates more than 20 elements because we use the extra elements to make the polygon to wrap around itself. We use bnd_pt[1] to bnd_pt[20] to store all of the boundary points, then set bnd_pt[0] equal to bnd_pt[20] and bnd_pt[21] equal to bnd_pt[1]. Doing this eases our work with polygons because it reduces the likelihood that we will go past the end of the points defined.

ALGORITHM

An algorithm can be developed from the specific example. It consists of the following steps:

Read the coordinates of the point to be evaluated.
Read all of the points on the boundary of the plume.
Establish the endpoints of the probe (the long horizontal dashed line of Fig. 8.51).
Loop over all of the boundary points:
    Establishing a line segment from two boundary points
    Calling a function (which we call *cross*) to determine if the line segment and probe intersect
    Incrementing the counter if the line segment and probe intersect
Print the result.

FIG. 8.52

Referencing the x and y coordinates of line segment 1 using the defined structures

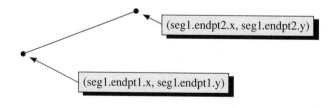

The function cross, for determining the intersection of the line segments, has the general algorithm:

Calculate the slopes and intercepts of the two line segments.
Find the intersection point of the two lines representing the two line segments.
Determine if the intersection point of the lines lies on the line segments.
Return 0 if the line segments do not intersect and return 1 if they do intersect.

Some of the details of cross are important, as they handle conditions that must be considered:

1. A line segment is vertical. This is a special condition because a vertical line has an infinite slope. Unless we make adjustments for this case, we will encounter a division by zero in applying equation (8.1).
2. The two line segments are parallel. This is a special condition because no inter-section point exists for the lines when line segments are parallel. Unless we account for this case, we will divide by zero in using equation (8.3).

To keep it reasonably simple, we have decided to not present a general function for determining whether two line segments intersect. In the function, we use the fact that the probe is horizontal. If the other line segment is vertical, the intersection point is at the *x* value of the line segment and the *y* value of the probe. Also, if the first line segment has a zero slope, the two line segments are parallel.

In Chapter 1 we did not go through all of the details, but many decimal numbers cannot be represented exactly using a limited number of bits of memory. For instance, the decimal 3.7 can be represented most closely with 8 bits as 11.111001. However, this representation translates to only 3.6953125 decimal. Therefore, with bits, it is common that we only approximately represent real decimal numbers. Because of this, when we use the double data type in a program, we may not be able to use $==$ in a comparison as we have with int.

For instance, it may not be sufficient to use

```
if (seg1.slope == 0.0 ) return(0);
```

in our function. Instead, we use

```
if (fabs(seg1.slope) <1.0E-15) return(0);
```

At this point you are an experienced programmer, so we do not go through the development of each step of the code. Read the details of the program in the annotations. Pay particular attention to the way we have used the dot operator with the structures.

SOURCE CODE

```
#define LARGE_NUM 1.0E10
#include<stdio.h>
#include<math.h>
struct Point
        {
        double x;
        double y;
        };


struct Line_seg
        {
        struct Point endpt1;
        struct Point endpt2;
        double slope;
        double intcpt;
        };


int cross(struct Line_seg seg1, struct Line_seg seg2);


void main(void)
{
        int i, check, numcross=0;

        struct Point sample, bnd_pt[22];


        struct Line_seg probe, bnd_line;
        FILE *infile;

        if((infile=fopen("bndary.dat","r"))==NULL) printf("Cannot open file\n");

        printf("Enter the x and y coordinates of the sample point.\n");
        scanf("%lf %lf", &sample.x, &sample.y);

        for (i=1; i<=20; i++) fscanf(infile,"%lf %lf",&bnd_pt[i].x, &bnd_pt[i].y);

        bnd_pt[0]=bnd_pt[20];
        bnd_pt[21]=bnd_pt[1];


        probe.endpt1=sample;
        probe.endpt2.x=LARGE_NUM;
        probe.endpt2.y=sample.y;
```

We define a large number for the the second *x* coordinate of the probe.

Point structure.

Line segment structure.

The function cross determines whether the two line segments (seg1 and seg2) intersect.

The purpose of the program is to determine if the point `sample` is within the polygon.

The plume boundary points. Because this is an array of the structure Point, it contains both the *x* and *y* coordinates for each point.

One line segment of the boundary.

The probe line segment.

Reading the coordinates of the sample point. Note the use of the dot operator.

Making the plume boundary points "wrap around."

Reading both the *x* and *y* coordinates of all the boundary points.

Making the left endpoint of the probe equal to the sample point. This statement makes probe.end pt1.x=sample.x and probe.end pt1.y=sample.y

Making the right endpoint of the probe a large horizontal distance from the left endpoint.

```
                                        Looping over all points.
           for (i=1; i<=20; i++)
                          {                              Creating a line segment from two
                          bnd_line.endpt1=bnd_pt[i];     adjacent points on the boundary.
                          bnd_line.endpt2=bnd_pt[i+1];
```

The function cross returns 1 if the boundary line segment (bnd_line) and the probe intersect.

```
                        if (cross(bnd_line,probe)==1) numcross++;
                        }
```

If the two line segments intersect, we increment numcross.

numcross%2 is 0 if numcross is even; it is 1 if numcross is odd.

If numcross is even, the point is outside the plume.

```
           check=numcross%2;
           if (check==0) printf("The sample point is outside the plume.\n");
           if (check==1) printf("The sample point is inside the plume.\n");
}
```

If numcross is odd, the point is inside the plume.

```
int cross (struct Line_seg seg1, struct Line_seg seg2)
{
           double x, y, x_low_1, x_high_1, x_low_2, x_high_2;
           double y_low_1, y_high_1, y_low_2, y_high_2;
```

From equation (8.5)

This is true if seg1 is a vertical line.

```
           if (fabs(seg1.endpt2.x-seg1.endpt1.x)<1.0E-15)
                          {
                          x = seg1.endpt1.x;
                          y = seg2.endpt1.y;
                          }
```

When seg1 is a vertical line, the lines represented by the probe and seg1 intersect at seg1's value of x and the probe's (seg2's) value of y.

Slope of the boundary line segment using equation (8.1).

```
           else
                          {
                          seg1.slope=(seg1.endpt2.y-seg1.endpt1.y)/(seg1.endpt2.x-seg1.endpt1.x);
                          seg2.slope=0.0;
```

The probe has a slope of 0.

```
                          seg1.intcpt=seg1.endpt2.y-seg1.slope*seg1.endpt2.x;
                          seg2.intcpt=seg2.endpt2.y;
```

Intercept of the first line segment.

The probe is a horizontal line so it has an intercept equal to the y value of its endpoints.

If the boundary line segment is horizontal, the probe and line segment do not intersect.

```
                          if(fabs(seg1.slope)<1.0E-15) return(0);
                          x=(seg2.intcpt-seg1.intcpt)(seg1.slope-seg2.slope);
                          y=seg1.slope*x + seg1.intcpt;
                          }
```

Equations (8.3) and (8.4).

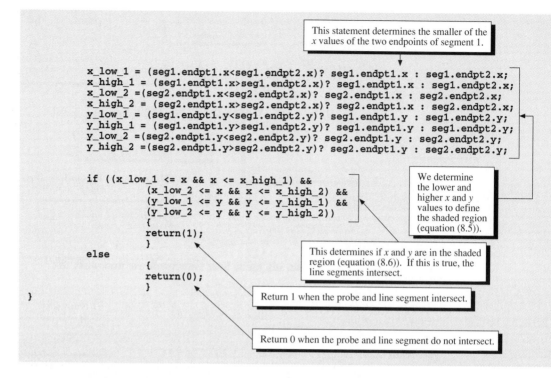

This statement determines the smaller of the x values of the two endpoints of segment 1.

```
x_low_1 = (seg1.endpt1.x<seg1.endpt2.x)? seg1.endpt1.x : seg1.endpt2.x;
x_high_1 = (seg1.endpt1.x>seg1.endpt2.x)? seg1.endpt1.x : seg1.endpt2.x;
x_low_2 =(seg2.endpt1.x<seg2.endpt2.x)? seg2.endpt1.x : seg2.endpt2.x;
x_high_2 = (seg2.endpt1.x>seg2.endpt2.x)? seg2.endpt1.x : seg2.endpt2.x;
y_low_1 = (seg1.endpt1.y<seg1.endpt2.y)? seg1.endpt1.y : seg1.endpt2.y;
y_high_1 = (seg1.endpt1.y>seg1.endpt2.y)? seg1.endpt1.y : seg1.endpt2.y;
y_low_2 =(seg2.endpt1.y<seg2.endpt2.y)? seg2.endpt1.y : seg2.endpt2.y;
y_high_2 =(seg2.endpt1.y>seg2.endpt2.y)? seg2.endpt1.y : seg2.endpt2.y;

    if ((x_low_1 <= x && x <= x_high_1) &&
         (x_low_2 <= x && x <= x_high_2) &&
         (y_low_1 <= y && y <= y_high_1) &&
         (y_low_2 <= y && y <= y_high_2))
         {
         return(1);
         }
    else
         {
         return(0);
         }
}
```

We determine the lower and higher x and y values to define the shaded region (equation (8.5)).

This determines if x and y are in the shaded region (equation (8.6)). If this is true, the line segments intersect.

Return 1 when the probe and line segment intersect.

Return 0 when the probe and line segment do not intersect.

OUTPUT

Keyboard input
```
                   Enter the x and y coordinates of the sample point.
                   12 27
                   The sample point is outside the plume.
```

COMMENTS

Note that we have used the ?: operator to determine the lower and higher of two points. This helps us determine the shaded region. The ?: operator is very convenient for comparing two values and selecting just one of them. Also, this is not the most elegant way of determining whether a point is in a polygon. However, the concept behind it is simple.

Modification exercises

1. Modify the program to make the cross function a general function for determining whether or not any two line segments intersect. In other words, use a probe that is not horizontal and long.

2. In this program, we have the probe as a horizontal line going far in the positive x direction. However, the probe can go in any direction. Modify the program so that the probe goes from the sample point to a point far in both the positive x and y directions.

3. This program is designed to handle 20 points for defining the plume boundary. Modify the program so that it can use 100 points to define the plume boundary.

4. Modify the program so that it can handle a variable number of points on the plume boundary.

5. Modify the program so that it can read ten sample points and determine which are contained in the plume boundary.

▮ APPLICATION PROGRAM 8.2 RECURSIVE FUNCTION— CALCULATION OF PI (NUMERICAL METHOD EXAMPLE)

Problem statement

Use a recursive function to estimate the value of π. No input is needed to the program. The output should be the value of π printed to the screen.

Solution

RELEVANT EQUATIONS

The French mathematician Vieta (1540–1603) developed the following equation to calculate the value of π:

$$\frac{2\sqrt{2}}{\pi} = \left[\sqrt{\left(\frac{1}{2} + \frac{1}{2}\sqrt{\frac{1}{2}}\right)}\right]\left[\sqrt{\frac{1}{2} + \frac{1}{2}\sqrt{\left(\frac{1}{2} + \frac{1}{2}\sqrt{\frac{1}{2}}\right)}}\right]\left[\sqrt{\frac{1}{2} + \frac{1}{2}\sqrt{\frac{1}{2} + \frac{1}{2}\sqrt{\left(\frac{1}{2} + \frac{1}{2}\sqrt{\frac{1}{2}}\right)}}}\right] [\text{etc.}]$$

This equation can be used in a recursive function to calculate π to a large number of decimal places. There are an infinite number of terms on the right hand side of the equation. Observe the pattern of the terms.

To successfully develop a recursive function, it is necessary to identify the base case. The base case becomes the one evaluated at the reversal and forms the False block of the if control structure in the recursive function.

From the equation, one can see that the base case (or simplest case) is represented by the portion of the equation

$$\sqrt{\left(\frac{1}{2} + \frac{1}{2}\sqrt{\frac{1}{2}}\right)}$$

With each succeeding term of the equation, the base case gets more deeply embedded within the equation structure.

In our representation, we show the first three terms of the expression. In our hand calculation, we will calculate the first four terms. In our computer calculation, we will take the first 30 terms.

SPECIFIC CALCULATION

The hand calculation will be done by first calculating an individual term and then getting an accumulation by taking the product of the individual terms previously calculated. If we work from left to right in the equation, we get the following.

First term,

$$\sqrt{\left(\frac{1}{2} + \frac{1}{2}\sqrt{\frac{1}{2}}\right)} = 0.923879532$$

Accumulated value $= 0.923879532$

Second term,

$$\sqrt{\left(\frac{1}{2} + \frac{1}{2}(0.923879532)\right)} = 0.98078528$$

Accumulated value $= 0.98078528 \times 0.923879532 = 0.906127445$

Third term,

$$\sqrt{\left(\frac{1}{2} + \frac{1}{2}(0.98078528)\right)} = 0.995184726$$

Accumulated value $= 0.995184726 \times 0.906127445 = 0.901764193$

Fourth term,

$$\sqrt{\left(\frac{1}{2} + \frac{1}{2}(0.995184726)\right)} = 0.998795456$$

Accumulated value $= 0.998795456 \times 0.901764193 = 0.900677978$

This is all of the terms we will carry; therefore,

$$\frac{2\sqrt{2}}{\pi} \approx 0.913612265$$

Solving for π we get

$$\pi \approx 3.140331165$$

Recall that π is roughly 3.14159, making our calculated value not a very good approximation. We would need more terms to get a better result.

ALGORITHM

A major difference between the recursive function that we write here and the recursive function that we wrote earlier is that, with each recursive call, we need to pass two values, not just one, on the return. These are

1. The value of the single term
2. The accumulated value

Since we need to return more than one value, we need to use the parameter list to transfer the information. We could return both values through the parameter list. However, for illustration purposes, we choose to use a return statement to return the value of the single term and the parameter list to return the accumulated value.

Keeping this in mind, we can write an algorithm for the recursive function. Our test expression will be the value of n being less than or equal to 30 (to carry 30 terms).

The algorithm for recursive pi function is

```
increment n
if n <=30
        call recursive function
        calculate single term
        calculate accumulated value
        return single term, pass accumulated value
```

```
else
        calculate base case
        create accumulated value
        return base case, pass accumulated value
```

After repeated calls to the recursive function, the accumulated value can be passed to main where the final step of dividing the accumulated value into $2(2)^{0.5}$ can give the value of π (see Fig. 8.53).

SOURCE CODE
This source code has been written from the algorithm and the specific calculation.

> Through the parameter list, *tot is returned.

> A double value is returned.

> Prototype for recursive function.

```
#include <stdio.h>
#include <math.h>


double recurs_pi (int n, double *tot);
```

> Calling the recursive function. The variable *n* is passed, and accum is returned.

```
void main (void )
{
        int n;
        double a=0.5, b=2.0, pi, two_over_pi, accum;
        n=1;
        recurs_pi(n,&accum);
        two_over_pi = sqrt (a) * accum;
        pi = b/two_over_pi;
        printf ("The value of pi is approximately %1.251f \n", pi);
}
```

> Pi is calculated from accum.

```
double recurs_pi (int n, double *tot)
{
        double z=0.5, accum, single_term;
        n++;
        if (n <= 30)
                {
                single_term = sqrt (z + z * recurs_pi(n, &accum));
                *tot = accum * single_term;
                return (single_term);
                }
        else
                {
                single_term = sqrt(z + z * sqrt(z));
                *tot = single_term;
                return (single_term);
                }
}
```

> Recursive function call. The return value represents the base case or simplest case.

> At the reversal, the value of *tot goes into accum, and single_term is returned to the location shown.

FIG. 8.53
Flow of recurs_pi function for four calls

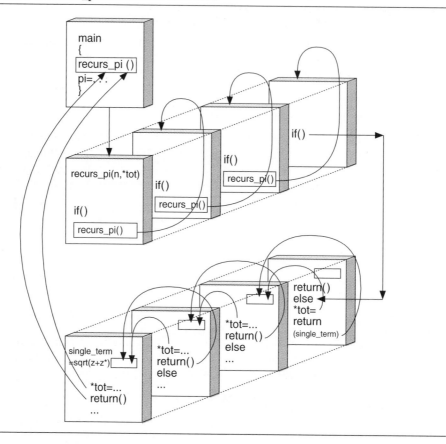

OUTPUT

```
The value of pi is approximately 3.141592653897931200000000
```

COMMENTS

Passing values. The feature that we have not yet seen for recursive functions is using the parameter list to pass values. It becomes a little confusing, but remember the fundamentals of using the parameter list. Remember,

1. In the function call, the address of operator (&) should be used in front of the variable value to be returned.
2. In the function prototype and header, the * should be used on variables corresponding with those having & in the function call (the * indicating that these variables are pointer variables).
3. In the recursive function itself, variables to be returned are pointer variables and all calculations with them should have the unary * operator.

Look at the function recurs_pi in the source code. As indicated by the three rules, in the header for the function, *tot is used as a variable to be returned. Also, *tot is used in calculations in the function body. However, in the recursive function call in the True block of the function, &accum is used. This fits with rule 1. Note that, immediately on return from the call, the value that is in accum is the value that was in *tot in the previous step, because of the correspondence between accum in the function call and *tot in the function declarator. Remember your & and * operators. Essentially, through this entire process, *tot and accum are exactly the same thing. When we put something into *tot we are putting it into accum.

This is illustrated in Fig. 8.54. The figure shows a different table of values for each function call (as is created with recursive functions). The *tot operation is shown. It illustrates that the value of *tot is the same as accum on the previous call. Not all of the variable values and addresses are shown in this figure for the purpose of clarity. The case illustrated is for only four recursive calls, like the hand calculation.

Debugging. You also need to debug your recursive functions. To do this, note that you can evaluate the performance of a recursive function in a manner similar to that of evaluating the performance of a loop structure. By making a table of the values and how they change with each recursive call, you can trace the operations of the function.

For debugging, set a small *n*. That way you can do a hand calculation of all of the numbers. You then can compare your hand calculation to the values your program gives on each step. Use either printf statements or the compiler's debugger to get the values on each step.

In tracing the flow of the program, we first realize that, during the calling phase, the only calculation done is the incrementation of *n*. So, we do not need to trace any values during the calling phase, we can start at the reversal. This puts us into the False block of the if control structure. If you are using printf statements to trace the flow of the program, put one into the False block.

The returning phase begins and control shifts to the point of the function call in the True block of the if control structure. With the value returned, the expression containing the function call can be evaluated. The single term and total are the same and easily calculated by hand. The assignment statement gives a new value for the single term. This also can be calculated by hand. The following table results:

Return step from reversal	Value of single term	Accumulated value
Reversal ($n = 4$)	0.923879532	0.923879532
1	0.980785280	0.906127445
2	0.995184726	0.901764193
3	0.998795456	0.900677978

Since this program works correctly, it need not be debugged. But the procedure just described can be used to debug similar programs.

Developing your own programs with recursive functions. You will find that developing your own recursive functions is not simple. We recommend that you take the following steps to get going:

1. Recognize the base case or simplest case.
2. Do a simple hand calculation, working backward from the base case. As you do the hand calculation, recognize how many values you must return from each call.
3. If you need to return more than one value from each call, you must use the parameter list to transfer values.
4. Start writing the recursive function with the False block in the if control structure. Put the base case into the False block.
5. Decide on your test condition. Put in the test condition and a statement that changes the test condition for each function call. Make sure that the condition is tested and changed for each function call. This is important. Without it, your recursive functions could go on forever.
6. Write the True block of the if control structure. In it should be a call to the function, the recursive call. You can get this part correct if you think about the transfer of information (what is returned) during the return phase of the recursive process. Begin your logic process by thinking about the reversal. In other words, ask yourself, "What value is returned from the function at the reversal?" That is the base case value. That value will be returned to the location of the recursive call. Realizing this, write your True block around that value (also use any other values that the recursive function returns).
7. As you are beginning, use just a few recursive calls. You can debug the program by making a table of the values as shown in Fig. 8.54.

Following these steps will get you on your way to writing a large number of programs with recursive functions.

Modification exercises

1. Rewrite the program using an iterative structure (loop) rather than a recursive structure.

2. Modify the program to include 50 terms instead of 30.

■ APPLICATION PROGRAM 8.3 SORTING—QUICKSORT ALGORITHM

Problem statement

Write a program that can sort a list of integers using the quicksort algorithm.

Solution

The quicksort algorithm (C.A.R. Hoare, "QuickSort," *Computer Journal,* Vol. 5, 1962, pp. 10–15) performs substantially better than the bubble sort algorithm and

FIG. 8.54

The operation of *tot putting values into accum on the previous function call. Note that there is a table of values for each function call because a different region of memory is reserved with each call. Recursion causes passing of values from one table to the other. Action shown with arrows is *tot.

Variable name	Variable type	Variable address		Variable value	
n	int			1	*Variables*
accum	double	FFE0		0.901764193	*in memory region*
single_term	double		*		*for first call to*
tot	address to double	FFD0			*recurs_pi*
n	int			2	*Variables*
accum	double	FFC0		0.906127445	*in memory region*
single_term	double		*		*for second call to*
tot	address to double	FFB0		FFE0	*recurs_pi*
n	int			3	*Variables*
accum	double	FFA0		0.923879532	*in memory region*
single_term	double		*		*for third call to*
tot	address to double	FAD0		FFC0	*recurs_pi*
n	int			4	*Variables*
accum	double	FAB0			*in memory region*
single_term	double				*for fourth call to*
tot	address to double	FA90		FFA0	*recurs_pi*

has been found to be an efficient sorting method for a number of different types of sorting problems. It is used in commercial software.

Quicksort uses a partitioning approach to sorting. It breaks a list into two parts (left and right), where the left part is made to contain only values less than a certain value in the list and the right part is made to contain only values greater than that value. Between the two parts, the value is inserted. When quicksort has accomplished partitioning, the value is in the correct sorted location (so it does not need to be worked on further) and it has created two separate lists, which if sorted individually without regard to the other, would result in a completely sorted single list. Quicksort then works with the two separate lists, repeating the process until all of the individual values are in their proper locations.

SPECIFIC EXAMPLE

We begin with an unordered list of numbers, which is shown in boxes. First, a value from the list is chosen, called the *pivot value*. Our algorithm uses the rightmost value (29) as the pivot to get things going. With this value we begin from both the left and right. Our goal with this pass is to get the number 29 in its proper position. From the right we look for values less than 29 (because any values less than 29 on this portion of the list are out of place) and from the left we look for values greater than 29 (because any values greater than 29 on this portion of the list are out of place). When we encounter an out-of-place value on both sides, we swap them.

From the left, we hit 98 and from the right we hit 18.

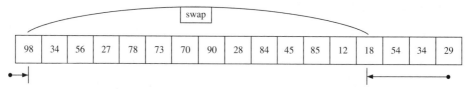

We now swap these two values and proceed from them, giving

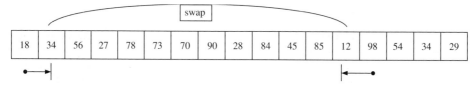

We hit 34 and 12. We swap them and proceed, giving

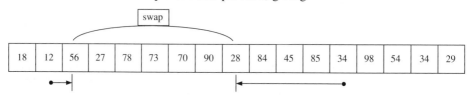

We hit 56 and 28. We swap them and proceed, giving

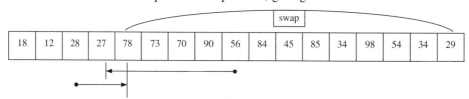

We hit 78 and 27. At this point our left position has gone past our right position making the arrows overlap. This indicates that we stop and swap our pivot (29) with the left position value (78), giving

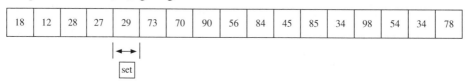

At this point, our pivot value, 29, is in its proper location with all values to the left of it less than 29 and all values to the right of it greater than 29. As a result, we need not touch the location with 29 in it any more. In addition, we have created two new lists: to the left of 29 and to the right of 29. If we sort these two lists independently, we end up with a single sorted list.

In our algorithm, when given a choice, we will work with the left list first; therefore, we now operate on the left list. We set the pivot value as 27 (the rightmost value in the list) and look for numbers less than 27 and greater than 27 as we proceed from the right and left, respectively. The arrow from the left goes to 28 and the arrow from the right goes to 12.

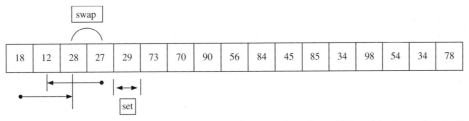

Since the two arrows overlap, we stop and swap the pivot (27) with the value indicated by the left arrow (28), giving

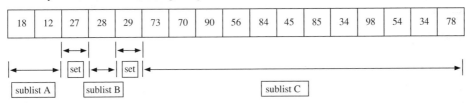

At this point, we have the values 27 and 29 in their proper locations and have created three sublists. Since we always choose the left list, we work on sublist A. Without going into detail, we switch the 18 and 12, putting both in their proper locations. Then we work on sublist B. We check to see if the starting locations for the right and left are the same (meaning only one value is present in the sublist) then we do nothing. Then sublist C is addressed using a pivot of 78.

In abbreviated form, we show these steps in Fig. 8.55:

1. Find the next values >78 and <78 (being 90 and 34).
2. Swap 90 and 34.
3. Find the next values >78 and <78 (being 84 and 54).
4. Swap 84 and 54.
5. Find the next values >78 and <78 (being 85 and 34).
6. Swap 85 and 34.
7. Find the next values >78 and <78 (being 85 and 34 again).
8. Since the arrows have crossed, stop, and swap the location indicated by the left arrow (85) with the far right value (78).
9. The result of this pass, indicated by the lowest numbers in each column in Fig. 8.55, follows:

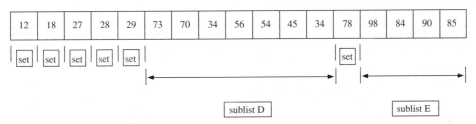

FIG. 8.55
The steps in quicksort, right list

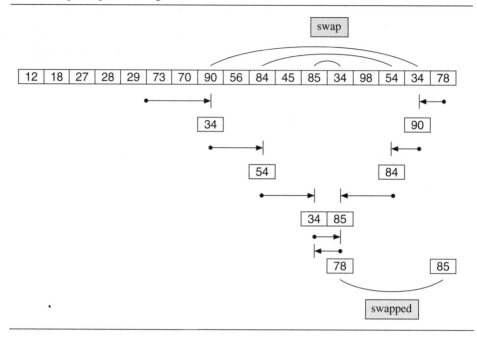

At this point, the value 78 is in its proper location and two sublists are created. Since we always take the left list, we would work with sublist D next. We will not go through the rest of the procedure in detail; however, we encourage you to follow the steps and complete the sort by hand. A schematic illustration of the operations of a list is shown in Fig. 8.56.

ALGORITHM
From the hand example and Fig. 8.56, we can see that the overall procedure is the following:

1. Establish the starting points for the left and right arrows.
2. Set the pivot equal to the value at the right starting point.
3. Move from the left to find a value greater than the pivot.
4. Move from the right to find a value less than the pivot.
5. If the arrows do not overlap, swap the values in 3 and 4 and repeat steps 3 through 5.
6. If the arrows overlap, stop and swap the pivot and the value indicated by the left arrow.
7. At this point the pivot value has been put into its proper location and two new sublists created. Work on the left sublist created (using steps 1–6) and create a new left and right sublist. Work with the new left sublist. Repeat this until the last left

■ **FIG. 8.56**

Pattern of operation of quicksort in a list. Read this diagram from the top line, then follow down each succeeding line. Each time a list is broken into sublists, the left sublist is first to be operated on. This continues until no left sublists are unsorted. Then, the right sublist is evaluated. You can see from this diagram that first the low numbers are sorted (roughly) and as the sort proceeds the higher numbers are sorted. The last portion of the list to be sorted is the far right end.

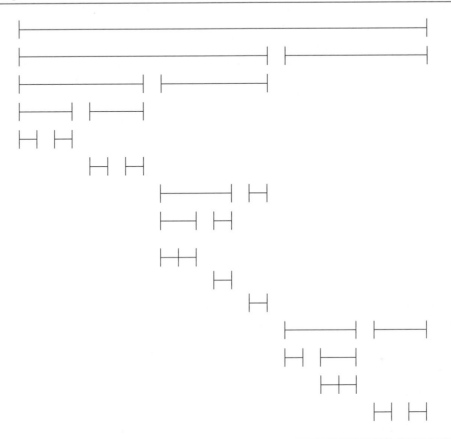

sublist has only three values that have been sorted. Then work with the last right sublist created. In other words, do not work on a right sublist until the last trailing left sublist has been sorted.

One other step should be done before all of this: Check that the right starting point indeed is to the right of the left starting point. If it is not, then the sublist has already been sorted and none of the steps need be done.

SOURCE CODE

We develop the source code step by step from the algorithm. We call the list array a[ ]. The variable i is the index value that moves from the left, and j is the index that moves from the right.

To move from the left to a value greater than the pivot, we can use a while loop. For instance, the loop

```
while (a[++i] < pivot);
```

causes the value of the index i to be incremented continuously until a value in the array a is found to be greater than or equal to the pivot. At the end of executing this loop, the value of i is equal to the subscript of the first value from the left that is greater than or equal to the pivot. This represents the location of the arrowhead of the left arrow.

Similarly, the loop

```
while (a[--j] > pivot);
```

causes the value of the index j to be decremented continuously until a value in the array a is found to be less than or equal to the pivot. At the end of executing this loop, the value of j is equal to the subscript of the first value from the right that is less than or equal to the pivot. This represents the location of the arrowhead of the right arrow.

Note that, at the end of this, if the value of i is greater than or equal to the value of j then the arrowheads overlap and we should not swap any values. The statement

```
if (i >= j) break;
```

takes us out of any loop that does the swapping.

To swap the values of a[i] and a[j], we use the following code (with the variable swap as a variable used for temporary storage of a value so the swap can be completed):

```
swap = a[i];
a[i] = a[j];
a[j] = swap;
```

We have not illustrated this particular sequence of instructions previously; however, the situation of swapping values comes up frequently in programming. Since two actions cannot be done simultaneously, an intermediate storage location (here represented by the variable swap) must be used. The action of these three statements (using a[i] = 90 and a[j] = 34 as statements executed before the three previously listed statements) is

a[i]	90
a[j]	34
swap	90

Action of　`swap = a[i];`

a[i]	34
a[j]	34
swap	90

Action of　`a[i] = a[j];`

a[i]	34
a[j]	90
swap	90

Action of   `a[j] = swap;`

The result is that the values of a[i] and a[j] have been swapped.

We can put all these statements together within a loop to form one complete pass through a sublist:

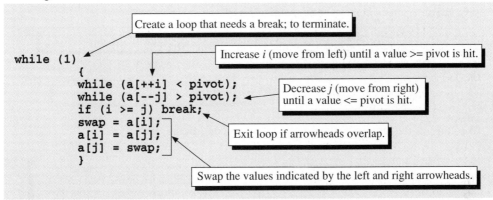

The loop created by the line while(1) can be an infinite loop, since the value 1 always is True. However, the statement if(i >= j) break; causes the loop to terminate. Be careful when you write this type of loop. If *i* never becomes greater than or equal to *j,* the loop will not terminate and your computer (if you are running it on a PC) will appear to have locked up.

If the initial indices of the tails of the left and right arrows are the variables left_position and right_position, we can swap the pivot (a[right_position]) and the value at the left arrowhead (a[i]) with the following sequence of statements:

```
swap = a[i];
a[i] = a[right_position];
a[right_position] = swap;
```

Adding this to the previous code gives the following:

```
while (1)
        {
        while (a[++i] < pivot);
        while (a[--j] > pivot);
        if (i >= j) break;
        swap = a[i];
        a[i] = a[j];
        a[j] = swap;
        }
swap = a[i];
a[i] = a[right_position];
a[right_position] = swap;
```

Loop to move arrowheads from left and right and swap the values at their arrowheads. Exit the loop when the arrowheads overlap.

On exiting the loop, swap the values at the left arrowhead and the far right position.

Before we get this process going, we must check that the left position is not to the right of the right position and that we initialize the values of *i* and *j* and the pivot. These statements would be

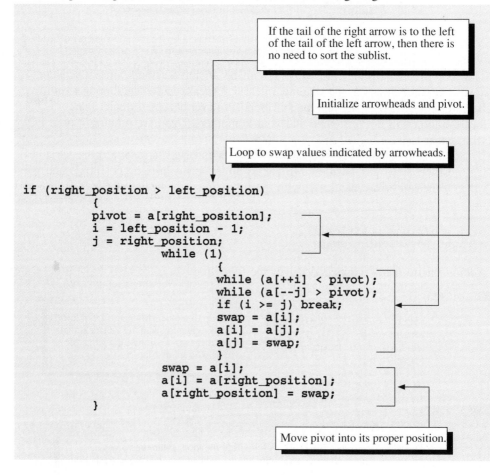

If the tail of the right arrow is to the right of the tail of the left arrow, set the pivot to be the tail of the right arrow.

```
if (right_position > left_position)
{
pivot = a[right_position];
i = left_position - 1;
j = right_position;
}
```

Initialize *i* to be the tail of the left arrow − 1.

Initialize *j* to be the tail of the right arrow.

We can put our previous statements into this conditional, giving

If the tail of the right arrow is to the left of the tail of the left arrow, then there is no need to sort the sublist.

Initialize arrowheads and pivot.

Loop to swap values indicated by arrowheads.

```
if (right_position > left_position)
        {
        pivot = a[right_position];
        i = left_position - 1;
        j = right_position;
                while (1)
                        {
                        while (a[++i] < pivot);
                        while (a[--j] > pivot);
                        if (i >= j) break;
                        swap = a[i];
                        a[i] = a[j];
                        a[j] = swap;
                        }
                swap = a[i];
                a[i] = a[right_position];
                a[right_position] = swap;
        }
```

Move pivot into its proper position.

These statements form one complete pass. At the end of this pass, we have the pivot value in its proper location, a left sublist, and a right sublist. We now need to repeat the procedure for the left and right sublists. This can be done in different ways, but it lends itself to using recursion. Because we need to repeat the procedure for both the left and right sublists, we need two recursive calls instead of just one, as done previously.

As we had with the single recursive call, we have to consider both the calling and returning phases. However, with two recursive calls, the situation is not as straightforward. Figure 8.57 illustrates the flow of a function with two recursive calls (with the reversal for each call occurring after just a couple calls). This is an important figure. Follow the numbers in it and reason out why a function with two recursive calls would operate in this manner. Note that each function, unless it has hit the reversal (meaning that an if control structure has caused bypassing both calls), calls itself twice.

For our function, we first put in the recursive calls and then follow the flow of the program. The source code to call the function qksort follows. Note the two recursive calls in this function. The first call works on the left sublist and the second call works on the right sublist.

> Parameters are the array, and the locations of the tails of the left and right arrows.

```
int qksort(int a[ ], int left_position, int right_position)◄

{
        int pivot, i, j, swap;


        PREVIOUS CODE
```

> Call to qksort to sort left sublist.

```
        qksort(a, left_position, i-1);  ◄
        qksort(a, i+1, right_position); ◄
        }
}
```

> Call to qksort to sort right sublist.

The flow of this function is illustrated in Fig. 8.58 for 39 recursive calls. Each square in the figure is meant to represent the function. The small squares within the function square are meant to represent the two recursive calls. The code prior to the recursive calls performs one complete pass of a sublist. Therefore, each call to the function sets one value in its proper position and creates two new sublists (except at the reversal). Since the first recursive call works with the left sublist and the second recursive call works with the right sublist, the squares from those calls in this figure have been positioned to the left and right of their calling function.

The flow is quite complicated. In the 6, 7, 8, and 9 squares, the function call for evaluating the left part of the list is performed repeatedly (from 6 to 7 to 8 to 9), then, from 9, the left part of the list is finished (since right_position > left_position). On returning to 8, the function is called to evaluate the right sublist and returned. If you compare this with the hand calculation we did, you will find that this figure is

FIG. 8.57

The order of operation for a function that has two recursive calls for the case where there are a small number of calls. Trace the numbered lines to understand how the program flows.

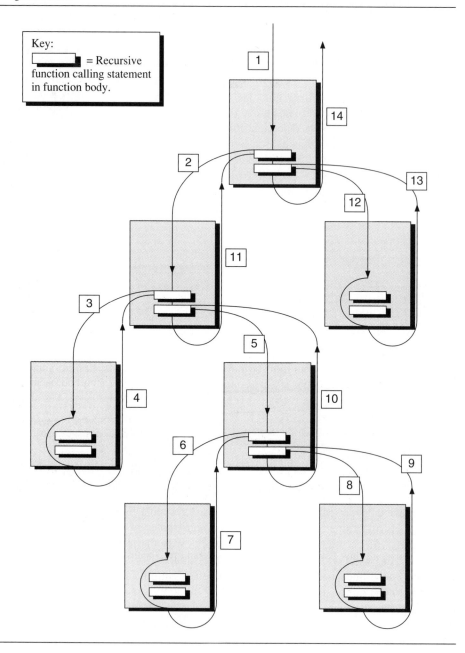

FIG. 8.58

Flow of the qksort program for 39 recursive calls. Compare this to Fig. 8.56. Note that, in both figures, the operations are first performed on the left (all of the left sublists are first sorted), then the right sublists are sorted.

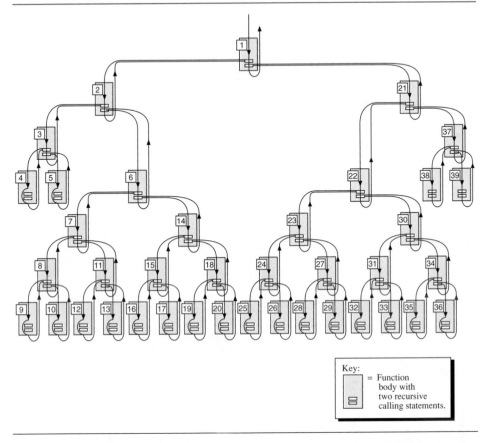

Key:

= Function body with two recursive calling statements.

very similar (compare this to Fig. 8.56). Also, you can see the similarity between what is shown in Fig. 8.58 and a binary tree. Because of this similarity, we will use a function with two recursive calls in working with a binary tree in the next Application Program.

   The complete source code, including the main function, follows. It sorts the list given by the vector list[ ].

```
#include <stdio.h>
#define SIZE 17
int qksort(int a[ ], int left_position, int right_position);

void main (void)
{
```

```
        int
list[]={98,34,56,27,78,73,70,90,28,84,45,85,12,18,54,34,29};
        int i;

        qksort(list, 0, SIZE-1);
        for (i=0; i<SIZE; i++)
                printf("%d ", list[i]);
        printf("\n");
}

int qksort(int a[ ], int left_position, int right_position)
{
        int pivot, i, j, swap;
        if (right_position > left_position)
                {
                pivot = a[right_position];
                i = left_position-1;
                j = right_position;
                while (1)
                        {
                        while (a[++i] < pivot);
                        while (a[--j] > pivot);
                        if (i >= j) break;
                        swap = a[i];
                        a[i] = a[j];
                        a[j] = swap;
                        }
                swap = a[i];
                a[i] = a[right_position];
                a[right_position] = swap;
                qksort(a, left_position, i-1);
                qksort(a, i+1, right_position);
                }
return 0;
}
```

OUTPUT

```
12 18 27 28 29 34 34 45 54 56 70 73 78 84 85 90 98
```

COMMENTS

One method of evaluating the performance of sorting algorithms is to examine the number of comparisons needed to complete a sort. For a given number of items, N, in a list, the number of comparisons needed for a particular method is not necessarily constant, because some lists may begin in a favorable order. For instance, the most favorable initial arrangement for quicksort is one in which, each time two sublists are created, both are of the same size. The analysis goes beyond the scope of this book; however, we can say that the number of comparisons on average for the quicksort is on the order of $N \log_2 N$. For the bubble sort it is on the order of N^2. So, for 10,000 values in a list, quicksort on average requires on the order of $10,000(\log_2 10,000) = 132,900$ comparisons, whereas the bubble sort requires on the order of $(10,000)^2 = 100$ million comparisons. Clearly, quicksort is far superior to the bubble sort.

A number of suggestions for improving the quicksort algorithm have been made since the method was first developed. A popular one is to use a pivot value different from the far right value. If a median of the left, middle, and right values is used, the likelihood of a more even split in the creation of the sublists increases. This improves the efficiency of the method.

Last, to use a quicksort algorithm you also can use the C library function qsort described in Lesson 8.15.

■ APPLICATION PROGRAM 8.4 PERFORMING ARITHMETIC OPERATIONS; USING A BINARY TREE

Problem statement

Write a program that can perform the operation

`((5+6)/(7-9))*3`

using a binary tree as shown in Fig. 8.59. Enter the operation using postfix (reverse Polish) notation as

`5 6 + 7 9 - / 3 *`

Use the program in Lesson 8.9 for creating the tree. Write the result of the expression to the screen. The program should be general enough to be able to perform similar postfix type operations. Please review Lesson 8.9 if you do not remember it well.

Solution

RELEVANT EQUATIONS

The only equation involved is the arithmetic expression as given in the Problem Statement.

■ **FIG. 8.59**
Binary tree of arithmetic expression

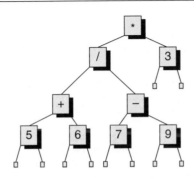

SPECIFIC EXAMPLE

The result of the expression is

```
((5+6)/(7-9))*3 = -16.5
```

DATA STRUCTURES

As described in Lesson 8.9, we use both a tree and a stack structure. However, we find that we need to add to the tree structure a member that can store the result of the operation involving the left and right children. For instance, consider the subtree shown in Fig. 8.60 of the full tree shown in Fig. 8.59. The result of the operation indicated is $5 + 6 = 11$. We can store the number 11 in a member of the structure for the node $+$. To do this, we create a member of the tree structure, cum, that we did not have in Lesson 8.9. This gives us the following tree structure:

```
struct Tree_block
        {
        char type, operat;
        double operand, cum;
        struct Tree_block *left_child, *right_child;
        };
```

The member cum is of type double because it will hold the real result of the arithmetic expression. The value of cum for the node that is the root is the final result of the expression. Fig. 8.61 shows the value of cum for each node.

ALGORITHM

What we did not show in Lesson 8.9 is performing the operation indicated by the expression. To perform the operation, we must traverse the tree. There are three standard ways of traversing a binary tree: preorder, inorder, and postorder. Because we created the tree using a postfix expression, we should traverse the tree in postorder. Postorder involves beginning at the root and following these steps in the order listed:

1. Traverse the left subtree of the node in postorder.
2. Traverse the right subtree of the node in postorder.
3. Visit the node.

FIG. 8.60

Subtree of the larger tree. The result of this expression, 11, can be stored in the block of memory reserved for the $+$ node. To do this, create a member of the tree structure, cum, to hold the cumulative result of the operation involving the left and right children of a node.

FIG. 8.61
The value of cum for each node

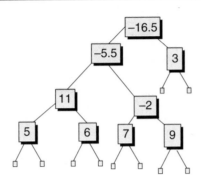

Although we did not label it as such, we followed the postorder method of traversing a tree in Application Program 8.3. Figures 8.57 and 8.58 show a tree and the order in which a postorder traversal takes place. We did not call it *tree traversal* in Application Program 8.3, because we did not formally create a tree. However, essentially, we traversed a tree using postorder in that example. If you examine the program, you can see that we used a function with two recursive calls to do the traversal. Rather than describe the process in detail here, we refer you to Application Program 8.3. If you do not understand that program, please also review recursion in Lesson 8.10.

   We call the recursive function for this program *traverse*. If we wanted to simply print the cumulative value of each node in postorder within the recursive function traverse, we would have the following types of statements:

```
traverse ( current->left_child );
traverse ( current->right_child);
printf ("The cumulative value is: %lf\n", current->cum);
```

In other words, for a postorder traversal of a tree, we have a recursive call with the left child, a recursive call with the right child, and an operation. The statements are not exact; however, we create our function traverse with the realization that we need this general form.

   You may recall that, in creating a recursive function, we need to recognize the base case or simplest case to help us recognize the reversal point. For a tree, the reversal is reached when we encounter a terminating node. For our tree, if either of the children of a node is a terminating node, then we have reached a reversal point.

In this book, we created recursive functions by using an if control structure with one block being the base case (thus, the reversal) and the other block having the recursive calls. Thus, the form of the function traverse becomes

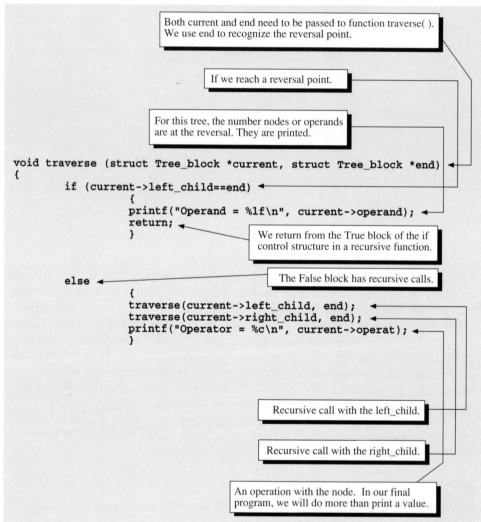

Both current and end need to be passed to function traverse( ). We use end to recognize the reversal point.

If we reach a reversal point.

For this tree, the number nodes or operands are at the reversal. They are printed.

```
void traverse (struct Tree_block *current, struct Tree_block *end)
{
        if (current->left_child==end)
            {
            printf("Operand = %lf\n", current->operand);
            return;
            }

        else
            {
            traverse(current->left_child, end);
            traverse(current->right_child, end);
            printf("Operator = %c\n", current->operat);
            }
```

We return from the True block of the if control structure in a recursive function.

The False block has recursive calls.

Recursive call with the left_child.

Recursive call with the right_child.

An operation with the node. In our final program, we will do more than print a value.

Note that with this code, we enter the False block of the function only when current is a node containing an operator and not a number (or operand). This is because none of the operator nodes has a terminating node as a child. As a result, within the False block, we perform the operation indicated by the operator and store the result in the structure member cum.

The following if-else control structure uses the operator and cumulative value for the left and right children of the node to perform the operation:

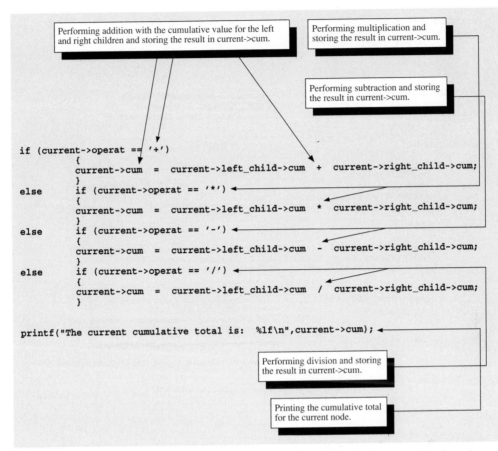

```
if (current->operat == '+')
        {
        current->cum  =  current->left_child->cum  +  current->right_child->cum;
        }
else    if (current->operat == '*')
        {
        current->cum  =  current->left_child->cum  *  current->right_child->cum;
        }
else    if (current->operat == '-')
        {
        current->cum  =  current->left_child->cum  -  current->right_child->cum;
        }
else    if (current->operat == '/')
        {
        current->cum  =  current->left_child->cum  /  current->right_child->cum;
        }

printf("The current cumulative total is:  %lf\n",current->cum);
```

Boxes (annotations):
- Performing addition with the cumulative value for the left and right children and storing the result in current->cum.
- Performing multiplication and storing the result in current->cum.
- Performing subtraction and storing the result in current->cum.
- Performing division and storing the result in current->cum.
- Printing the cumulative total for the current node.

Putting this if-else control structure into the false block of the recursive function leads to the code shown. Note the location in main where we call the function, traverse. Other than that, the functions main, pop, and push are identical to those shown for Lesson 8.9.

SOURCE CODE

The structure member cum accumulates the result of the tree calculations. Each node has a cum. For a number node, cum is equal to the number. For an operator node, cum is the result of the operation involving the left and right child.

```
#include <stdio.h>
#include <stdlib.h>


struct Tree_block
        {
        char type, operat;
        double operand, cum;
        struct Tree_block *left_child, *right_child;
        };
```

```
struct Stack_block
        {
        struct Tree_block *tree_address;
        struct Stack_block *next_node;
        };


void push(struct Tree_block *address,  struct Stack_block *head);
struct Tree_block * pop(struct Stack_block *head);
void traverse (struct Tree_block *current, struct Tree_block *end);


void main(void)
{
        int i;
        double number;
        char optr, tp;
        struct Tree_block *current, *end;
        struct Stack_block *head, *tail;



        head=(struct Stack_block *)calloc(1,sizeof(struct Stack_block));
        tail=(struct Stack_block *)calloc(1,sizeof(struct Stack_block));
        end=(struct Tree_block *) calloc (1, sizeof(struct Tree_block));


        head->next_node=tail;
```

> The function traverse is a recursive function used to visit each node of the tree. It works with a current node (the node being visited) and the terminating (end) node.

```
for (i=1; i<=9; i++)
                {
                current=(struct Tree_block *)calloc(1, sizeof(struct Tree_block));

                printf ("Enter 'n' for number or 'o' for operator then return\n");
                fflush(stdin);
                tp = getchar();
                fflush(stdin);

                if (tp == 'n')
                        {
                        printf("Enter a number\n");
                        scanf ("%lf", &number);
                        current->type='n';
                        current->operand=number;
                        current->cum=number;
                        current->left_child = end;
                        current->right_child = end;
                        }

                if (tp == 'o')
                        {
                        printf("Enter an operator\n");
                        optr=getchar();
                        current->type='o';
                        current->operat=optr;
                        current->cum=0.0;
                        current->right_child = pop(head);
                        current->left_child = pop(head);
                        }
                push(current, head);
                }

                traverse (current, end);
}
```

> Traversing the tree.

> Creating the tree.

```
void push(struct Tree_block *address,  struct Stack_block *head)
{
        struct Stack_block *new_node;


        new_node=(struct Stack_block *)calloc(1,sizeof(struct Stack_block));
        new_node->tree_address=address;
        new_node->next_node=head->next_node;
        head->next_node=new_node;
}

struct Tree_block *pop(struct Stack_block *head)
{
        struct Stack_block *dummy;
        struct Tree_block *address;


        address=head->next_node->tree_address;
        dummy=head->next_node;
        head->next_node=head->next_node->next_node;
        free(dummy);
        return(address);
}

void traverse (struct Tree_block *current, struct Tree_block *end)
{
        if (current->left_child==end)
                {
                printf("Operand = %lf\n", current->operand);
                return;
                }


        else
                {
                traverse(current->left_child, end);
                traverse(current->right_child, end);
                printf("Operator = %c\n", current->operat);
                if (current->operat == '+')
                        {
                        current->cum=current->left_child->cum+current->right_child->cum;
                        }
                else    if (current->operat == '*')
                        {
                        current->cum=current->left_child->cum*current->right_child->cum;
                        }
                else    if (current->operat == '-')
                        {
                        current->cum=current->left_child->cum-current->right_child->cum;
                        }
                else    if (current->operat == '/')
                        {
                        current->cum=current->left_child->cum/current->right_child->cum;
                        }
                printf("The current cumulative total is:  %lf\n",current->cum);
                }
}
```

> Traversing the tree.

> If the node has the end node for a child, return.

> Traversing the tree using postorder.

> Performing the operation with the children of the node with the operator.

OUTPUT

	Enter 'n' for number or 'o' for operator then return
Keyboard input	n
	Enter a number
Keyboard input	5
	Enter 'n' for number or 'o' for operator then return
Keyboard input	n
	Enter a number

```
Keyboard input    6
                  Enter 'n' for number or 'o' for operator then return
Keyboard input    o
                  Enter an operator
Keyboard input    +
                  Enter 'n' for number or 'o' for operator then return
Keyboard input    n
                  Enter a number
Keyboard input    7
                  Enter 'n' for number or 'o' for operator then return
Keyboard input    n
                  Enter a number
Keyboard input    9
                  Enter 'n' for number or 'o' for operator then return
Keyboard input    o
                  Enter an operator
Keyboard input    -
                  Enter 'n' for number or 'o' for operator then return
Keyboard input    o
                  Enter an operator
Keyboard input    /
                  Enter 'n' for number or 'o' for operator then return
Keyboard input    n
                  Enter a number
Keyboard input    3
                  Enter 'n' for number or 'o' for operator then return
Keyboard input    o
                  Enter an operator
Keyboard input    *
                  Operand  = 5.000000
                  Operand  = 6.000000
                  Operator = +
                  The current cumulative total is:  11.000000
                  Operand = 7.000000
                  Operand = 9.000000
                  Operator = -
                  The current cumulative total is:  -2.000000
                  Operator = /
                  The current cumulative total is:  -5.500000
                  Operand = 3.000000
                  Operator = *
                  The current cumulative total is: -16.500000
```

COMMENTS

Because of the similarity of this program and Application Program 8.3, Figs. 8.57 and 8.58 illustrate the program flow. Although we would not have obtained the correct arithmetic answer, we could have traversed the tree in preorder or inorder. These have the following steps.

Preorder.
1. Visit the node.
2. Traverse the node's left subtree in preorder.
3. Traverse the node's right subtree in preorder.

These steps lead to the following False block in the function traverse, if when we visit the node, we simply print the operator:

```
printf("Operator = %c\n", current->operat);
traverse(current->left_child, end);
traverse(current->right_child, end);
```

This would lead to the output being printed in the order:

`* / + 5 6 - 7 9 3`

Inorder.
1. Traverse the node's left subtree in inorder.
2. Visit the node.
3. Traverse the node's right subtree in inorder.

This leads to the following statements in the False block of the function traverse:

```
traverse(current->left_child, end);
printf("Operator = %c\n", current->operat);
traverse(current->right_child, end);
```

This would lead to the output being printed in the order

`5 + 6 / 7 - 9 * 3`

Had we wanted to perform an arithmetic calculation with preorder or inorder traversal, we would have had to change the if-else control structure in the function traverse.

Modification exercises

1. Modify the program so that it can handle 21 binary tree nodes.

2. Modify the program so that the user can enter the number of nodes that are to be in the tree.

3. Modify the program to perform the calculation $5 + 6 / 7 - 9 \times 3$ without any consideration of parentheses using an inorder traversal of the tree.

4. Modify the program so that the tree is created by entering the data in inorder.

5. Modify the program so that the tree is created by entering the data in preorder.

▪ APPLICATION EXERCISES (SEE THE DESCRIPTION BEFORE APPLICATION EXERCISE 6.16 TO REVIEW RANDOM NUMBER GENERATION)

1. Write a program that deletes a node anywhere in a linked list 30 long.

2. Write a program that inserts a node anywhere into a linked list 30 long.

3. Write a program that creates a queue, 12 long. In the queue put playing cards. Have a user remove a card from the top of the queue and accept the card or return it to the bottom of the queue. Have the player quit when the sum total of his or her cards is 21.

4. In industrial engineering, it is sometimes necessary to simulate manufacturing and service operations to improve efficiency and optimize the system. In this problem, model a rapid oil change and lubrication business. The queue data structure can be used to do this. Assume that the station has the following:

There is one bay for changing oil.
It takes 15 minutes to service each car.
One car arrives randomly between every 2 and 20 minutes.
At the end of a 12-hour day, all waiting cars are sent away.

Write your program to model 30 days of operation. Determine the average waiting time for the cars and the total amount of idle time for the bays. Run your program 12 times to get an understanding of the variability of the results. Report all of your answers.

5. Have the program automatically execute 50 times and average the results of all 50 runs.

6. Modify the program of the previous problem to handle five bays. How does this change the answers?

7. The company is thinking of issuing a preferred customer card to minimize the waiting for certain customers. Allow these customers to move to the front of the queue. Have 15 of these customers appear each day. What is the average waiting time for the nonpreferred customers?

8. Change the number of preferred customers in the previous program until the waiting time for the other customers increases 10, 30, and 100 percent. How many preferred customers can be created if the other customers are to wait no more than an additional 25 percent if 0.5 percent of all preferred customers arrive each day?

9. A small bicycle manufacturing company relies on its suppliers to deliver crucial parts in a timely manner for it to be able to complete a bicycle. However, its suppliers cannot deliver parts in a completely predictable manner. Use three different stacks to model the supply and use of chains, wheel rims, and spokes. Each bicycle uses 100 spokes, two wheel rims, and one chain. The suppliers offer the following reliability of delivery:

	Spokes	Wheel rims	Chains
Frequency of delivery	Every 30 to 60 days	Every 30 to 90 days	Every 40 to 150 days
Number delivered	3000 to 5000	400 to 700	180 to 400

Assume that all other parts are on hand and do not affect production. If the parts are delivered randomly in these ranges, determine how many bicycles can be made in a three-year period. At the end of this time, how many of each type of parts are on hand? Have your program automatically execute 50 times, so that you get an understanding of the effects of the randomness.

10. Modify the program to consider the supply of brakes. Assume that 200 to 900 brakes can be delivered every 20 to 100 days. How many bikes can be manufactured now?

11. Have the program print a weekly report of the number of parts on hand. Have it compute the number of days for which no bicycle is manufactured.

12. The exact boundaries of a city are being determined by locating individual points on its boundaries. Each point is given x and y coordinates according to a predetermined coordinate system. As each individual point is located, the boundaries of the city are updated. Initially the city's boundaries are determined with just four points:

x (ft.)	y (ft.)
58	93
3597	198
3498	78965
87	97866

The location of a new point on the list is important because the boundary is formed by connecting the adjacent points on the list. Write a program that creates a list of boundary points that can be updated continuously. Begin with the four listed points. Have points added by first inputting the coordinates of the new boundary point and then the coordinates of the boundary point that it succeeds. Use a linked list structure to do this. Each new data point should be inserted in the linked list. Print the final list of boundary points.

13. Modify the program so that it calculates the length of the boundary each time a new point is added.

14. On complex construction projects, a sequence of steps must be taken in a particular order to complete the project on schedule and reliably. This is called a *critical path*. The steps in a critical path sometimes are modified as the project progresses because some actions are not foreseeable. Write a program that uses a linked list to represent a critical path. Read the description of the activity from the keyboard and the name of the activity that it follows. Add this activity to the critical path and print out the new critical path. The first activity should follow "head."

15. Transportation engineers coordinate bus traffic in major cities. Consider ten different stops on one bus line. Use a queue for each stop. Assume that the patrons arrive randomly at each stop at a rate of one every three to six minutes. Each time a passenger boards a bus, the bus is delayed ten seconds. The time for the bus to drive between stops is four minutes. Write a program that simulates driving two buses. The buses begin at evenly spaced intervals along the route. In other words, the buses begin at the first and sixth stops. Determine the average waiting time for the patrons at the stops.

16. Modify the program to use three buses, evenly spaced. What is the average waiting time now?

17. Modify the program to limit the number of passengers allowed on a bus. Allow no more than ten passengers at a time. Determine the average waiting time of passengers able to get on a bus and the number of passengers left stranded.

18. Modify the program to determine the number of buses necessary to assure that only one passenger is stranded.

19. Modify the passenger arrival rates of the bus programs and the spacing between bus stops to gain an understanding of the effects of these parameters.

20. Products that have limited shelf life are put onto shelves in a supermarket in a manner similar to that of the stack data structure. New products arriving are simply put in front of older products. Customers remove the newer products until the older ones become exposed. Write a program that simulates this behavior for four different products. Assume that new products arrive every day according to the following schedule:

	Bread	**Milk**	**Eggs**	**Chicken**
Supply	8:00 AM, 300 loaves	11:00 AM, 900 liters	2:00 PM, 600 boxes	4:00 PM, 60 chickens
Removal	Random, every 3–15 minutes	Random, every 1–6 minutes	Random, every 4–8 minutes	Random, every 5–10 minutes

At the end of every third (12-hour) day, the leftover product is discarded. If no product exists when a customer wants it, register a complaint with management. Determine the amount of product discarded and the number of complaints registered over a 30-day period.

21. Have your program select randomly among the following supply and removal rates and compute the amount of product that is discarded and the number of complaints registered:

Supply arrival time	Supply number of units	Removal (minutes)
8:00 AM	2000	2–5
9:30 AM	40	10–30
11:00 AM	90	25–60
12:00 PM	3500	20–90
1:00 PM	200	4–9
2:00 PM	4000	12–20
4:00 PM	7000	5–15
6:00 PM	120	8–30

Appendix A: C Operators

Operator	Name	Associativity	Prece-dence	Example	Notes
()	Parentheses/function	Left to right	1	sin(x)	Call sin function with argument x
[]	Array indexing	Left to right	1	a[5]	Array of five elements
->	Member selection	Left to right	1	ptr->x	Member named x in a structure that ptr points to
.	Member selection	Left to right	1	str.x	Member x in structure str
++	Postincrement	Left to right	2	x++	Increment x after execution
--	Postdecrement	Left to right	2	x--	Decrement x after execution
*	Pointer indirection	Right to left	2	*ptr	Content of location whose address is stored in ptr
&	Address	Right to left	2	&x	Address of x
!	Logical NOT	Right to left	2	!x	If x is True (1), !x is False (0) and vice versa
~	Bitwise negation	Right to left	2	~x	Toggle 1 bits to 0 and 0 bits to 1
-	Negation	Right to left	2	-x	Negate the value of x
+	Plus sign	Right to left	2	+x	Unary plus operator
++	Preincrement	Right to left	2	++x	Increment x before execution
--	Predecrement	Right to left	2	--x	Decrement x before execution
sizeof	Size of data	Right to left	2	sizeof(x)	Return size of x in bytes

Operator	Name	Associativity	Prece-dence	Example	Notes
(type)	Type cast	Right to left	2	int(x)	Convert x to int type
*	Multiple	Left to right	3	x*y	Multiply x and y
/	Divide	Left to right	3	x/y	Divide x by y
%	Modulus	Left to right	3	x%y	Find remainder of dividing x by y
+	Addition	Left to right	4	x+y	Add x and y
−	Subtraction	Left to right	4	x−y	Subtract y from x
<<	Left shift	Left to right	5	x<<3	x shifted to left by 3 bit position, which is equal to $x = x \times 2^3$
>>	Right shift	Left to right	5	x>>3	x shifted to right by 3 bit position, which is equal to $x = x/2^3$
<	Less than	Left to right	6	x<y	True (1) if x is less than y, else False (0)
<=	Less than or equal to	Left to right	6	x<=y	True (1) if x is less than or equal to y, else False (0)
>	Greater than	Left to right	6	x>y	True (1) if x is greater than y, else False (0)
>=	Greater than or equal to	Left to right	6	x>=y	True (1) if x is greater than or equal to y, else False (0)
==	Equal to	Left to right	7	x==y	True (1) if x is equal to y, else False (0)
!=	Not equal to	Left to right	7	x!=y	True (1) if x is not equal to y, else False (0)
&	Bitwise AND	Left to right	8	x&y	Bits become ones at bits where corresponding bits of x and y are ones, else zeros
^	Bitwise exclusive OR	Left to right	9	x^y	Bits become ones at bits where corresponding bits of x and y differ, else zeros
|	Bitwise OR	Left to right	10	x|y	Bits become ones at bits where corresponding bits of x and/or y is ones, zeros if both x and y are zero
&&	Logical AND	Left to right	11	x&&y	1 if both x and y are 1, else 0

Operator	Name	Associativity	Prece-dence	Example	Notes
‖	Logical OR	Left to right	12	x‖y	1 if either x or y is 1, 0 if both x and y are 0
?:	Conditional	Right to left	13	x?y:z	If x is not 0, y is evaluated; else if x is 0, z is evaluated. For example, $z =$ (x>y)?x:y is equivalent to $z = \max(x,y)$
=	Compound assignment	Right to left	14	x=y	Assign x to y
+=	Compound assignment	Right to left	14	x+=y	Equivalent to $x = x + y$
−=	Compound assignment	Right to left	14	x−=y	Equivalent to $x = x - y$
*=	Compound assignment	Right to left	14	x*=y	Equivalent to $x = x * y$
/=	Compound assignment	Right to left	14	x/=y	Equivalent to $x = x/y$
%=	Compound assignment	Right to left	14	x%=y	Equivalent to $x = x \% y$
>>=	Compound assignment	Right to left	14	x>>=y	Equivalent to $x = x >> y$
<<=	Compound assignment	Right to left	14	x<<=y	Equivalent to $x = x << y$
&=	Compound assignment	Right to left	14	x&=y	Equivalent to $x = x \& y$
‖=	Compound assignment	Right to left	14	x‖=y	Equivalent to $x = x‖y$
^=	Compound assignment	Right to left	14	x^=y	Equivalent to $x = x \wedge y$
,	Comma/ sequential evaluation	Left to right	15	int x,y;	First declare x, then y

B

Appendix B: ASCII Table

Decimal	Octal	Hex	Character	Key	Escape sequence
0	0	0	NUL	Ctrl/1	
1	1	1	SOH	Ctrl/A	
2	2	2	STX	Ctrl/B	
3	3	3	ETX	Ctrl/C	
4	4	4	EOT	Ctrl/D	
5	5	5	ENQ	Ctrl/E	
6	6	6	ACK	Ctrl/F	
7	7	7	BEL	Ctrl/G	bell '\a'
8	10	8	BS	Ctrl/H	backspace '\b'
9	11	9	HT	Ctrl/I	horizontal tab '\t'
10	12	A	LF	Ctrl/J	newline '\n'
11	13	B	VT	Ctrl/K	vertical tab '\v'
12	14	C	FF	Ctrl/L	formfeed '\f'
13	15	D	CR	Ctrl/M	carriage return '\r'
14	16	E	SO	Ctrl/N	
15	17	F	SI	Ctrl/O	
16	20	10	DLE	Ctrl/P	
17	21	11	DC1	Ctrl/Q	
18	22	12	DC2	Ctrl/R	
19	23	13	DC3	Ctrl/S	
20	24	14	DC4	Ctrl/T	
21	25	15	NAK	Ctrl/U	
22	26	16	SYN	Ctrl/V	
23	27	17	ETB	Ctrl/W	
24	30	18	CAN	Ctrl/X	
25	31	19	EM	Ctrl/Y	
26	32	1A	SUB	Ctrl/Z	
27	33	1B	ESC	Esc	
28	34	1C	FS	Ctrl\	
29	35	1D	GS	Ctrl/]	
30	36	1E	RS	Ctrl/=	
31	37	1F	US	Ctrl/−	

Decimal	Octal	Hex	Character	Key	Escape sequence
32	40	20	SP	Spacebar	
33	41	21	!		
34	42	22	"		
35	43	23	#		
36	44	24	$		
37	45	25	%		
38	46	26	&		
39	47	27	'		
40	50	28	(		
41	51	29	)		
42	52	2A	*		
43	53	2B	+		
44	54	2C	,		
45	55	2D	−		
46	56	2E	.		
47	57	2F	/		
48	60	30	0		
49	61	31	1		
50	62	32	2		
51	63	33	3		
52	64	34	4		
53	65	35	5		
54	66	36	6		
55	67	37	7		
56	70	38	8		
57	71	39	9		
58	72	3A	:		
59	73	3B	;		
60	74	3C	<		
61	75	3D	=		
62	76	3E	>		
63	77	3F	?		
64	100	40	@		
65	101	41	A		
66	102	42	B		
67	103	43	C		
68	104	44	D		
69	105	45	E		
70	106	46	F		
71	107	47	G		
72	110	48	H		
73	111	49	I		
74	112	4A	J		
75	113	4B	K		
76	114	4C	L		
77	115	4D	M		
78	116	4E	N		
79	117	4F	O		
80	120	50	P		
81	121	51	Q		
82	122	52	R		

Decimal	Octal	Hex	Character	Key	Escape sequence
83	123	53	S		
84	124	54	T		
85	125	55	U		
86	126	56	V		
87	127	57	W		
88	130	58	X		
89	131	59	Y		
90	132	5A	Z		
91	133	5B	[		
92	134	5C	\		
93	135	5D	]		
94	136	5E	^		
95	137	5F	_		
96	140	60	`		
97	141	61	a		
98	142	62	b		
99	143	63	c		
100	144	64	d		
101	145	65	e		
102	146	66	f		
103	147	67	g		
104	150	68	h		
105	151	69	i		
106	152	6A	j		
107	153	6B	k		
108	154	6C	l		
109	155	6D	m		
110	156	6E	n		
111	157	6F	o		
112	160	70	p		
113	161	71	q		
114	162	72	r		
115	163	73	s		
116	164	74	t		
117	165	75	u		
118	166	76	v		
119	167	77	w		
120	170	78	x		
121	171	79	y		
122	172	7A	z		
123	173	7B	{		
124	174	7C	|		
125	175	7D	}		
126	176	7E	~		
127	177	7F	DEL Del		

Appendix C: ANSI Standard Functions and Macros

Diagnostics <assert.h> (function prototype is in assert.h)

Function name	Description
assert	Puts debugging statements into program

Character handling <ctype.h>

Function name	Description
isalnum	Tests for alphanumeric character
isalpha	Tests for alphabetic character
iscntrl	Tests for control character
isdigit	Tests for decimal digit character
isgraph	Tests for printable character except space character
islower	Tests for lowercase character
isprint	Tests for printable character
ispunc	Tests for punctuation character
isspace	Tests for white-space character
isupper	Tests for uppercase character
isxdigit	Tests for hexadecimal digit character
tolower	Converts an uppercase character to the corresponding lowercase one
toupper	Converts a lowercase character to the corresponding uppercase one

Localization <locale.h>

Function name	Description
setlocale	Sets the appropriate locale data members for the program
localeconv	Sets appropriate locale values for formatting numeric quantities

Mathematics <math.h>

Function name	Description
acos	Calculates the arc cosine
asin	Calculates the arc sine
atan	Calculates the arc tangent (required one argument)
atan2	Calculates the arc tangent (required two arguments)
ceil	Calculates the integer ceiling
cos	Calculates the cosine
cosh	Calculates the hyperbolic cosine
exp	Calculates the exponential function
fabs	Calculates the absolute value of a floating-point number
floor	Calculates the largest integral value no greater than the argument
fmod	Calculates the floating-point remainder
frexp	Calculates an exponential value
ldexp	For ldexp(x, y), calculates the value of x times 2 raised to the power y
log	Calculates the natural logarithm
log10	Calculates the base-10 logarithm
modf	Breaks the argument value into integral and fractional parts
pow	Calculates a value raised to a power
sin	Calculates the sine
sinh	Calculates the hyperbolic sine
sqrt	Calculates the square root of the argument value
tan	Calculates the tangent
tanh	Calculates the hyperbolic tangent

Nonlocal jumps <setjump.h>

Function name	Description
setjump	Saves its calling environment for later use by the longjump function
longjump	Restores a stack environment saved by the setjump function

Signal handling <signal.h>

Function name	Description
signal	Allows a process to choose how to handle a signal from the operating system
raise	Sends a signal to the executing program

Variable arguments <stdarg.h>

Function name	Description
va_start	Initializes a special argument pointer to be used to access the variable arguments
va_arg	Processes argument from the list using pointer set by va start
va_end	Resets pointer and ends variable arguments processing

Input/output <stdio.h>

Function name	Description
clearerr	Clears the error indicator for a stream
fclose	Closes a stream
feof	Tests for EOF (end of file) on a stream
ferror	Tests for error on a stream
fflush	Flushes a stream
fgetc	Reads a character from a stream
fgetpos	Gets the file position indicator of a stream
fgets	Reads a string from a stream
fopen	Opens a stream
fprintf	Writes formatted information to a stream
fputc	Writes a character to a stream
fputs	Writes a string to a stream
fread	Reads unformatted information from a stream
freopen	Reassigns a file pointer (to be used to redirect a preopened file)
fscanf	Reads formatted data from a stream
fseek	Moves file position to a specified location
fsetpos	Sets the file position indicator of a stream
ftell	Finds the current file position
fwrite	Writes unformatted data to a stream
getc	Reads a character from a stream
getchar	Reads a character from stdin
gets	Reads a line from stdin
perror	Prints an error message to stderr
printf	Writes formatted data to stdout
putc	Writes a character to a stream
putchar	Writes a character to stdout
puts	Writes a line to a stream
remove	Deletes a file
rename	Renames a file
rewind	Moves file position to the beginning of a stream
scanf	Reads formatted data from stdin
setbuf	Controls buffering for stream
setvbuf	Controls buffering for a stream with a specified buffer size
sprintf	Writes a formatted data to a string
sscanf	Reads a formatted data from a string
tmpfile	Creates a temporary file
tmpnam	Creates a temporary file name
ungetc	Puts a character back to the stream
vfprintf	Writes formatted data to a stream using a pointer to a list of arguments
vprintf	Writes formatted data to stdout using a pointer to a list of arguments
vsprintf	Writes formatted data to a string using a pointer to a list of arguments

General utilities <stdlib.h>

Function name	Description
abort	Aborts the current process
abs	Calculates absolute value
atexit	Schedules functions to be executed in last in, first out order
atof	Converts string to double
atoi	Converts string to int
atol	Converts string to long
bsearch	Performs binary search in a sorted array
calloc	Allocates memory for an array with all elements initialized to 0
div	Calculates the quotient and the remainder of two int values
exit	Terminates the calling process after flushing all file buffers
free	Frees allocated memory
getenv	Finds the value of a specified environment variable
labs	Calculates the absolute value of a long number
ldiv	Calculates the quotient and the remainder of two long numbers
malloc	Allocates memory block
mblen	Finds the size of a multibyte character (such as Chinese)
mbtowc	Finds the size of a multibyte character and determines the code for the corresponding multibyte character
mbstowcs	Converts multibyte characters to wide characters
qsort	Uses a quicksort algorithm to sort an array
rand	Generates a pseudorandom number
realloc	Reallocates memory block
srand	Generates a starting point for random number generation
strtod	Converts a string to a double number
strtol	Converts a string to a long number
strtoul	Converts a string to an unsigned integer
system	Executes an operating system command
wctomb	Finds the size of a wide character
wcstombs	Converts wide characters to multibyte characters

String Handling

Function name	Description
memchr	Finds a character in a buffer
memcmp	Compares characters in two buffers
memcpy	Copies characters between two buffers
memicmp	Similar to memcmp but is case insensitive
memmove	Copies characters between two buffers, overlapping region is copied before overwritten
memset	Sets buffer to a specified character
strcat	Concatenates two strings
strchr	Finds a character in a string
strcmp	Compares two strings
strcoll	Compares two strings using locale-specific information
strcpy	Copies a source string to a destination string
strcspn	Finds a substring in a string
strerror	Finds system error message
strlen	Finds the size of a string
strncat	Concatenates two strings with a specified number of characters appended
strncmp	Compares two strings with a specified number of characters compared
strncpy	Copies one string to another with a specified number of characters copied
strpbrk	Scans a source string for the first occurrence of any character specified in a character set
strrchr	Scans a source string for the last occurrence of a character
strspn	Finds the maximum length of initial segment of the first string that consists entirely of characters from the second string
strstr	Finds a substring from a string to be searched
strtok	Finds the next token in a string
strxfrm	Transforms a string to a new form based on locale-specific information

Date and time <time.h>

Function name	Description
asctime	Converts time from a structure form to a character string
clock	Gets the elapsed time required for a process
ctime	Converts time from a long integer form (the number of seconds elapsed since 00:00:00 Greenwich mean time) to a character string
difftime	Calculates the difference between two times
gmtime	Converts time from long integer form to a structure form
localtime	Converts time from long integer to a structure form with local correction
mktime	Converts local time in structure form to a calendar value
strftime	Formats a time string
time	Finds the number of seconds elapsed since 00:00:00 Greenwich mean time

I N D E X